FED-BATCH CULTURES

Many, if not most, industrially important fermentation and bioreactor operations are carried out in fed-batch mode, producing a wide variety of products. Despite this, until now, no single book has dealt with fed-batch operations. This is the first book that puts together all the necessary background material regarding the what, why, and how of optimal and suboptimal fed-batch operations. Numerous examples are provided to illustrate the application of optimal fed-batch cultures. This unique book, by world experts with decades of research and industrial experience, is a must for researchers and industrial practitioners of fed-batch processes (modeling, control, and optimization) in the biotechnology, fermentation, food, pharmaceutical, and waste treatment industries.

Henry C. Lim gained industrial experience by working for Pfizer for five years in reaction engineering and separation and purification. He taught for 21 years at Purdue University and initiated biochemical engineering research in 1970. He was recruited to the University of California, Irvine, where he was founding chair of biochemical engineering and chemical engineering for 10 years and taught for 22 years. He has extensive consulting experience with Leeds and Northrup, Novo Enzyme Corporation, Pharmacontrol Inc., Eli Lilly, Merck, Monsanto, LG Biotech Inc., CJ Biotechnology, and Zander Renewable Systems, LLC. He has studied bioreactions and bioreactor engineering, modeling, optimization and control of bioreactors, cellular regulation, recombinant DNA technology, and bioremediation. Dr. Lim has supervised more than 50 PhD dissertations and has published more than 160 journal articles, as well as two books. He received the Food, Pharmaceutical, and Bioengineering Division Award of the AIChE for his work on bioreactor and enzyme engineering.

Hwa Sung Shin received formal education at Postech (Korea) and the University of California (UC), Irvine. After postdoctoral positions at UC Irvine and the University of Michigan, he joined the Department of Biological Engineering of Inha University. Dr. Shin studies the optimization and control of fed-batch fermentation, control of stem cells and neurons in microfluidic cell culture systems, and application of optimization theory to stem cells and neural tissues in defined macro- and microenvironments.

Fed-Batch Cultures

PRINCIPLES AND APPLICATIONS OF SEMI-BATCH BIOREACTORS

Henry C. Lim
University of California, Irvine

Hwa Sung Shin
Inha University

CAMBRIDGE
UNIVERSITY PRESS

CAMBRIDGE UNIVERSITY PRESS
Cambridge, New York, Melbourne, Madrid, Cape Town,
Singapore, São Paulo, Delhi, Mexico City

Cambridge University Press
32 Avenue of the Americas, New York, NY 10013-2473, USA

www.cambridge.org
Information on this title: www.cambridge.org/9780521513364

First published 2013

Printed in the United States of America

A catalog record for this publication is available from the British Library.

Library of Congress Cataloging in Publication Data

Lim, Henry C., 1935–
Fed-batch cultures : principles and applications of semi-batch bioreactors /
Henry C. Lim, University of California, Irvine, Hwa Sung Shin, Inha University.
    pages cm. – (Cambridge series in chemical engineering)
Includes bibliographical references and index.
ISBN 978-0-521-51336-4 (hardback)
1. Bioreactors. I. Shin, Hwa Sung, 1974– II. Title.
TP248.25.B55L56 2013
660′.6–dc23   2012033203

ISBN 978-0-521-51336-4 Hardback

To my wife, Sun Boo Lim; children, David, Carol, Michael,
Tom, and Melia; and grandchildren, Natalie and Lanie
and
To my parents, Mr. and Mrs. Shin; wife, Jung Hye Hyun; and
children, Alyssa and Claire

Contents

Preface

Fed-batch operations are semi-batch operations in which one or more streams of feed containing nutrient sources, precursors, inducers, and mineral sources are fed either continuously or intermittently during the course of otherwise batch operations. The culture content is harvested either fully or partially at the end of the run and is used as the inoculum for the next cycle. By regulating the feed rates, it is possible to regulate the bioreactor environment to maximize the total rate of production, the reactor productivity, or the product yield.

Many industrially important bioreactor operations involving microbial and animal cells are carried out in fed-batch mode. These so-called fed-batch cultures have been found to be particularly effective for fermentation processes and cell cultures in which it is desirable to overcome such common phenomena as substrate inhibition, catabolite repression, product inhibition, and glucose effects to achieve high cell density for efficient fermentation, to minimize high viscosity effects, and to take advantage of auxotrophic mutants. Products produced by fed-batch cultures include amino acids, antibiotics, enzymes, microbial cells, organic chemicals, polysaccharides, proteins, tissue culture products, and various recombinant DNA products.

Despite the long history of the industrial use of fed-batch cultures, only recently have we gained a thorough understanding of them, and industrial practices have tended to be empirical in nature and most often based on experiences gained through bench, pilot plant, or production-level operations. Theoretical analyses and experimental studies of fed-batch operation have received considerable attention in the past 10 years. A better understanding of principles and applications of modern optimal control theory and geometric interpretation has opened the door for improving the performance of fed-batch fermentation processes and cell cultures. As a result of the surge in research activity, an abundance of information can now be effectively utilized in improving the performance of industrial fed-batch processes. Until now, however, no single book or monograph has been fully devoted to fed-batch processes, and therefore, the information has remained largely scattered and untapped. Textbooks dealing with biochemical engineering principles provide at best either very little coverage of fed-batch operations or unconventional treatments. Therefore, there was a strong need for a book that would put together all principles, guide the reader through the up-to-date theoretical developments in fed-batch processes, and describe step-by-step procedures leading to the optimization of fed-batch processes.

Such a book, dealing with the what, why, and how of fed-batch operation, would provide basic principles of fed-batch cultures in the simplest terms and practical means for optimizing the processes.

This book is a first attempt to provide all the necessary background materials regarding the what, why, and how of fed-batch operations. A simple conception of a fed-batch operation is a one-bioreactor operation mimicking a two-bioreactor operation, a continuous-stirred tank reactor followed by a batch reactor (a temporal equivalence of a tubular reactor) that maximizes the overall reaction rate, the product yield, or a combination of the two.

The systematic coverage presented in this work is an attempt to include elementary principles, theoretical developments in optimization and control, and practical implementation of optimal strategies for fed-batch processes. This book can be used fully or partially as a textbook and/or reference by advanced undergraduate and graduate students in biochemical engineering, environmental engineering, chemical engineering, biotechnology, and other related areas. It is also useful as a reference or guide for practitioners and research and development personnel in biotechnology, fermentation, food, pharmaceuticals, and industrial waste treatment.

Acknowledgments

The basis and details of this book were developed by the first author while teaching bioreactor optimization and bioreactor engineering courses to senior undergraduate and graduate students at Purdue University and the University of California, Irvine. We wish to thank a number of people, including former students and research associates of the first author. A formal study of fed-batch culture was initiated in his MS thesis by Professor Ravindra Waghmare of T. S. Engineering College of Mumbai, India, and by Yves Tayeb and Peter Bonte of Rhone Poulenc of France, followed by Professor Satish Parelukar of the Illinois Institute of Technology. Special thanks go to Professor Jayant Modak of the Indian Institute of Science, Bangalore, who made substantial progress toward formal fed-batch optimization. Experimental studies of fed-batch cultures made significant contributions: penicillin fermentation by Dr. Venkatesh K. Chittur of Exxon-Mobil, with help from Eli Lilly and Co.; recombinant yeast fermentation for invertase by Dr. Jon Hansen of Martek Biosciences; baker's yeast fermentation by Dr. Kyusung Lee of Samsung Biologics; and recombinant *E. coli* fermentation for poly-β-hydroxybutyric acid by Professor Jung Heon Lee of Korea University. Professor Tammy Chan of California State University, Los Angeles, and Professor Yong K. Chang of Korea Advance Science and Technology contributed to the genetic algorithm and neural network approach to the optimization of fed-batch cultures. Finally, the work of the second author in his PhD thesis contributed the most to completing this book.

We wish to thank Cambridge University Press, especially Peter Gordon, for his patience in extending a number of deadlines so that we could finish the book – it has taken much more time than we anticipated. We also express our appreciation to Purdue University and the University of California, Irvine, for the opportunity to initiate and complete this book.

1 Introduction to Fed-Batch Cultures

A living cell of a microbial, plant, or animal source is essentially an expanding and dividing biochemical reactor in which a large number of enzyme-catalyzed biochemical reactions take place. Microbial cultures involve live microbial cells, while tissue cultures involve live plant or animal cells. These cultures can be run, as in the case of chemical and biochemical reactions, in three classical operational modes: batch, continuous, or semi-batch (semi-continuous). For the past three decades, there has been tremendous growth in the use of semi-batch reactors in the fermentation, biotechnology, chemical, and waste-treatment industries owing to increasing demands for specialty chemicals and products and to certain advantages semi-batch reactors provide. Batch and semi-batch processes are used to handle usually low-volume, high-value products such as fermentation products, including amino acids and antibiotics, recombinant DNA products, and specialty chemicals. Owing to high values of these products, profitability can be improved greatly even with marginal improvements in yield and productivity. Therefore, there are incentives to optimize batch or semi-batch reactor operations.

For a batch or semi-batch process, the objective is to maximize the profit that can be realized at the end of the run, at which time the reactor content is harvested for further processing such as separation and purification. Thus, the problem is called *end point optimization* as only the end, not the intermediate, results are relevant to the overall profit.

1.1 Batch Cultures

In a batch operation, all necessary medium components and the inoculum are added at the beginning and not during period of fermentation. Therefore, their concentrations are not controlled but are allowed to vary as the living cells take them up. The products, be they intra- or extracellular, are harvested only at the end of the run. Basic controls for pH, temperature, dissolved oxygen, and foam are applied during the course of batch culture. The pH, dissolved oxygen, and temperature are normally held constant during the course of batch reactor operation. The only optimization parameters are the initial medium composition. However, profile optimizations of temperature and pH may lead to improved performance over the operations carried out at constant temperature and constant pH.

1.2 Continuous Cultures

In a continuous operation, one or more feed streams containing the necessary nutrients are fed continuously, while the effluent stream containing the cells, products, and residuals is continuously removed. A steady state is established by maintaining an equal volumetric flow rate for the feed and effluent streams. In so doing, the culture volume is kept constant, and all nutrient concentrations remain at constant steady state values. Continuous reactor operations are common in chemical industries. With the exception of single-cell protein production, certain beer production, and municipal waste treatment processes, continuous cultures have not been adopted widely by industry. It is not a dominant mode of industrial operation primarily because of the difficulty in maintaining sterility (contamination by other organisms) and protecting against phage attacks or mutations and because often, steady state operations are found to yield poorer results than dynamic operations, for reasons not yet fully understood.

1.3 Fed-Batch Cultures

A fed-batch culture is a semi-batch operation in which the nutrients necessary for cell growth and product formation are fed either intermittently or continuously via one or more feed streams during the course of an otherwise batch operation. The culture broth is harvested usually only at the end of the operational period, either fully or partially (the remainder serving as the inoculum for the next repeated run). This process may be repeated (repeated fed-batch) a number of times if the cells are fully viable and productive. Thus, there are one or more feed streams but no effluent during the course of operation. Sources of carbon, nitrogen, phosphates, nutrients, precursors, or inducers are fed either intermittently or continuously into the culture by manipulating the feed rates during the run. The products are harvested only at the end of the run. Therefore, the culture volume increases during the course of operation until the volume is full. Thereafter, a batch mode of operation is used to attain the final results. Thus, the fed-batch culture is a dynamic operation. By manipulating the feed rates, the concentrations of limiting nutrients in the culture can be manipulated either to remain at a constant level or to follow a predetermined optimal profile until the culture volume reaches the maximum, and then a batch mode is used to provide a final touch. In so doing, the concentration of the desired product or the yield of product at the end of the run is maximized. This type of operation was first called a *fed-batch culture* or *fed-batch fermentation*.[1,2] It is also known as *Zulaufverfahren* in German or *ryukaho*[2] (a flow addition method) in Japanese. Obviously, this type of operation is a semi-batch reactor operation that is used for chemical and biochemical reactions. In environmental engineering dealing with toxic waste, this type of operation is known as a *fill and draw operation* or as a *sequencing batch reactor*. In biomedical engineering, the breathing process in and out of the lung is known as *stick and balloon*, as the volume of the lung increases as we inhale and decreases as we exhale, which is a form of fed-batch process.

1.3.1 Reasons for Fed-Batch Cultures

The fed-batch culture has been practiced since the early 1900s, when it was recognized in yeast production from malt wort that the malt concentration in the medium had to be kept low enough to suppress alcohol formation and maximize the yield[3] of yeast cells. High malt concentration would accelerate the cell growth, which in turn would cause anaerobic conditions that favored ethanol formation and lowered the yield of yeast cells. Additional wort was added at a rate that was always less than the rate at which the yeast cells could use it. Intermittent or incremental feeding of nutrients to an initially dilute medium was introduced thereafter in large-scale yeast production to improve the yeast yields while obviating the production of ethanol.[4] However, there is some speculation that a small amount of ethanol may be necessary to ensure the quality of the baker's yeasts.

Through the manipulation of one or more feed rates, the fed-batch operation can provide unique means of regulating the concentration of compounds that control the key reaction rates and, therefore, can provide a definite advantage over the batch or continuous operation. If it is advantageous to control independently the concentration of more than one species, more than one feed stream may be used.

The fed-batch operation described previously can be also shown to yield a superior performance (a higher yield or higher productivity) for certain chemical and biochemical reactions that exhibit a maximum in the overall reaction rate or to maximize the selectivity of a specific product in systems of multiple reactions such as biological reactions and polymerization reactions. Examples include inhibited enzyme reactions and autocatalytic reactions, reactions in which one of the products acts as a catalyst, certain adiabatic reactions, and series parallel reactions. Fermentation and cell cultures are autocatalytic reactions in the sense that the cells produced are in turn producing additional cells, and it is not surprising to find that most of the industrially important fermentations are carried out in fed-batch mode. Extensive reviews of fed-batch techniques[5–7,145,146] are available elsewhere.

1.3.2 Applications of Fed-Batch Cultures

The oldest and first well-known industrial application of a fed-batch operation was introduced after the end of World War I. It was the yeast cell production in which sugar (glucose) was added incrementally during the course of fermentation to maintain a low sugar concentration to suppress alcohol formation.[3] The manufacture of yeast by fed-batch culture has gone through a series of improvements and is an industrially important fed-batch process. This process was historically followed by penicillin fermentation, in which the energy source (e.g., glucose) and precursors (e.g., phenyl acetic acid) were added incrementally during the course of fermentation[8] to improve penicillin production. Prior to this practice, a slowly metabolized but more expensive substrate, lactose, was used in place of glucose in a batch culture. Oversupply of a carbon source resulted in more mycelial growth and low penicillin formation, while undersupply resulted in slower mycelial growth and, eventually, slower penicillin formation.

The most important questions to ask in fed-batch culture operations are what compounds(s) should be fed and how they should be added. The answers depend on the characteristics of the organisms used. The primary candidates in the list of compounds that may be fed during the course of the operation include the limiting substrate, inducers, precursors, a carbon source, a nitrogen source, a phosphate source, inducers, and other nutrient sources. The feeding patterns are open loop or feedback controlled to maintain some key variables at constant optimum values such as the specific growth rate, respiratory quotient, pH, partial pressure of carbon dioxide, dissolved oxygen, substrate concentration, and some metabolite concentrations. The optimum feed rates sometimes require keeping these parameters to follow certain optimum profiles rather than keeping them at constant values.

To maximize the cell formation rate for the case of constant cell mass yield, it is obvious that the substrate concentration should be maintained at the value that maximizes the specific growth rate, S_m, until the reactor is full. Therefore, it is also obvious that the initial substrate concentration should be S_m, that is, $S(0) = S_m$, and that the substrate concentration should be maintained at S_m throughout the course of fermentation. This will lead to the maximum cell concentration at the end of the run. To achieve this, the feed rate must be regulated properly to hold the substrate concentration constant at S_m. If it is not possible to set the initial substrate concentration to S_m for one reason or another, the substrate concentration should be brought to this value as soon as possible by applying at the beginning the maximum substrate feed rate ($S(0) < S_m$) or a batch period ($S(0) > S_m$) and then regulating thereafter to maintain the substrate concentration to remain at S_m until the fermentor is full. Once the fermentor is full, it is run in a batch mode to reduce the substrate concentration to a desired level, and the cells containing the product are harvested. This is the basis for the simplest case of a fed-batch culture.

Fed-batch cultures with the addition of nutrients, precursors, inducers, or other additives have been tested in laboratories, pilot plants, and industrial plants for production of various products[2,5,6,7,9] such as yeasts; antibiotics; amino acids; fine organic acids; enzymes; alcoholic solvents; recombinant DNA products; proteins; tissue cultures, including hybridoma and Chinese hamster ovaries (CHO) cells; insect cell cultures; and others. These are listed in Table 1.1.

1.3.3 A Simple Example of a Fed-Batch Culture

Let us consider a simple example to illustrate the advantage of controlling the concentration of the limiting nutrient. Consider the simplest case of growing cells, or a fermentation process in which the intracellular metabolite concentration is proportional to the cell concentration so that the metabolite production is maximized by maximizing the cell mass production. Assume that the specific growth rate of cells is substrate inhibited. In other words, the rate increases first with the substrate concentration reaching a maximum value of μ_m at the substrate concentration of S_m and then decreases with further increase in the substrate concentration. In other words, the rate is a nonmonotonic function of the substrate concentration, as shown in Figure 1.1.

If the specific growth rate increases asymptotically with the substrate concentration to a maximum, as in the case of Monod-type monotonic kinetics, as shown in Figure 1.2, then it is obvious to keep the substrate concentration as high as possible to maximize the specific growth rate. Because the total growth rate is the product of

Table 1.1. *Various products produced or attempted to be produced by fed-batch techniques*

Product	References	Product	References
Amino Acids		**Solvents**	
DOPA	10, 148	Acetone and butanol	125, 164–166
Glutamic Acid	11–23, 149	Glycerol	126, 167–168
Lysine	24, 25, 150, 151	1,3-Propanediol	123, 124
Tyrosine	25	**Vitamins**	
Tryptophan	26–28, 152	Riboflavin	127–129, 169
Alanine	225	Vitamin B_{12}	130–133, 170
Antibiotics		**Others**	
Candidin and candihexin	29	Acetic acid	134, 195–196
Cephalosporin C	30–32, 153–155	Citric acid	135, 197–198
Chlorotetracycline	33	Gibberellic acid	136, 199–201
Griseofulvin	34–36	Gibberellins	137–140
Novobiocin	37	Neutral lipids	141
Oxytetracycline	38–39, 156	Sorbose from sorbitol	142
Penicillin	40–75, 157–158		
Rifamycin	76, 159–160		
Streptomycin	77–79, 161–162		
Tetracycline	80–81, 163		
Thuringiensin	222		
Baker's yeasts	4, 82–91		
Enzymes		**Others**	
Cellulase	92–96, 171–172	5-Aminolevulinic acid (ALA)	216
Galactosidase	97, 173–174	Monoclonal antibody	217
Isoamylase	98–99, 175–176	Sophorolipid	218
Penicillin amidase	100, 177	Polyhydroxyalkanoate (PHA)	220
Polygalacturonic acid	101	Dihydroxyacetone (DHA)	221
Trans-eliminase		Human interferon-γ	223
Protease	102–104, 178–180	Glutathione	224
β–amylase	105, 181	Clavulanic acid	226
α–amylase	182–187	Poly-β-hydroxybutyrate (PHB)	227
β–Galactosidase	106–107, 188–191		
β–Glucanase	108, 192–194		
β–Glucosidase	219		
Microbial Cell Mass		**Animal Cell Culture**	
Bacteria		Non-GS NS0 cell line	230
A bacterium	109	Hybridoma	228, 229, 231
Cellulomonas sp.	110, 202	CHO	232
Protaminobact ruber	111, 112, 203		
Pseudomonas	113, 204, 205		
Pseudomonas AM-1	114, 206–208		
Yeasts			
Candida boidinii	115		
Candida brassicae	116–118, 209		
Candida utilis	119–120, 210–211		
Methylomonas L3	212		
Pichia farinosa	121		
Pichia methanothermo	122, 213		
Saccharomyces cerevisiae	147, 214–215		

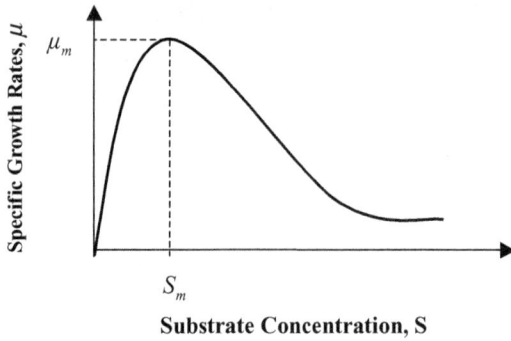

Figure 1.1. Nonmonotonic specific growth rate.

the growth rate and fermentor volume, μXV, it is advantageous to have the largest working volume as possible from the start. In other words, a batch culture with a full working volume is the best for this situation.

1.4 Alternatives to Fed-Batch Cultures

There are potential alternative operations. Why not run as a continuous culture with a proper dilution rate to maintain the substrate concentration at S_m, corresponding to the maximum specific growth rate? First, associated with a continuous culture is the problem of contamination by other microorganisms, attack by phages, and potential mutations over a long period of operation. In addition, the residual substrate concentration S_m may be substantial and may need to be further reduced for easier separation or a better yield. Therefore, an additional reactor is required to reduce the substrate concentration such as a batch or a large volume continuous fermentor. Also, there is evidence that the maximum production rate for some products can be achieved under a dynamic environment in which cells must continue to grow, however small it may be. For example, an attempt to replace the fed-batch fermentation by a continuous culture has not been successful for penicillin[144] and bacterial antigens.[145] Fed-batch cultures provide transient growth conditions for cells (oftentimes exponential growth of cells). Along this line of thought, it is well known that experimental kinetic data from one type of bioreactor may not be used to design another type of bioreactor, in particular, batch or fed-batch data may not be valid for continuous bioreactors, and vice versa. It is best to use experimental data obtained from the type of bioreactor that is to be used eventually to avoid this difficulty. This is presumed to be due to simplifying assumptions we make for a complex bioreactor.

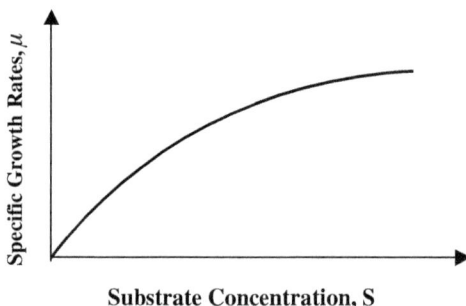

Figure 1.2. Monotonic specific growth rate, Monod type.

Conversely, chemical reactions are relatively simple in terms of our understanding, and therefore, the kinetic data obtained from one type of chemical reactor usually hold for any other type of reactor.

REFERENCES

1. Burrows, S. 1970. Baker's yeast, in *The Yeast*, vol. 3, ed. Rose, A. S., and Harrison, J. S., p. 350. Academic Press.
2. Yamane, T. 1978. Kinetic studies on fed-batch fermentations: A monograph. *Hakko Kogaku Kaishi* 56: 310–322.
3. White, J. 1954. *Yeast Technology*. Vol. 31. John Wiley.
4. Reed, G., and Peppler, H. J. 1973. *Yeast Technology*. Avi.
5. Yamane, T., and Shimizu, S. 1984. Fed-batch techniques in microbial processes. *Advances in Biochemical Engineering/Biotechnology* 30: 147–202.
6. Parelukar, S., and Lim, H. 1986. Modeling, optimization and control of semi-batch bioreactors. *Advances in Biochemical Engineering/Biotechnology* 32: 1207–1258.
7. Minihane, B. J., and Brown, D. E. 1986. Fed-batch culture technology. *Biotechnology Advances* 4: 207–218.
8. Hosler, P., and Johnson, M. J. 1953. Penicillin from chemically defined media. *Industrial and Engineering Chemistry* 45: 871–874.
9. Whitaker, A. 1980. Fed-batch culture. *Process Biochemistry* 15: 10–12, 14–15, 32.
10. Yamada, H., Kumagai, J., Enei, H., Matsui, H., and Okumura, S. 1972. Microbiological synthesis of L-tyrosine and 3,4-dihydroxyphenyl-L-alanine, in *Fermentation Technology Today*, ed. Terui, G., p. 445. Society for Fermentation Technology.
11. Nakamura, T., Kuratani, T., and Morita, Y. 1985. Fuzzy control application to glutamic acid fermentation. *Proceedings of IFAC Modeling and Control of Biotechnology Processes* Noordwijkerhout, The Netherlands, December.
12. Tsao, J. H., Chaung, H. L., and Wu, W. T. 1991. On-line state estimation and control in glutamic acid production. *Bioprocess Biosystem Engineering* 7: 35–39.
13. Su, Y. C., and Yamada, K. 1960. Studies on L-glutamic acid fermentation. Part III. Pilot plant scale test of the fermentative production of L-glutamic acid by *Brevibacterium divaricatum nov.* sp. *Bulletin Agricultural Chemistry Society Japan* 24: 525–529.
14. Kishimoto, M., Alfafara, C. G., Nakajima, M., Yoshida, T., and Taguchi, H. 1989. The use of the maharanobis and modified distances for the improvement of simulation of glutamic acid production. *Biotechnology and Bioengineering* 33: 191–196.
15. Kishimoto, M., Moo-Young, M., and Allsop, P. 1991. A fuzzy expert system for the optimization of glutamic acid production. *Bioprocess Engineering* 6: 173–177.
16. Kishimoto, M., Kitta, Y., Takeuchi, S., Nakajima, M., and Yoshida, T. 1991. Computer control of glutamic acid production based on fuzzy clusterization of culture phases. *Journal of Fermentation and Bioengineering* 72: 110–114.
17. Wu, J. Y., and Wu, W. T. 1992. Glutamic acid production in an airlift reactor with net draft tube. *Bioprocess Engineering* 8: 183–187.
18. Sanraku Ocean Company Ltd and Ajinomoto Company Ltd. 1968. French Patent 1,556,808.
19. Tanaka, K., and Kimura K. 1970. Method for producing L-glutamic acid. U.S. Patent 3,511,752.
20. Yamamoto, M., Nishida, H., Inui, T., and Ozaki, A. 1972. Microbial production of amino acids from aromatic compounds. II. Production of L-glutamic acid from benzoate. *Hakko Kogaku Zasshi* 50: 876–883.

21. Yamamoto, M., Nishida, H., Inui, T., and Ozaki, A. 1972. Microbial production of amino acids from aromatic compounds. III. Metabolic pathway of benzoate utilized by *Brevibacterium* species No. 6, a typical L-glutamate producer. *Hakko Kogaku Zasshi* 50: 884–892.

22. Kishimoto, M., Yoshida, T., and Taguchi, H. 1981. Simulation of fed-batch culture for glutamic acid production with ethanol feeding by use of regression analysis. *Journal of Fermentation Technology* 59: 43–48.

23. Kishimoto, M., Yoshida, T., and Taguchi, H. 1981. On-line optimal control of fed-batch culture of glutamic acid production. *Journal of Fermentation Technology* 59: 125–129.

24. Fr. Demande (Kyowa Fermentation Industries Company Ltd). 1969. Bacterial production of L-lysine. U.S. Patent 3,595,751.

25. Ohno, H., Nakanishi, F., and Takamatsu, T. 1976. Optimal control of a semi-batch fermentation. *Biotechnology and Bioengineering* 18: 847–864.

26. Ikeda, M., and Katsumata, R. 1999. Hyperproduction of tryptophan by *Corynebacterium glutamicum* with the modified pentose phosphate pathway. *Applied Environmental Microbiology* 65: 2497–2502.

27. Aiba, S., Imanaka, T., and Tsunekawa, H. 1980. Enhancement of tryptophan production by *Escherichia coli* as an application of genetic engineering. *Biotechnology Letters* 2: 525–532.

28. Aiba, S., Tsunekawa, H., and Imanaka, T. 1982. New approach to tryptophan production by *Escherichia coli*: Genetic manipulation of composite plasmids in vitro. *Applied Environmental Microbiology* 43: 289–297.

29. Martin, J. F., and McDaniel, L. E. 1974. Submerged culture production of the polyene antifungal antibiotics candidin and candihexin. *Developments in Industrial Microbiology* 15: 324–337.

30. Pirt, S. J. 1974. The theory of fed-batch culture with reference to the penicillin fermentation. *Journal of Applied Chemistry and Biotechnology* 24: 415–424.

31. Trilli, A., Michelini, V., Mantovani, V., and Pirt, S. J. 1977. Estimation of productivities in repeated fed batch cephalosporin fermentation. *Journal of Applied Chemistry and Biotechnology* 27: 219–224.

32. Matsumura, M., Imanaka, T., Yoshida, T., and Taguchi, T. 1981. Modeling of cephalosporin C production and its application to fed-batch culture. *Journal of Fermentation Technology* 59: 115–123.

33. Avanzini, F. 1963. Preparation of tetracycline antibiotics. British Patent 939,476.

34. Hockenhull, D. J. D. 1959. Griseofulvin. British Patent 868,958.

35. Hockenhull, D. J. D. 1962. Griseofulvin in submerged aerobic culture. British Patent 934,527.

36. Calam, C. T., Ellis, S. H., and McCann, M. J. 1971. Mathematical models of fermentations and a simulation of the griseofulvin fermentation. *Journal of Applied Chemistry and Biotechnology* 21: 181–189.

37. Smith, C. G. 1956. Fermentation studies with *Streptomyces niveus*. *Applied Microbiology* 4: 232–236.

38. Bosnjak, M., Stroj, A., Curcic, M., Adamovic, V., Gluncic, Z., and Bravar, D. 1985. Application of scale-down experiments in the study of kinetics of oxytetracycline biosynthesis. *Biotechnology and Bioengineering* 27: 398–408.

39. Ettler, P. 1987. Oil feeding during the oxytetracycline biosynthesis. *Acta Biotechnologie* 7: 3–8.

40. Johnson, M. J. 1952. Recent advances in penicillin fermentation. *Bulletin of the World Health Organization* 6: 99–121.

41. Moyer, A., and Coghill, R. D. 1946. Penicillin: VIII. Production of penicillin in surface cultures. *Journal of Bacteriology* 51: 57–78.

42. Singh, K., and Johnson, M. J. 1948. Evaluation of the precursors for penicillin G. *Journal of Bacteriology* 56: 339–355.

43. Coghill, R. D., and Moyer, A. J. 1947. Production of increased yields of penicillin. U.S. Patent 2,423,873.
44. Brown, W. E., and Peterson, W. H. 1950. Factors affecting production of penicillin in semi-pilot-plant equipment. *Industrial and Engineering Chemistry* 42: 1769–1775.
45. Brown, W. E., and Peterson, W. H. 1950. Penicillin fermentations in a Waldhof-type fermenter. *Industrial and Engineering Chemistry* 42: 1823–1826.
46. Foster, J. W., and McDaniel, L. E. 1952. Fermentation process. U.S. Patent 2,584,009.
47. Owen, S. P., and Johnson, M. J. 1955. Effect of temperature changes on the production of penicillin by *Penicillium chrysogenum. Applied Microbiology* 3: 375–379.
48. Kolachov, P. J., and Schneider, W. C. 1952. Continuous process for penicillin production. U.S. Patent 2,609,327.
49. Soltero, F. V., and Johnson, M. 1953. The effect of the carbohydrate nutrition on penicillin production by *Penicillium chrysogenum* Q-176. *Applied Microbiology* 1: 52–57.
50. Davey, V. F., and Johnson, M. 1953. Penicillin production in corn steep media with continuous carbohydrate addition. *Applied Microbiology* 1: 208–211.
51. Anderson, R. F., Whitmore, L. M. J., Brown, W. E., Peterson, W. H., Churchill, B. W., Roegner, F. R., Campbell, T. H., Backus, M. P., and Stauffer, J. F. 1953. Penicillin production by pigment-free molds. *Industrial and Engineering Chemistry* 45: 768–773.
52. Soltero, F. V., and Johnson, M. J. 1954. Continuous addition of glucose for evaluation of penicillin-producing cultures. *Applied Microbiology* 2: 41–44.
53. Anderson, R. F., Tornqvist, E. G. M., and Peterson, W. H. 1956. Effect of oil in pilot plant fermentations for penicillin production. *Journal of Agricultural Food Chemistry* 4: 556–559.
54. Freaney, T. E. 1958. Penicillin. U.S. Patent 2,830,934.
55. Pan, S. C., Bonanno, S., and Wagman, G. H. 1959. Efficient utilization of fatty oils as energy source in penicillin fermentation. *Applied Microbiology* 7: 176–180.
56. Tornqvist, E. G. M., and Peterson, W. H. 1956. Penicillin production by high-yielding strains of *Penicillium chrysogenum. Applied Microbiology* 4: 277–283.
57. Chaturbhuj, K., Gopalkrishnan, K. S., and Ghosh, D. 1961. Studies on the feed rate for precursor and sugar in penicillin fermentation. *Hindustan Antibiotic Bulletin* 3: 144–151.
58. Chaturbhuj, K., and Ghosh, D. 1961. Semi-continuous penicillin fermentation. *Hindustan Antibiotic Bulletin* 4: 50–51.
59. Hockenhull, D. J. D., and Mackenzie, R. M. 1968. Preset nutrient feeds for penicillin fermentation on defined media. *Chemistry and Industry* 19: 607–610.
60. Noguchi, Y., Kurihara, S., and Arao, O. 1960. Influence of feeding of nitrogen source on the penicillin production. *Hakko Kogaku Zasshi* 38: 514–517.
61. McCann, E. P., and Calam, C. T. 1972. Metabolism of *Penicillium chrysogenum* and the production of penicillin using a high yielding strain, at different temperatures. *Journal of Applied Chemistry and Biotechnology* 22: 1201–1208.
62. Matelova, V., Brecka, A., and Matouskova, J. 1972. New method of intermittent feeding in penicillin biosynthesis. *Applied Microbiology* 23: 669–670.
63. Pan, C. H., Hepler, L., and Elander, R. P. 1972. Control of pH and carbohydrate addition in the penicillin fermentation. *Developments in Industrial Microbiology* 13: 103–112.
64. Fishman, V. M., and Birokov, V. V. 1974. Kinetic model of secondary metabolite production and its use in computation of optimal conditions. *Biotechnology and Bioengineering Symposium* 4: 647–662.

65. Verkhovtseva, T. P., Lur'e, L. M., Itsygin, S. B., Stepanova, N. E., and Levitov, M. M. 1975. Carbon metabolism under conditions of controlled penicillin biosynthesis. *Antibiotiki* 20: 102–106.
66. Noguchi, Y., Miyakawa, R., and Arao, O. 1960. Penicillin fermentation: Relation of the amount of glucose and soybean oil added and the potency of the culture in the fermentation by feeding technique. *Journal of Fermentation Technology* 38: 511–514.
67. Rodrigues, J. A. D., and Maciel Filho, R. 1999. Production optimization with operating constraints for a fed-batch reactor with DMC predictive control. *Chemica Engineering Science* 54: 2745–2751.
68. Badino, A. C., Jr., Barboza, M., and Hokka, C. O. 1994. Power input and oxygen transfer in fed-batch penicillin production process. *Advances in Bioprocess Engineering* , Kluwer Academic Publishers, 157–162.
69. Menezes, J. C., Alves, S. S., Lemos, J. M., and Feyo de Azevedo, S. 1994. Mathematical modeling of industrial pilot-plant penicillin-G fed-batch fermentations. *Journal of Chemical Technology and Biotechnology* 61: 123–138.
70. Nicolai, B. M., Van Impe, J. F., Vanrolleghem, P. A., and Vandewalle, J. 1991. A modified unstructured mathematical model for the penicillin G fed-batch fermentation. *Biotechnology Letters* 13: 489–494.
71. Squires, R. W. 1972. Regulation of the penicillin fermentation by means of a submerged oxygen-sensitive electrode. *Developments in Industrial Microbiology* 13: 128–135.
72. Calam, C. T., and Russell, D. W. 1973. Microbial aspects of fermentation process development. *Journal of Applied Chemistry and Biotechnology* 23: 225–237.
73. Moyer, A. J., and Coghill, R. D. 1947. Penicillin. X. The effect of phenylacetic acid on penicillin production. *Journal of Bacteriology* 53: 329–341.
74. Hegewald, E., Wolleschensky, B., Guthke, R., Neubert, M., and Knorre, W. A. 1981. Instabilities of product formation in a fed-batch culture of *Penicillium chrysogenum*. *Biotechnology and Bioengineering* 23: 1563–1572.
75. Mou, D.-G., and Cooney, C. L. 1983. Growth monitoring and control through computer-aided on-line mass balancing in a fed-batch penicillin fermentation. *Biotechnology and Bioengineering* 25: 225–255.
76. Bapat, P. M., Padiyar, N. U., Dave, N. N., Bhartiya, S., and Wangikar, P. P. 2006. Model-based optimization of feeding recipe for rifamycin fermentation. *Journal of American Institute of Chemical Engineers* 52: 4248–4257.
77. Singh, A., Bruzelius, E., and Heding, H. 1976. Streptomycin: A fermentation study. *European Journal of Applied Microbiology* 3: 97–101.
78. Jackson, C. J., and Milner, J. 1950. Improvements in the production of streptomycin. British Patent 644,078.
79. Inoue, S., Nishizawa, Y., and Nagai, S. 1983. Stimulatory effect of ammonium on streptomycin formation by *Streptomyces griseus* growing on a glucose minimal medium. *Journal of Fermentation Technology* 61: 7–12.
80. Avanzith, F. 1963. Preparation of tetracycline antibiotics. British Patent 939,476.
81. Makarevich, V. G., Slugina, M. D., Upiter, G. D., Zaslavskaia, P. L., and Gerasimova, T. M. 1976. Regulation of tetracycline biosynthesis by controlling the growth of the producer. *Antibiotiki* 21: 205–210.
82. Wang, H. Y., Cooney, C. L., and Wang, D. I. C. 1979. Computer control of bakers' yeast production. *Biotechnology and Bioengineering* 21: 975–995.
83. Finn, B., Harvey, L., and McNeil, B. 2006. Near-infrared spectroscopic monitoring of biomass, glucose, ethanol and protein content in a high cell density baker's yeast fed-batch bioprocess. *Yeast* 23: 507–517.
84. Hospodka, J. 1966. Oxygen-absorption rate-controlled feeding of substrate into aerobic microbial cultures. *Biotechnology and Bioengineering* 8: 117–134.
85. Mikiewicz, T., Leniak, W., and Ziobrowski, J. 1975. Control of nutrient supply in yeast propagation. *Biotechnology and Bioengineering* 17: 1829–1832.

86. Aiba, S., Nagai, S., and Nishizawa, Y. 1976. Fed batch culture of *Saccharomyces cerevisiae*: A perspective of computer control to enhance the productivity in baker's yeast cultivation. *Biotechnology and Bioengineering* 18: 1001–1016.

87. Wang, H. Y., Cooney, C. L., and Wang, D. I. C. 1977. Computer-aided baker's yeast fermentations. *Biotechnology and Bioengineering* 19: 69–86.

88. Dairaku, K., Yamasaki, Y., Kuki, K., Shioya, S., and Takamatsu, T. 1981. Maximum production in a bakers' yeast fed-batch culture by a tubing method. *Biotechnology and Bioengineering* 23: 2069–2081.

89. Dairaku, K., Izumoto, E., Morikawa, H., Shioya, S., and Takamatsu, T. 1983. An advanced microcomputer coupled control system in a bakers' yeast-fed batch culture using a tubing method. *Journal of Fermentation Technology* 61: 189–196.

90. Bach, H. P., Woehrer, W., and Roehr, M. 1978. Continuous determination of ethanol during aerobic cultivation of yeasts. *Biotechnology and Bioengineering* 20: 799–807.

91. Nanba, A., Hirota, F., and Nagai, S. 1981. Microcomputer-coupled baker's yeast production. *Journal of Fermentation Technology* 59: 383–389.

92. Yamane, K., Suzuki, H., Hirotani, M., Ozawa, H., and Nishizawa, K. 1970. Effect of nature and supply of carbon sources on cellulase formation in *pseudomonas fluorescens* var. cellulose. *Journal of Biochemistry* 67(1):9–18.

93. Hulme, M. A., and Stranks, D. W. 1970. Induction and the regulation of production of cellulase by fungi. *Nature* 226: 469–470.

94. Waki, T., Suga, K., and Ichikawa, K. 1981. Production of cellulase in fed-batch culture. *Advances in Biotechnology* 1: 359–364.

95. Allen, A. L., and Mortensen, R. E. 1981. Production of cellulase from *Trichoderma reesei* in fed-batch fermentation from soluble carbon sources. *Biotechnology and Bioengineering* 23: 2641–2645.

96. Gottvaldova, M., Kucera, J., and Podrazky, V. 1982. Enhancement of cellulase production by *Trichoderma viride* using carbon/nitrogen double-fed-batch. *Biotechnology Letters* 4: 229–231.

97. Yang, X. M. 1992. Optimization of a cultivation process for recombinant protein production by *Escherichia coli*. *Journal of Biotechnology* 23: 271–289.

98. Fujio, Y., Kojima, S., Sambuichi, M., and Ueda, S. 1972. Isoamylase production by *Aerobacter aerogenes*. IV. Controlled feeding of the culture by dissolved oxygen and its analysis. *Journal of Fermentation Technology* 50: 546–552.

99. Fujio, Y., Sambuichi, M., and Ueda, S. 1971. Isoamylase production by *Aerobacter aerogenes*. III. Isoamylase formation in continuous cultures. *Journal of Fermentation Technology* 49: 626–631.

100. Carrington, T. R., Savidge, T. A., and Walmsley, M. F. 1966. Penicillin-splitting enzymes. British Patent 1,015,554.

101. Hsu, E. J., and Vaughn, R. H. 1969. Production and catabolite repression of the constitutive polygalacturonic acid trans-eliminase of *Aeromonas liquefaciens*. *Journal of Bacteriology* 98: 172–181.

102. Gutellberg, A. V. 1954. A method for the production of the plakalbumin-forming proteinase from *Bacillus subtilis*. *Comptes rendus des travaux du Laboratoire Carlsberg* 29: 27–35.

103. Bergman, D. E., Denison, F. W. J., and Friedland, W. C. 1962. Culture process for gibberellic acid. U.S. Patent 3,021,261.

104. Kalabokias, G. 1071. Microbial alkaline protease by fermentation with *Bacillus subtilis var licheniformis*. U.S. Patent 3,623,956.

105. Yamane, T., and Tsukano, M. 1977. Kinetic studies on fed-batch cultures. V. Effect of several substrate-feeding modes on production of extracellular-amylase by fed-batch culture of *Bacillus megaterium*. *Journal of Fermentation Technology* 55: 233–242.

106. Clark, D. J., and Marr, A. G. 1964. Studies on the repression of beta-galactosidase in *Escherichia coli*. *Biochemistry and Biophysics Acta* 92: 85–94.

107. Gray, P. P., Dunnill, P., and Lilly, M. D. 1973. Effect of controlled feeding of glycerol on β-galactosidase production by *Escherichia coli* in batch culture. *Biotechnology and Bioengineering* 15: 1179–1188.

108. Markkanen, P., Reinwall, A., and Linko, M. 1976. Increase of β-glucanase production by *Bacillus subtilis* by use of starch feeding during fermentor cultivation. *Journal of Applied Chemistry and Biotechnology* 26: 41–46.

109. Yamane, T., Kishimoto, M., and Yoshida, F. 1976. Semi-batch culture of methanol-assimilating bacteria with exponentially increased methanol feed. *Journal of Fermentation Technology* 54: 229–240.

110. Srinivasan, V. R., Fleenor, M. B., and Summers, R. J. 1977. Gradient-feed method of growing high cell density cultures of cellulomonas in a bench-scale fermentor. *Biotechnology and Bioengineering* 19: 153–155.

111. Yano, T., Kobayashi, T., and Shimizu, S. 1978. Fed-batch culture of methanol-utilizing bacterium with DO-stat. *Journal of Fermentation Technology* 56: 416–420.

112. Kobayashi, T., Yano, T., Mori, H., and Shimizu, S. 1979. Cultivation of microorganisms with a DO-stat and a silicone tubing sensor. *Biotechnology and Bioengineering Symposium* 9: 73–83.

113. Harrison, D. E. F., Topiwala, H., and Hamer, G. 1972. *Fermentation Technology Today*, ed. Terui, G. Society for Fermentation Technology. Japan, 491.

114. Nishio, N., Tsuchiya, Y., Hayashi, M., and Nagai, S. 1977. Studies on methanol metabolism. VI. A fed-batch culture of methanol-utilizing bacteria with pH stat. *Journal of Fermentation Technology* 55: 151–155.

115. Reuss, M., Gnieser, J., Reng, H. G., and Wagner, F. 1975. Extended culture of *Candida boidinii* on methanol. *European Journal of Applied Microbiology* 1: 295–305.

116. Woehrer, W., and Roehr, M. 1981. Regulatory aspects of bakers' yeast metabolism in aerobic fed-batch cultures. *Biotechnology and Bioengineering* 23: 567–581.

117. Mori, H., Yano, T., Kobayashi, T., and Shimizu, S. 1979. High density cultivation of biomass in fed-batch system with DO-stat. *Journal of Chemical Engineering of Japan* 12: 313–319.

118. Matsumura, M., Umemoto, K., Shinabe, K., and Kobayashi, J. 1982. Application of pure oxygen in a new gas entraining fermentor. *Journal of Fermentation Technology* 60: 565–578.

119. Watteeuw, C. M., Armiger, W. B., Ristroph, D. L., and Humphrey, A. E. 1979. Production of single cell protein from ethanol by fed-batch process. *Biotechnology and Bioengineering* 21: 1221–1237.

120. Edwards, V. H., Cottschalk, M. J., Noojin, A. Y., III, Tuthill, L. B., and Tannahill, A. L. 1970. Extended culture: Growth of *Candida utilis* at controlled acetate concentrations. *Biotechnology and Bioengineering* 12: 975–999.

121. Yamane, T., Matsuda, M., and Sada, E. 1981. Application of porous teflon tubing method to automatic fed-batch culture of microorganisms. II. Automatic constant-value control of fed substrate (ethanol) concentration in semi-batch culture of yeast. *Biotechnology and Bioengineering* 23: 2509–2524.

122. Minami, K., Yamamura, M., Shimizu, S., Ogawa, K., and Sekine, N. 1978. Single cell production from methanol. II. Methods and apparatus for methanol feeding with less growth-inhibition. *Journal of Fermentation Technology* 56: 35–40.

123. Yang, G., Tian, J., and Li, J. 2007. Fermentation of 1,3-propanediol by a lactate deficient mutant of *Klebsiella oxytoca* under micro-aerobic conditions. *Applied Microbiology and Biotechnology* 73: 1017–1024.

124. Zhao, Y. N., Chen, G., and Yao, S. J. 2006. Microbial production of 1,3-propanediol from glycerol by encapsulated *Klebsiella pneumoniae*. *Biochemical Engineering Journal* 32: 93–99.

125. Richard, S., Allenet, and CIE Soc (Societe Richard, Allenet & Cie) 1921. Acetone; butyl alcohol. British Patent 176,284.
126. Eoff, J. R., Linder, W. V., and Beyer, G. F. 1919. The production of glycerol from sugar by fermentation. *Journal of Industrial and Engineering Chemistry* 11: 842–845.
127. Moss, A. R., and Klien, R. 1949. Riboflavin. British Patent 615,847.
128. Pridham, T. G. 1951. Biological production of riboflavin. U.S. Patent 2,578,738.
129. Kojima, I., Yoshikawa, H., Okazaki, M., and Terui, G. 1972. Riboflavine production by *Eremothecium ashbyii*. I. Inhibiting factors of riboflavine production and their control. *Journal of Fermentation Technology* 50: 716–723.
130. Speedie, J. D., and Hull, G. W. 1960. Cobalamin preparation by fermentation. British Patent 829,232.
131. Jackson, P. W. 1960. Cobalamins. British Patent 846,149.
132. Hoffman-La Roche, F., Co., Akt-Ges. 1960. Vitamins of the B12 group. British Patent 866,488.
133. Kojima, I., Sato, H., and Fujiwara, Y. 1993. Vitamin B12 production by isopropanol assimilating microorganisms. *Journal of Fermentation and Bioengineering* 75: 182–186.
134. Masai, H., Kawamura, Y., and Yamada, K. 1978. Development of a new production process and use of fermentation vinegar. *Bulletin of the Agricultural Chemical Society, Japan* 52: R103–R109.
135. Shepard, M. W. 1963. Citric acid. U.S. Patent 3,083,144.
136. Darken, M. A., Jensen, A. L., and Shu, P. 1959. Production of gibberellic acid by fermentation. *Applied Microbiology* 7: 301–303.
137. Borrow, A., Jefferys, E. G., and Nixon, I. S. 1960. Gibberellic acid. British Patent 838,033.
138. Giordano, W., and Domenech, C. E. 1999. Aeration affects acetate destination in *Gibberella fujikuroi*. *FEMS Microbiology Letters* 180: 111–116.
139. Serzedello, A., Simao, S., and Whitaker, N. 1958. Effect of gibberellic acid on lettuce crops (*Lactuca sativa* Linneu). *Revista de Agricultura* 33: 117–122.
140. Bergman, D. E., Denison, F. W. Jr., and Friedland, W. C. 1962. Culture process for gibberellic acid. U.S. Patent 3,021,261.
141. Yamauchi, H., Mori, H., Kobayashi, T., and Shimizu, S. 1983. Mass production of lipids by *Lipomyces starkeyi* in microcomputer-aided fed-batch culture. *Journal of Fermentation Technology* 61: 275–280.
142. Mori, T., Kobayashi, T., and Shimizu, S. 1981. High density production of sorbose from sorbitol by fed-batch culture with DO-stat. *Journal of Chemical Engineering Japan* 14: 65–70.
143. Wright, D. G., and Calam, C. T. 1968. Importance of the introductory phase in penicillin production, using continuous flow culture. *Chemistry and Industry* 38: 1274–1275.
144. Pirt, S. J., Thackeray, E. J., and Harris-Smith, R. J. 1961. The influence of environment on antigen production by *Pasteurella pestis* studied by means of the continuous flow culture technique. *Journal of General Microbiology* 25: 119–130.
145. Lee, J., Lee, S. Y., Park, S., and Middelberg, A. P. J. 1999. Control of fed-batch fermentations. *Biotechnology Advances* 17: 29–48.
146. Yee, L., and Blanch, H. W. 1992. Recombinant protein expression in high cell density fed-batch cultures of *Escherichia coli*. *Bio/technology* 10: 1550–1556.
147. Namdev, P. K., Thompson, B. G., and Gray, M. R. 1992. Effect of feed zone in fed-batch fermentations of *Saccharomyces cerevisiae*. *Biotechnology and Bioengineering* 40: 235–246.
148. Lee, J.-Y., and Xun, L. 1998. Novel biological process for L-DOPA production from L-tyrosine by p-hydroxyphenylacetate 3-hydroxylase. *Biotechnology Letters* 20: 479–482.

149. Kitsuta, Y., and Kishimoto M. 1994. Fuzzy supervisory control of glutamic acid production. *Biotechnology and Bioengineering* 44: 87–94.
150. Tada, K., Kishimoto, M., Omasa, T., Katakura, Y., and Suga, K. 2001. Constrained optimization of L-lysine production based on metabolic flux using a mathematical programming method. *Journal of Bioscience and Bioengineering* 91: 344–351.
151. Drysch, A., El Massaoudi, M., Wiechert, W., de Graaf, A. A., and Takors, R. 2004. Serial flux mapping of *Corynebacterium glutamicum* during fed-batch L-lysine production using the sensor reactor approach. *Biotechnology and Bioengineering* 85: 497–505.
152. Dodge, T. C., and Gerstner, J. M. 2002. Optimization of the glucose feed rate profile for the production of tryptophan from recombinant *E coli*. *Journal of Chemical Technology and Biotechnology* 7: 1238–1245.
153. Cruz, A. J. G., Silva, A. S., Araujo, M. L. G. C., Giordano, R. C., and Hokka, C. O. 1999. Modelling and optimization of the cephalosporin C production bioprocess in a fed-batch bioreactor with invert sugar as substrate. *Chemical Engineering Science* 54: 3137–3142.
154. Kim, N. R., Lim, J. S., Hong, S. I., and Kim, S. W. 2005. Optimization of feed conditions in a 2.5-l fed-batch culture using rice oil to improve cephalosporin C production by *Cephalosporium acremonium* M25. *World Journal of Microbiology and Biotechnology* 21: 787–789.
155. Kim, J. H., Lim, J. S., and Kim, S. W. 2004. The improvement of cephalosporin C production by fed-batch culture of *Cephalosporium acremonium* M25 using rice oil. *Biotechnology and Bioprocess Engineering* 9: 459–464.
156. Papapanagiotou, P. A., Quinn, H., Molitor, J.-P., Nienow, A. W., and Hewitt, C. J. 2005. The use of phase inversion temperature (PIT) microemulsion technology to enhance oil utilisation during *Streptomyces rimosus* fed-batch fermentations to produce oxytetracycline. *Biotechnology Letters* 27: 1579–1585.
157. Wang, Z., Lauwerijssen, M. J. C., and Yuan, J. 2005. Combined age and segregated kinetic model for industrial-scale penicillin fed-batch cultivation. *Biotechnology and Bioprocess Engineering* 10: 142–148.
158. Tiller, V., Meyerhoff, J., Sziele, D., Schuegerl, K., and Bellgardt, K.-H. 1994. Segregated mathematical model for the fed-batch cultivation of a high-producing strain of *Penicillium chrysogenum*. *Journal of Biotechnology* 34: 119–131.
150. El-Tayeb, O. M., Salama, A. A., Hussein, M. M. M., and El-Sedawy, H. F. 2004. Optimization of industrial production of rifamycin B by *Amycolatopsis mediterranei*. III. Production in fed-batch mode in shake flasks. *African Journal of Biotechnology* ONLINE 3: 387–394.
160. Jin, Z. H., Lin, J. P., Xu, Z. N., and Cen, P. L. 2002. Improvement of industry-applied rifamycin B-producing strain, *Amycolatopsis mediterranei*, by rational screening. *Journal of General and Applied Microbiology* 48: 329–334.
161. Zhang, Q., Jin, X., Wang, S., Rong, G., and Chen, Y. 2000. Modeling for streptomycin fermentation process based on principal component analysis and fuzzy model. *Wuxi Qinggong Daxue Xuebao* 19: 446–450.
162. Xiaoming Jin, X., Wang, S., and Chu, J. 2003. Hybrid modeling and monitoring of streptomycin fermentation process. *Proceedings 42nd IEEE Conference on Design and Control* 5: 4765–4769.
163. Ross, A., and Schuegerl, K. 1998. Tetracycline production by *Streptomyce aureofaciens*: The time lag of production. *Applied Microbiology and Biotechnology* 29: 174–180.
164. Qureshi, N., Ezeji, T. C., Blaschek, H. P., and Cotta, M. A. 2004. A novel biological process to convert renewable biomass to acetone and butanol (AB). AIChE Annual Meeting, Austin, TX, Nov. 7–12.
165. Tashiro, Y., Takeda, K., Kobayashi, G., Sonomoto, K., Ishizaki, A., and Yoshino, S. 2004. High butanol production by *Clostridium saccharoperbutylacetonicum*

N1–4 in fed-batch culture with pH-stat continuous butyric acid and glucose feeding method. *Journal of Bioscience and Bioengineering* 98: 263–268.

166. Ezeji, T. C., Qureshi, N., and Blaschek, H. P. 2004. Acetone butanol ethanol (ABE) production from concentrated substrate: Reduction in substrate inhibition by fed-batch technique and product inhibition by gas stripping. *Applied Microbiology and Biotechnology* 63: 653–658.

167. Xie, D., Liu, D., Zhu, H., and Zhang, J. 2002. Optimization of glycerol fed-batch fermentation in different reactor states: A variable kinetic parameter approach. *Applied Biochemistry and Biotechnology* 101: 131–151.

168. Xie, D., Liu, D., Zhu, H., and Liu, T. 2001. Multipulse feed strategy for glycerol fed-batch fermentation: A steady-state nonlinear optimization approach. *Applied Biochemistry and Biotechnology* 95: 103–112.

169. Lu, W., Zhang, K., and Wu, P. 2000. Fed-batch fermentation of *Eremothecium ashbyii* for riboflavin production. *Wuxi Qinggong Daxue Xuebao* 19: 240–243.

170. Andriantsoa, M., Laget, M., Cremieux, A., and Dumenil, G. 1984. Constant fed-batch culture of methanol-utilizing corynebacterium producing vitamin B12. *Biotechnology Letters* 6: 783–788.

171. Yong, Q., Hong, F., Ding, Y., and Yu, S. 1998. Production of cellulase in fed-batch cultures. *Linchan Huaxue Yu Gongye* 18: 73–77.

172. McLean, D., and Podruzny, M. F. 1985. Further support for fed-batch production of cellulases. *Biotechnology Letters* 7: 683–688.

173. Prytz, I., Sanden, A. M., Nystroem, T., Farewell, A., Wahlstroem, A., Foerberg, C., Pragai, Z., Barer, M., Harwood, C., and Larsson, G. 2003. Fed-batch production of recombinant β-galactosidase using the universal stress promoters uspA and uspB in high cell density cultivations. *Biotechnology and Bioengineering* 83: 595–603.

174. Nor, Z. M., Tamer, M. I., Mehrvar, M., Scharer, J. M., Moo-Young, M., and Jervis, E. J. 2001. Improvement of intracellular-galactosidase production in fed-batch culture of *Kluyveromyces fragilis*. *Biotechnology Letters* 23: 845–849.

175. Lai, J.-T., and Liu, H.-S. 1996. Production enhancement of *Pseudomonas amyloderamosa* isoamylase. *Bioprocess Engineering* 15: 139–144.

176. Wu, D. H., Wen, C. Y., Chu, W. S., Lin, L. L., and Hsu, W. H. 1993. Selection of antibiotic-resistant mutants with enhanced isoamylase activity in *Pseudomonas amyloderamosa*. *Biotechnology Letters* 15: 883–888.

177. Rao, K. J., and Panda, T. 1996. Effect of mode of operation of bioreactors on the biosynthesis of penicillin amidase in *Escherichia coli*. *Bioprocess Engineering* 14: 317–321.

178. Beshay, U., and Moreira, A. 2005. Production of alkaline protease with *Teredinobacter turnirae* in controlled fed-batch fermentation. *Biotechnology Letters* 27: 1457–1460.

179. Dutta, J. R., Dutta, P. K., and Banerjee, R. 2005. Modeling and optimization of protease production by a newly isolated *Pseudomonas* sp. using a genetic algorithm. *Process Biochemistry* 40: 879–884.

180. Mao, W., Pan, R., and Freedman, D. 1992. High production of alkaline protease by *Bacillus licheniformis* in a fed-batch fermentation using a synthetic medium. *Journal of Industrial Microbiology* 11: 1–6.

181. Ahmad, R., Qadeer, M. A., Jafri, R. H., Baig, M. A., and Khan, J. 1995. Kinetic studies on the biosynthesis of *Bacillus polymyxa* PCSIR-90 β-amylase. *Pakistan Journal of Scientific Research* 47: 55–61.

182. Shiina, S., Ohshima, T., and Sato, M. 2007. Extracellular production of α-amylase during fed-batch cultivation of recombinant *Escherichia coli* using pulsed electric field. *Journal of Electrostatics* 65: 30–36.

183. Huang, H., Ridgway, D., Gu, T., and Moo-Young, M. 2004. Enhanced amylase production by *Bacillus subtilis* using a dual exponential feeding strategy. *Bioprocess Biosystem Engineering* 27: 63–69.

184. Skolpap, W., Scharer, J. M., Douglas, P. L., and Moo-Young, M. 2004. Fed-batch optimization of α-amylase and protease-producing *Bacillus subtilis* using Markov chain methods. *Biotechnology and Bioengineering* 86: 706–717.

185. Enayati, N., Tari, C., Parulekar, S. J., Stark, B. C., and Webster, D. A. 1999. Production of α-amylase in fed-batch cultures of vgb+ and vgb⁻ recombinant *Escherichia coli*: Some observations. *Biotechnology Progress* 15: 640–645.

186. Lee, J., and Parulekar, S. J. 1993. Enhanced production of α-amylase in fed-batch cultures of *Bacillus subtilis* TN106[pAT5]. *Biotechnology and Bioengineering* 42: 1142–1150.

187. Yoo, Y. J., Cadman, T. W., Hong, J., and Hatch, R. T. 1998. Fed-batch fermentation for the production of α-amylase by *Bacillus amyloliquefaciens*. *Biotechnology and Bioengineering* 31: 426–432.

188. Prytz, I., Sanden, A. M., Nystroem, T., Farewell, A., Wahlstroem, A., Foerberg, C., Pragai, Z., Barer, M., Harwood, C., and Larsson, G. 2003. Fed-batch production of recombinant β-galactosidase using the universal stress promoters uspA and uspB in high cell density cultivations. *Biotechnology and Bioengineering* 83(5): 595–603.

189. Nor, Z. M., Tamer, M. I., Mehrvar, M., Scharer, J. M., Moo-Young, M., and Jervis, E. J. 2001. Improvement of intracellular β-galactosidase production in fed-batch culture of *Kluyveromyces fragilis*. *Biotechnology Letters* 23: 845–849.

190. Patnaik, P. R. 1999. Neural control of an imperfectly mixed fed-batch bioreactor for recombinant β-galactosidase. *Biochemical Engineering Journal* 3: 113–120.

191. Bedard, C., Perret, S., and Kamen, A. A. 1997. Fed-batch culture of Sf-9 cells supports 3×10^7 cells per ml and improves baculovirus-expressed recombinant protein yields. *Biotechnology Letters* 19: 629–632.

192. Shene, C., Andrews, B. A., and Asenjo, J. A. 1999. Fedbatch fermentations of *Bacillus subtilis* ToC46 (pPFF1) for the synthesis of a recombinant β-1,3-glucanase: Experimental study and modeling. *Enzyme and Microbial Technology* 24: 247–254.

193. Chuen-Im, S., and Lynch, H. C. 1998. Production of a yeast cell wall degrading enzyme, β-1,3-glucanase by recombinant *Bacillus subtilis*. *Biochemical Society Transaction* 26: S175.

194. Argyropoulos, D., and Lynch, H. C. 1997. Recombinant β-glucanase production and plasmid stability of *Bacillus subtilis* in cyclic fed-batch culture. *Biotechnology Techniques* 11: 187–190.

195. Ozadali, F., Glatz, B. A., and Glatz, C. E. 1996. Fed-batch fermentation with and without online extraction for propionic and acetic acid production by *Propionibacterium acidipropionici*. *Applied Microbiology and Biotechnology* 44: 710–716.

196. Eaton, D. C., and Gabelman, A. 1995. Fed-batch and continuous fermentation of *Selenomonas ruminantium* for natural propionic, acetic and succinic acids. *Journal of Industrial Microbiology* 15: 32–38.

197. Crolla, A., and Kennedy, K. J. 2004. Fed-batch production of citric acid by *Candida lipolytica* grown on n-paraffins. *Journal of Biotechnology* 110: 73–84.

198. Rakshit, S. K., Khan, N. H., and Lakshmi, K. V. 1994. A fed-batch surface culture process for the production of citric acid with reuse of *Aspergillus niger* mycelia. *Bioprocess Engineering* 11: 199–201.

199. Shukla, R., Chand, S., and Srivastava, A. K. 2005. Improvement of gibberellic acid production using a model based fed-batch cultivation of *Gibberella fujikuroi*. *Process Biochemistry* (Oxford) 40: 2045–2050.

200. Bandelier, S., Renaud, R., and Druand, A. 1996. Production of gibberellic acid by fed-batch solid state fermentation in an aseptic pilot-scale reactor. *Process Biochemistry* (Oxford) 32: 141–145.

201. Lale, G., Jogdand, V. V., and Gadre, R. V. 2006. Morphological mutants of *Gibberella fujikuroi* for enhanced production of gibberellic acid. *Journal of Applied Microbiology* 100: 65–72.

202. Chapman, B. D., Schleicher, M., Beuger, A., Gostomski, P., and Thiele, J. H. 2006. Improved methods for the cultivation of the chemolithoautotrophic bacterium *Nitrosomonas europaea. Journal of Microbiological Methods* 65: 96–106.

203. Rodriguez, H., and Gallardo, R. 1993. Single-cell protein production from bagasse pith by a mixed bacterial culture. *Acta Biotechnology* 13: 141–149.

204. Yano, T., Kurokawa, M., and Nishizawa, Y. 2001. Optimum substrate feed rate in fed-batch culture with the DO-stat method. *Journal of Fermentation and Bioengineering* 71: 345–349.

205. Sato, K., Maruyama, K., Mori, H., Suzuki, T., and Shimizu, S. 1984. Protein production by lysis with glycine of a facultative methylotroph, *Protaminobacte ruber*, cultivated in a fed-batch system. *Journal of Fermentation Technology* 62: 301–303.

206. Diniz, S. C., Taciro, M. K., Gomez, J. G. C., Pradella, J. G. da Cruz. 2004. High-cell-density cultivation of *Pseudomonas putida* IPT 046 and medium-chain-length polyhydroxyalkanoate production from sugarcane carbohydrates. *Applied Biochemistry and Biotechnology* 119: 51–69.

207. Thuesen, M. H., Nørgaard, A., Hansen, A. M., Caspersen, M. B., and Christensen, H. E. M. 2003. Expression of recombinant *Pseudomonas stutzeri* di-heme cytochrome c4 by high-cell-density fed-batch cultivation of *Pseudomonas putida. Expression and Purification* 27: 175–181.

208. Tocaj, A., Hof, A., Hagander, P., and Holst, O. 1993. Fed-batch cultivation of *Pseudomonas cepacia* with on-line control of the toxic substrate salicylate. *Applied Microbiology and Biotechnology* 38: 463–466.

209. Vandeska, E., Amartey, S., Kuzmanova, S., and Jeffries, T. W. 1996. Fed-batch culture for xylitol production by *Candida boidinii. Process Biochemistry* (Oxford) 31: 265–270.

210. Yano, T., Endo, T., Tuji, T., and Nishizawa, Y. 1991. Fed-batch culture with a modified DO-stat method. *Journal of Fermentation and Bioengineering* 71: 35–38.

211. Huang, H. P., Chang, L. L., Chao, Y. C., and Huang, S. Y. 1988. On-line optimal feed of substrate during fed-batch culture of cell mass. *Chemical Engineering Communication* 68: 221–236.

212. Cheng, C., and Ma, J.-H. 1996. Enantioselective synthesis of S-(-)-1-phenylethanol in *Candida utilis* semi-fed-batch cultures. *Process Biochemistry* (Oxford) 31: 119–124.

213. Vijaikishore, P., and Karanth, N. G. 1987. Glycerol production by fermentation: A fed-batch approach. *Biotechnology and Bioengineering* 30: 325–328.

214. Finn, B., Harvey, L. M., and McNeil, B. 2006. Near-infrared spectroscopic monitoring of biomass, glucose, ethanol and protein content in a high cell density baker's yeast fed-batch bioprocess. *Yeast* 23: 507–517.

215. Belo, I., Pinheiro, R., and Mota, M. 2003. Fed-batch cultivation of *Saccharomyces cerevisiae* in a hyperbaric bioreactor. *Biotechnology Progress* 19: 665–671.

216. Qin, G., Lin, J., Liu, X., and Cen, P. 2006. Effects of medium composition of 5 aminolevulinic acid by recombinant *Escherichia coli. Journal of Bioscience and Bioengineering* 102: 316–322.

217. Choo, C. Y., Tian, Y., Kim, W. S., Blatter, E., Conary, J., and Brady, C. P. 2007. High-level production of a monoclonal antibody in murine melanoma cells by perfusion culture using a gravity settler. *Biotechnology Progress* 23: 225–231.

218. Felse, P. A., Shah, V., Chan, J., Rao, K. J., and Gross, R. A. 2007. Sophorolipid biosynthesis by *Candida bombicola* from industrial fatty acid residues. *Enzyme Microbial Techniques* 40: 316–323.

219. Amouri, B., and Gargouri, A. 2006. Characterization of a novel β-glucosidase from a *Stachybotrys* strain. *Biochemical Engineering Journal* 32: 191–197.
220. Koller, M., Horvat, P., Hesse, P., Bona, R., Kutschera, C., Atlic, A., and Braunegg, G. 2006. Assessment of formal and low structured kinetic modeling of polyhydroxy-alkanoate synthesis from complex substrates. *Bioprocess Biosystem Engineering* 29: 367–377.
221. Hekmat, D., Bauer, R., and Neff, V. 2007. Optimization of the microbial synthesis of dihydroxyacetone in a semi-continuous repeated-fed-batch process by in situ immobilization of *Gluconobacter oxydans*. *Process Biochemistry* 42: 71–76.
222. Zhou, J. W., Chang, Y. F., Xu, Z. H., Yu, Z. N., and Chen, S. W. 2007. Production of thuringiensin by fed-batch culture of *Bacillus thuringiensis* subsp. Darmstadiensis 032 with an improved pH-control glucose feeding strategy. *Process Biochemistry* 42: 52–56.
223. Babaeipour, V., Shojaosadati, S. A., Robatjazi, S. M., Khalilzadeh, R., and Maghsoudi, N. 2007. Over-production of human interferon-γ by HCDC of recombinant *Escherichia coli*. *Process Biochemistry* 42: 112–117.
224. Wen, S., Zhang, T., and Tan, T. 2006. Maximizing production of glutathione by amino acid modulation and high-cell-density fed-batch culture of *Saccharomyces cerevisiae*. *Process Biochemistry* 41: 2424–2428.
225. Smith, G. M., Lee, S. A., Reilly, K. C., Eiteman, M. A., and Altman, E. 2006. Fed-batch two phase production of alanine by a metabolically engineered *Escherichia coli*. *Biotechnology Letters* 28: 1695–1700.
226. Teodoro, J. C., Baptista-Neto, A., Cruz-Hernandez, I. L., Hokka, C. O., and Badino, A. C. 2006. Influence of feeding conditions on clavulanic acid production in fed-batch cultivation with medium containing glycerol. *Applied Microbiology and Biotechnology* 72: 450–455.
227. Patnaik, P. R. 2006. Enhancement of PHB biosynthesis by *Ralstonia eutropha* in fed-batch cultures by neural filtering and control. *Food Bioprocesses* 84: 150–156.
228. Xie, L., and Wang, D. C. 1994. Fed-batch cultivation of animal cells using different medium design concepts and feeding strategies. *Biotechnology and Bioengineering* 43: 1175–1189.
229. Zhou, W., Rehm, J., and Hu, W. S. 1995. High viable cell concentration fed-batch cultures of hybridoma cells through online nutrient feeding. *Biotechnology and Bioengineering* 46: 579–587.
230. Burky, J. E., Wesson, M. C., Young, A., Farnsworth, S., Dionne, B., Zhu, Y., Hartman, T. E., Qu, L., Zhou, W., and Sauer, P. W. 2007. Protein-free fed-batch culture of non-GS NS0 cell lines for production of recombinant antibodies. *Biotechnology and Bioengineering* 96: 281–293.
231. Dhir, S., Morrow, K. J., Jr., Rhinehart, R. R., and Wiesner, T. 2000. Dynamic optimization of hybridoma growth in a fed-batch bioreactor. *Biotechnology and Bioengineering* 67: 197–205.
232. Wong, D. C. F., Wong, K. T. K., Goh, L. T., Heng, C. K., and Yap, M. G. S. 2005. Impact of dynamic online fed-batch strategies on metabolism, productivity and N-glycosylation quality in CHO cell cultures. *Biotechnology and Bioengineering* 89: 164–177.

2 Idealized Reactors and Fed-Batch Reactors

Prior to presenting the principles of fed-batch operations, it is instructive to review quickly various forms of idealized reactors: batch, continuous, and semi-batch reactors. This will help in understanding the simplest form of fed-batch operation as equivalent to a continuous-stirred tank reactor (CSTR) followed by a batch reactor (BR). It becomes simple and easy to understand the operation of a fed-batch bioreactor as mimicking a dynamic CSTR followed by a BR to maximize the reaction rate or product yield. In other words, for a single reaction, the sufficient condition for superior performance of fed-batch operation is that the reaction rate and/or product yield show a maximum or decrease with the substrate concentration.

Basically, a fed-batch operation is preferred when the main or a side reaction rate exhibits a maximum or if the side reaction is more sensitive than the main reaction to reactant concentrations. Fed-batch operations can take advantage of the maximum rate or the sensitivity of side reactions (yield) by manipulating the substrate (reactant) concentration in the reactor.

2.1 Material Balances

A quantitative description of a reactor operation requires a material balance for each independent species and an energy balance for heat removal or addition calculations. For isothermal reactions, as most fermentation processes are run isothermally at temperatures slightly above or at room temperature, an energy balance is needed to determine only the heat removal rate; the material balances describe the process. Therefore, henceforth, we assume isothermal operations.

A material balance for each key species is written for a control volume, taking into account the species entering and leaving the control volume and the species generated or consumed by reactions within the control volume, as depicted in Figure 2.1. Although we begin with one input stream for simplicity, an extension to the case of multiple inputs is straightforward by tracking the same species in all of the input streams.

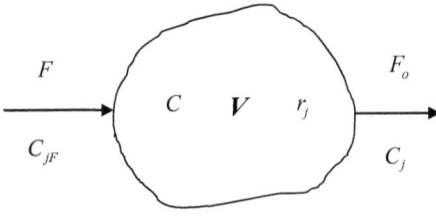

Figure 2.1. A control volume with input and output for mass balance.

$$
\begin{bmatrix} \text{Flow rate of species} \\ j \text{ into the control} \\ \text{volume (moles/time} \\ \text{or grams/time)} \end{bmatrix} - \begin{bmatrix} \text{Flow rate of species} \\ j \text{ out of the control} \\ \text{volume (moles/time} \\ \text{or grams/time)} \end{bmatrix}
$$

(2.1)

$$
+ \begin{bmatrix} \text{Generation (or } - \text{consumption)} \\ \text{rate of species } j \text{ within the} \\ \text{control volume (moles/time or} \\ \text{grams/time)} \end{bmatrix} = \begin{bmatrix} \text{Accumulation rate} \\ \text{of species } j \text{ within} \\ \text{the control volume} \\ \text{(moles/time or} \\ \text{grams/time)} \end{bmatrix}
$$

Notice the units in Eq. (2.1). The normal units are moles/time for chemical reactions in which the molecular weights of the reactants, intermediates, and products are precisely known. For reactions involving living cells, that is not the case, as we cannot assign molecular weights to cells. Therefore, it is usually more convenient to use the weight unit rather than the mole unit. In symbols, Eq. (2.1) may be written for the j species as

$$
FC_{jF} - F_o C_j + \int^V r_j dV = \frac{d(C_j V)}{dt}
$$

(2.2)

where F and F_o are the volumetric flow rates of the influent and effluent, respectively; C_{jF} and C_j are the concentrations of species j in the influent and the effluent (or reactor), respectively; and r_j is the rate of generation of species j per unit control volume per unit time. When a species is consumed, the generation term takes on a negative value. In general, the generation rate r_j is a function of concentrations of one or more reactant and product species as well as the temperature and pH.

In an idealized batch or a continuous-flow stirred tank reactor, the mixing is assumed to be perfect so that any material supplied into the reactor is mixed instantaneously and homogeneously. The latter implies that samples taken from any parts of the reactor would have the same composition, that is, that there is no spatial variation. When the mixing within the control volume is perfect so that there is no spatial variation within the control volume, the rate can be taken out of the integral in Eq. (2.2), and the volume is integrated to yield

$$
FC_{jF} - F_o C_j + r_j V = \frac{d(C_j V)}{dt}
$$

(2.3)

Equation (2.3) is a general mass balance equation for a species j that goes through a reaction in the control volume. If the reaction rate r_j depends on the concentrations of other species, then additional mass balances for those species are required.

The reactor volume may change due to the difference between the influent and effluent rates and also due to evaporation caused by aeration. The overall material balance, ignoring the evaporation loss, is

$$F\rho - F_o\rho_o = \frac{d(V\rho_o)}{dt} \quad V(0)\rho_o(0) = V_0\rho_{o0} \tag{2.4}$$

where ρ and ρ_o represent the densities of the feed and effluent streams, respectively. Because perfect mixing is assumed, the physical properties of the effluent are the same as those of the reacting fluid in the reactor.

2.2 Various Types of Ideal Reactors

It is appropriate at this point to review very briefly various types of ideal reactor operations prior to using the preceding equation and to introduce the concept of semi-batch operation. Reactors can be classified into three types: batch, continuous flow, and semi-batch. Continuous-flow reactors include CSTRs and tubular reactors. Tubular reactors include the idealized plug-flow reactors (PFRs) and packed-bed reactors (PBRs).

2.2.1 Batch Reactor Operation

In a BR operation, all nutrients and the inoculum (or reactants) are placed in the reactor at the beginning, and no further supply or withdrawal is made during the course of reactor operation. In other words, there is neither a feed (influent) nor a harvest stream (effluent). Controls necessary to keep constant such variables as the temperature, dissolved oxygen, and pH and to handle foaming problems are applied during the course of batch operation.

A BR does not require much supporting equipment compared to a continuous reactor and is therefore used for small-scale operations, including experimental studies of reaction kinetics, the production of expensive products, and processes that are not amenable to continuous operations.

The material balance on species j in an ideal BR (Figure 2.2) is obtained from Eq. (2.3). Because there is no input and output, the species mass balance becomes

$$r_j V = \frac{d(C_j V)}{dt} \tag{2.5}$$

If reactor volume remains constant, it may be taken out of the right-hand side and canceled by the volume on the left-hand side to yield

$$r_j = \frac{dC_j}{dt} \quad C_j(0) = C_{j0} \tag{2.6}$$

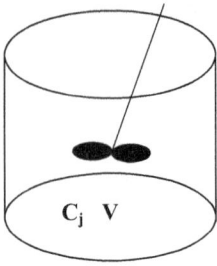

Figure 2.2. A batch reactor.

Equation (2.6) can be rearranged and integrated with the initial condition if the rate is known as a function of the reactant concentration:

$$t = \int_{C_{j0}}^{C_j(t)} \frac{dC_j}{r_j(C_j)} \quad r_j = \frac{dC_j}{dt} \quad C_j(0) = C_{j0} \tag{2.7}$$

Equation (2.7) yields the time necessary to obtain the final reactant concentration, or it gives the reactant concentration at any given time. The time to achieve the desired conversion is determined from Eq. (2.7) if the rate expression $r_j(C_j)$ is known, or it can be obtained from experimental rate data. The reciprocal of the rate of formation, $1/r_j(C_j)$, is plotted against the reactant concentration C_j and the area under the curve between the initial concentration C_{j0}, and the desired concentration $C_j(t)$ represents the reaction time necessary to convert the jth species from C_{j0} to $C_j(t)$. This is shown in Figure 2.3.

For a BR in which microbial cells are grown on a limiting substrate, the mass balances for cell mass and substrate are obtained from Eq. (2.3) by recognizing that the cellular growth rate is the product of the specific growth rate μ and cell concentration X, $r_X = \mu X$, and that the substrate consumption rate is the product of the specific substrate consumption rate σ and cell concentration, $r_S = -\sigma X$:

$$r_X V = \mu X V = \frac{d}{dt}(XV) \tag{2.8}$$

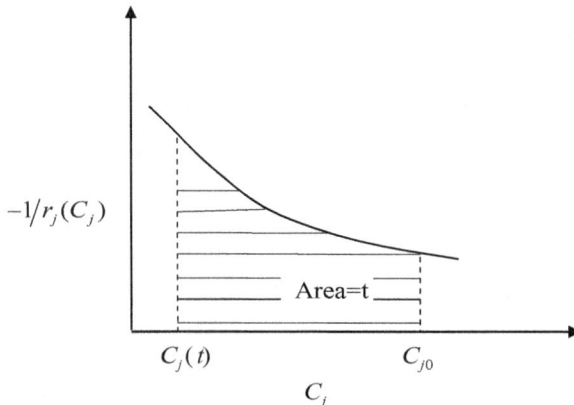

Figure 2.3. A graphical representation of conversion-time for a batch reactor.

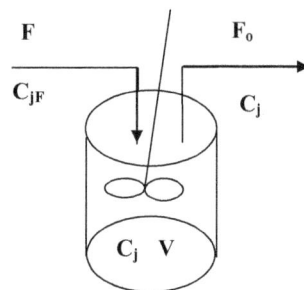

Figure 2.4. A continuous-stirred tank reactor (CSTR).

and

$$r_S V = -\sigma X V = \frac{d}{dt}(SV) \tag{2.9}$$

Here the specific rates of cell growth and substrate consumption, $\mu(S) = r_X/X$ and $\sigma(S) = -r_S/X$, are usually dependent on the substrate concentration S. Under a constant-volume assumption, these equations reduce to

$$r_X = \mu(S)X = \frac{dX}{dt}, \quad X(0) = X_0 \tag{2.10}$$

and

$$r_S = -\sigma(S)X = \frac{dS}{dt}, \quad S(0) = S_0 \tag{2.11}$$

respectively. When the functional forms of $\mu(S)$ and $\sigma(S)$ are known, Eqs. (2.10) and (2.11) can be integrated simultaneously to obtain the time profiles of $X(t)$ and $S(t)$ and therefore the time necessary to obtain the desired cell concentration or the residual substrate concentration.

2.2.2 Continuous-Flow Reactor Operation

Continuous-flow reactors include a CSTR, a PFR, and a tubular packed reactor (TPR) packed with immobilized cells, catalyst, or other packing.

2.2.2.1 Continuous-Stirred Tank Reactor

A very common type of reactor with a continuous feed and a continuous withdrawal is called by various names such as a "continuous-stirred tank reactor," a "continuous-flow stirred-tank reactor" (CFSTR), a "back-mix reactor," or a "mixed reactor." As shown in Figure 2.4, reactants are introduced via the feed stream(s), and the products and unreacted reactants are withdrawn by the effluent stream. Because of the influent and effluent streams, the CSTR can be operated at unsteady state, or it may be operated at steady state after a short start-up period of unsteady state operation. The CSTR is usually operated at steady state and therefore provides constant reaction conditions. CSTRs are ideal for processing large quantities of reaction materials. The common homogeneous liquid-phase flow reactors are CSTRs. One of the disadvantages of CSTRs is that for reactions whose rates decrease with reactant concentration, the conversion of reactant per reactor volume is the smallest among all continuous reactors and therefore requires a larger volume reactor. The material balance for species j in the CSTR shown in Figure 2.4 is obtained from Eq. (2.3):

$$FC_{jF} - F_o C_j + r_j V = \frac{d(C_j V)}{dt} \quad C_j(0)V(0) = C_{j0}V_0 \tag{2.12}$$

For aqueous processes involving microbial, plant, and mammalian cells, the density difference may be small, $\rho = \rho_o$, and may not change appreciably so that the overall mass balance (Eq. (2.4)) reduces to

$$F - F_o = \frac{d(V)}{dt} \quad V(0) = V_0 \tag{2.13}$$

There is a period of unsteady state operation due to the start-up of the reactor before the reactor operation reaches steady state. If the volume remains constant due to equal feed and withdrawal rates ($F = F_o$), the species balance (Eq. (2.12)) reduces to

$$F(C_{jF} - C_j) + r_j V = V \frac{dC_j}{dt} \quad C_j(0) = C_{j0} \tag{2.14}$$

At steady state, the influent and effluent volumetric flow rates are the same, $F = F_o$; the species concentration remains constant, and the reactor volume remains constant so that Eq. (2.14) reduces to

$$\frac{F}{V}(C_{jF} - C_j) + r_j = 0 \tag{2.15}$$

where F/V is called the *dilution rate*. Rearranging Eq. (2.15) yields

$$\frac{F}{V} = D = \frac{-r_j(C_j)}{C_{jF} - C_j} = \frac{-r_1(C_1)}{C_{1F} - C_1} = \frac{-r_2(C_2)}{C_{2F} - C_2} = \cdots = \frac{-r_n(C_n)}{C_{nF} - C_n} \tag{2.16}$$

The dilution rate fixes the concentration of each species in the reactor and, therefore, in the effluent. The space time, τ_{CSTR}, which is the reciprocal of the dilution rate, necessary to convert the reactant concentration from C_{jF} to C_{jf}, is obtained from Eq. (2.15):

$$\tau_{CSTR} = \frac{V}{F} = \frac{1}{D} = \left(\frac{1}{-r_{jf}}\right)(C_{jF} - C_{jf}) \tag{2.17}$$

A graphical interpretation of Eq. (2.17) can be obtained readily by recognizing that the right-hand side is a rectangle whose height is $1/-r_{jf}$ and whose base is the difference between the feed and the effluent concentrations, $(C_{jF} - C_{jf})$. This is shown in Figure 2.5.

It is apparent that for reaction rates that decrease with the reactant concentration (thus, the reciprocal rates increase monotonically), the clock time for a BR (black hatched area) is larger than the space time of a CSTR (black lined area). Therefore, the reactor volume required to achieve a given conversion is smaller for a CSTR than for a BR.

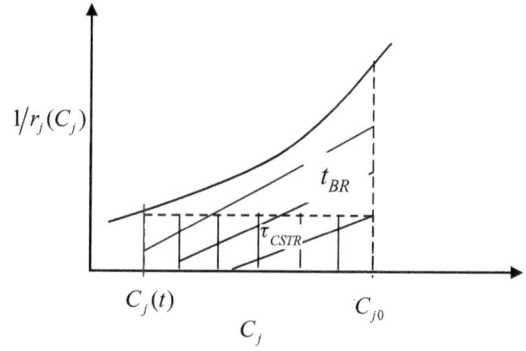

Figure 2.5. A graphical representation of space time for a CSTR.

2.2.2.2 Continuous-Stirred Tank Biological Reactor

For a CSTR in which cells are growing on a single substrate, one can write unsteady state mass balances for cell mass and substrate, as we did for a BR. This continuous-stirred tank biological reactor (CSTBR) can be operated with either a sterile feed (feed with no cells) or nonsterile feed (feed containing cells). We shall develop mass balance equations for a nonsterile feed ($X_F \neq 0$) that contains cells first. The cell balance is

$$FX_F - F_oX + \mu XV = \frac{d}{dt}(XV) \tag{2.18}$$

while the substrate balance is

$$FS_F - F_oS - \sigma XV = \frac{d}{dt}(SV) \tag{2.19}$$

If the inlet and outlet flow rates are equal so that the reactor volume remains constant, then Eqs. (2.18) and (2.19) reduce to

$$D(X_F - X) + \mu X = \frac{dX}{dt} \tag{2.20}$$

and

$$D(S_F - S) - \sigma XV = \frac{dS}{dt} \tag{2.21}$$

respectively. When the feed is sterile (no cells), then Eq. (2.20) reduces to

$$-DX + \mu X = \frac{dX}{dt} \tag{2.22}$$

When the steady state is reached so that the concentrations of the cells and substrate remain time invariant, then

$$-DX + \mu X = 0 \quad \Rightarrow D = \mu \tag{2.23}$$

$$D(S_F - S) - \sigma XV = 0 \quad \Rightarrow D(S_F - S) = \sigma XV \tag{2.24}$$

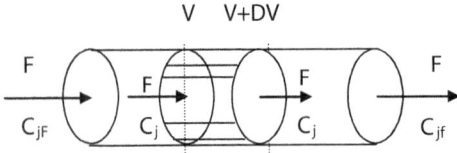

Figure 2.6. A plug-flow reactor (PFR).

2.2.2.3 Tubular Reactors

Another type of reactor that finds wide application in industry is the tubular reactor, either free of any packing or packed with immobilized enzymes and cells, catalysts, or other packing.

2.2.2.3.1. PLUG-FLOW REACTORS.

A PFR is a tubular reactor without packing in which the reacting mixture is continually put into the inlet and pushed out at the other end. Ideally, there is neither axial mixing nor dispersion, and therefore, the flow pattern is assumed to be in the form of a plug or a piston, hence the name "plug-flow reactor." The PFR usually requires very little maintenance as it has no moving parts, and usually, the conversion of reactant per reactor volume is highest. However, the disadvantage is that it is difficult to control temperature or pH within the reactor. The PFR may be used either as a single, long tube or in a bank of tubes. Many homogeneous gas phase flow reactors are tubular and find wide application in the chemical industry. In addition, aqueous phase bioreactors are found in wastewater treatment plants.

The mass balance equation for species j is written around an infinitesimal control volume between V and $V + \Delta V$ (see Figure 2.6):

$$FC_j|_V - FC_j|_{V+\Delta V} + r_j\Delta V = \frac{\partial (C_j\Delta V)}{\partial t} \qquad (2.25)$$

By dividing both sides of Eq. (2.25) by ΔV and then taking the limit as $\Delta V \to 0$, we obtain

$$\lim_{\Delta V \to 0}\left[\frac{FC_j|_V - FC_j|_{V+\Delta V}}{\Delta V}\right] + r_j = \frac{\partial (C_j)}{\partial t} \quad\Rightarrow\quad -\frac{\partial (FC_j)}{\partial V} + r_j = \frac{\partial C_j}{\partial t} \quad (2.26)$$

which reduces at steady state ($\partial C_j/\partial t = 0$) to an ordinary differential equation:

$$-\frac{d(FC_j)}{dV} + r_j = 0 \qquad (2.27)$$

When there is no flow rate variation along the axial distance, Eq. (2.27) can be rearranged to yield

$$r_j = F\frac{dC_j}{d(V)} = \frac{dC_j}{d(V/F)}, \qquad C_j(0) = C_{j0} \qquad (2.28)$$

Because V/F represents the space time, the PFR material balance equation can be written in terms of the space time, $\tau = V/F$:

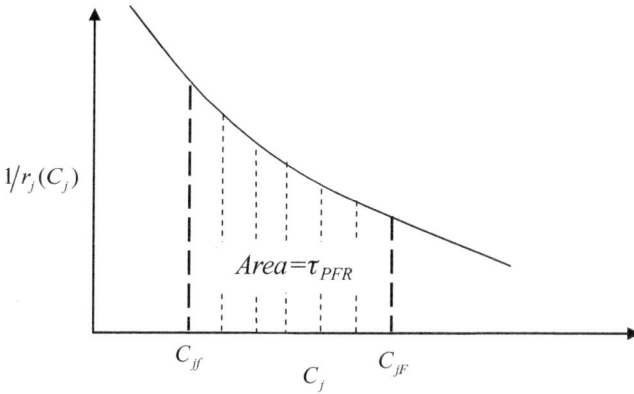

Figure 2.7. A graphical representation of space time for a PFR.

$$r_j = \frac{dC_j}{d\tau_{\text{PFR}}} \quad C_j(0) = C_{j0} \tag{2.29}$$

The steady state PFR material balance equation in terms of space time τ is identical to the BR mass balance equation in which the time is the running clock time t (Eq. (2.7)). Thus, a PFR is a spatial dual of a BR, or a BR is a temporal dual of a PFR.

Equation (2.29) can be rearranged to yield the space time necessary to reduce the feed reactant concentration from C_{jF} to the effluent concentration C_{jf}:

$$\tau_{\text{PFR}} = \int_{C_{jf}}^{C_{jF}} \frac{1}{-r_j(C_j)} dC_j \tag{2.30}$$

Equation (2.30) states that the space time for the PFR is the area under the curve of the reciprocal rate $1/r_j(C_j)$ plotted against the reactant concentration C_j between the feed concentration C_{jF} and the desired effluent concentration C_{jf}. This is shown in Figure 2.7.

Equation (2.30) can be rewritten as

$$\frac{V}{F} = \tau_{\text{PFR}} = \int_{C_{jf}}^{C_{jF}} \frac{1}{-r_j} dC_j = \left[\int_{C_{jf}}^{C_{jF}} \frac{1}{-r_j} dC_j \Big/ \int_{C_{jf}}^{C_{jF}} dC_j \right] \int_{C_{jf}}^{C_{jF}} dC_j = \left[\frac{1}{-r_j} \right]_{\text{mean}} (C_{jF} - C_{jf}) \tag{2.31}$$

where the mean rate is the concentration-averaged reaction rate. Equation (2.31) states that the PFR space time can be interpreted as the area of a rectangle formed by the mean reciprocal rate as the height and the difference between the feed and desired concentrations as the base.

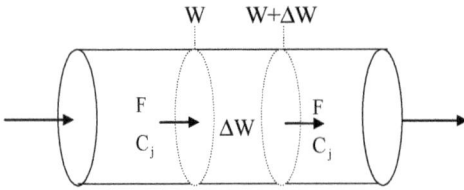

Figure 2.8. A packed-bed reactor (PBR).

As is done previously, the mass balance equations for a PFR in which cells are grown on a single substrate are obtained from Eq. (2.29):

$$\frac{dX}{d(V/F)} = \frac{dX}{d\tau} = r_X = \mu X, \quad X(0) = X_F \tag{2.32}$$

$$\frac{dS}{d(V/F)} = \frac{dS}{d\tau} = r_S = -\sigma X, \quad S(0) = S_F \tag{2.33}$$

Because of the nature of the PFR, the feed to the reactor must contain cells (non-sterile feed, $X_F \neq 0$). Otherwise, there would be washout of cells, and no cells would remain in the reactor.

2.2.2.3.2. PACKED-BED REACTORS (PBRs). A PBR, or a fixed-bed reactor, is a tubular reactor that is packed with catalyst pellets, immobilized enzyme pellets, immobilized cells, or other packing material. This heterogeneous reactor is most useful for gas reactions and also for liquid reactions. For reactions whose rate increases with the reactant concentration, it usually yields the highest conversion of reactant per catalyst weight of any catalytic reactor. However, it is possible for channeling to occur, which can result in a part of the packed-bed catalyst not being effectively utilized for the reaction.

For the idealized case in which the mass transfer resistance is negligible and any radial and axial dispersion can be ignored, we can write, as in the case of PFR, the mass balance of PBR over an infinitesimal volume (see Figure 2.8). Because of the packing, the linear velocity in a PBR is much faster than that in a PFR by the ratio of V_{empty}/V_{void}. The difference here is that the rate depends on the amount of the catalyst for this heterogeneous reaction. Defining by r'_j the rate of formation of species j per unit weight of catalyst (or immobilized cells), we can write the mass balance over an infinitesimal weight of catalyst, W and $W + \Delta W$ (or V and $V + \Delta V$),

$$FC_j \big|_W - FC_j \big|_{W+\Delta W} + r'_j \Delta W = \frac{\partial(C_j \Delta V)}{\partial t} \tag{2.34}$$

and at steady state, the usual limiting process after the division by ΔW leads to

$$-F\frac{dC_j}{dW} + r'_j = \frac{V}{W}\frac{\partial C_j}{\partial t} = 0 \tag{2.35}$$

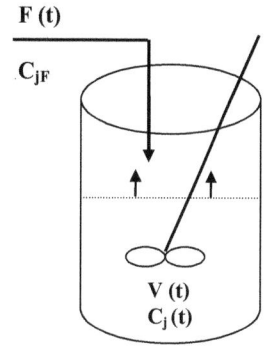

Figure 2.9. A fed-batch reactor (semi-batch reactor).

At steady state, the weight of catalyst (immobilized cells) required to achieve the desired concentration is obtained by integrating Eq. (2.35):

$$W = F \int_{C_{jF}}^{C_{jf}} \frac{dC_j}{r_j'} = F \int_{C_{jf}}^{C_{jF}} \frac{dC_j}{-r_j'} \tag{2.36}$$

2.2.3 Semi-Batch Reactor Operation

A semi-batch reactor (SBR) or a fed-batch reactor (FBR) is a variation of a batch reactor in the sense that one or more feed streams are introduced but there is no effluent stream so that the reactor volume changes with time. The operation begins with a specified initial volume and reactant concentration. The feed containing reactants, inducers, precursors, or nutrients is fed either intermittently or continuously into the BR, and as a result, the reactor volume increases with time. When the reactor volume is full, the feeding is ceased, and the operation continues in batch mode until a desired conversion is achieved. Thus, the SBR or FBR is a BR of variable volume with one or more feed streams but without an effluent stream. The SBR is a flexible reactor that offers an excellent means to manipulate the reaction conditions in the reactor by programming the feed streams that contain the reactant, inducers, or precursors. The SBR is more commonly known in fermentation and biotechnology industries as a *fed-batch culture* and finds wide applications. Indeed, much of the industrially important fermentation processes and cell cultures are carried out in fed-batch mode. As will be shown later, under certain conditions, the SBR is theoretically equivalent to a combination of a CSTR followed by a BR (a temporal dual of PFR).

The mass balance for species *j* for a SBR with a feed stream containing *j* species (see Figure 2.9) and no effluent stream is obtained from Eq. (2.3):

$$FC_{jF} + r_j V = \frac{d(C_j V)}{dt} \tag{2.37}$$

The overall balance, assuming an aqueous system so that there is no variation in density, is obtained from Eq. (2.4):

$$F = \frac{dV}{dt} \tag{2.38}$$

It is interesting to expand the right-hand side of Eq. (2.37) and substitute into it Eq. (2.38) to obtain

$$FC_{jF} + r_jV = V\frac{dC_j}{dt} + C_j\frac{dV}{dt} = V\frac{dC_j}{dt} + C_jF \tag{2.39}$$

or

$$F(C_{jF} - C_j) + r_jV = V\frac{dC_j}{dt} \tag{2.40}$$

Comparing Eq. (2.40) with Eq. (2.14), it is apparent that the SBR is equivalent in form to that of the constant-volume (feed rate is equal to withdrawal rate) unsteady state CSTR. However, one has to keep in mind that the semi-batch operation lasts only until the reactor volume is full, whereas the CSTR operation lasts indefinitely. It should be kept in mind that in the SBR, the volume increases with time. It is also clear that if the feed rate $F(t)$ is manipulated to keep the species concentration constant so that $dC_j/dt = 0$, the SBR operation reduces to

$$F(C_{jF} - C_j) + r_jV = 0 \tag{2.41}$$

which is identical to the steady state CSTR operation of Eq. (2.15). The SBR operation can mimic the steady state CSTR. In addition, when no feed is supplied, the SBR reduces to a constant-volume BR. Therefore, the SBR can mimic a CSTR followed by a BR (or a temporal equivalent of PFR).

When a species k is not supplied into the feed, as in the case of feed that does not contain product, the mass balance is obtained from Eq. (2.37),

$$r_kV = \frac{d(C_kV)}{dt} \tag{2.42}$$

or from Eq. (2.40),

$$-FC_k + r_kV = V\frac{dC_k}{dt} \tag{2.43}$$

It is interesting to note that Eq. (2.43) is equivalent to a constant-volume unsteady state CSTR mass balance equation for a product (no product in the feed). From the preceding observation, one can deduce that the SBR operation is equivalent to an unsteady state constant-volume CSTR operation until the reactor is full, followed by a BR operation.

2.2.4 Fed-Batch Cultures

As noted earlier, the SBR operation is traditionally known in the fermentation industry as a fed-batch culture or fed-batch operation. Taking the simplest case of growing cells from a single limiting substrate in a single feed stream, $S \rightarrow X$, we begin by writing the material balances for the cells and the limiting substrate and the overall material balance.

The overall material balance is

$$F\rho_F = \frac{d(V\rho)}{dt} \tag{2.44}$$

where ρ_F and ρ are the densities of the feed and culture, respectively. The mass added to the culture by pH adjustments and the potential mass (volatile substrate) lost due to aeration are assumed negligible compared to the mass added by the feed stream. Because an aqueous medium is employed and the apparent cell density is very close to that of water, it is normally safe to assume that the densities are the same and do not change with time. Therefore, the overall mass balance can be simplified to

$$F = \frac{dV}{dt} \tag{2.45}$$

Because the feed stream is free of cells (sterile feed) and there is no effluent, the cell balance states that the rate of generation is equal to the rate of accumulation,

$$r_X V = \mu X V = \frac{d(XV)}{dt} \tag{2.46}$$

where μ is the specific growth rate of cells, X is the cell concentration, V is the culture volume, and $r_X = \mu X$ is the rate of generation of cells per unit culture volume. Unlike other ideal reactors, the reactor volume changes with time because there are one or more feed streams that supply the limiting substrate and necessary medium components. By expanding the right-hand side of Eq. (2.46) and substituting Eq. (2.45) into it, we obtain

$$r_X V = \mu X V = \frac{d(XV)}{dt} = V \frac{dX}{dt} + \frac{dV}{dt} X = V \frac{dX}{dt} + FX \tag{2.47}$$

or

$$-FX + \mu X V = V \frac{dX}{dt} \tag{2.48}$$

We note here that the cell balance equation for the cell-free (sterile) feed is identical to the product balance equation of the SBR (Eq. (2.43)). In other words, the fed-batch cell balance equation is identical in form to that of the product for a constant-volume unsteady state CSTR.

The mass balance on the limiting substrate is

$$FS_F + r_S V = FS_F - \sigma X V = \frac{d(SV)}{dt} \tag{2.49}$$

where $F(t)$ is the volumetric feed rate, S_F is the limiting substrate concentration of the feed, and σ is the specific substrate consumption rate. Expanding the right-hand side of Eq. (2.49), substituting overall mass balance equation (2.45) into it, and rearranging yields

$$FS_F - FS - \sigma X V = V \frac{dS}{dt} \tag{2.50}$$

Equation (2.50) is identical in form to that of the substrate for a constant-volume unsteady state CSTR. One must keep in mind that the CSTR balance equation holds for an indefinite time, whereas the fed-batch balance holds only until the fed-batch culture volume is full.

From the preceding developments, it is apparent that the fed-batch operation can mimic a CSTR operation until the reactor volume is full. Once the reactor volume is full, a batch operation continues until the desired yield or conversion is achieved. Therefore, the fed-batch operation is one reactor that mimics the roles of two reactors in series, CSTR + BR, over the operational time and not indefinitely.

REFERENCES

1. Hill, Charles G., Jr. 1977. *Introduction to Chemical Engineering Kinetics and Reactor Design*. John Wiley.
2. Harriott, P. 2002. *Chemical Reactor Design*. Marcel Dekker.
3. Levenspiel, O. 1999. *Chemical Reaction Engineering*. 3rd ed. John Wiley.
4. Fogler, H. Scott. 2006. *Elements of Chemical Reaction Engineering*. 4th ed. Prentice Hall.

3 Maximization of Reaction Rates and Fed-Batch Operation

In this chapter, we consider simple chemical reactions in idealized reactors. Simplicity of such systems helps us understand more complex microbial reactions. With the brief description of each type of idealized reactor operation presented in Chapter 2, we consider now the simplest case of maximizing the reaction rate of a single isothermal reaction in a single reactor. This will be followed by the problem of maximizing the reaction rate using two reactors. Finally, the problem of maximizing the reaction rate of a single reaction by a single reactor with an inlet stream and an outlet stream is considered. This procedure leads naturally to the concept of fed-batch operation, a form of semi-batch operation in which there is only an inlet stream. The simplest fermentation, growing of microbial cells on a single limiting substrate, is used to illustrate the concept of fed-batch culture.

The primary objective of optimization of chemical and biochemical reactions is to maximize the reaction rate and/or yield (selectivity). For single reactions, maximization of the product formation rate or substrate consumption rate leads to the minimum reactor size or maximum productivity for a given reactor size. The minimum reactor size leads to minimum investment costs, and maximum productivity results in minimum maintenance costs. For multiple reactions, maximization of the product yield results in a minimum raw material cost, and maximization of the reaction rate leads to a minimum reactor size or a maximum productivity for a given reactor size.

We begin by examining the simplest case of a single reaction with a rate expression that is a function of the single reactant concentration. This is done first using an intuitive argument.

3.1 Intuitive Maximization of a Single Reaction Rate

It is instructive to take a single reaction whose rate exhibits a maximum with respect to the reactant concentration and perform an intuitive optimization to show that this procedure leads naturally to the concept of semi-batch (fed-batch) reactor operation.

The reaction rate (or product formation rate) may increase or decrease with the reactant concentration, may be independent of reactant concentration, or may go through a maximum. Figures 3.1a–3.1d illustrate these situations. When

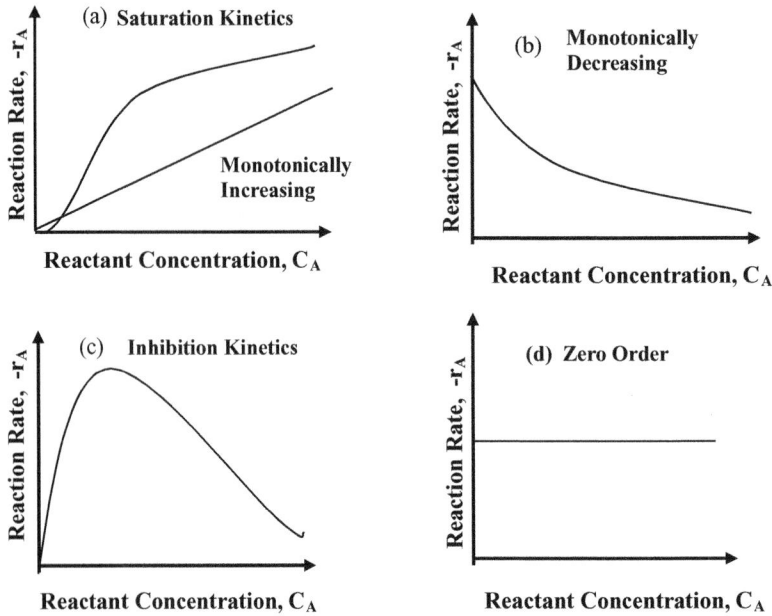

Figure 3.1. Various forms of reaction rates for a single reaction $A \rightarrow B$, $-r_A(C_A)$.

the reaction rate is independent of the reactant concentration, as in the case of zero-order reactions (Figure 3.1d), the reactant concentration does not affect the rate so that any type of idealized reactor operation would suffice, be it a batch reactor (BR), a continuous-stirred tank reactor (CSTR), or a plug flow reactor (PFR).

Conversely, if the reaction rate increases with the reactant concentration (Figure 3.1a), as in the case of an nth-order power law, $-r_A = kC_A^n$, and saturation kinetics such as Michaelis–Menten or Langmuir–Hinshelwood, $-r_A = k_1 C_A / (k_2 + C_A)$, it is obvious that the reactant concentration in the reactor should be kept as high as possible. This is accomplished by introducing the reactant all at once into the reactor, that is, a batch reactor or its spatial equivalent, a PFR. If the reaction rate decreases with the reactant concentration (Figure 3.1b), the reaction rate is maximized by keeping the reactant concentration as low as possible. Thus, a CSTR operation with a low dilution rate (feed rate/reactor volume) is the clear choice. These choices are made clear in Figure 3.2, where the reciprocal rate is plotted against the reactant concentration. The rectangular area (vertical lines) represents the space time (reactor volume/feed rate) of CSTR, τ_{CSTR}, while the area under the curve (horizontal lines) represents the space time of PFR, τ_{PFR}. Smaller space times represent a smaller reactor volume.

If the rate goes through a maximum (Figure 3.1c), first increasing with the reactant concentration reaching a maximum value, r_{Am}, at C_{Am} and then decreasing with further increases in reactant concentration, then the choice of reactor operation depends on the final desired concentration, C_{Af}. There are three possible cases depending on the desired final reactant concentration, C_{Af}, relative to the concentration at which the rate is maximum, C_{Am}: (Case I) $C_{Af} = C_{Am}$, (Case II) $C_{Af} < C_{Am}$, and (Case III) $C_{Af} > C_{Am}$. These three cases are depicted in Figure 3.3.

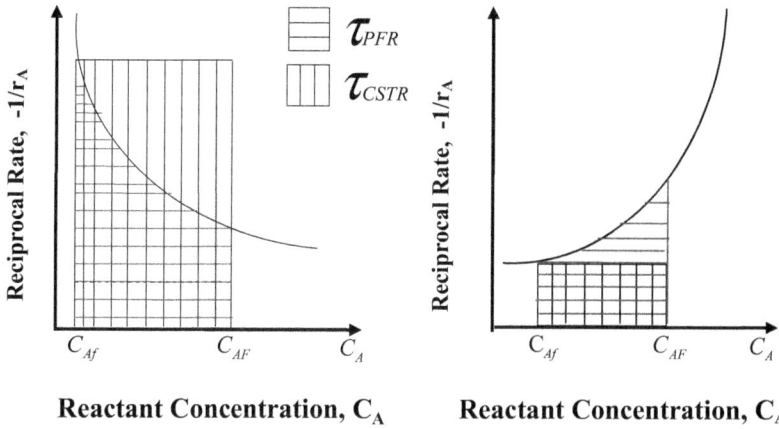

Figure 3.2. Space times of continuous-stirred tank reactor (CSTR) and plug flow reactor (PFR).

3.1.1 Optimum One-Reactor Operations

Case I: The final reactant concentration equals that at which the reaction rate is maximum, $C_{Af} = C_{Am}$.

Since the final concentration coincides with the concentration at which the rate is maximum, one should maintain the reactant concentration at C_{Am} throughout the

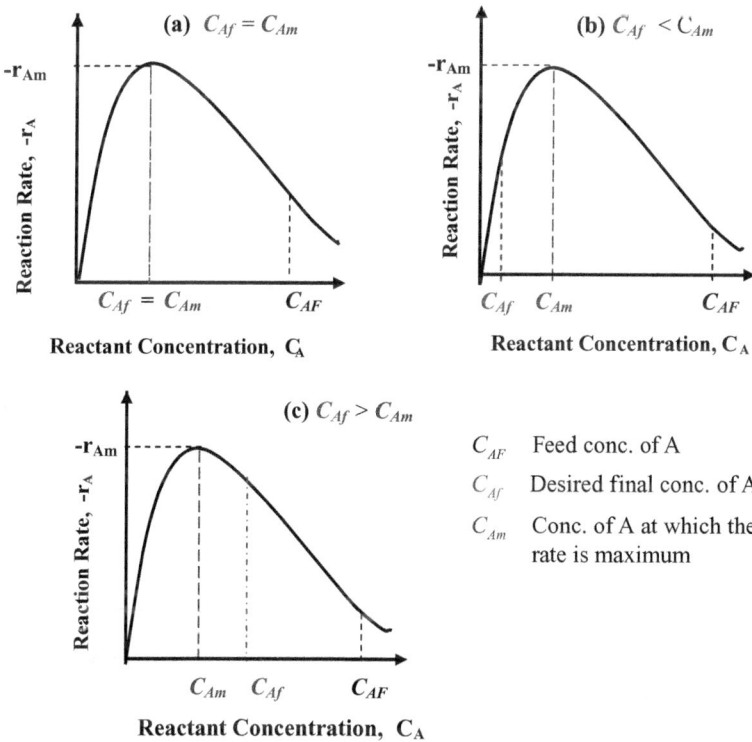

C_{AF} Feed conc. of A

C_{Af} Desired final conc. of A

C_{Am} Conc. of A at which the rate is maximum

Figure 3.3. Final reactant concentration C_{Af} relative to the concentration at which the reaction rate is at maximum C_{Am}.

course of reactor operation to obtain the maximum reaction rate. Obviously, a steady state CSTR operation with a proper dilution rate would maintain the concentration at C_{Am}. This is done by choosing a proper flow rate, F, for a given reactor size, which is obtained from the material balance of Eq. (2.15):

$$\frac{F}{V} = \frac{-r_{Am}(C_{Am})}{C_{AF} - C_{Af}} \tag{3.1}$$

where C_{AF} is the feed concentration and $C_{Am} = C_{Af}$ is the desired final concentration, respectively, of species A.

In a PFR, the reactant concentration varies throughout the length of the reactor from C_{AF} to C_{Af} so that the average rate in the PFR is less than the maximum rate in the CSTR:

$$(-\overline{r_A})_{PFR} = \left[\int_{C_{Am}}^{C_{AF}} (-r_A)dC_A \middle/ \int_{C_{Am}}^{C_{AF}} dC_A \right] < (-r_{Am})_{CSTR} \tag{3.2}$$

Another way to look at the result is in terms of space times. Rearrangement of Eq. (3.1) yields

$$\frac{V}{F} = \tau_{CSTR} = \frac{1}{D} = \left(\frac{1}{-r_{Am}(C_{Am})} \right)(C_{AF} - C_{Am}) \tag{3.3}$$

Equation (3.3) states that the area of the rectangle formed with the minimum reciprocal rate (the maximum rate) as the height (ordinate) and $(C_{AF} - C_{Am})$ as the base (abscissa), as shown in Figure 3.4a, represents the space time[2] for the CSTR, τ_{CSTR}. In terms of the rate, Eq. (3.3) is rearranged to give

$$-r_{Am} = (C_{AF} - C_{Am})/\tau_{CSTR} \tag{3.4}$$

Conversely, the area under the curve between the feed concentration, C_{AF}, and the final concentration, $C_{Af} = C_{Am}$ (Figure 3.4b), represents the space time[2] of the PFR:

$$\frac{V}{F} = \tau_{PFR} = \left[\int_{C_{Am}}^{C_{AF}} \frac{1}{-r_A}dC_A \middle/ \int_{C_{Am}}^{C_{AF}} dC_A \right] \int_{C_{Am}}^{C_{AF}} dC_A = \left[\frac{1}{-r_A} \right]_{mean} (C_{AF} - C_{Am}) \tag{3.5}$$

The mean reaction rate for a PFR is obtained from Eq. (3.5):

$$(-r_A)_{mean, PFR} = (\overline{-r_A})_{PFR} = (C_{AF} - C_{Am})/\tau_{PFR} \tag{3.6}$$

Therefore, the ratio of the maximum reaction rates is inversely proportional to the ratio of space times:

$$\frac{(-r_{Am})_{CSTR}}{(-\overline{r_{Am}})_{PFR}} = \frac{(C_{AF} - C_{Am})/\tau_{CSTR}}{(C_{AF} - C_{Am})/\tau_{PFR}} = \frac{\tau_{PFR}}{\tau_{CSTR}} = \frac{V_{PFR}}{V_{CSTR}} > 1 \tag{3.7}$$

Equation (3.7) states that the maximization of reaction rate is equivalent to the minimization of space time (minimization of reactor volume or maximization of throughputs) and that the required reactor volume of the CSTR is smaller than that of the PFR. Thus, the use of the CSTR leads to a higher reaction rate or maximization of the mean reaction rate.

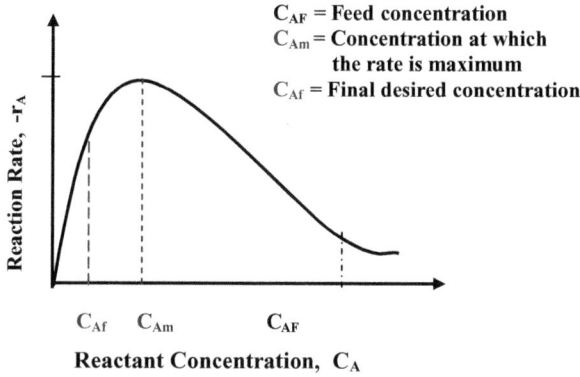

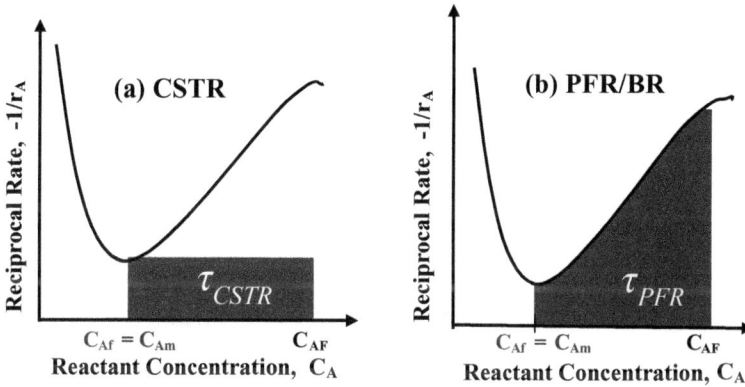

Figure 3.4. Space times of CSTR and PFR/batch reactor, $C_{Af} = C_{Am}$.

Case II: The final reactant concentration is less than that at which the rate is maximum, $C_{Af} < C_{Am}$.

As shown in Figure 3.5, the CSTR must be operated to maintain the effluent concentration at C_{Af}, which is smaller than C_{Am}. Therefore, in this case, one cannot take advantage of the maximum rate at C_{Am}. The space time of the CSTR may be smaller, equal to, or larger than that of the PFR, depending on the exact value of C_{Af}

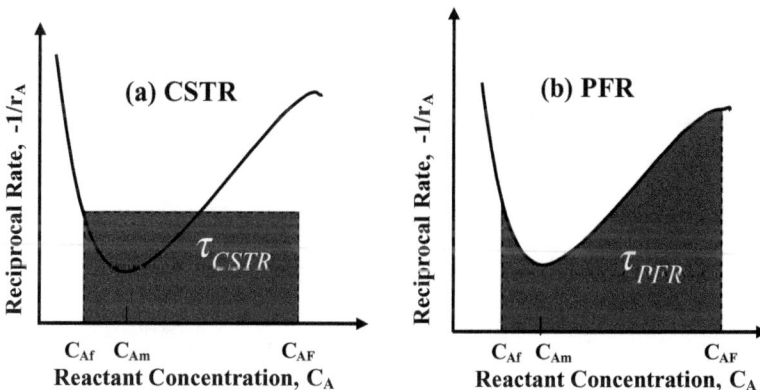

Figure 3.5. Space times of CSTR and PFR, $C_{Af} < C_{Am}$.

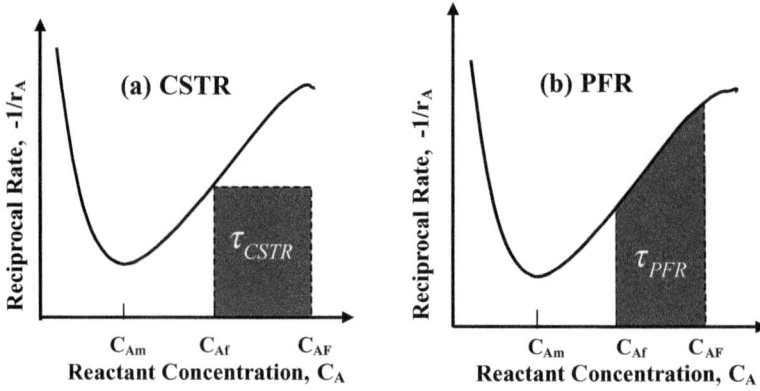

Figure 3.6. Space times of CSTR and PFR, $C_{Af} > C_{Am}$.

relative to C_{Am}. As C_{Af} gets smaller, the space time of the CSTR becomes larger and will surpass that of the PFR. Hence, the PFR is favored over the CSTR. Conversely, if C_{Af} is closer to C_{Am} (see the square formed by dotted lines), the space time of the CSTR gets smaller than that of the PFR, thus favoring the choice of the CSTR.

Case III: The final concentration is greater than that at which the reaction rate is maximum, $C_{Af} > C_{Am}$.

As shown in Figure 3.6, the space time of the CSTR is smaller than that of the PFR. In this case, one can still take advantage of running the reaction at the maximum rate, r_{Am}, at C_{Am} and bypass a part of the feed and mix with the effluent to meet the desired value, C_{Af}. Thus, the choice is a CSTR with a bypass. Figure 3.7 illustrates a CSTR with a bypass.

At the junction point, the material balance is

$$F_1 C_{Am} + F_2 C_{AF} = FC_{Af} = (F_1 + F_2)C_{Af} \qquad (3.8)$$

where F_1 and F_2 represent the flow rate to the CSTR and the bypass flow rate, respectively. Thus, Eq. (3.8) can be solved for the ratio of flow rates:

$$F_1/F_2 = (C_{AF} - C_{Af})/(C_{Af} - C_{Am}) \qquad (3.9)$$

Because

$$F = F_1 + F_2 \qquad (3.10)$$

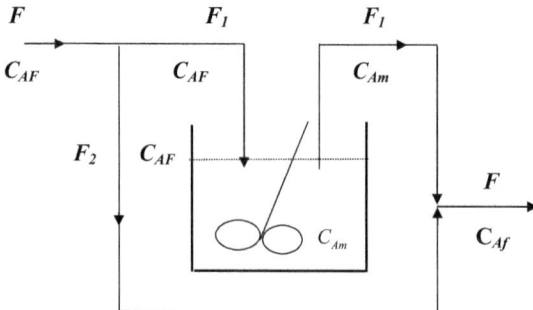

Figure 3.7. A CSTR with a bypass.

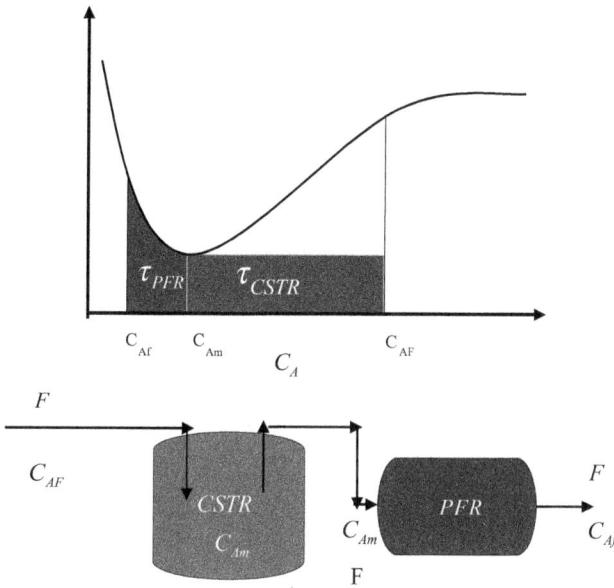

Figure 3.8. Two-reactor system (CSTR + PFR).

Equations. (3.9) and (3.10) are combined to obtain the fraction of feed to the CSTR, F_1/F, and the fraction to bypass, F_2/F:

$$F_1/F = (C_{AF} - C_{Af})/(C_{AF} - C_{Am}) \qquad (3.11)$$
$$F_2/F = (C_{Af} - C_{Am})/(C_{AF} - C_{Am})$$

It should be noted that Case III is an unusual situation. For single reactions, the desired conversion is usually very high because the raw material and separation costs depend on the unreacted reactant concentration. Thus, the desired final reactant concentration is usually very low, lower than the reactant concentration at which the reaction rate is at maximum.

It is also instructive to reconsider Case II, $C_{Af} < C_{Am}$. This time, we will allow a two-reactor configuration. We may assume that C_{Af} is much smaller than C_{Am} so that one PFR is optimal over one CSTR.

3.1.2 Optimum Two-Reactor Operations

Let us reconsider Case II, represented by Figure 3.5 through the use of two reactors instead of one. It is obvious that a two-reactor system, a CSTR followed by a PFR, would maximize the reaction rate or minimize the space time. The feed at C_{AF} is sent to a CSTR operating at C_{Am} to take advantage of operating at the maximum rate, and because the final concentration, C_{Af}, is less than the CSTR effluent concentration, C_{Am}, the effluent from the CSTR is sent to a PFR. In the PFR, the reactant concentration is reduced from C_{Am} to the final concentration C_{Af}. Thus, a two-reactor configuration consisting of a CSTR followed by a PFR would maximize the reaction rate and therefore yield the minimum reactor volume or a shorter operating time. This is shown in Figure 3.8.

The next question that arises naturally is, can we use only one reactor to mimic the operation of these two reactors, a CSTR followed by a PFR?

3.1.3 One-Reactor Operation Mimicking a Two-Reactor Operation

The objective is to maximize for all time the total rate, $-r_A V$, the product of the reaction rate per unit volume, and the reactor volume. In the case of batch, continuous flow, and tubular reactors, the reactor volume usually does not change so that maximization of total rate, $-r_A V$, is achieved by maximizing the reaction rate per unit volume, $-r_A$, alone. However, the volume of semi-batch (fed-batch) reactors increases during the operation so that the total rate, $-r_A V$, must be maximized. Thus, maintaining the reactant concentration constant at C_{Am}, the value at which the reaction rate is maximum, $-r_A(C_{Am}) = -r_{Am}$, maximizes the reaction rate, $-r_{Am}$, but does not necessarily maximize the total rate, $-r_A V$. Let us denote the reactant concentration that is to be maintained constant as C_{As}, which is yet to be determined. Then, the feed rate, $F(t)$, to keep the reactant concentration at C_{As} is obtained from the mass balance equation

$$F(t)C_{AF} + r_A(C_{As})V(t) = C_{As}\frac{dV}{dt} = C_{As}F \qquad (3.12)$$

or solving for the feed rate, we obtain

$$F(t) = \frac{-r_A(C_{As})}{C_{AF} - C_{As}}V(t) = \alpha V(t), \quad \alpha = \frac{-r_A(C_{As})}{C_{AF} - C_{As}} \qquad (3.13)$$

According to Eq. (3.13), the feed rate, $F(t)$, to keep the reactant concentration constant at C_{As} is proportional to the reactor volume, $V(t)$. With this feed rate, the volume increases exponentially:

$$F = \frac{dV}{dt} = \alpha V \quad \Rightarrow \quad V(t) = V(0)\exp(\alpha t) \qquad (3.14)$$

Therefore, the feed rate obtained by substituting Eq. (3.14) into Eq. (3.13) is also exponential:

$$F(t) = \alpha V(t) = \alpha V(0)\exp(\alpha t) = \frac{-r_A(C_{As})V(0)}{C_{AF} - C_{As}}\exp\left[\frac{-r_A(C_{As})}{C_{AF} - C_{As}}\right]t \qquad (3.15)$$

The total reaction rate obtained by maintaining the reactant concentration at C_{Am} corresponding to the maximum rate is

$$-r_A(C_{Am})V_m = -r_A(C_{Am})V(0)\exp\left[\frac{-r_A(C_{Am})}{C_{AF} - C_{Am}}t\right] \qquad (3.16)$$

while the corresponding total rate obtained by holding the reactant concentration constant at a value of C_{As} is

$$-r_A(C_{As})V_s = -r_A(C_{As})V(0)\exp\left[\frac{-r_A(C_{As})}{C_{AF} - C_{As}}t\right] \qquad (3.17)$$

The ratio of the total rate obtained by holding at C_{As} to that obtained by holding at the peak concentration, C_{Am}, is

$$\frac{-r_A(C_{As})V_s}{-r_A(C_{Am})V_m} = \frac{-r_A(C_{As})}{-r_A(C_{Am})}\exp\left\{\left[\frac{-r_A(C_{As})}{C_{AF} - C_{As}} - \frac{-r_A(C_{Am})}{C_{AF} - C_{Am}}\right]t\right\} \qquad (3.18)$$

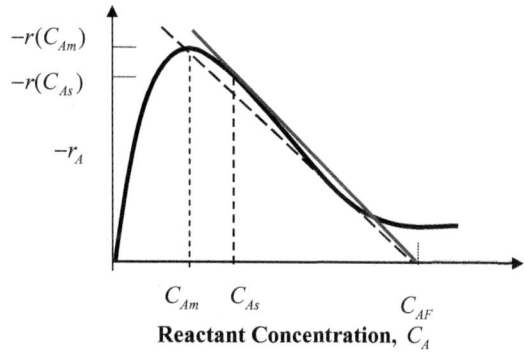

Figure 3.9. C_{Am} versus C_{As}.

Therefore, if the ratio is greater than one, the operation holding the reactant concentration at the value C_{As} yields a higher total rate than holding the reactant concentration at the value C_{Am}. It is therefore necessary for the exponent to be positive:

$$\left[\frac{-r_A(C_{As})}{C_{AF} - C_{As}} - \frac{-r_A(C_{Am})}{C_{AF} - C_{Am}} \right] > 0 \qquad (3.19)$$

Because the first term in the bracket represents the negative of the tangent to the rate curve at C_{As} drawn from C_{AF} and the second term represents the negative of the tangent to the rate curve at C_{Am} drawn from C_{AF}, it is apparent that the concentration C_{As} that is to be maintained must be larger than the peak concentration C_{Am}, $C_{As} > C_{Am}$, for the exponent to be positive. This is shown in Figure 3.9. As shown in Chapter 12, the concentration C_{As} obtained rigorously by variational calculus satisfies

$$\frac{-r_A(C_{As})}{(C_{AF} - C_{As})} = -r'_A(C_{As}) \qquad (3.20)$$

The solution to this equation yields the concentration to be maintained during the filling operation, which is greater than the concentration at which the rate is at maximum, $C_{As} > C_{Am}$. The feed rate to achieve this condition is obtained from Eq. (3.15):

$$F(t) = \left[\frac{-r_A(C_{As})}{C_{AF} - C_{As}} \right] V(0) \exp \left[\frac{-r_A(C_{As})}{C_{AF} - C_{As}} t \right] \qquad (3.21)$$

Thus, the initial reactant concentration should be C_{As} so that the total rate $-r(C_{As})V$ is at the maximum from the beginning. Then, to continue maximizing the total rate as long as possible, this concentration C_{As} should be maintained by supplying the feed given by Eq. (3.21). Once the reactor volume is full, then the reactor should be run as batch without a feed until the desired concentration is attained. This is a semi-batch operation. This is illustrated graphically in Figure 3.10. If the initial concentration is less than C_{As}, it should be brought instantaneously to C_{As}. This is done by supplying the feed at the maximum rate, that is, an impulse type of feeding. If the initial concentration is greater than C_{As}, it should be brought down to C_{As} as soon as possible by running as a batch reactor without any feed.

This type of semi-batch operation is also known as the *extended fed-batch* operation.[1] It should be emphasized that the exponent in the exponential feed,

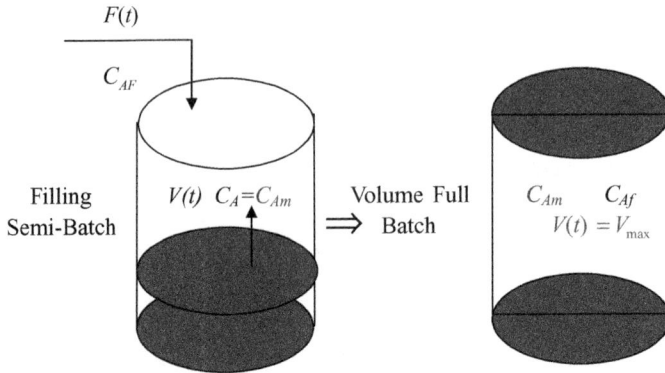

Figure 3.10. A fed-batch reactor operating scheme: a semi-batch and a batch reactor.

a, is not any arbitrary constant. It is given by Eq. (3.13) and depends on the reactant concentration being held constant, C_{As}; the feed concentration, C_{AF}; and the reaction rate at C_{As}, $-r_{As}$. This exponential feed rate would continue until the reactor is full so that the filling time is obtained by integrating Eq. (3.14) from zero time ($V = V_0$) to $V = V_{max}$ and solving the resultant for the time at which the reactor volume is full, t_{full}:

$$t_{full} = \frac{1}{\alpha} \ln \left(\frac{V_{max}}{V_0} \right) \tag{3.22}$$

The semi-batch reactor operation described earlier, mimicking the two-reactor configuration, can now be summarized.

3.1.3.1 The Optimal Operational Sequence, a Semi-Batch Operation
The optimal operational sequence is summarized as follows:

1. Determine the concentration C_{As} in the reactor to satisfy Eq. (3.20):

$$-r_A(C_{As}) = -r'_A(C_{As})(C_{AF} - C_{As})$$

2. Set $C_A(0) = C_{As}$ and the optimum initial volume (yet to be determined):

$$V(0) = V(t_f) = \beta V_{max}$$

3. Supply the feed exponentially using the rate given by Eq. (3.21):

$$F(t) = \left[\frac{-r_A(C_{As})V(0)}{(C_{AF} - C_{As})} \right] \exp \left[\frac{-r_A(C_{As})}{(C_{AF} - C_{As})} \right] t$$

4. When the reactor becomes full, $V = V_{max}$, stop feeding and operate as a batch reactor until the desired final concentration is attained, $C_A(t_f) = C_{Af}$.
5. Draw out the reactor content instantaneously, retaining the original initial volume for the next cycle, $V(t_f) = \beta V_{max} = V(0)$.
6. Fill the reactor instantaneously to attain $C_A = C_{As}$.
7. Repeat steps 3–6.

The time profiles resulting from the preceding semi-batch operation are sketched in Figure 3.11.

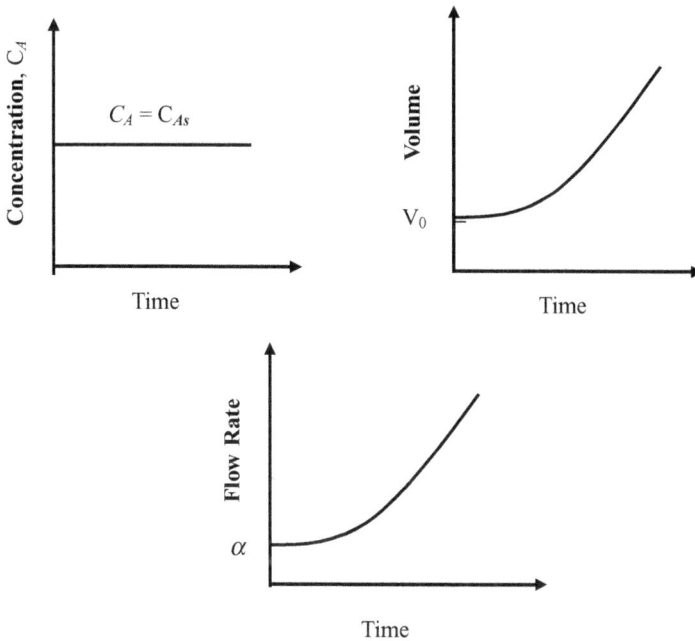

Figure 3.11. Time profiles of reactant concentration, feed rate, and reactor volume.

3.1.4 Rationale for Mimicking Optimal Two-Reactor Operations

Why do we need a single reactor that mimics the optimal two-reactor system, a CSTR followed by a PFR (BR)? There are a number of reasons for preferring one semi-batch operation over a continuous operation of two reactors in series. Let us consider chemical reactions separately from biological reactions.

For chemical reactions, the major reason for preferring a single reactor operation over a two-reactor operation is rather obvious: there are fewer maintenance costs and lower equipment costs associated with one reactor than with two reactors in series. Another reason is that for nonmonotonic reaction rates that exhibit a maximum rate, it is difficult to operate the continuous reactor at near the maximum rate because of the potential instability of the steady state. However, for large production volume, continuous operations are advantageous over repeated semi-batch operations.

For biological reactors with living cells that normally require oxygen, operations of plug-flow-type reactors are difficult to mimic and require a supply of cells into the reactor or a cell recycle. But the major reason for a semi-batch operation is a potential contamination in continuous reactor operation. A reactor system, even for a single reactor with a continuous feed stream and a continuous withdrawal stream, is subject to potential contamination much more so than a semi-batch operation that has a single feed stream that is sterile. Continuous operation over a long period can increase potential contamination. It is difficult to maintain sterile conditions on an industrial scale for a long period of operation. In addition, continuous operation requires sterile equipment backup, which can add additional capital costs. When recombinant cells are used, a segregational instability (reversion to host cells on cell division) may lead to eventual washout of the recombinant cells, leaving only the nonproductive host cells, which grow faster than the recombinant cells, in the reactor. Therefore, a semi-batch operation with a fresh inoculum each time can reduce this type of problem. As in the case of chemical reactors, a CSTR operation at near the maximum specific growth rate can lead to a washout of cells if the specific growth rate is less than the dilution rate. Thus, it can lead to a reactor instability problem. Another problem is that attempts to produce secondary metabolites in continuous reactors have not been successful in most reported situations, apparently requiring dynamic operations. In addition to these contamination and physiological problems, there are also reasons for preferring single-reactor operation: simpler operation and lower equipment costs associated with one reactor than with two reactors in series. Additional physiological and physical reasons for fed-batch operation are discussed in much more in detail in Chapter 4.

3.2 Optimization of Multiple Reactions

Consider a simple multiple reaction scheme of two parallel reactions,

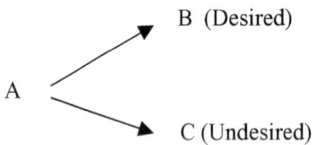

where the desired product is B and the side reaction yields the undesired product C. Both rates of formation r_B and r_C are functions of the reactant concentration C_A only. Figure 3.12 depicts simple examples of constant and variable yields. If the order of reaction or the rate expressions were same for the desired and undesired products, then the yield would be constant, independent of the reactant concentration. Conversely, if the order of the desired reaction were higher than that of the undesired reaction, the yield would increase with the reactant concentration. For example, if the orders are second and first for the desired and undesired reactions, respectively, then the yield will increase linearly with the reactant concentration. Conversely, if the order of the desired reaction is lower than that of the undesired reaction, the yield will decrease with the reactant concentration.

The yield may also depend on other environmental factors such as temperature and pH. The objective is to maximize the total rate of formation of desired species B, $r_B(C_A)V$, by controlling the reactant concentration C_A in the reactor. The instantaneous yield for B as a function of the reactant concentration is defined as

$$Y_{B/A}(C_A) = \frac{r_B(C_A)}{-r_A(C_A)} = \frac{r_B(C_A)}{r_B(C_A) + r_C(C_A)} = \frac{1}{1 + r_C/r_B} \qquad (3.23)$$

so that the total rate of formation of the desired product is

$$r_B(C_A)V = [-r_A(C_A)V][Y_{B/A}(C_A)] \qquad (3.24)$$

Thus, maximization of the total rate of formation of B is equivalent to maximizing the right-hand side of Eq. (3.24), that is, maximizing the product of the total rate of disappearance of A, $(-r_A(C_A)V)$, and the yield coefficient for B, $(Y_{B/A}(C_A))$. This process is considered in the following sections for the general cases of constant and variable yields.

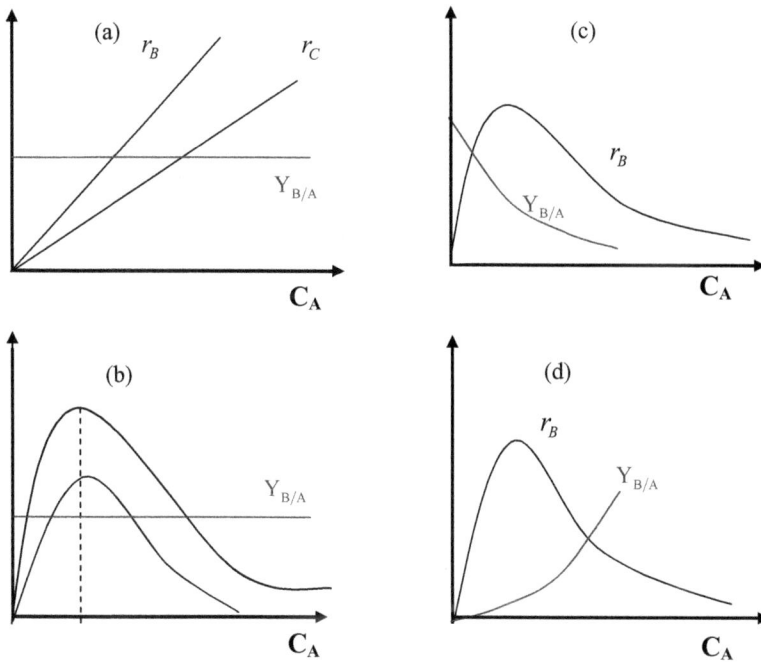

Figure 3.12. Various forms of yield coefficients as a function of reactant concentration.

3.2.1 Case of Constant Yields

When the reaction rate expressions or the reaction order are the same but differ in the numerator constants, the yield coefficient is constant (Figure 3.12a), that is, it is independent of the reactant concentration. When the yield coefficient is constant, the reactant concentration does not affect the yield, and thus the problem is equivalent to the case of a single reaction. Therefore, the total rate is maximized by maximizing the product of the rate of disappearance of species A and the reactor volume, $(-r_A)V$. Thus, when the rate of disappearance of species A increases monotonically and the yield coefficient is constant, then the rate of formation of the desired species B is maximized by operating the reactor at the maximum reactant concentration, that is, a batch reactor.

Conversely, when both reaction rates are nonmonotonic but peak at the same reactant concentration so that the yield coefficient is constant (see Figure 3.12b), then the rate of formation of the desired species B is maximized by maximizing the product of the rate for species A and the reactor volume, $(-r_A)V$, thus reducing to the single reaction optimization that was covered earlier in Section (3.1). The solution is to maintain the reactant concentration constant at the value C_{As} that satisfies Eq. (3.20), that is, a semi-batch reactor operation.

When the rate for species A is nonmonotonic and the yield coefficient is not a constant but a function of reactant concentration, as depicted in Figures 3.12c and 3.12d, then the situation is much more complex, and it is no longer optimal to maintain the reactant concentration constant. Therefore, ordinary calculus cannot be applied, and a more rigorous approach[3] is through variational calculus, as shown in Chapter 10. The reactant concentration must be varied during the course of the reaction to maximize the product of the yield coefficient and the total rate of disappearance of A.

Alternatively to using a fed-batch reactor with the feed flow rate manipulated to keep the reactant concentration at C_{As}, a conventional steady state CSTR may be used to maintain the reactant concentration constant, not at C_{As}, but at C_{Am}. However, the reactant concentration C_{Am} may be higher than the desired final concentration and may require an additional reactor to reduce it to the desired conversion. However, the fermentation industry is very reluctant to use a continuous process because of potential contamination and mutation problems and owing to very poor results they have experienced with continuous fermentation, contrary to theoretical predictions based on simplifying assumptions. With a batch process, the volume remains at the maximum, but the reactant concentration, which affects the rate and/or yield, cannot be regulated. A fed-batch process provides the means to manipulate the rate and/or yield.

3.2.2 Case of Variable Yields

When the yield depends on the reactant concentration, the right-hand side of Eq. (3.24) must be maximized. The reactant concentration that maximizes the right-hand side of Eq. (3.24) is obtained by differentiating it with respect to the reactant concentration and setting it to zero:

$$\frac{\partial (r_B V)}{\partial C_A} = \frac{\partial Y_{B/A}}{\partial C_A}[(-r_A)V] + Y_{B/A}\frac{\partial (-r_A V)}{\partial C_A} = 0 \tag{3.25}$$

which may be rearranged to yield

$$\frac{\partial (-r_A V)}{\partial C_A} = \left[\frac{(-r_A V)}{Y_{B/A}}\right]\left[\frac{-\partial Y_{B/A}}{\partial C_A}\right] \tag{3.26}$$

where, for convenience, the dependence of $-r_A$ and $Y_{B/A}$ on C_A has been suppressed.

Inspection of Eq. (3.26) yields some insight into the nature of the optimal reactant concentration profile. Because the volume changes with time, the optimal concentration changes with time, $C_A(t)$, and therefore, ordinary calculus cannot be applied; instead, variational calculus must be used to determine the optimal reactant concentration profile. The first term on the right-hand side is positive, and therefore, the sign of the left-hand side is determined by the sign of the second eight-hand-side term. Taking the sign function of both sides' yields, we obtain

$$\text{sign}\left[\frac{\partial (-r_A V)}{\partial C_A}\right] = -\text{sign}\left[\frac{dY_{B/A}}{dC_A}\right] \tag{3.27}$$

Thus, the reactant concentration C_A that maximizes the rate of formation of B varies in the region where the slope of the total rate, $d(-r_A V)/dC_A$, is opposite in sign to the slope of the yield curve, $dY_{B/A}/dC_A$. This implies that the optimum reactant concentration is not a constant but varies with time, that is, $C_A(t)$. Thus, the optimum reactant concentration profile cannot be determined by ordinary calculus. Instead, variational calculus must be used to obtain the optimal time profile of the reactant concentration. A formal treatment of this optimization will be covered in detail in Chapters 9–13. It will be also shown that in semi-batch operation, the reactor content is removed fully or partially only at the end of the run, indicating that *any withdrawal during the course of operation other than at the end is not optimal.* It is also shown that the necessary condition for singular operation (fed-batch) is that the reaction rate go through a maximum, and feed concentration must be sufficiently larger than the concentration at which the rate is maximum, $C_{AF} \gg C_{Am}$.

So far, we have considered maximizing the amount of product at the end of the reaction time. However, numerous other objective functions could be optimized. For example, it may be desirable to minimize the time required to obtain a specified final product concentration, or it may be desirable to maximize the productivity (final amount of product/final time). Ultimately, the engineering objective is to minimize the production costs (operational costs plus the raw material costs) or maximize the profit. Various forms of the objective functions will be considered in Chapter 9, which deals with fed-batch optimization.

3.3 Maximization of Cell Mass of a Simple Microbial Fed-Batch Culture

We now consider the simplest case of growing cells on a single substrate, say, glucose. This case corresponds to the preceding section on a single reaction involving a single reactant. Writing the mass balance on the cells (the feed contains no cells, and there is no removal of cells),

$$0 - 0 + \mu XV = \frac{d(XV)}{dt} \tag{3.28}$$

where the specific growth rate may depend on other variables besides the limiting substrate concentration, but we shall assume for simplicity that it depends on only the limiting substrate concentration, $\mu(S)$.

Because the objective is to maximize the total amount of cells in the reactor at a specified final time, $(XV)(t_f)$, we integrate Eq. (3.28) to obtain

$$(XV)(t_f) = (XV)(0) \exp \int_0^{t_f} \mu dt = (XV)(0) \exp \left[\frac{\int_0^{t_f} \mu dt}{\int_0^{t_f} dt} \right] \int_0^{t_f} dt = XV(0) \exp \overline{\mu} t_f \tag{3.29}$$

where $\overline{\mu}$ is the mean specific growth rate. Maximization of $(XV)(t_f)$ implies that the integral of the specific growth rate, or the mean specific growth rate, must be maximized. The integral is maximized by maximizing the integrand (specific growth rate) during the fixed time interval of 0 to t_f, $\text{Max}_S \mu \Rightarrow \text{Max}_S \overline{\mu}$. In other words, the specific growth rate must be maximized at all times during the time interval 0 to t_f by properly controlling the substrate concentration S to maximize the total amount of cells at the final time. Thus, we need to consider various functional forms of specific growth rates μ, just as we have considered various forms of reaction rates for single reactions. Figure 3.13 illustrates various forms of specific growth rates μ as functions of the substrate concentration S. We see here that there is a parallel

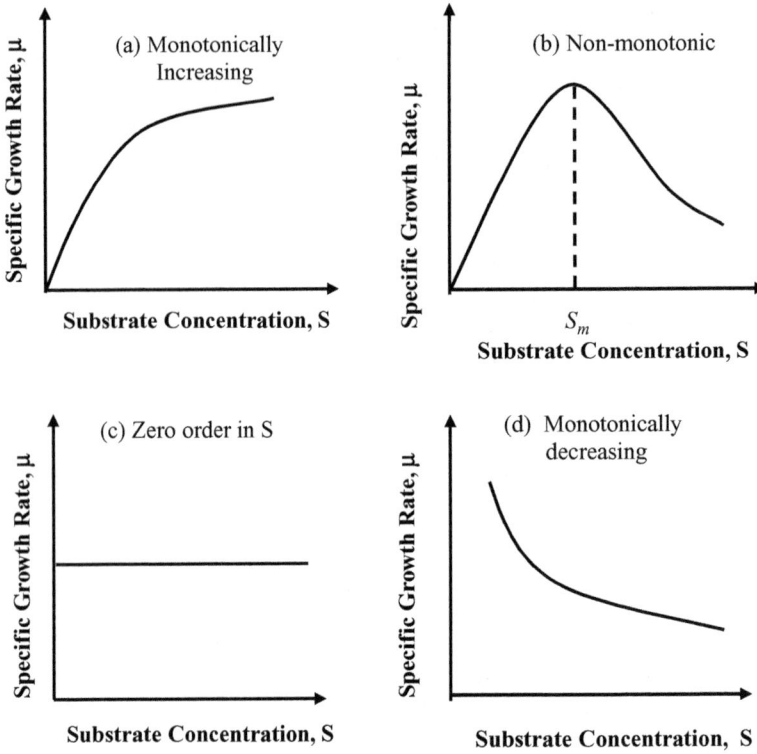

Figure 3.13. Various forms of specific growth rates as a function of substrate concentration.

between the reaction rates of single reactions and the specific growth rates of cell mass formation.

When the specific growth rate increases monotonically, as in the case of Monod form (Figure 3.13a), the specific growth rate is maximized by maintaining the substrate concentration as high as possible. Therefore, a batch mode of operation is optimal. Conversely, if the specific growth rate is a nonmonotonic function of the substrate concentration, increasing with the substrate concentration, going through a maximum, and then decreasing with further increases in the substrate concentration (Figure 3.13b), then the substrate concentration should be brought to and held as long as possible at the value S_m, the substrate concentration corresponding to the maximum specific growth rate, μ_m. If the initial substrate concentration is less than S_m, $S_0 < S_m$, then it should be brought up to S_m instantaneously by feeding at the maximum rate (theoretically an impulse feeding but practically a maximum feed rate over a very short time period). If the initial substrate concentration exceeds S_m, $S_0 > S_m$, then a batch mode of operation without any feed should be used to reduce the substrate concentration to S_m.

3.3.1 Feed Rate to Maintain the Substrate Concentration That Maximizes the Specific Growth Rate, $S = S_m$

Suppose that the initial substrate concentration is at S_m, $S_0 = S_m$. What should be the feed rate profile that would keep the substrate concentration constant at S_m?

With the initial substrate concentration chosen as S_m, the amount of cells obtained by maintaining the substrate concentration at S_m is obtained by integrating the cell balance equation, Eq. (3.28), with $S = S_m$ and $\mu(S_m) = \mu_m$:

$$X(t)V(t) = X_0 V_0 \exp(\mu_m t) \tag{3.30}$$

The overall balance, assuming equal and constant density for the feed and the reactor content, is

$$\frac{dV}{dt} = F \tag{3.31}$$

The substrate balance equation with $S = S_m$ so that $\mu = \mu_m$ and $\sigma = \sigma_m$ is

$$FS_F - \sigma_m XV = \frac{d(S_m V)}{dt} = S_m \frac{dV}{dt} = S_m F \tag{3.32}$$

where Eq. (3.31) is used. Equation (3.30) is substituted into Eq. (3.32), and the resultant is solved for the required feed rate:

$$F(t) = \frac{\sigma_m XV}{S_F - S_m} = \left[\frac{\sigma_m X_0 V_0}{S_F - S_m}\right] \exp(\mu_m t) = \gamma \exp(\mu_m t) \tag{3.33}$$

where the constant γ is

$$\gamma = \sigma_m X_0 V_0 / (S_F - S_m) = V_0[\sigma_m X_0 / (S_F - S_m)] = V_0[\mu_m X_0 / Y_{X/S}(S_F - S_m)] \tag{3.34}$$

Equation (3.33) states that the feed rate is an exponential function as in the case of a single reaction. An exponential feed rate is applied to support the exponential growth of cells at the constant substrate concentration, S_m. If we choose the initial cell concentration properly as

$$X_0 = Y_{X/S}(S_F - S_m) \tag{3.35}$$

then the constant γ reduces to

$$\gamma = V_0 \mu_{\max}$$

and the feed rate is

$$F(t) = \frac{\sigma_m XV}{S_F - S_m} = V_0 \mu_m \exp(\mu_m t) \tag{3.36}$$

The reactor volume increases exponentially also and is obtained by substituting Eq. (3.36) into Eq. (3.31) and integrating the resultant,

$$V(t) = V_0 + V_0[\exp(\mu_m t) - 1] = V_0[\exp(\mu_m t)] \tag{3.37}$$

or the time to fill the reactor completely is

$$t_{\text{full}} = \left(\frac{1}{\mu_m}\right) \ln \left(\frac{V_{\max}}{V_0}\right) \tag{3.38}$$

When the reactor volume is full, the feed should be shut off and the reactor should be run in batch mode until the desired final substrate concentration is attained. The cell concentration during the period of exponential feed is obtained from Eqs. (3.30) and (3.35):

$$X = \frac{XV}{V} = \frac{X_0 V_0 \exp(\mu_m t)}{V_0 \exp(\mu_m t)} = X_0 = Y_{X/S}(S_F - S_m) \qquad (3.39)$$

Thus, the cell concentration and substrate concentration remain constant (remain the same as the initial concentrations), X_0 and S_m, respectively, during the exponential feed period, when the initial cell concentration is chosen according to Eq. (3.35) and the initial substrate concentration is chosen properly, $S_0 = S_m$. Simply, the volume increases exponentially and thus the total amount of cells increases exponentially, while the cell and substrate concentrations remain the same as the initial values, $X = X_0$ and $S = S_m$. Once the reactor volume is full, the reactor may be operated as a batch without any feed, until the desired substrate concentration is reached.

Thus, we have a sequence of operations that would be described as a semi-batch or *fed-batch* operation. This result parallels exactly the result obtained for single reactions. The only difference is that the specific growth $\mu (= r_X / X)$ is maximized here instead of the reaction rate $-r_A$. As was done with single reactions, we can summarize the fed-batch operation to maximize the total amount of cells produced at the end of the final time.

3.3.2 Optimal Feed Rate Sequence for a Fed-Batch Culture

The optimal feed rate sequence for the fed-batch described previously is summarized as follows:

1. Choose the initial substrate concentration to equal the value at which the specific growth rate is maximum, $S(0) = S_m$. The inoculum concentration should be selected in accordance with Eq. (3.35) and the initial volume.
2. Supply the feed exponentially according to Eq. (3.36) so that the substrate concentration remains constant at S_m and cell concentration remains at X_0 until the reactor is full.
3. Run the reactor as a batch until the desired substrate concentration is reached.

In the preceding scheme, we paid no attention to the amount of substrate consumed because it is directly proportional to the amount of cells formed owing to the assumption of a constant-yield coefficient. However, if the yield coefficient is a function of the substrate concentration, then one should also consider the amount of substrate consumed as it may turn out that the substrate concentration that maximizes the specific growth rate may lead to a substantially high consumption of the substrate (a low cell yield). For example, if the substrate consumption rate at S_m corresponding to the maximum specific growth rate is twice as high as that corresponding to 80 percent of the maximum specific growth rate, $\sigma(\mu = \mu_m) = 2\sigma(\mu = 0.8\mu_m)$, then the fed-batch operation at the maximum specific growth rate would result in a 25 percent decrease in fermentation time, but at the expense of doubling the substrate consumption. Thus, if the substrate cost is substantial relative to the price of the cell, the optimization should take into account the cost of the substrate and the operating cost and optimize the profit instead of the cell mass.

Optimization of cell mass production under various conditions and performance indices is covered in great detail in Chapter 12, where the problems are formulated and rigorously solved using Pontryagin's maximum principle.

REFERENCES

1. Edward, V. H., Gottschalk, M. J., Noojii, A. Y., III, Tuthill, L. B., and Tannaholl, A. L. 1970. Extended culture: The growth of *Candida utilis* at controlled acetate concentration. *Biotechnology and Bioengineering* 7: 975–999.
2. Levenspiel, O. 1962. *Chemical Reaction Engineering.* John Wiley.
3. Wagmare, R., and Lim, H. C. 1981. Optimal operation of isothermal reactors. *Industrial Engineering Chemistry Fundamentals* 20: 361–369.

4 Phenomena That Favor Fed-Batch Operations

One of the primary objectives of reaction engineering is to optimize the rate of formation of the product (productivity) and/or the relative rates (selectivity or yield). For an existing plant, a faster product formation rate implies a higher productivity and corresponding reductions in plant operating time and operating cost. For a new plant to be built, the increased rate implies, in addition to improved productivity, a smaller reactor and therefore a lower capital investment cost. Likewise, an improved yield implies a lower raw material cost and a lower capital investment for existing and new plants.

Fed-batch operations are well suited for situations in which the cell growth and/or product formation rates are sensitive to the concentration of the limiting substrate, an intermediate, or a product so that the overall rate increases with the limiting substrate concentration, reaches a maximum, and decreases with further increases in the substrate concentration. Fed-batch operation finds wide applications in bioindustry as it takes advantages of various biochemical and physiological phenomena of cell cultures and is also able to overcome adverse physical effects. Fed-batch operations provide potential advantages for autocatalytic reactions, to which cellular processes belong, and multiple reactions in which the relative rates vary with the reactant concentration.

The specific rates usually depend on the limiting substrate concentration due to such common phenomena as activation, inhibition, induction, and repression. Thus, one or more rates may be nonmonotonic functions of the limiting substrate concentration, exhibiting a maximum. In these situations, it is advantageous through manipulation of the feed rate to regulate the limiting substrate concentration to remain constant or to follow the time profile that maximizes a weighted sum of the specific rate and yield. Therefore, a relatively simple preliminary appraisal can be obtained by checking if the specific rates or yield coefficients depend on the limiting substrate concentration. The goal is to provide rapidly and as long as possible the best conditions for cell growth and product formation.

In general, when a certain medium component concentration affects significantly a rate or rates nonmonotonically, a fed-batch culture operation may be superior to the traditional batch culture. We will first look at various situations in which fed-batch operations provide advantages. Although various situations are listed separately in the following, in practice, these situations often are found in combinations in industrial fermentations. Various physiological and biochemical phenomena can be cited for which fed-batch operations may be desirable or beneficial. They can be classified as either chemical or physical phenomena.

4.1 Chemical Phenomena

4.1.1 Substrate Inhibition[1–10]

Relatively low concentrations of substrates such as methanol, ethanol, acetic acid, and aromatic compounds and moderate concentrations of such substrates as glucose inhibit directly or via an intermediate the growth of various microorganisms and the rate at which the product is formed. For example, overfeeding of glucose may lead to acetic acid formation by *Escherichia coli*, which slows down the growth; overfeeding of methanol to methylotroph can lead to the formation of formaldehyde, which is detrimental to cell growth; and overfeeding of glucose to yeasts can lead to the formation of ethanol, resulting in lower yeast yields. Overfeeding of glucose can also lead to rapid cell growth at the expense of lower antibiotic yields.

By varying the feed rates of limiting substrates, it is possible to control the concentrations of substrates in the medium so that the growth rate and/or product formation rate is maximized and therefore the productivity and/or yield can be improved substantially.

4.1.2 Glucose (Crabtree) Effect[11–15]

Overfeeding of nutrients often leads to formation of undesired by-products. For example, in the production of baker's yeast with glucose as the substrate, when glucose concentration exceeds a critical level (0.02–0.85 mM), glucose is partially metabolized to ethanol, thus lowering the yield for baker's yeast. This is the main cause of low cell yield and is known as the glucose or Crabtree effect. Fed-batch culture provides an excellent means of regulating the concentration of glucose to improve the yield or productivity. For recombinant DNA products based on yeast cells as hosts, one observes a similar phenomenon. The same is true when recombinant cells of *E. coli* are utilized to produce metabolites using glucose as the substrate.

4.1.3 Catabolite Repression[16–41]

Syntheses of many metabolites are repressed when cells are grown rapidly on readily utilizable carbon sources such as glucose due to the resultant increase in intracellular concentration of adenosine monophosphate (AMP), which causes repression of biosynthesis of certain enzymes. Such repression, *catabolite repression*, occurs in the production of antibiotics and enzymes. In producing recombinant products utilizing *E. coli* as host cells, high glucose concentration in the medium leads to the formation of acetic acid, which reduces the cell growth and product formation rate. With *Saccharomyces cerevisiae* as host cells, high glucose concentrations lead to the formation of ethanol, which reduces the growth rate and product formation. One powerful way to circumvent the depressed formation of desired products is to limit the growth rate by keeping the concentration low by slow feeding of the carbon source.

When methanol is used as the substrate, as in the case of producing single-cell proteins, methanol is oxidized first to formaldehyde, which plays havoc in the cellular process, and as a result, it is essential to keep the methanol concentration in an optimally low range. A similar phenomenon is observed when ethanol is used as a substrate in amino acid production or a single-cell protein as it is first oxidized to acetaldehyde, although the effect is much milder.

4.1.4 Utilization of Auxotrophic Mutants[42–44]

By treating with mutagens such as UV light and nitrosoguanidine, it is possible to isolate mutants that require one or more nutrients for growth. These auxotrophic mutants requiring a certain nutrient for growth, such as amino acids, purine, pyrimidine, or vitamins, are often utilized in industrial amino acid production. These auxotrophs can be grown rapidly in the presence of an excess of the required nutrient to a high concentration without any accumulation of the desired product owing to feedback inhibition and/or end product repression. After growing the cells to a high concentration, a medium lacking or extremely low in the required nutrient concentration can be fed into a fed-batch culture to force the grown cells to begin synthesizing the desired product, that is, amino acids. In this way, the production of amino acid is maximized. Various auxotrophs have been used to produce commercially such amino acids as arginine, lysine, glutamic acid, valine, alanines, homoserine, phenylalanine, threonine, ornithine, citrulline, and proline. A tyrosine-requiring auxotroph produces phenylalanine, the raw material for the artificial sweetener aspartame. Similarly, tyrosine is produced by a phenylalanine-requiring auxotroph. Literature utilizing auxotrophic mutants is abundant. Pyruvate is also produced by auxotrophic mutants.

4.2 Physical Phenomena

4.2.1 High Cell Density[45–52,90]

In cell cultures, be they microbial or animal cells, it is desirable to achieve the highest possible cell concentration (e.g., 100 g/L or more of microbial cells) because the total rate is proportional to the total number of cells (concentration times the reactor volume). Thus a high cell concentration in the reactor leads to a high metabolite production rate and resultant high productivity. In addition, the purification cost may be reduced substantially for media high in metabolite concentration owing to the reduction in the volume to be processed and high product concentration. Additional reduction in cost may be realized owing to reductions in equipment and operating costs. In a batch operation, a high cell concentration calls for a high initial substrate concentration, which may be inhibitory to the growth or may result in a lower yield. For instance, a glucose concentration above 50 g/L, an ammonia concentration above 3 g/L, a phosphorus concentration above 10 g/L, and a zinc concentration above 38 mg/L are known to be inhibitory.

4.2.2 Extension of Operational Period[53–55]

At the end of batch operation, if the cells continue to produce the desired metabolite, then by adding the nutrients, the culture can be operated for an extended period of time, and the concentration of the metabolite can be increased at the time of harvest. A fed-batch operation is then effectively utilized to extend the culture time if the cells are capable of producing the desired metabolite.

In antibiotic production, such as for penicillin, microorganisms initially in the growth phase (idiophase) utilize rapidly the carbon energy source for growth but not for penicillin synthesis, which is followed by a period in which the cell growth is slow but penicillin synthesis takes place actively (tropophase). A fed-batch operation is used to accommodate these phenomena by providing relatively high substrate concentration in the idiophase and low substrate concentration to extend the tropophase in which penicillin synthesis takes place.

4.2.3 Alleviation of High Broth Viscosity[56,57]

In the production of microbial biopolymers such as dextran, pullulan, and xanthan gum, the broth viscosity increases tremendously, accompanied by increased agitation power consumption and decreased oxygen transfer efficiency. This high viscosity–derived problem can be substantially improved by fed-batch operation by continuous feeding of nutrient solutions. By so doing, the viscosity increases gradually and obviates the problems caused by high viscosity until the end of fermentation.

4.2.4 Makeup for Lost Water by Evaporation[59–61]

In extended aerobic processes for certain antibiotics and animal feed additives, where the fermentation period may last a week to a month, the water loss through the exhaust gas is substantial. Left uncorrected, this loss leads to a considerable concentration of the broth, which in turn changes the rates of growth and product formation or rheological properties of the broth and causes problems. The water loss due to evaporation caused by aeration is very significant when aerobically cultivating hyperthermophiles[54] at elevated temperature. Regulating the feed rate to counterbalance the amount of water loss can eliminate this concentration effect caused by evaporation.

4.2.5 Better Plasmid Stability of Recombinant Cells[64,68–86]

There have been conflicting reports on the plasmid stability of recombinant cells in continuous and batch cultures. Plasmid stability in bacteria and yeasts has been observed to increase[68–70] or decrease[71,72] growth rates. A variety of host–vector systems in continuous or batch cultures have been shown either to increase,[73–75] to decrease,[76–78] to show the optimum,[79–82] or to not be related[83–85] at all with respect to the specific growth rate. Plasmid stability has been reported to depend on the limiting substrate concentration[69,74] and on reactor operation mode.[64,79,80] Conversely, plasmid stability in fed-batch cultures has been reported to be high.[65,86]

4.3 Other Phenomena[87-89]

4.3.1 Experimental Kinetic Studies[63-67]

Another application of fed-batch operation is in obtaining accurate reaction rates experimentally.[62] It has been reported that for fast and slow reactions, more accurate rate data can be obtained using a fed-batch reactor with a constant feed than can be obtained using a continuous-stirred tank reactor. The idea is based on the fact that by monitoring reactions with constant feed rates and noting the concentration at the point at which the reactant concentrations reach their peak, it is possible to calculate the rate without taking a time derivative of the concentration profile. This method of generating reaction rate data without taking time derivatives is covered in detail in Chapter 8. In addition, fed-batch cultures have been used as a tool for experimental kinetic studies.[63,65,66] In particular, an exponential-feed fed-batch was used to obtain a complete kinetic characterization of the effect of growth rate on recombinant penicillin acylase production.[66]

4.3.2 Various Other Situations

In the production of riboflavin by *Eremethecium ashbyii*, it was reported that intermittently slow feeding of glucose or inositol, or both, increased the yield.[87] Plant cell cultures also have been reported[88] to see increased specific product formation rates of chlorophyll and higher potential photosynthesis by *Ocimum basilicum* by the use of a glucose fed-batch. In ethanol production from xylose[89] by *Pachysolen tannophilus*, an improved yield of ethanol was obtained by slow feeding of xylose and also by adding glucose to inhibit the respiration of ethanol.[58] Fed-batch fermentation of molasses by *S. cerevisiae*[59] was studied as used in commercial ethanol plants.

REFERENCES

1. Edwards, V. H., Cottschalk, M. J., Noojin, A. Y., III, Tuthill, L. B., and Tannahill, A. L. 1970. Extended culture: The growth of *Candida utilis* at controlled acetate concentrations. *Biotechnology and Bioengineering* 12: 975–999.
2. Yamane, T., Kishimoto, M., and Yoshida, F. 1976. Semi-batch culture of methanol assimilating bacteria with exponentially increased methanol feed. *Journal of Fermentation Technology* 54: 229–240.
3. Shimizu, S., Tanaka, A., and Fukui, S. 1969. Utilization of hydrocarbons by microorganisms. VII. Production of coenzyme Q by *Candida tropicalis* pK 233 in hydrocarbon fermentation. Effect of aromatic compounds on coenzyme Q production. *Journal of Fermentation Technology* 47: 551–557.
4. Renard, J. M., Mansouri, A., and Cooney, C. L. 1984. Computer controlled fed-batch fermentation of the methylotroph Pseudomonas AM1. *Biotechnology Letters* 6: 577–580.
5. Silman, R. W. 1984. Ethanol production by *Zymomonas mobilis* in fed-batch fermentations. *Biotechnology and Bioengineering* 26: 247–251.
6. Kishimoto, M., Yoshida, T., and Taguchi, H. 1980. Optimization of fed-batch culture by dynamic programming and regression analysis. *Biotechnology Letters* 2: 403–408.

7. Kishimoto, M., Yoshida, T., and Taguchi, H. 1981. Simulation of fed-batch culture for glutamic acid production with ethanol feeding by use of regression analysis. *Journal of Fermentation Technology* 59: 43–48.
8. Kishimoto, M., Yoshida, T., and Taguchi, H. 1981. On-line optimal control of fed-batch culture of glutamic acid production. *Journal of Fermentation Technology* 59: 125–129.
9. Lee, S., and Wang, H. Y. 1982. Repeated fed-batch rapid fermentation using yeast cells and activated carbon extraction systems. *Biotechnology Bioengineering Symposium* 12: 221–231.
10. Dibiasio, D. 1980. An investigation of stability and multiplicity of steady states in a laboratory biological reactor. PhD dissertation, Purdue University.
11. Weigand, W. A., Lim, H. C., Creagan, C. C., and Mohler, R. D. 1979. Optimization of a repeated fed-batch reactor for maximum cell productivity. *Biotechnology and Bioengineering Symposium* 9: 335–348.
12. White, J. 1954. *Yeast Technology*. Vol. 31. John Wiley.
13. Crabtree, H. G. 1929. Observations on the carbohydrate metabolism of tumors. *Biochemistry Journal* 23: 536–545.
14. Whitaker, A. 1980. Fed-batch culture. *Process Biochemistry* 15: 10–12, 14–15, 32.
15. Moss, F. J., Rickard, P. A. D., Bush, F. E., and Caiger, P. 1971. The response by microorganisms to steady-state growth in controlled concentrations of oxygen and glucose. II. *Saccharomyces carlsbergensis. Biotechnology and Bioengineering* 13: 63–75.
16. Nanba, A., Hirota, F., and Nagai, S. 1981. Microcomputer-coupled baker's yeast production. *Journal of Fermentation Technology* 59: 383–389.
17. Hosler, P., and Johnson, M. J. 1953. Penicillin from chemically defined media. *Industrial and Engineering Chemistry* 45: 871–874.
18. Clark, D. J., and Marr, A. G. 1964. Studies on the repression of β-galactosidase in *Escherichia coli. Biochemistry Biophysics Acta* 92: 85–94.
19. Hsu, E. J., and Vaughn, R. H. 1969. Production and catabolite repression of the constitutive polygalacturonic acid trans-eliminase of *Aeromonas liquefaciens. Journal of Bacteriology* 98: 172–181.
20. Yamane, K., Suzuki, H., and Nishizawa, K. 1970. Purification and properties of extracellular and cell-bound cellulase components of *Pseudomonas fluorescens var. cellulosa. Journal of Biochemistry* 67: 19–35.
21. Hulme, M. A., and Strank, D. W. 1970. Induction and the regulation of production of cellulose by fungi. *Nature* 226: 469–470.
22. Hockenhull, D. J. D., and Mackenzie, R. M. 1968. Present nutrient feeds for penicillin fermentation on defined mediums. *Chemistry and Industry* 19: 607–610.
23. Gray, P. P., Dunnill, P., and Lilly, M. D. 1973. The effect of controlled feeding of glycerol on β-galactosidase production by *Escherichia coli* in batch culture. *Biotechnology and Bioengineering* 15: 1179–1188.
24. Allen, A., and Andreotti, R. E. 1982. Cellulase production in continuous and fed-batch culture by *Trichoderma reesei* MCG 80. *Biotechnology and Bioengineering Symposium* 12: 451–459.
25. Gottvaldova, M., Kucera, J., and Podrazky, V. 1982. Enhancement of cellulase production by *Trichoderma viride* using carbon/nitrogen double-fed-batch. *Biotechnology Letters* 4: 229–231.
26. Gottvaldova, M., Kucera, J., and Podrazky, V. 1982. Fed-batch cultivation of *Trichoderma viride*: Effect of pH and nitrogen source supplementation on cellulase production. *Biotechnology Letters* 4: 645–646.
27. Hendy, N., Wilke, C., and Blanch, H. 1982. Enhanced cellulase production using Solka floc in a fed-batch fermentation. *Biotechnology Letters* 4: 785–788.

28. Hendy, N., Wilke, C., and Blanch, H. 1984. Enhanced cellulase production in fed-batch culture of *Trichoderma reesei* C-30. *Enzyme and Microbial Technology* 6: 73–77.

29. McLean, D., and Podruzny, M. F. 1985. Further support for fed-batch production of cellulases. *Biotechnology Letters* 7: 683–688.

30. Matsumura, M., Imanaka, T., Yoshida, T., and Taguchi, H. 1981. Modeling of cephalosporin C production and its application to fed-batch culture. *Journal of Fermentation Technology* 59: 115–123.

31. Rutkov, A. B. 1984. Microbiological production of L-lysine by employing fed batch cultivation. *Doklady Bolgarskoi Akademii Nauk* 37: 1677–1680.

32. Vu-Trong, K., and Gray, P. P. 1984. Stimulation of enzymes involved in tylosin biosynthesis by cyclic feeding profiles in fed batch cultures. *Biotechnology Letters* 6: 435–440.

33. Vu-Trong, K., and Gray, P. P., eds. 1982. *Proceedings of the 5th Australian Biotechnology Conference*, University of New South Wales.

34. Vu-Trong, K., and Gray, P. P. 1982. Stimulation of tylosin productivity resulting from cyclic feeding profiles in fed batch cultures. *Biotechnology Letters* 4: 725–728.

35. Matsumura, M., Imanaka, T., Yoshiuda, T., and Taguchi, H. 1978. Effect of glucose and methionine consumption rates on cephalosporin C production by *Cephalosporium acremonium*. *Journal of Fermentation Technology* 56: 345–353.

36. Matsumura, M., Imanaka, T., Yoshiuda, Y., and Taguchi, H. 1981. Optimal conditions for production of cephalosporin C in fed-batch culture. *Advances in Biotechnology* 1: 297–302.

37. Galliher, P. M., Cooney, C. L., Langer, R., and Lindhart, R. J. 1981. Heparinase production by *Flavobacterium heparinum*. *Applied Environmental Microbiology* 41: 360–365.

38. Yamane, T., and Tsukano, M. 1977. Kinetic studies on fed-batch cultures. V. Effect of several substrate-feeding modes on production of extracellular α-amylase by fed-batch culture of *Bacillus megaterium*. *Journal of Fermentation Technology* 55: 233–242.

39. Shin, S. B., Kitagawa, Y., Suga, K., and Ichikawa, K. 1978. Cellulase biosynthesis by *Trichoderma viride* on soluble substrates. *Journal of Fermentation Technology* 56: 396–402.

40. Waki, T., Suga, K., and Ichigawa, K. 1981. Production of cellulase in fed-batch culture. *Advances in Biotechnology* 1: 359–364.

41. Ohno, H., Nakanishi, E., and Takamatsu, T. 1976. Optimal control of a semi-batch fermentation. *Biotechnology and Bioengineering* 18: 847–864.

42. Suzuki, T., Mori, H., Yamane, T., and Shimizu, S. 1985. Automatic supplementation of minerals in fed-batch culture to high cell mass concentration. *Biotechnology and Bioengineering* 27: 192–201.

43. Demain, A. L. 1971. Overproduction of microbial metabolites and enzymes due to alteration of regulation. *Advances in Biochemical Engineering* 1: 113–142.

44. Bailey, E. J., and Ollis, D. F. 1986. *Biochemical Engineering Fundamentals*. 2nd ed. McGraw-Hill.

45. Bauer, S., and Schiloach, J. 1974. Maximal exponential growth rate and yield of *E. coli* obtainable in a bench-scale fermentor. *Biotechnology and Bioengineering* 16: 933–941.

46. Mori, H., Yano, T., Kobayashi, T., and Shimizu, S. 1979. High density cultivation of biomass in fed-batch system with DO-Stat. *Journal of Chemical Engineering, Japan* 12: 313–319.

47. Riesenberg, D., and Guthke, R. 1999. High-cell-density cultivation of microorganisms. *Applied Microbiology and Biotechnology* 51: 422–430.

48. Kobayashi, T., Yano, T., Mori, H., and Shimizu, S. 1979. Cultivation of microorganisms with a DO-stat and a silicone tubing sensor. *Biotechnology and Bioengineering Symposium* 9: 73–83.
49. Mori, H., Kobayashi, T., and Shimizu, S. 1981. High density production of sorbose from sorbitol by fed-batch culture with DO-stat. *Journal of Chemical Engineering, Japan* 14: 65–70.
50. Nishio, N., Tsuchiya, Y., Hayashi, M., and Nagai, S. 1979. Studies on methanol metabolism. VI. A fed-batch culture of methanol-utilizing bacteria with pH stat. *Journal of Fermentation Technology* 55: 151–155.
51. Yamauchi, H., Mori, H., Kobayashi, T., and Shimizu, S. 1983. Mass production of lipids by *Lipomyces starkeyi* in microcomputer-aided fed-batch culture. *Journal of Fermentation Technology* 61: 275–280.
52. Park, C. B., and Lee, S. B. 1997. Constant-volume fed-batch operation for high density cultivation of hyperthermophilic aerobes. *Biotechnology Techniques* 11: 277–281.
53. Huang, W.-B., and Chu, S. Y. 1984. Ethanol production by *Zymomonas mobilis* in fed-batch fermentations. *Biotechnology and Bioengineering* 26: 247–251.
54. Lim, H. C., Tayeb, Y. J., Modak, J. M., and Bonte, P. 1986. Computational algorithms for optimal feed rates for a class of fed-batch fermentation: Numerical results for penicillin and cell mass production. *Biotechnology and Bioengineering* 28: 1408–1420.
55. Vicik, S. M., Fedor, A. J., and Swartz, R. W. 1990. Defining an optimal carbon source/methionine feed strategy for growth and cephalosporin C formation by *Cephalosporium acremonium*. *Biotechnology Progress* 6: 333–340.
56. Bajpai, R. K., and Reub, M. 1981. Evaluation of feeding strategies in carbon-regulated secondary metabolite production through mathematical modeling. *Biotechnology and Bioengineering* 23: 717–738.
57. Bhargava, S., Nandakumar, M. P., Roy, A., Wenger, K. S., and Marten, M. R. 2003. Pulsed feeding during fed-batch fungal fermentation leads to reduced viscosity without detrimentally affecting protein expression. *Biotechnology and Bioengineering* 81: 341–347.
58. De Swaaf, M. E., Sijtsma, L., and Pronk, J. T. 2003. High-cell-density fed-batch cultivation of the docosahexaenoic acid producing marine alga *Crypthecodinium cohnii*. *Biotechnology and Bioengineering* 81: 666–672.
59. Jeffries, T. W., Fady, J. H., and Lightfoot, E. N. 1985. Effect of glucose supplements on the fermentation of xylose by *Pachysolen tannophilus*. *Biotechnology and Bioengineering* 27: 171–176.
60. Koshimizu, L. H., Valdeolivas Gomez, E. I., Bueno Netto, C. L., Regina de Melo Cruz, M., Vairo, M. L. R., and Borzani, W. 1984. Constant fed-batch ethanol fermentation of molasses. *Journal of Fermentation Technology* 62: 205–210.
61. Park, C. B., and Lee, S. B. 1997. Constant-volume fed-batch operation for high density cultivation of hyperthermophilic aerobes. *Biotechnology Techniques* 11: 277–281.
62. Lo Curto, R. B., and Tripodo, M. M. 2001. Yeast production from virgin grape marc. *Bioresource Technology* 78: 5–9.
63. Rangel-Yagui, C. de O., Danesi, E. D., de Carvalho, J. C., and Sato, S. 2004. Chlorophyll production from *Spirulina platensis*: Cultivation with urea addition by fed-batch process. *Bioresource Technology* 92: 133–141.
64. Lee, H. H., and Yau, B. O. 1981. An experimental reactor for kinetic studies: Continuously fed batch reactor. *Chemical Engineering Science* 36: 483–488.
65. Hardjito, L., Greenfield, P. F., and Lee, P. L. 1993. Recombinant protein production via fed-batch culture of the yeast *Saccharomyces cerevisiae*. *Enzyme and Microbial Technology* 15: 120–126.
66. Keller, R., and Dunn, I. J. 1978. Fed-batch microbial culture: Models, errors and applications. *Journal of Applied Chemistry and Biotechnology* 28: 508–514.

67. Esener, A. A., Roels, J. A., and Kossen, N. W. F. 1981. Fed-batch culture: Modeling and application in the study of microbial energetics. *Biotechnology and Bioengineering* 23: 1851–1871.

68. Ramirez, O. T., Zamora, R., Quintero, R., and Lopez-Munguia, A. 1994. Exponentially fed-batch cultures as an alternative to chemostats: The case of penicillin acylase production by recombinant *E. coli. Enzyme and Microbial Technology* 16: 895–903.

69. Sterkenburg, A., Prozee, G. A. P., Leegwater, P. A. J., and Wouters, J. T. M. 1984. Expression and loss of the pBR322 plasmid in *Klebsiella aerogenes* NCTC 418, grown in chemostat culture. *Antonie van Leeuwenhoek* 50: 397–404.

70. Chew, L. C. K., Tacon, W. C. A., and Cole, J. A. 1988. Effect of growth conditions on the rate of loss of the plasmid pAT153 from continuous cultures of *Escherichia coli* HB101. *FEMS Microbiology Letters* 56: 101–104.

71. Tottrup, H. V., and Carlsen, S. 1990. A process for the production of human proinsulin in *Saccharomyces cerevisiae. Biotechnology and Bioengineering* 35: 339–348.

72. Siegel R., and Ryu, D. D. Y. 1985. Kinetic study of instability of recombinant plasmid pPLc23 trpA1 in *E. coli* using two-stage continuous culture system. *Biotechnology and Bioengineering* 27: 28–33.

73. Nancib, N., and Boudrant, J. 1992. Effect of growth rate on stability and gene expression of a recombinant plasmid during continuous cultures of *Escherichia coli* in non-selective medium. *Biotechnology Letters* 14: 643–648.

74. Enberg, B., and Nordstrom, K. 1975. Replication of R-factor R1 in *Escherichia coli* K-12 at different growth rates. *Journal of Bacteriology* 123: 179–186.

75. Curless, C., Pope, J., and Tsai, L. 1990. Effect of pre-induction specific growth rate on recombinant alpha consensus interferon synthesis in *Escherichia coli. Biotechnology Progress* 6: 149–152.

76. Bentley, W. E., Mirjalili, N., Andersen, D. C., Davis, R. H., and Kompala, D. S. 1990. Plasmid-encoded protein: The principal factor in the "metabolic burden" associated with recombinant bacteria. *Biotechnology and Bioengineering* 35: 668–681.

77. Seo, J.-H., and Bailey, J. E. 1985. Effects of recombinant plasmid content on growth properties and cloned gene product formation in *Escherichia coli. Biotechnology and Bioengineering* 27: 1668–1674.

78. Riesenberg, D., Menzel, K., Schulz, V., Schumann, K., Veith, G., Zuber, G., and Knorre, W. A. 1990. High cell density fermentation of recombinant *Escherichia coli* expressing chuman interferon alpha 1. *Applied Microbiology and Biotechnology* 34: 77–82.

79. Ryan, W., and Parulekar, S. J. 1991. Recombinant protein synthesis and plasmid instability in continuous cultures of *Escherichia coli* JM103 harboring a high copy number plasmid. *Biotechnology and Bioengineering* 37: 415–429.

80. Seo, J.-H., and Bailey, J. E. 1986. Continuous cultivation of recombinant *Escherichia coli*: Existence of an optimum dilution rate for maximum plasmid and gene product concentration. *Biotechnology and Bioengineering* 28: 1590–1594.

81. Park, S., Ryu, D. D. Y., and Kim, J. Y. 1990. Effect of cell growth rate on the performance of a two-stage continuous culture system in a recombinant *Escherichia coli* fermentation. *Biotechnology and Bioengineering* 36: 493–505.

82. Ramirez, O. T., Zamora, R., Espinosa, G., Merino, E., Bolivar, F., Quintero, R. 1994. Kinetic study of penicillin acylase production by recombinant *E. coli* in batch cultures. *Process Biochemistry* 29: 197–206.

83. Curless, F. K., Swank, R., Menjares, A., Fieschko, J., and Tsai, L. 1991. Design and evaluation of a two-stage, cyclic, recombinant fermentation process. *Biotechnology and Bioengineering* 38: 1082–1090.

84. Jensen, E. B., and Carlsen, S. 1990. Production of recombinant human growth hormone in *Escherichia coli*: Expression of different precursors and physiological effects of glucose, acetate, and salts. *Biotechnology and Bioengineering* 36: 1–11.

85. Zabriskie, D. W., Wareheim, D. A., and Plansky, M. J. 1987. Effects of fermentation feeding strategies prior to induction of expression of a recombinant malaria antigen in *Escherichia coli*. *Journal of Industrial Microbiology* 2: 87–95.

86. Nasri, M., Sayadi, S., Barbotin, J. N., and Thomas, D. 1987. The use of the immobilization of whole living cells to increase stability of recombinant plasmids in *Escherichia coli*. *Journal of Biotechnology* 6: 147–157.

87. Horn, U., Krug, M., and Sawistoski, J. 1990. Effect of high density cultivation on plasmid copy number in recombinant *Escherichia coli* cells. *Biotechnology Letters* 12: 191–196.

88. Kojima, I., Yoshikawa, H., Okazaki, M., and Terui, Z. 1972. Studies on riboflavin production by *Eremothecium ashbyii*. *Journal of Fermentation Technology* 50: 716–723.

89. Dalton, C. C. 1983. Chlorophyll production in fed-batch cultures of *Ocimum basilicum* (sweet basil). *Plant Science Letters* 32: 263–270.

90. Woods, M., and Millis, N. F. 1985. Effect of slow feeding of xylose on ethanol yield by *Pachysolen tannophilus*. *Biotechnology Letters* 7: 679–682.

91. Glick, R. G., and Pasternak, J. J. 1998. *Molecular Biotechnology: Principles and Applications of Recombinant DNA*. ASM Press.

Classification and Characteristics of
Fed-Batch Cultures

As for any reactor operation, the primary purpose of fed-batch operation is to maximize the rates of cell growth and product formation so that the total rate of product formation (productivity) or product yield (selectivity) is maximized. The desired product may be classified into (1) high volume–high margin products, (2) high volume–low margin products, (3) low volume–high margin products, and (4) low volume–low margin products. For high-margin products, the raw material cost may be negligible in comparison to the price of the product. In this case, there is little incentive to minimize the raw material cost. However, the production cost, which is roughly inversely proportional to the productivity, may be reduced by increasing the rate (productivity). For low-margin products, there is much incentive to minimize both the raw material and the processing costs.

For those processes for which the raw material is relatively inexpensive, one may wish to maximize the productivity (rate), while one may wish to improve the yield (selectivity) if the cost of the raw material and/or the product is relatively high. Ultimately, one must minimize the total production costs. This is achieved by regulating the feed rates of the limiting substrates, nitrogen and phosphate sources, inducers, precursors, or intermediates and by the selection of proper initial conditions. Through the manipulation of the feed rates of the medium containing the substrate and nutrients, the fed-batch operation allows regulation of the concentration of key substances that control the cell growth and/or product formation rate.

5.1 Classification Based on Feeding Patterns

A number of feeding patterns have been tried for various purposes, including constant feed rates, linearly increasing (or decreasing) feed rates, exponential feed rates, intermittent feed rates, feed rates to maintain the limiting nutrient concentration constant (extended fed-batch), empirical feed rates, optimal feed rates that optimize various objectives, and closed-loop feedback-controlled feed rates. To understand the basics of various fed-batch operations, it is best to utilize a simple example, involving the growth of cells and metabolite production on a limiting substrate. We begin by examining the material balance equations that describe the fed-batch operation.

5.1.1 Mass Balance Equations

For convenience, we deal with only one feed stream. Extension to the case of multiple feed streams is straightforward; simply add additional terms for the extra feed streams. The overall mass balance that keeps track of all materials introduced into and removed from the bioreactor is obtained from the general mass balance equation (5.1):

$$\frac{d(V\rho_c)}{dt} = F\rho_F + m_a + m_b + m_f - m_e \tag{5.1}$$

where ρ_c and ρ_F are the densities of the culture and the feed, respectively; V is the culture volume; F is the feed rate; m_a, m_b, and m_f are the mass flow rate of the acid, base, and antifoam, respectively; and m_e is the rate of mass loss, mainly for volatile components and water, due to evaporation caused by aeration.

5.1.2 Cell Mass Balance

The material balance for the cells is obtained by applying Eq. (5.1) to the cell mass, recognizing that there are no cells in the feed and effluent, so that the rate of accumulation, $d(XV)/dt$, is equal to the rate of generation, which is the product of the net specific growth rate, $\mu(S)$, and the total cell mass, XV:

$$\frac{d(XV)}{dt} = \mu(S)XV \quad XV(0) = X_0V_0 \tag{5.2}$$

where $\mu(S)$ denotes that the specific growth rate is a function of the limiting substrate concentration S. The specific growth rate may be a function of not only the substrate concentration but also the product concentration,

5.1.3 Substrate Balance

For the limiting substrate, the rate of accumulation is equal to the rate of feed $F(t)S_F$ minus the rate of consumption by cells $\sigma(S)XV$, which is the product of the net specific substrate consumption rate, $\sigma(S)$, which, for now, is assumed to be a function of substrate concentration, and the total mass of cells, XV:

$$\frac{d(SV)}{dt} = F(t)S_F - \sigma(S)XV \quad SV(0) = S_0V_0 \tag{5.3}$$

where S_F is the feed substrate concentration and $F(t)$ is the substrate feed rate.

5.1.4 Product Balance

Because there is no product in the feed and no product is removed, the rate of accumulation of metabolite is equal to the rate of generation, which is equal to the net specific product formation rate, $\pi(S)$, times the total cell mass, XV:

$$\frac{d(PV)}{dt} = \pi(S)XV \tag{5.4}$$

where P is the product concentration. For now, the specific rate is assumed to be a function only of the limiting substrate concentration S.

The specific rates include the usual specific rates as well as the maintenance and decay terms:

$$\mu(S) - k_x \tag{5.5}$$

and

$$\sigma(S) - m_X \tag{5.6}$$

where m_X is the usual maintenance coefficient for cells and k_x is the decay constant for cell mass.

5.1.5 Overall Mass Balance

Assuming the mass lost due to evaporation into the exhaust gas and the mass gained by the additions of a base, acid, and antifoam to be insignificant as compared to the mass flow rate of the feed, the overall material balance relates the mass flow rate of the feed to the mass accumulation rate in the bioreactor:

$$\frac{d(V\rho_c)}{dt} = F\rho_F \quad V(0) = V_0 \tag{5.7}$$

Assuming that there is no appreciable density difference between the feed and the culture broth, $\rho_C = \rho_F$, and that the densities remain time-invariant – reasonable assumptions for aqueous fermentations – we can eliminate the density to obtain

$$\frac{dV}{dt} = F(t) \quad V(0) = V_0 \tag{5.8}$$

The feed flow rate $F(t)$ affects directly the culture volume (Eq. (5.1)), and the total amount of substrate in the culture (Eq. (5.3)) or the substrate concentration, which affects the cell growth rate through Eq. (5.2). It is also clear that even in the simplest situation, there are three initial conditions that should be chosen appropriately: the initial volume, the total amount of inoculum, and the total amount of substrate initially, V_0, X_0V_0, and S_0V_0. These initial conditions must be chosen optimally, in addition to the optimal feed flow rate profile, to maximize a profit function. However, practical limitations put an upper limit on the initial concentration and therefore on the amount of inoculum.

With the preceding mass balance equations, it is possible to describe quantitatively the characteristics of various forms of fed-batch cultures.

5.1.6 Fed-Batch Cultures with Constant Feed Rates

This is perhaps the simplest fed-batch operation, in which the feed rate is held constant until the bioreactor is full, and as such, it has received extensive coverage in mathematical analyses by Yamane and colleagues,[1,2] Pirt,[3,4] and Dunn and colleagues.[5,6] Yamane and Hirano[7] made experimental studies. Many classical practices belong to this category. The feed rate is held constant, $F = F_c$, and the defining mass balance equations are obtained from Eqs. (5.2)–(5.4) and Eq. (5.8):

$$\frac{dV}{dt} = F_c \quad V(0) = V_0 \tag{5.9}$$

$$\frac{d(XV)}{dt} = \mu XV \quad XV(0) = X_0 V_0 \tag{5.10}$$

$$\frac{d(SV)}{dt} = F_c S_F - \sigma XV = F_c S_F - \frac{\mu XV}{Y_{X/S}} \quad SV(0) = S_0 V_0 \tag{5.11}$$

$$\frac{d(PV)}{dt} = \pi XV = \frac{\mu XV}{Y_{P/S}} \quad PV(0) = P_0 V_0 \tag{5.12}$$

Because of the constant feed rate, the culture volume as obtained by integrating Eq. (5.9) increases in the form of a ramp function until it reaches the maximum:

$$V(t) = V_0 + \int_0^t F_c d\tau = V_0[1 + (F_c/V_0)t] = F_c[1 + D_0 t] \quad 0 \le t \le (V_f - V_0)/F_c \tag{5.13}$$

Therefore, the dilution rate, which decreases with time, is

$$D(t) \triangleq \frac{F}{V}(t) = \frac{F_c}{V_0 + F_c t} = \frac{F_c/V_0}{1 + (F_c/V_0)t} = \frac{D_0}{1 + D_0 t} \tag{5.14}$$

The total cell mass in the culture at any time is obtained by integrating Eq. (5.10) with the initial condition:

$$XV(t) = X_0 V_0 \exp \int_0^t \mu d\tau \tag{5.15}$$

The specific growth rate in Eq. (5.15) is not constant but varies with the substrate concentration and therefore also with time. Thus, the total cell mass XV increases not exponentially but semiexponentially with time. The cell concentration as obtained by dividing Eq. (5.15) by Eq. (5.13) can increase, decease, or remain constant, depending on the initial dilution rate D_0:

$$X = XV/V = X_0 V_0 \exp \int_0^t \mu d\tau \Big/ F_c(1 + D_0 t) \tag{5.16}$$

Taking as an example the Monod form as the specific growth rate, $\mu = kS/(K + S)$, the time profiles of total cell mass and cell mass concentration are shown in Figure 5.1. It is clear that both the amount and concentration of cell mass increase semiexponentially.

It is now convenient to introduce a moving average (a time average) of the specific growth rate:

$$\bar{\mu}(t) = \int_0^t \mu d\tau \Big/ \int_0^t d\tau = \int_0^t \mu d\tau \Big/ t \tag{5.17}$$

The moving average is a function of time, $\bar{\mu}(t)$. The cell mass concentration can be expressed in terms of the moving average of the specific growth rate:

$$X(t) = \frac{XV}{V} = \frac{X_0 V_0 \exp(\bar{\mu} t)}{V_0 + F_c t} = \frac{X_0 \exp(\bar{\mu} t)}{(1 + D_0 t)} \tag{5.18}$$

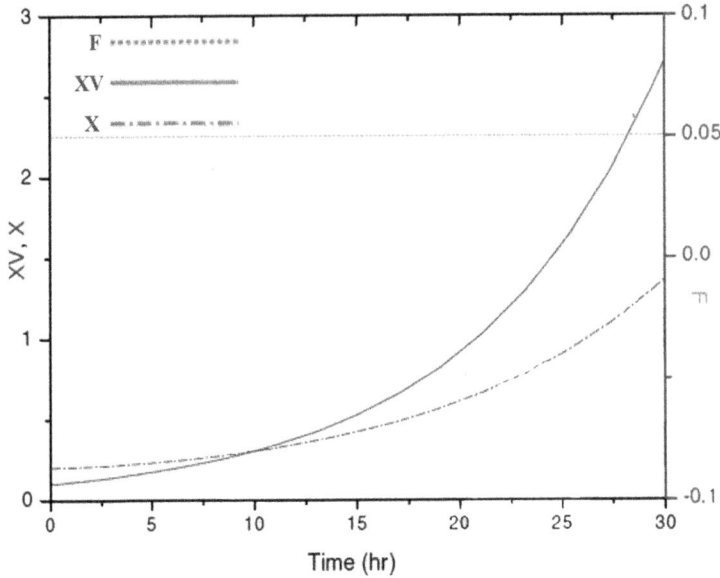

Figure 5.1. Time profiles of cell concentration and total cell mass for fed-batch culture with a constant feed rate: $\mu = 0.11s/(s + 0.006)$; $\pi = \mu/1.2$; $\sigma = \mu/4.7$; $S_F = 100\,\text{g/L}$; $t_f = 30\,\text{hr}$; $F_c = 0.05\,\text{L/hr}$; $[XV, SV, PV, V](t = 0) = [0.1\,\text{g/L}, 1\,\text{g/L}, 0\,\text{g/L}, 0.5\,\text{L}]$.

That the cell concentration can increase, decrease, or remain constant can be made clearer by expanding the left-hand side of the cell balance equation (Eq. (5.10)) and substituting into it Eq. (5.9),

$$\frac{d(XV)}{dt} = \frac{dX}{dt}V + X\frac{dV}{dt} = V\frac{dX}{dt} + XF_c = \mu XV \tag{5.19}$$

and then solving for dX/dt,

$$\frac{dX}{dt} = \left(\mu - \frac{F_c}{V}\right)X = (\mu - D)X = \left(\mu - \frac{F_c}{V_0 + F_c t}\right)X = \left(\mu - \frac{D_0}{1 + D_0 t}\right)X \tag{5.20}$$

From Eq. (5.20), we conclude that the cell concentration can increase, decrease, or remain constant with respect to time, depending on the value of the specific growth rate relative to the time-variant dilution rate, $D = F_c/(V_0 + F_c t)$:

$$\frac{dX}{dt} \begin{cases} < 0 & \text{if} \quad D > \mu(S) \\ = 0 & \text{if} \quad D = \mu(S) \\ > 0 & \text{if} \quad D < \mu(S) \end{cases} \tag{5.21}$$

If the time-variant dilution rate is greater than the specific growth rate, the dilution effect is greater than the growth effect, and consequently, the cell concentration decreases, whereas if the dilution rate is less than the growth rate, the dilution effect is overcome by the growth of cells, and the cell concentration increases with time. This is a general characteristic with a constant flow rate with any arbitrary specific growth rate. Because the time-variant dilution rate decreases with time, the cell concentration would eventually increase with time. In fact, the cell concentration can go through a local valley or peak during the course of a run. This phenomenon is utilized in Chapter 8 to estimate the specific rates.

It should be noted that to maintain the cell concentration constant over a period of time, the dilution rate, which decreases with time, must be matched exactly with the specific growth rate,

$$\mu(S) = D = \frac{F_c}{V} = \frac{F_c}{V_0 + F_c t} = \frac{D_0}{1 + D_0 t} \tag{5.22}$$

which implies that the specific growth rate must also decrease with time. Because μ is an arbitrary function of the substrate concentration, it is not apparent whether the conditions imposed by Eq. (5.22) can be met for any time interval. The time behavior of the specific growth rate may be examined by differentiating it with respect to time:

$$\frac{d\mu}{dt} = \frac{d\mu}{dS}\frac{dS}{dt} = -\frac{D_0^2}{(1 + D_0 t)^2} < 0$$

or

$$\text{sign}\left[\frac{d\mu}{dt}\right] = \text{sign}\left[\frac{d\mu}{dS}\right]\text{sign}\left[\frac{dS}{dt}\right] < 0$$
$$\Rightarrow \text{sign}\left[\frac{d\mu}{dS}\right] = -\text{sign}\left[\frac{dS}{dt}\right] \tag{5.23}$$

Equation (5.23) states that the specific growth rate must decrease with time and that the signs of the two terms $d\mu/dS$ and dS/dt must be opposite to each other because the sign of the product is negative. Therefore, to maintain the cell concentration constant, the substrate concentration must decrease with time, that is, $dS/dt < 0$ for the specific growth rate that increases with the substrate concentration $d\mu/dS > 0$, as in the case of the Monod model. Conversely, for nonmonotonic specific growth rates that can increase as well as decrease with the substrate concentration so that $d\mu/dS > 0$ or $d\mu/dS < 0$, the substrate concentration can increase or decrease depending on the slope of the μ curve; that is, if μ is on the rising side ($d\mu/dS > 0$), the substrate concentration decreases with time, whereas it increases with time if μ is on the diminishing side ($d\mu/dS < 0$) of the peak. All this implies that it would be difficult to maintain the cell concentrations constant over any appreciable time period. This phenomenon has been named *quasi steady state* and has been numerically demonstrated[5] to take place asymptotically after a long time if the constant feed rate is small enough relative to the maximum bioreactor volume, $V/F_c = (V_0 + F_c t)/F_c = (1 + D_0 t)/D_0 \gg 1$.

To assess the substrate concentration profile, we begin with the substrate balance equation. Equation (5.11) is rearranged and Eq. (5.9) is substituted into it to obtain

$$\frac{d(SV)}{dt} = F_c S_F - \frac{1}{Y_{X/S}}\frac{d(XV)}{dt} \tag{5.24}$$

and then each term is integrated under the assumption that the yield coefficient $Y_{X/S}$ is constant:

$$SV(t) - S_0 V_0 = F_c S_F t - \frac{XV - X_0 V_0}{Y_{X/S}} = F_c S_F t - \frac{X_0 V_0 (e^{\overline{\mu} t} - 1)}{Y_{X/S}} \tag{5.25}$$

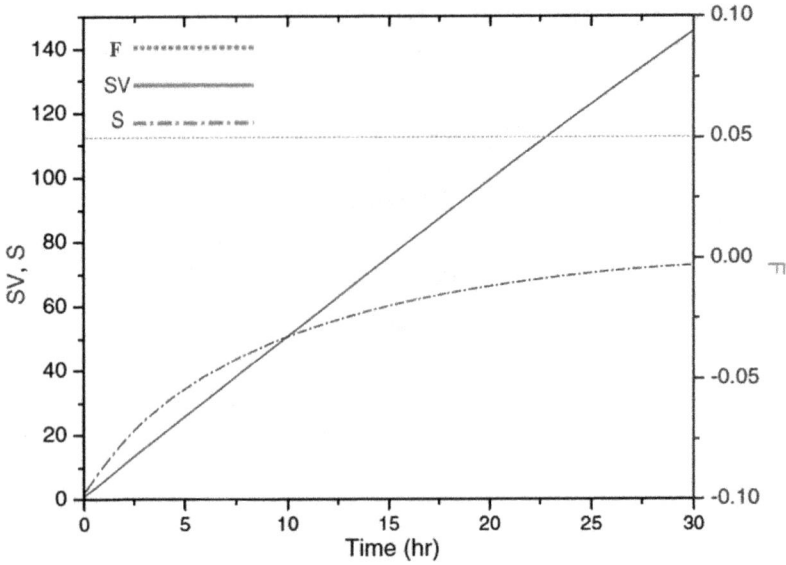

Figure 5.2. Time profiles of substrate concentration and total amount of substrate for fed-batch culture with a constant feed rate: $\mu = 0.11s/(s + 0.006)$; $\pi = \mu/1.2$; $\sigma = \mu/4.7$; $S_F = 100 \, \text{g/L}$; $t_f = 30 \, \text{hr}$; $F_c = 0.05 \, \text{L/hr}$; $[XV, SV, PV, V](t_0) = [0.1 \, \text{g/L}, 1 \, \text{g/L}, 0 \, \text{g/L}, 0.5 \, \text{L}]$.

The total amount of substrate remaining in the reactor can increase or decrease, depending on the amount of substrate supplied $(F_c S_F t)$ relative to the amount of substrate consumed to form cells $(X_0 V_0 (e^{\overline{\mu}t} - 1)/Y_{X/S})$. Because the volume increases in the form of a ramp function, the substrate concentration can increase or decrease, just as the total amount of substrate can increase or decrease:

$$S = \frac{S_0 + X_0/Y_{X/S} + D_0 S_F t}{1 + D_0 t} - \frac{X}{Y_{X/S}} \tag{5.26}$$

The time profiles of substrate concentration and total amount of substrate are shown in Figure 5.2. The metabolite balance equation (Eq. (5.12)) may be integrated if the product yield coefficient is constant. Equation (5.10) is substituted into Eq. (5.12) and integrated to obtain

$$PV = P_0 V_0 + (XV - X_0 V_0)/Y_{P/S} \tag{5.27}$$
$$= P_0 V_0 + X_0 V_0 [\exp(\overline{\mu}t) - 1]/Y_{P/S}$$

The metabolite concentration is obtained by dividing Eq. (5.27) by Eq. (5.13):

$$P = \frac{P_0 V_0 + X_0 V_0 [\exp(\overline{\mu}t) - 1]/Y_{P/S}}{V_0 + F_c t} \tag{5.28}$$

For constant feed rate operation, two parameters can be selected: the initial bioreactor volume, V_0, and the constant feed rate, F_c. By a proper selection of these parameters, one can regulate the dilution rate as well as the time to fill the culture completely, which is

$$t_{\text{full}} = \frac{V_{\text{max}} - V_0}{F_c} \tag{5.29}$$

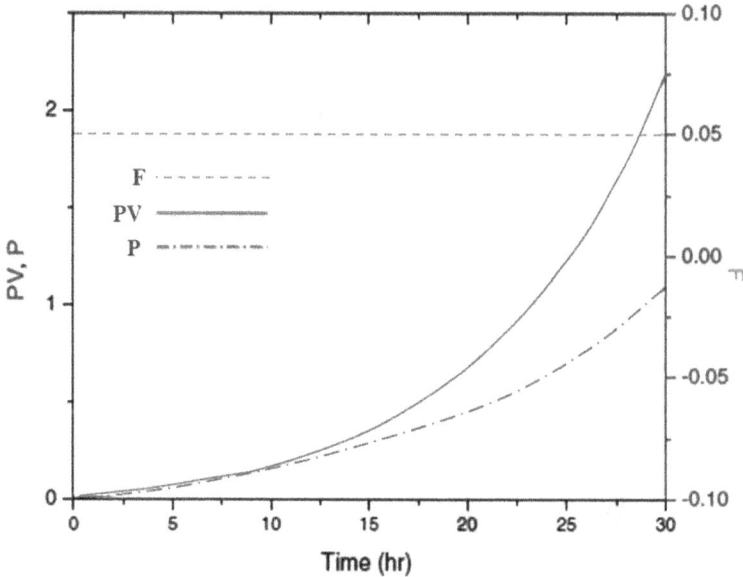

Figure 5.3. Time profiles of product concentration and total amount of product for a fed-batch culture with a constant feed rate: $\mu = 0.11s/(s + 0.006)$; $\pi = \mu/1.2$; $\sigma = \mu/4.7$; $S_F = 100\,\text{g/L}$; $t_f = 30\,\text{hr}$; $F_c = 0.05\,\text{L/hr}$; $[XV, SV, PV, V](t_0) = [0.1\,\text{g/L}, 1\,\text{g/L}, 0\,\text{g/L}, 0.5\,\text{L}]$.

When the bioreactor volume is full, the operation may continue in batch mode until the substrate concentration drops to a desired level. Time profiles of product concentration and the total amount of product are given in Figure 5.3.

5.1.6.1 Early Growth Phase or Very Small Specific Growth Rate

Because the exponential function can be expressed in a Taylor series,

$$\exp(\overline{\mu}t) = 1 + \overline{\mu}t + (\overline{\mu}t)^2/2! + (\overline{\mu}t)^3/3! + \cdots \tag{5.30}$$

the total cell mass (Eq. (5.18)) can be represented by

$$XV(t) = X_0V_0[1 + \overline{\mu}t + (\overline{\mu}t)^2/2! + (\overline{\mu}t)^3/3! + \cdots] \tag{5.31}$$

From Eq. (5.18), we note that the total cell mass increases exponentially with time. However, it is also apparent from Eq. (5.31) that when the values of $\overline{\mu}t$ are small, the second- and higher-order terms are negligible with respect to the linear term $(1 + \overline{\mu}t)$, and the exponential function may be well approximated by the linear function. Therefore, the total mass appears to increase linearly:

$$XV(t) = X_0V_0(1 + \overline{\mu}t) \tag{5.32}$$

In other words, if the mean specific growth rate is small, as would be the case if the initial substrate concentration is low and the feed rate is small, the total cell mass would appear to increase linearly in the early stage (for small t) of fed-batch operation. However, the cell mass is actually growing exponentially. Eventually, as the time progresses, the remaining terms weigh heavily, and the total mass can no longer be represented by the linear terms. In other words, even when the cell mass appears to increase linearly in the early stage, it eventually goes into a full-fledged exponential phase as time increases. This phenomenon occurs when the mean specific growth rate $\bar{\mu}$ is very small so that its product with the time, $\bar{\mu}t$, is small. Indeed, when $\bar{\mu}t$ is less than 0.4, the difference between the exponential function, $\exp(\mu_m t) = 1.492$, and its linear approximation, $1 + \bar{\mu}t = 1.4$, is 6%, which is well within the experimental error. Therefore, for all practical purposes, the exponential increase would appear to be linear. These phenomena have been observed experimentally as well as in simulations using the Monod form of specific growth rate.[1,7] However, it should be noted here that these observations should hold for any form of the specific growth rate, not just for the Monod form, as long as the value of $\bar{\mu}t$ is small.

As shown earlier, for small values of $\bar{\mu}t$, the exponential function may be approximated by a linear function (Eq. (5.32)) and with a proper choice of the constant feed rate so that the cell concentration appears to remain constant at the level of the initial inoculum concentration for slowly growing cells in the early stage of operation:

$$X(t) = \frac{X_0(1 + \bar{\mu}t)}{1 + (F_c/V_0)t} = X_0 \quad \text{if } F_c/V_0 = \bar{\mu} = \int_0^t \mu \, d\tau / t \tag{5.33}$$

5.1.6.2 Quasi Steady States

The state of a fed-batch culture with a constant feed rate, $\mu = F/V$, and the concentrations of cell mass and substrate do not change appreciably, that is, $dX/dt = dS/dt \cong 0$. This phenomenon is termed the *quasi steady state*.[3] This phenomenon has received some attention[8] and is also utilized advantageously.[9,10] From Eq. (5.20), one can deduce that for the cell concentration to remain time invariant, the relationship

$$\mu = \frac{F_c}{V} = \frac{F_c}{V_0 + F_c t} = \frac{D_0}{1 + D_0 t} \tag{5.34}$$

must be satisfied, and a steady state in substrate concentration can be reached if the following holds:

$$\mu/\sigma = Y_{X/S} = \frac{X}{(S_F - S)} \tag{5.35}$$

Equation (5.34) implies that the specific growth rate must decrease with time or remain approximately constant if the reactor volume remains approximately constant, that is, the increase in volume $F_c t$ relative to the initial volume V_0 is negligible. Thus, the initial volume must be large and the feed rate must be small. Equation (5.35) imposes an additional constraint on the operational conditions. For a feed with a specified substrate concentration S_F, the concentrations of cell mass and substrate, $X(t)$ and $S(t)$, must satisfy Eq. (5.35). It is instructive to calculate the yield coefficient for a fed-batch culture from the definition (the yield coefficient is equal to the amount of cell mass formed divided by the amount of substrate consumed)

$$Y_{X/S} = \frac{XV - X_0 V_0}{(V - V_0)S_F + S_0 V_0 - SV} = \frac{X - \frac{X_0 V_0}{V}}{(S_F - S) + (S_0 - S_F)\frac{V_0}{V}} \tag{5.36}$$

To match the condition imposed by Eq. (5.35) with that of Eq. (5.36) so that the quasi steady state can be observed over a finite time interval, it is necessary to pick

$$S_0 = S_F \tag{5.37}$$

and

$$X \gg \frac{X_0 V_0}{V} \Rightarrow XV \gg X_0 V_0 \tag{5.38}$$

In other words, the initial substrate concentration must be equal to the feed substrate concentration, and the total amount of cells must be much greater than the amount of initial inoculum. Because the initial volume has to be large (Eq. (5.34)), the initial cell concentration must be very small. Thus, it is possible to observe the quasi steady state over *a short time interval* by matching the feed concentration with the initial substrate concentration and by choosing a large initial volume with a very slow feed rate and a negligibly small inoculum ($S_0 = S_F$, $V \cong V_0$, $X_0 \cong 0$). Another way to look at the situation is to accept the conditions necessary to achieve $dX/dt = 0$ (Eq. (5.34)), that is, $F_c/V = \mu$, and apply this feed rate to the substrate balance equation (Eq. (5.11)) and substitute Eq. (5.36) into the resultant:

$$\frac{dS}{dt} = \frac{F_c}{V}(S_F - S) - \sigma X = \mu(S_F - S) - \sigma X = \mu \left[(S_F - S) - \frac{X}{Y_{X/S}} \right]$$

$$= \frac{\mu S_F (X - X_0) V_0}{XV - X_0 V_0} > 0 \tag{5.39}$$

Even if the cell concentration approaches asymptotically a constant value after some time, Eq. (5.39) states that the substrate concentration must increase, not remain constant. Thus, the quasi steady state is possible only for the cell concentration. It is obvious that to maintain two variables constant would require, in general, two manipulated variables, not just one: $F_c = \mu V$. This conclusion was also reported[5] by others. Because the quasi steady state implies that both the cell and substrate concentrations must be kept constant, it would in general require two manipulated variables to rapidly force the system toward the quasi steady state.[8]

To determine precisely the conditions under which the quasi steady state with respect to both cell and substrate concentrations occurs, it is necessary to take a specific form of the specific growth rate and make the material balance equations dimensionless and to examine the small parameters that multiply the time derivatives, $\varepsilon_1 dX/dt$ and $\varepsilon_2 dS/dt$, and to look for the conditions under which the small parameters ε_1 and ε_2 vanish. In other words, one can apply a singular perturbation analysis.[26] However, this process is beyond the scope of this book and is therefore omitted here.

Another application is to generate more accurate reaction rate data than those obtained using a continuous-stirred tank reactor (CSTR) for fast and slow reactions. This is the use of fed-batch culture using constant feed rates.

5.1.6.3 Rate Data Acquisition

We begin with the cell, substrate, and product balances by expanding the accumulation terms (the left-hand side) and rearranging into convenient forms by recognizing that $dV/dt = F_c$:

$$\frac{d(XV)}{dt} = \frac{dX}{dt}V + X\frac{dV}{dt} = \frac{dX}{dt}V + XF_c = \mu XV \tag{5.10}$$

$$\Rightarrow \frac{dX}{dt} = \left(\mu - \frac{F_c}{V}\right)X$$

$$\frac{d(SV)}{dt} = \frac{dS}{dt}V + S\frac{dV}{dt} = F_c S_F - \sigma XV \tag{5.11}$$

$$\Rightarrow \frac{dS}{dt} = F_c(S_F - S) - \sigma XV$$

and

$$\frac{d(PV)}{dt} = \frac{dP}{dt}V + P\frac{dV}{dt} = \frac{dP}{dt}V + PF_c = \pi XV \tag{5.12}$$

$$\Rightarrow \frac{dP}{dt} = \left(\pi - P\frac{F_c}{V}\right)X$$

Equations (5.10)–(5.12) can be further reduced if we can eliminate the time derivatives. If we can locate the times at which the concentrations go through their peak or valley, then the time derivatives would vanish, and the specific rates would be obtained from the concentrations of cells, substrate, and product at the points of their peaks and valleys, the constant feed rate, and the culture volume. By setting to zero the time derivatives in Eqs. (5.10)–(5.12), we obtain the specific rates as

$$\frac{dX}{dt}(t_1) = 0 = \left(\mu_1 - \frac{F_c}{V}\right)X \Rightarrow \mu_1 = \frac{F_c}{V} = \frac{1}{V_0/F_C + t_1} = \frac{1}{\tau_0 + t_1} \tag{5.40}$$

$$\frac{dS}{dt}(t_2) = 0 = F_c(S_F - S_2) - \sigma_2 XV \Rightarrow \sigma_2 = \frac{(S_F - S_2)}{X_2(\tau_0 + t_2)} \tag{5.41}$$

$$\frac{dP}{dt}(t_3) = 0 = \left(\pi_3 - P_3\frac{F_c}{V_3}\right)X_3 \Rightarrow \pi_3 = \frac{P_3}{\tau_0 + t_3} \tag{5.42}$$

where $t_1, t_2,$ and t_3 are the times at which the peaks and valleys in concentration appear for the cell, the substrate, and the product, respectively. It is apparent that if we can locate the times at which these concentrations go through their peaks and valleys, then the specific rates μ, σ, and π can be calculated algebraically from Eqs. (5.40)–(5.42), that is, without taking time derivatives of experimental data, which is the major source for large errors. Having done this once, we repeat the process with a number of different constant feed rates to generate a data set of specific rates with corresponding concentrations of cells, substrates, and product concentrations, as shown in Table 5.1. The constant feed rates must be chosen to cover a wide range of substrate and product concentrations because the objective is to correlate the specific rates as functions of substrate concentrations, both substrate and product concentrations, or (but rarely) substrate, product, and cell concentrations. The

Table 5.1. *Fed-batch kinetic data*

Run	F_{ci}	D_{0i}	t_{1i}	μ_i(5.40)	t_{2i}	X_{2i}	S_{2i}	σ_i(5.41)	t_{3i}	P_{3i}	π_i(5.42)
1	F_{c1}	D_{01}	t_{11}	μ_1	t_{21}	X_{21}	S_{21}	σ_1	t_{31}	P_{31}	π_1
2	F_{c2}	D_{02}	t_{12}	μ_2	t_{22}	X_{22}	S_{22}	σ_2	t_{32}	P_{32}	π_2
3	F_{c3}	D_{03}	t_{13}	μ_3	t_{23}	X_{23}	S_{23}	σ_3	t_{33}	P_{33}	π_3
.	.	.	.	.	.	.	.	.	.	.	.
.	.	.	.	.	.	.	.	.	.	.	.
.	.	.	.	.	.	.	.	.	.	.	.
n	F_{cn}	D_{0n}	t_{1n}	μ_n	t_{2n}	X_{2n}	S_{2n}	σ_n	t_{3n}	P_{3n}	π_n

results are then a table of calculated specific rates of cell growth, substrate consumption, and product formation, μ, σ, and π, versus the concentrations of cells, substrate, and product, X, S, and P. The remaining task is to correlate the specific rates with S, S, and P, or S, P, and X. The method outlined here avoids the need to take time derivatives of rate data. Of course, the rate data can be obtained using a steady state CSTR, requiring no derivatives of experimental data. However, for biological reactors, the rate data obtained at steady state do not necessarily hold up for unsteady state operations such as batch and fed-batch cultures. Even if the steady state rate data are applicable to unsteady state operations, it takes a long time for CSTRs to reach a steady state, especially when the dilution rate is small (rates are low). When the dilution rate is high (rates are high), the substrate concentration approaches that of the feed concentration, and the required difference between these two is subject to a higher degree of error. In fact, fed-batch reactors with constant feed rates have been reported to yield more accurate rate data for fast and slow reactions than the CSTR.[12]

5.1.7 Fed-Batch Cultures with Linearly Varying Feed Rates

The feed is in the form of increasing ramp or decreasing ramp, that is, $F = F_0 (1 + at)$, where $a > 0$ for the linearly increasing feed rate and $a < 0$ for the linearly decreasing feed rate. With the feed rate given by

$$F = F_0(1 + at) \quad \begin{bmatrix} a > 0 \\ a < 0 \end{bmatrix} \tag{5.43}$$

the mass balance equations are as follows:

$$\frac{dV}{dt} = F_0(1 + at) \quad V(0) = V_0 \tag{5.44}$$

$$\frac{d(XV)}{dt} = \mu XV \quad XV(0) = X_0 V_0 \tag{5.45}$$

$$\frac{d(SV)}{dt} = F_0(1 + at)S_F - \frac{\mu XV}{Y_{X/S}} \quad SV(0) = S_0 V_0 \tag{5.46}$$

and

$$\frac{d(PV)}{dt} = \pi XV = \frac{\mu XV}{Y_{P/S}} \quad PV(0) = P_0 V_0 \tag{5.47}$$

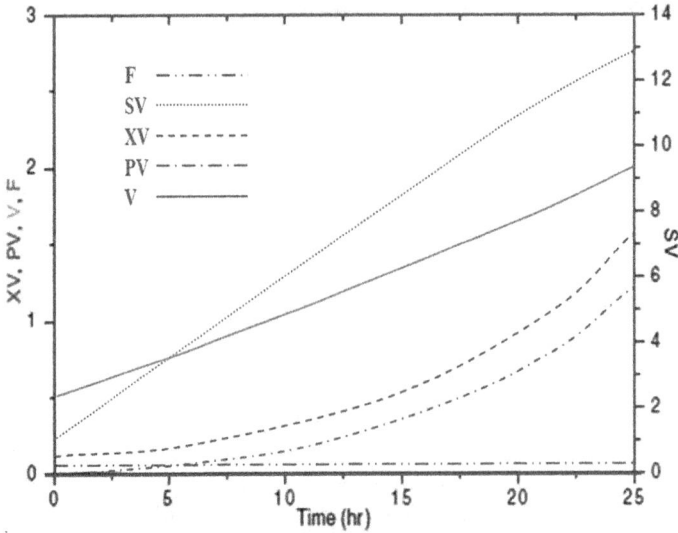

Figure 5.4. Time profiles of a fed-batch culture with a linearly increasing feed rate: $F = 0.05(1 + 0.016t)$; $t_f = 25$; $S_F = 10$ g/L; $V_{max} = 2$ L; $\mu = 0.11s/(s + 0.006)$; $\pi = \mu/1.2$; $\sigma = \mu/4.7$; $[XV, SV, PV, V](t_0) = [0.1$ g/L, 1 g/L, 0 g/L, 0.5 L$]$.

Integration of Eq. (5.44) shows that the culture volume changes quadratically in time:

$$V = V_0 + F_0(t + at^2/2) \tag{5.48}$$

The time-variant dilution rate increases $(a > 0)$ or decreases $(a < 0)$, and it determines whether the cell concentration and the total amount of substrate increase or decrease with time:

$$\frac{F}{V} = \frac{(F_0/V_0)(1 + at)}{1 + (F_0/V_0)(t + at^2/2)} \tag{5.49}$$

The time required to fill culture volume completely for the case of increasing flow rate $(a > 0)$ is obtained from Eq. (5.48) by noting that $V = V_f$ at $t = t_f$:

$$t_f = \frac{1}{a}\left\{-1 + \left[1 + \frac{2a(V_f - V_0)}{F_0}\right]^{1/2}\right\} \quad a > 0 \tag{5.50}$$

Typical time profiles of culture volume; concentrations of cell mass, substrate, and product; and the total amounts of cell mass substrate and product are shown in Figure 5.4 for the case of linearly increasing feed rate.

　　When the feed rate decreases with time $(a < 0)$, the feed rate may become zero before the culture volume is completely filled. This would be an ineffective use of the bioreactor. Thus, it may be assumed that this does not happen, and in that case, the time to fill the culture volume is

$$t_f = \frac{1}{a}\left\{-1 - \left[1 + \frac{2a(V_f - V_0)}{F_0}\right]^{1/2}\right\} \quad a < 0 \tag{5.51}$$

Typical time profiles of culture volume; concentrations of cell mass, substrate, and product; and the total amounts of cell mass, substrate, and product are shown in

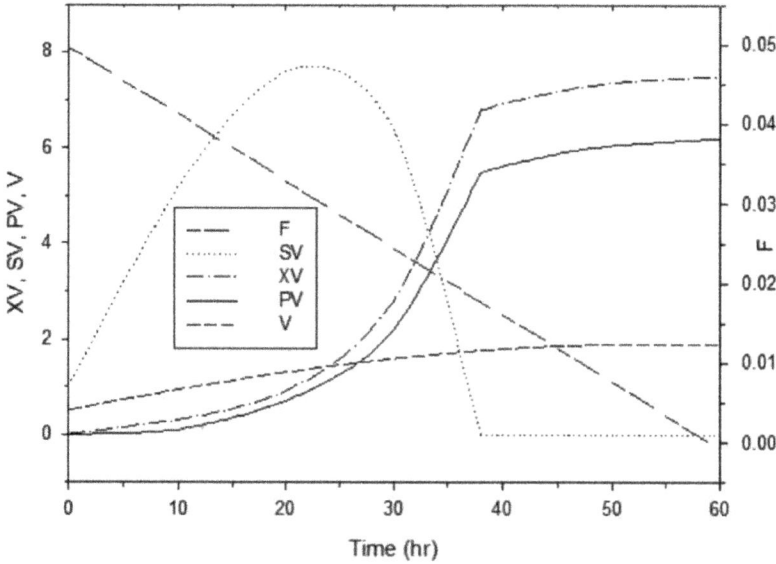

Figure 5.5. Time profiles of fed-batch culture with a linearly decreasing feed rate: $F = 0.05(1-0.0167t)$; $t_f = 60$; $SF = 10$ g/L; $V_{max} = 2$ L; $\mu = 0.11s/(s + 0.006)$; $\pi = \mu /1.2$; $\sigma = \mu/0.47$; $[XV, SV, PV, V](t_0) = [0.1$ g/L, 1 g/L, 0 g/L, 0.5 L].

Figure 5.5. Conceptually, it is difficult to envision a case in which a linearly decreasing feed rate is applied. Because the cells are growing exponentially or semiexponentially, the substrate must be supplied to keep up with the cell growth.

5.1.8 Fed-Batch Cultures with Exponential Feed Rates

This mode of operation is common and has received much attention.[14–17] Theoretical analyses were made by two groups.[14,15] The basic idea was based on the traditional yeast fermentation in which an amount of sugar was needed to obtain cell mass in a fixed time interval. In this operation, the feed rate increases exponentially with time,

$$F(t) = F_0 \exp(\beta t) \tag{5.52}$$

where F_0 and β are the initial feed rate and an arbitrary positive constant, respectively.

The mass balance equations are as follows:

$$\frac{dV}{dt} = F_0 \exp(\beta t) \quad V(0) = V_0 \tag{5.53}$$

$$\frac{d(XV)}{dt} = \mu XV \quad XV(0) = X_0 V_0 \tag{5.54}$$

$$\frac{d(SV)}{dt} = F_c S_F - \frac{\mu XV}{Y_{X/S}} \quad SV(0) = S_0 V_0 \tag{5.55}$$

and

$$\frac{d(PV)}{dt} = \pi XV = \frac{\mu XV}{Y_{P/S}} \quad PV(0) = P_0 V_0 \tag{5.56}$$

As a consequence of the exponential feed (Eq. (5.52)), the volume also increases exponentially from its initial value and is obtained by integrating the overall mass balance equation (Eq. (5.53)):

$$V(t) = V_0 + \frac{F_0}{\beta}[\exp(\beta t) - 1] = \left(V_0 - \frac{F_0}{\beta}\right) + \frac{F_0}{\beta}\exp(\beta t) \qquad (5.57)$$

The dilution rate is obtained from Eqs. (5.52) and (5.57):

$$\frac{F}{V}(t) = \frac{F_0 \exp(\beta t)}{V_0 - F_0/\beta + (F_0/\beta)\exp(\beta t)} = \frac{F_0/V_0}{(1 - F_0/V_0\beta)\exp(-\beta t) + F_0/V_0\beta} \qquad (5.58)$$

According to Eq. (5.58), the dilution rate varies with time, starting at F_0/V_0 and approaching β. Thus, the dilution rate can increase, decrease, or remain constant at F_0/V_0, depending on the value of β relative to F_0/V_0:

$$\frac{d(F/V)}{dt} \begin{bmatrix} > 0 & \text{if} & \beta > F_0/V_0 \\ = 0 & \text{if} & \beta = F_0/V_0 \\ < 0 & \text{if} & \beta < F_0/V_0 \end{bmatrix} \qquad (5.59)$$

The cell mass balance equation is same as before, and therefore, the total amount of cells grows exponentially with time,

$$XV = X_0 V_0 \exp(\bar{\mu} t) \qquad (5.60)$$

and the cell concentration is obtained by dividing Eq. (5.60) by Eq. (5.57):

$$X = \frac{X_0 \exp(\bar{\mu} t)}{1 - \dfrac{F_0}{V_0 \beta} + \dfrac{F_0}{V_0 \beta}\exp(\beta t)} \qquad (5.61)$$

The total amount of substrate in the culture is obtained by integrating Eq. (5.55),

$$SV(t) = S_0 V_0 + \frac{F_0}{\beta}S_F[\exp(\beta t) - 1] - \frac{X_0 V_0[\exp(\bar{\mu} t) - 1]}{Y_{X/S}} \qquad (5.62)$$

and the substrate concentration is obtained by dividing Eq. (5.62) by (5.57):

$$S(t) = 1 - \frac{(S_F - S_0)}{1 + \dfrac{F_0}{V_0 \beta}\exp(\beta t)} - \frac{X_0[\exp(\bar{\mu} t) - 1]}{Y_{X/S}\left[1 + \dfrac{F_0}{V_0 \beta}\exp(\beta t)\right]} \qquad (5.63)$$

Typical time profiles of concentrations and the total amounts of cell mass, substrate, and culture volume are shown in Figure 5.6.

5.1.8.1 Constant Dilution Rate

A proper selection of the initial feed rate F_0 in Eq. (5.58) can lead to a special case of constant dilution rate:

$$F_0 = V_0 \beta \qquad (5.64)$$

With this choice of initial feed rate, the feed rate is obtained from Eq. (5.52):

$$F(t) = V_0 \beta \exp(\beta t) \qquad (5.65)$$

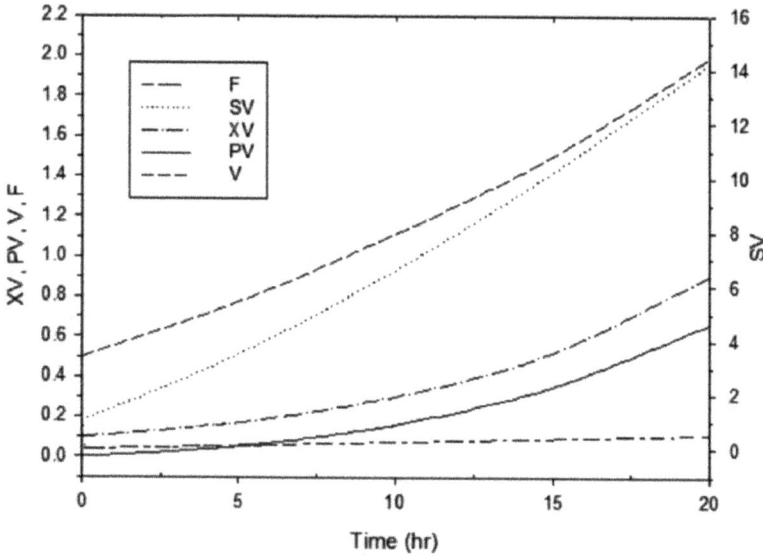

Figure 5.6. Time profiles of fed-batch culture with an exponential feed rate: $F = 0.05\exp(0.0381t)$; $t_f = 20$; $S_F = 10$ g/L; $V_{max} = 2$ L; $\mu = 0.11s/(s + 0.006)$; $\pi = \mu/1.2$; $\sigma = \mu/4.7$; $[XV, SV, PV, V](t_0) = [0.1$ g/L, 1 g/L, 0 g/L, 0.5 L$]$.

The culture volume is obtained by integrating Eq. (5.53):

$$V(t) = V_0 \exp(\beta t) \tag{5.66}$$

The dilution rate given by Eq. (5.58) reduces to

$$F/V = F_0/V_0 = \beta \tag{5.67}$$

Equations (5.66) and (5.67) show that with a proper selection of the initial rate of the exponential feed rate $F_0 = \beta V_0$, the volume can be made to increase exponentially, and the dilution rate can be held constant at β. The maximum time of fed-batch operation is given by

$$t_{max} = \frac{1}{\beta} \ln\left(\frac{V_{max}}{V_0}\right) \tag{5.68}$$

where V_{max} is the maximum culture volume. The operational time increases with smaller dilution rates (smaller F_0 and β) and initial volume and larger maximum culture volume. Typical time profiles of the total amount of substrate, cell mass, and products as well as the feed rate and culture volume are shown in Figure 5.7.

5.1.8.2 Constant Substrate and Cell Concentrations

Because of the exponential feed rate, the addition rate of the limiting substrate, FS_F, is also exponential, and this type of operation can be forced to keep up with an exponential growth of cells. If one can maintain the limiting substrate concentration constant throughout the operation, say, at $S = S_e$, by manipulation of the feed rate $F = \alpha\exp(\beta t)$, the specific growth rate, which depends on the limiting substrate concentration, would also be held constant, $\mu = \mu(S_e) = \mu_e$, and the cells would also

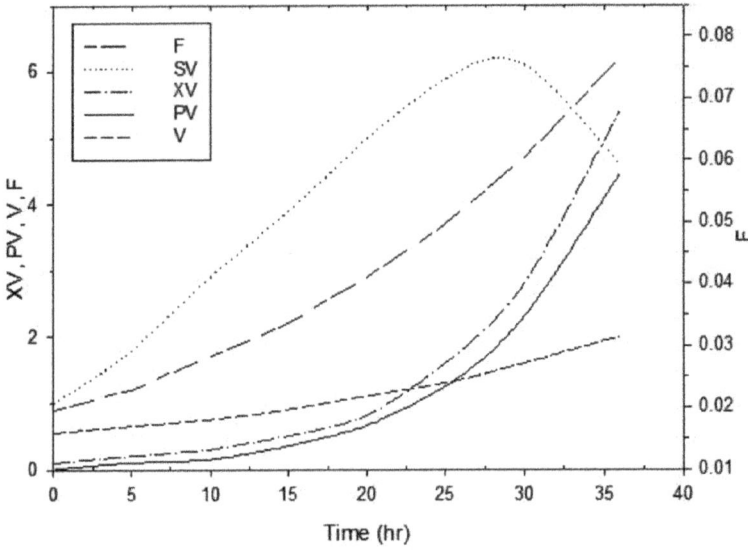

Figure 5.7. Time profiles of fed-batch culture with a constant dilution rate (exponential feed rate): $F = 0.0191\exp(0.0381t)$; $t_f = 36.3857$; $S_F = 10$ g/L; $V_{max} = 2$ L; $\mu = 0.11s/(s + 0.006)$; $\pi = \mu/1.2$;; $\sigma = \mu/4.7$; $[XV, SV, PV, V](t_0) = [0.1$ g/L, 1 g/L, 0 g/L, 0.5 L$]$.

grow exponentially according to the cell mass balance equation (Eq. (5.54)), which is integrated to yield

$$XV(t) = X_0V_0\exp(\mu_e t) \tag{5.69}$$

The substrate feed rate to maintain the substrate concentration constant at $S = S_e$ is obtained for the simplest case from the substrate mass balance equation (5.11) by setting $S = S_e$ and using Eq. 5.69,

$$F = [\sigma_e X_0V_0/(S_F - S_e)]\exp(\mu_e t) = \delta\exp(\mu_e t) \tag{5.70}$$

in which $\delta = [\sigma_e X_0V_0/(S_F - S_e)]$ and $\sigma_e = \sigma(S = S_e)$ is the specific consumption rate evaluated at $S = S_e$. It is clear that the feed rate necessary to keep the substrate concentration constant at an arbitrary value, $S = S_e$, is an exponential function and that the culture volume also increases exponentially:

$$V = V_0 + \int_0^t F d\tau = \left(V_0 - \frac{\delta}{\mu_e}\right) + \frac{\delta}{\mu_e}\exp(\mu_e t) \tag{5.71}$$

The dilution rate is

$$\frac{F}{V} = \frac{\delta\exp(\mu_e t)}{\left(V_0 - \dfrac{\delta}{\mu_e}\right) + \dfrac{\delta}{\mu_e}\exp(\mu_e t)} \tag{5.72}$$

To hold the dilution rate constant at μ_e, it is necessary in Eq. (5.72) to set

$$V_0 = \frac{\delta}{\mu_e} = \frac{[\sigma_e X_0V_0/(S_F - S_e)]}{\mu_e} = \frac{X_0V_0/(S_F - S_e)}{Y_{X/S}} \Rightarrow S_F = S_e + \frac{X_0}{Y_{X/S}} \tag{5.73}$$

so that the volume is a pure exponential function,

$$V = \frac{\delta}{\mu_e} \exp(\mu_e t) = V_0 \exp(\mu_e t) \tag{5.74}$$

and the feed rate becomes

$$F = [\sigma_e X_0 V_0 / (S_F - S_e)] \exp(\mu_e t) = V_0 \mu_e \exp(\mu_e t) \tag{5.75}$$

The feed substrate concentration must be selected to satisfy Eq. (5.73) to achieve a constant dilution rate:

$$\frac{F}{V} = \frac{V_0 \mu_e \exp(\mu_e t)}{V_0 \exp(\mu_e t)} = \mu_e \tag{5.76}$$

The resulting cell concentration is obtained from Eqs. (5.69) and (5.75):

$$X(t) = \frac{X_0 V_0 \exp(\mu_e t)}{V_0 \exp(\mu_e t)} = X_0 \tag{5.77}$$

Equation (5.77) states that the cell concentration remains constant at the initial concentration.

In summary, the exponential fed-batch can be characterized by the following equations:

$$F = \mu_e V_0 \exp(\mu_e t), \quad V = V_0 \exp(\mu_e t), \quad D = F/V, \quad S = S_e, \quad X = X_0 \tag{5.78}$$

With a proper choice of the feed substrate concentration (Eq. (5.73)) and the exponential feed rate of Eq. (5.74), the dilution rate remains constant at μ_e (Eq. (5.76)), and the culture volume increases exponentially (Eq. (5.75)), while the cell concentration remains constant at the initial value and the substrate concentration remains constant at the chosen value, S_e. Thus, this fed-batch operation mimics a steady state continuous culture. The difference is that the steady state continuous culture may be maintained indefinitely, while for the exponentially fed fed-batch culture, these conditions cease once the culture volume is full, t_{full}. The time profiles are shown in Figure 5.8. Note that the cell and substrate concentrations remain constant, while the total amounts increase with time.

5.1.9 Extended Fed-Batch Cultures

This mode of operation refers to the case in which the concentration of the limiting substrate concentration is maintained at a constant value throughout the course of fed-batch operation by manipulating its feed rate.[2,14,19,20] As seen earlier, when the limiting substrate concentration is kept constant, say, at $S = S_e$, then the specific growth rate is also kept constant, μ_e, so that the cells grow exponentially, $XV = X_0 V_0 \exp(\mu_e t)$. From the substrate balance (Eq. (5.3)), we seek the substrate feed rate that would keep the substrate concentration constant at $S = S_e$:

$$\frac{d(SV)}{dt} = S_e \frac{dV}{dt} = S_e F = FS_F - \sigma_e X \tag{5.79}$$

or

$$F = \sigma_e XV / (S_F - S_e) = [\sigma_e X_0 V_0 / (S_F - S_e)] \exp(\mu_e t) = \alpha \exp(\mu_e t) \tag{5.80}$$

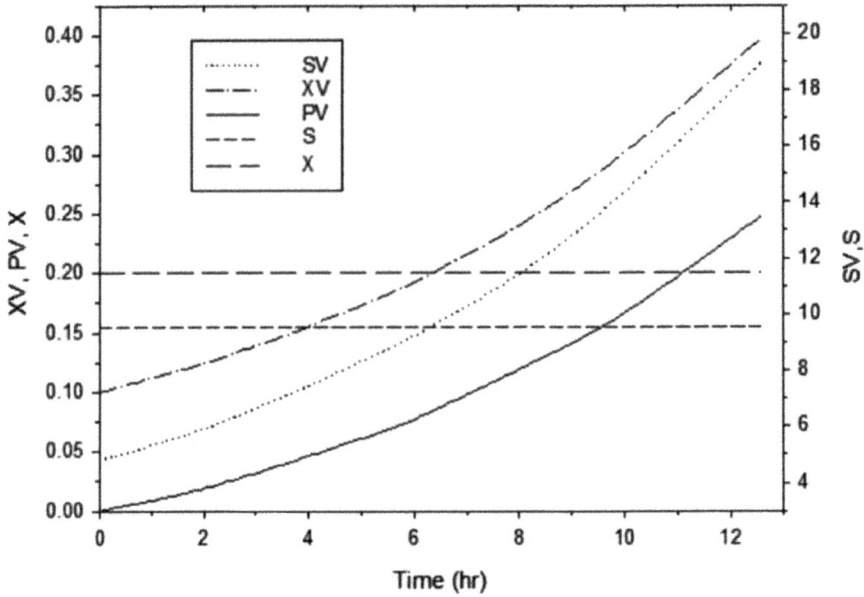

Figure 5.8. Time profile of fed-batch culture with constant substrate and cell concentrations (exponential feed rate): $F = 0.0549\exp(0.1099t)$; $t_f = 12.6141$; $S_e = 9.9872$; $\mu_e = 0.1099$; $S_F = 10$ g/L; $V_{max} = 2$ L; $\mu = 0.11s/(s + 0.006)$; $\pi = \mu/1.2$; $\sigma = \mu/4.7$; $[XV, SV, PV, V](t_0) = [0.1$ g/L, 1 g/L, 0 g/L, 0.5 L].

According to Eq. (5.80), the feed rate is an exponential function with a preexponential factor that depends on the initial amount of inoculum $X_0 V_0$, the specific substrate consumption rate σ_e, and the difference between the feed and culture substrate concentrations $S_F - S_e$. Thus, the extended culture is a form of exponentially fed fed-batch culture. The fed-batch volume also increases exponentially, which is confirmed by integrating Eq. (5.1) with Eq. (5.80):

$$V(t) = V_0 + \alpha[\exp(\mu_e t) - 1]/\mu_e \qquad (5.81)$$

However, if we select properly the values of S_F and S_e (therefore, σ_e and μ_e are also fixed) so as to force the preexponential factor equal to the product of specific growth rate and the initial volume,

$$\alpha = V_0 \mu_e \qquad (5.82)$$

the feed flow rate becomes

$$F_e = V_0 \mu_e \exp(\mu_e t) \qquad (5.83)$$

and the volume is also a pure exponential function:

$$V(t) = V_0 + V_0[\exp(\mu_e t) - 1] = V_0 \exp(\mu_e t) \qquad (5.84)$$

Thus, the bioreactor volume increases exponentially from its initial volume. As a consequence of extended fed-batch operation, which keeps the limiting substrate concentration at a constant value, the cell concentration is also maintained constant:

$$X = XV/V = X_0 V_0 \exp(\mu_e t)/V_0 \exp(\mu_e t) = X_0 \qquad (5.85)$$

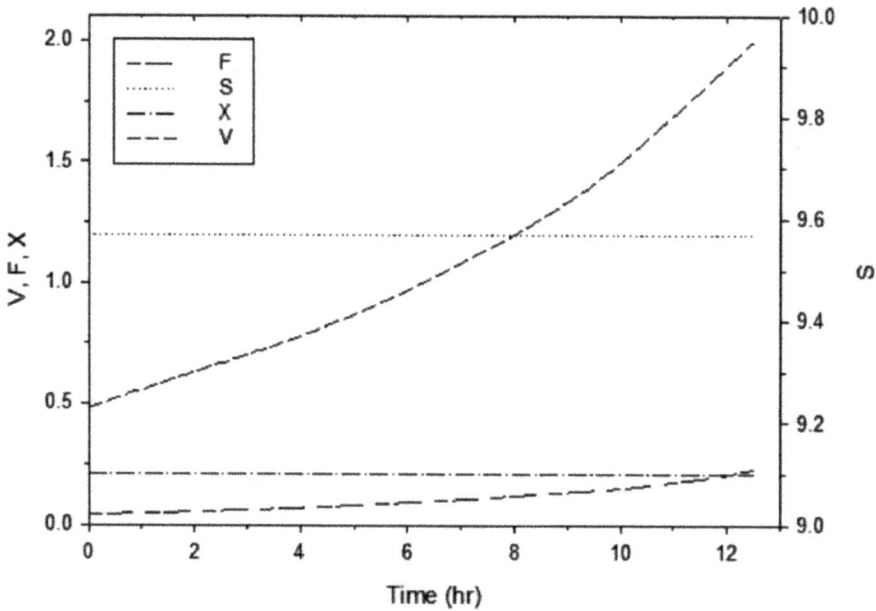

Figure 5.9. Time profile of feed flow rate and volume in the extended fed-batch culture with constant substrate and cell concentrations: $F = 0.0549\exp(0.1099t)$; $t_f = 12.6141$; $S_e = 9.9872$; $\mu_e = 0.1099$; $S_F = 10$ g/L; $V_{max} = 2$ L; $\mu = 0.11s/(s + 0.006)$; $\pi = \mu/1.2$; $\sigma = \mu/4.7$; $[XV, SV, PV, V](t_0) = [0.1$ g/L, 1 g/L, 0 g/L, 0.5 L$]$.

Conversely, the total amount of cell mass increases exponentially. The bioreactor size and the growth rate limit the maximum operating time:

$$t_{max} = \ln[(V_{max}/V_0)]/m_e \tag{5.86}$$

Because the concentrations of the limiting substrate and cells are kept constant, the extended fed-batch appears to be analogous to a continuous-flow bioreactor (chemostat). However, the total amounts of limiting substrate and cells increase exponentially in the extended fed-batch culture, whereas they remain constant in a continuous culture. In addition, there is no operational time limit in a chemostat.

It is obvious that only a particular form of the exponentially fed fed-batch, that is, with a proper preexponential factor and the exponent (Eq. (5.83)), is equivalent to the extended fed-batch culture. Time profiles of various dependent variables are shown in Figure 5.9. The concentrations of cell mass and substrate are held constant, while the feed rate and the reactor volume increase exponentially. Thus, it mimics the steady state chemostat with constant steady state cell and substrate concentrations.

5.1.10 Fed-Batch Cultures with Intermittent Feed Rates

For certain situations, such as when the feed is molasses, which is so thick and gooey that a live steam injection is necessary, the feed is applied intermittently. This type of fed-batch operation is necessary if the carbon source is very viscous and requires intermittent feed.[21,22]

5.1.11 Fed-Batch Cultures with Empirical Feed Rates

There are a number of fed-batch operations in which the feed rates are determined empirically. For commercial production of penicillin and other antibiotics and amino acids, a number of carbon and nitrogen sources and precursors are added, and in particular, the carbon source, glucose or molasses, is added intermittently through a steam-injection valve throughout the course of fermentation. The addition is made at a constant time interval or at predetermined times to keep the limiting substrate low.

5.1.12 Fed-Batch Cultures with Optimal Feed Rates

Fed-batch culture can be rigorously optimized if a mathematical model is available and if one knows the performance index to be optimized. Maximum principles[23] and singular control theory[24] can be applied to optimize the performance index for a model in the form of a set of ordinary differential equations with appropriate boundary conditions. The optimal performance index may include a maximum amount of product at the end of fed-batch operation, a maximum productivity, or a minimum time to achieve a desired conversion. If the raw material cost is high, it may pay to maximize the yield, whereas cheaper material costs may warrant a maximum productivity.

   In general, it is possible to obtain the optimal time profile of feed rate that would optimize the chosen performance index. However, the computational effort needed increases exponentially with the number of ordinary differential equations necessary to form the model of fed-batch culture. For a simple case considered in Chapter 2, growing cells on a limiting substrate, it is possible to obtain readily the optimal solution. However, if the number of differential equations exceeds four, an extensive numerical effort is required. The optimization of fed-batch cultures is treated extensively in Chapters 9, 10, 12, and 13.

   Another way of classifying fed-batch culture is based on whether a single cycle is used using a fresh inoculum for each cycle or a repeated fed-batch, in which a part of the culture is retained as the inoculum for the next cycle.

5.2 Classification Based on Number of Operational Cycles

A fed-batch operation is most frequently carried out using a new inoculum each time, or it can be repeated by retaining a part of the previous run serving as the inoculum for the subsequent run. A single cycle involving charging, inoculation, fermentation, and discharge of the entire bioreactor content is repeated. In other words, a single cycle of operation is repeated cycle after cycle. In other cases, the fed-batch content is partially removed, and the rest is left in the bioreactor to serve as the inoculum for the next cycle rather than using a new inoculum. This procedure is repeated a number of times, that is, multiple cycle operation, and is termed a *repeated fed-batch operation*.

5.2.1 Single-Cycle Operations

A single-cycle operation is the normal fed-batch operation, that is to say, the ideal run is repeatedly run until the production schedule is met. Therefore, the characteristics described earlier apply without any modification, and no additional information is necessary.

5.2.2 Multiple-Cycle Operations, Repeated Fed-Batch Operations

In a multiple-cycle operation, the assumption is that the cells at the end of one cycle of operation are just as healthy as the fresh inoculum and are totally viable so that they are capable of producing the product, more cells, or a metabolite. This type of operation is very feasible if cells themselves are the product, as in the case of a single cell protein and some intracellular products. Because yeast cells have a limit in the number of buds they can have in their life owing to scar formation left by breaking of daughter cell buds, the repeated fed-batch operation may be limited by the maximum number of buds the yeast cells can have.

In this case, the end of one run (cycle) serves as the beginning of the next run (cycle). For convenience, we shall assume that the time to pump out the fermentation broth is insignificant as compared to the fermentation time. The initial conditions are

$$X^1(t_0) = X_0^1, \quad S^1(t_0) = S_0^1, \quad V^1(t_0) = V_0^1 \tag{5.87}$$

and the cyclic boundary conditions are

$$X^{i+1}(t_0) = X^i(t_f), \quad S^{i+1}(t_0) = S^i(t_f), \quad V^{i+1}(t_0) = V^i(t_f) - V_p \tag{5.88}$$

where t_0 and t_f represent the start and end time, respectively, the superscript denotes the cycle number, and V_p stands for the volume that has been pumped out before beginning the next cycle. The cyclic conditions state that the initial conditions for the $(i + 1)$th cycle are the same as the final conditions of the ith cycle. It is apparent that there is a transient period before the operation reaches a pseudo steady state in its initial substrate and cell concentrations and bioreactor volume. It may take a few cycles to reach this point. In this case, not only does one have to determine the optimal feed rate profile as a function of time but the three parameters, the cyclic boundary conditions, must also be chosen properly. Thus, complexity of repeated fed-batch optimization is apparent. Repeated-cycle fed-batch operation has been analyzed fully using a Monod form of specific growth rate.[25]

REFERENCES

1. Yoshida, F., Yamane, T., and Nakamoto, K. 1973. Fed-batch hydrocarbon fermentation with colloidal emulsion feed. *Biotechnology and Bioengineering* 15: 257–270.
2. Yamane, T., and Hirano, S. 1977. Kinetic studies on fed-batch cultures. III. Semi-batch culture of microorganisms with constant feed of substrate: A mathematical simulation. *Journal of Fermentation Technology* 55: 156–165.
3. Pirt, S. J. 1974. The theory of fed-batch culture with reference to the penicillin fermentation. *Journal of Applied Chemistry and Biotechnology* 24: 415–424.
4. Pirt, S. J. 1979. Fed-batch culture of microbes. *Annals of the New York Academy of Science* 326: 119–125.
5. Dunn, I. J., and Mor, J.-R. 1975. Variable-volume continuous cultivation. *Biotechnology and Bioengineering* 17: 1805–1822.
6. Dunn, I. J., Shioya, S., and Keller, R. 1979. Analysis of fed-batch microbial culture. *Annals of the New York Academy of Science* 326: 127–139.

7. Yamane, T., and Hirano, S. 1977. Kinetic studies on fed-batch cultures. Part V. Semi-batch culture of microorganisms with constant feed of substrate: An experimental study. *Journal of Fermentation Technology* 55: 380–387.

8. Boyle, T. J. 1979. Control of the quasi-steady-state in fed-batch fermentation. *Biotechnology and Bioengineering Symposium* 9: 349–358.

9. Kalogerakis, N., and Boyle, T. J. 1981. Experimental evaluation of a quasi-steady-state controller for yeast fermentation. *Biotechnology and Bioengineering* 23: 921–938.

10. Kalogerakis, N., and Boyle, T. J. 1981. Implementation and demonstration of a QSS controller for yeast fermentations. *Canadian Journal of Chemical Engineering* 59: 377–380.

11. Panov, D. P., and Kristapsons, M. Z. 1991. An algorithm for operating a fed-batch fermentation using quasi-steady state. *Acta Biotechnologie* 11: 457–466.

12. Lee, H. H., and Yau, B. O. 1981. An experimental reactor for kinetic studies: Continuously fed batch reactor. *Chemical Engineering Science* 36: 483–488.

13. Esener, A. A., Roels, J. A., and Kossen, N. W. F. 1981. Fed-batch culture: Modeling and application in the study of microbial energetics. *Biotechnology and Bioengineering* 23: 1851–1871.

14. Lim, H. C., Chen, B. J., and Creagan, C. C. 1977. An analysis of extended and exponentially-fed-batch cultures. *Biotechnology and Bioengineering* 19: 425–433.

15. Yamane, T., Kishimoto, M., and Yoshida, F. 1976. Semi-batch culture of methanol-assimilating bacteria with exponentially increased methanol feed. *Journal of Fermentation Technology* 54: 229–240.

16. Yamane, T. 1978. Kinetic studies on fed-batch fermentations: A monograph. *Hakko Kogaku Kaishi* 56: 310–322.

17. Keller, R., and Dunn, I. J. 1978. Fed-batch microbial culture: Models, errors and applications. *Journal of Applied Chemistry and Biotechnology* 28: 508–514.

18. Edwards, V. H., Gottschalk, M. J., Noojin, A. Y., III, Tuthill, L. B., and Tannahill, A. L. 1970. Extended culture: The growth of *Candida utilis* at controlled acetate concentrations. *Biotechnology and Bioengineering* 12: 975–999.

19. Ramirez, A., Durand, A., and Blachere, H. T. 1981. Optimal bakers' yeast production in extended fed-batch culture by using a computer coupled pilot fermenter. *Biotechnology Letters* 3: 555–560.

20. Peringer, P., and Blachere, H. T. 1979. Modeling and optimal control of bakers' yeast production in repeated fed-batch culture. *Biotechnology and Bioengineering Symposium* 9: 205–213.

21. Limtong, S., Kishimoto, M., Seki, T., Yoshida, T., and Taguchi, H. 1987. Simulation and optimization of fed-batch culture for ethanol production from molasses. *Bioprocess Engineering* 2: 141–147.

22. Bae, S., and Shoda, M. 2004. Bacterial cellulose production by fed-batch fermentation in molasses medium. *Biotechnology Progress* 20: 1366–1371.

23. Pontryagin, L. S., Boltynskii, V. G., Gamkrelidze, R. V., and Mishchenko, E. F. 1962. *The Mathematical Theory of Optimal Processes.* Interscience.

24. Lewis, R. M. 1980. Definition of order and junction conditions in singular optimal control problems. *SIAM Journal of Control and Optimization* 18: 21–32.

25. Weigand, W. A. 1981. Maximum cell productivity by repeated fed-batch culture for constant yield case. *Biotechnology and Bioengineering* 23: 249–266.

26. O'Malley, R. E., Jr. 1974. *Introduction to Singular Perturbations.* Academic.

6 Models Based on Mass Balance Equations

The cellular behavior in a bioreactor has a complexity unparalleled in chemical reactors and consequently is very difficult to predict from the external bioreactor conditions. Cell growth is highly complex so that a drastic simplification is required before a manageable model can be developed.[1] In many cases, the objective of mathematical model development is explicitly aimed at providing the basis for predicting, controlling, and optimizing the performance of a bioreactor. Mathematical models are indispensable in determining the optimal operating conditions of bioreactors, in particular, fed-batch bioreactors. Nonequation models such as neural network models can be developed and used for optimization purposes when it is not possible to write mass balance equations with appropriate specific rate expressions. However, the results obtained are not as satisfactory as those obtained from equation-based models. Nevertheless, when it is not possible to write one or more mass balance equations and appropriate rate expressions are not possible for key variables, this statistical approach is indispensable.

In general, models can be divided into two classes: phenomenological (unstructured) models and mechanistic (structured) models.[2,3] Phenomenological models ignore various cellular processes that are involved, including specific enzymes and cellular structural components. These unstructured models are used to describe the overall observed microbial response in terms of readily measurable variables only, and no provision is made for changes in the internal components of the cells. Conversely, structured models are based on the cellular mechanisms and take into account various cellular processes that depend on the activities of enzymes and the structural components such as ribosomes, mitochondria, macromolecular components such as proteins, DNA and RNA, and carbohydrates. An advantage of such highly structured models is the potential for modeling the interrelationship of the metabolic processes and predicting the effects of internal genetic modification and external disturbances on the overall cellular process.[4] The disadvantages of the highly structured models are the difficulty in determining the kinetics and kinetic constants in individual reactions because many of the cellular components are not readily measurable and computational difficulty is posed by a large number of balance equations that result.[2] Although the complex nature of biological systems may require complicated models of large dimension, oversophistication of models should be avoided because it tends to defy the very purpose of modeling by obscuring the essence of the model and makes the prediction of cellular behavior exceedingly difficult.[5]

When a complete set of mass and energy balance equations with appropriate kinetic data is available, that is, transport phenomena, specific rates of cell growth, product formation, formation of intermediates, and substrate consumption, it is possible to use these balance equations, which may contain a large number of parameters that need to be determined from experimental data. In this situation, an estimation scheme is used to determine the parameters in the model. However, there are many situations in which it is not possible due to lack of knowledge and understanding to write a complete set of mass and energy balance equations that reflect the effect on the outcome of fermentation of potential manipulated variables such as the effects of medium components. Thus, there is a need to model the process by some statistical means.

In general, equation-based models may be classified as either *distributed* or *segregated*. In the distributed model, the number of cells is omitted; instead, an empirical measure of concentration, biomass, is used, and the reproduction process is not treated. Conversely, the segregated model considers the number of cells as a fundamental variable. Models may be classified as either *unstructured* or *structured*. The structured models take account of changes in the internal environment, which in turn change the rates of cell growth, substrate consumption, and product formation. Thus, more than the specification of a single quantity is required to specify the state of population. The unstructured model has no built-in concept of the internal structure. The most common models are distributed and unstructured.

Looking into the phenomena of cellular growth, one can list three major steps: (1) transport of substrates, nutrients, and other matter from the abiotic phase into the cellular biotic phase; (2) intracellular reactions to convert the substrates into cellular components and metabolic products; and (3) excretion of metabolic products to the extracellular abiotic phase. Some products are intracellular and are retained within the cells, whereas others are extracellular and are excreted into the abiotic phase. Therefore, a rigorous model must incorporate the transport into the cells of substrates and out of the cells of products. Thus, modeling of bioprocess is much more complicated than modeling a catalytic reaction because the intracellular reactions typically number in the thousands. It is neither possible nor necessary to track all these reactions, and therefore, one adapts the concept of limiting reactant(s). The bacterial cells and perhaps some yeast cells may be small enough so that diffusion in the cell may be assumed negligible. Therefore, the effectiveness factor concept may be omitted, although this may not be the case for large yeast cells and fungi. It is also commonly assumed that the time scale of the transport in and out of the cell is much shorter than that of the reaction of the limiting substrate in the cell. Therefore, the simplest models have been built on the assumption that a biological reactor can be treated as a homogenous reaction without any mass transfer effect.

6.1 Mass Balance Equations

As for modeling any process, one begins with unsteady state mass and energy balance equations. Most bioreactor operations are carried out isothermally around room temperature or slightly above. An energy balance equation is used to determine the heat exchange requirement, and the mass balance equations are used to model the reactor operation. A simple form of fed-batch cultures was introduced in Section 2.2.4. We present more detailed forms of fed-batch cultures here.

6.1.1 Total Mass Balance Equation

Because there is no outlet stream for fed-batch operations, the total mass balance equation states that the accumulation is due to the rate of input because no new mass is generated:

$$
\begin{bmatrix}
\text{Total mass flow rate} \\
\text{into the reactor} \\
\text{(grams/time)}
\end{bmatrix}
=
\begin{bmatrix}
\text{Accumulation rate} \\
\text{within the reactor} \\
\text{(grams/time)}
\end{bmatrix}
\tag{6.1}
$$

In addition to the feed stream that contains the limiting nutrient, a number of feed streams may be introduced into the reactor that contain other essential nutrients, precursors, and inducers. There is no outlet stream during the operation of fed-batch culture, and the harvesting is done only at the end of a run. Let us denote by $F_i(i = 1, 2, 3, \ldots, n)$ the volumetric flow rates of the feed streams, F_a, the acid addition stream; F_b, the base addition stream; F_{af}, the antifoam addition stream; F_{iw}, the rate of water carried in by the air or aeration gas; and F_{ow}, the rate of water carried out by the exhaust gas. The densities associated with these addition streams are denoted by ρ with appropriate subscripts. The accumulation rate is the time rate of change of the total mass, which is volume times density. The total mass balance equation, according to Eq. (6.1), is

$$
\sum_i F_i \rho_i + F_a \rho_a + F_b \rho_b + F_{af} \rho_{af} + F_{ia} \rho_{iw} - F_{oa} \rho_{ow} = \frac{d(V\rho)}{dt} \quad V\rho(0) = V_0 \rho_0 \tag{6.2}
$$

where V is the effective bioreactor volume and t is the clock time. In Eq. (6.2), usually the total mass increases owing to the addition of acid, base, and antifoam and of water introduced by the aeration and decreases owing to the water carried out by the aeration gas. However, these rates are relatively small as compared to the addition rate of the feed so that we can ignore them, $\sum_i F_i \rho_i \gg F_a, F_b, F_{af}, F_{iw}, F_{ow} \approx 0$.

In most cases, the densities of culture and feed streams may be assumed equal and constant, $\rho_i = \rho$, owing to the fact that aqueous solutions are involved.

6.1.2 Component Mass Balances

In addition to the total mass balance, each of the key components must be accounted for. There are a large number of components. We are concerned with key components among them. They are cells (viable, nonviable, product producing fraction, etc.), limiting substrates, key intermediate compounds (including precursors), and products. These component balances are considered in this section. Because

substrates are converted to various intermediates and products, the general component balance is

$$
\begin{bmatrix} \text{Total mass flow rate} \\ \text{into the reactor} \\ \text{(grams/time)} \end{bmatrix} = \begin{bmatrix} \text{Accumulation rate} \\ \text{within the reactor} \\ \text{(grams/time)} \end{bmatrix} \tag{6.3}
$$

6.1.2.1 Cell Balance

The cell mass balance states that the accumulation is due to the generation rate
since the feed streams are assumed to be free of cells (sterile) and there is no outlet
stream:

$$
0 - 0 + r_X V = \mu^{net} XV = (\mu - k_x)XV = \frac{d(VX)}{dt} \quad VX(0) = V_0 X_0 \tag{6.4}
$$

where X represents the cell concentration and μ^{net} represents the net specific growth
rate of cells, which is defined as the difference between the specific growth rate and
cell death rate constant.

6.1.2.2 Limiting Substrate Balance

The limiting substrate accumulation is due to the difference between the substrate
feed rate in and the substrate consumption rate and the rate of loss due to evaporation
if the substrate is volatile. The substrate consumption rate accounts for the formation
of cell mass and product and for maintenance of cell viability:

$$
FS_F - \sigma^{net} XV - GS_g = FS_F - (\sigma XV + mXV) - GS_g
$$

$$
= FS_F - \left(\frac{\mu XV}{Y_{X/S}} + \frac{\pi XV}{Y_{P/S}} + mXV \right) - GS_g = \frac{d(VS)}{dt} \quad VS(0) = V_0 S_0 \tag{6.5}
$$

where S stands for the limiting substrate concentration, S_F is the limiting substrate
concentration in the feed stream, $\sigma^{net} XV (= -r_S V)$ is the net substrate consumption
rate, G is the aeration rate, S_g is the concentration of the limiting substrate (if
volatile) in the exhaust gas, m is the maintenance coefficient, and $Y_{X/S}$ and $Y_{P/S}$ are
the yield coefficients for cell mass and product, respectively. The last term in Eq.
(6.4), GS_g, represents the loss of volatile substrates such as methanol, ethanol, and
hydrocarbons due to aerations. When the limiting substrate is not volatile, or there
is a condensation device to condense the volatile substrate and return back to the
reactor, then $S_g = 0$.

6.1.2.3 Key Component Balances

If intermediates are produced as in the case of yeast cells, which produce ethanol,
and ethanol is also utilized for yeast cell growth, it must be accounted for. Sometimes
there are certain intermediates that affect the cell growth and product formation,
such as ethanol in *Saccharomyces cerevisiae* and acetic acid in *Escherichia coli*. These
intermediates must be included in the model:

$$
0 - 0 + \iota XV = \frac{d(VI)}{dt} \tag{6.6}
$$

where I is the concentration of intermediate species I and $\iota X V (= r_I V)$ is the specific intermediate formation rate.

6.1.2.4 Product Balance

Because no product is introduced in the feed streams and no product is harvested until the end of operation, the product accumulation rate is due to the net product formation rate:

$$0 - 0 + \pi^{net} X V = \pi X V - k_p P V = \frac{d(VP)}{dt} \quad V(0)P(0) = V_0 P_0 \qquad (6.7)$$

where P represents the product or metabolite concentration, $\pi^{net} X V$ is the net product formation rate $(= r_P V)$ (the difference between the product formation rate and the product decay rate), and k_p is the product decay constant.

6.2 Unstructured Models

For unstructured models, the most widely used state variables, the minimal essential variables that describe the state of the process, are the concentrations of substrate, biomass, and product and the bioreactor volume. Of course, there may be any number of other variables such as the key intermediates, enzymes, and precursors, which are ignored. In other words, the unstructured models ignore the internal state of cells and its effects on cellular processes of growth and product formation. The rate expressions are empirical in nature, and no structural components are included.

Modeling is based on mass and energy balances. As stated earlier, most fermentation is carried isothermally, and energy balances are used to estimate the energy removal rates for exothermic processes, while the mass balances keep track of each species.

Example 6.E.1: Simplest Fed-Batch Operation, Growing Cells

For fed-batch operations, there is no outflow, except at the end, so that $F_o = 0$ and one feed stream contains the substrate and nutrients only and is free of cells and product. Assuming that the limiting substrate is nonvolatile and that the cell mass is the only product, Eqs. (6.4)–(6.7) reduce to

$$\frac{dV}{dt} = F \quad V(0) = V_0 \qquad (6.E.1.1)$$

$$\frac{d(VX)}{dt} = \mu(S)XV \quad V(0)X(0) = V_0 X_0 \qquad (6.E.1.2)$$

$$\frac{d(VS)}{dt} = FS_F - \sigma(S)XV \quad V(0)S(0) = V_0 S_0 \qquad (6.E.1.3)$$

where F is the feed rate and the net specific rates $\mu(S)$ and $\sigma(S)$ are assumed to be functions only of the limiting substrate concentration. Thus, Eqs. (6.E.1.1)–(6.E.1.3) represent the simplest fed-batch operation.

It is instructive to rearrange Eqs. (6.E.1.2) and (6.E.1.3) by expanding the left-hand sides and substituting Eq. (6.E.1.1) into them:

$$\frac{d(VX)}{dt} = X\frac{dV}{dt} + V\frac{dX}{dt} = XF + V\frac{dX}{dt} = \mu[S]XV \qquad (6.E.1.4)$$

or

$$\frac{dX}{dt} = (\mu[S] - F/V)X = (\mu[S] - D)X \quad X(0) = X_0 \qquad (6.E.1.5)$$

and

$$\frac{d(VS)}{dt} = S\frac{dV}{dt} + V\frac{dS}{dt} = SF + V\frac{dS}{dt} = FS_F - \sigma[S]XV \qquad (6.E.1.6)$$

or

$$\frac{dS}{dt} = D(S_F - S) - \sigma[S]X \quad S(0) = S_0 \qquad (6.E.1.7)$$

where $D = F/V$ is the dilution rate. *These balance equations are identical in form to the mass balance equations for a constant-volume unsteady state continuous bioreactor* (Eqs. (2.21) and (2.22)). However, it should be emphasized that $D = F(t)/V(t)$ is time variant in fed-batch operation and is limited in time as no overflow is allowed, whereas D may be constant or time variant and applies for all times in the case of continuous operation. It is also interesting to note in Eq. (6.E.1.5) that by maintaining the dilution rate equal to the specific growth rate, $D = F(t)/V(t) = \mu(S)$, it is possible to maintain the cell concentration constant, that is, $dX/dt = 0$. Conversely, the total amount of cells in the bioreactor, VX, increases with time owing to the increase in the volume caused by the feed stream.

Example 6.E.2: Simplest Fed-Batch Operation, Metabolite Production

Assuming that the limiting substrate is nonvolatile and that the cell mass is the only product, Eqs. (6.4)–(6.7) reduce to

$$\frac{dV}{dt} = F \quad V(0) = V_0 \qquad (6.E.2.1)$$

$$\frac{d(VX)}{dt} = \mu[S]XV \quad V(0)X(0) = V_0X_0 \qquad (6.E.2.2)$$

$$\frac{d(VS)}{dt} = FS_F - \sigma[S]XV \quad V(0)S(0) = V_0S_0 \qquad (6.E.2.3)$$

$$\frac{d(VP)}{dt} = \pi XV \quad V(0)P(0) = V_0P_0 \qquad (6.E.2.4)$$

Thus, Eqs. (6.E.2.1)–(6.E.2.4) represent the simplest fed-batch operation for product formation without any intermediate playing an important role.

6.2.1 Specific Growth Rate of Cells, μ

The specific growth rate is the rate of formation of cell mass per unit time per unit concentration of cells, $\mu = r_X/X$, and has the unit of grams of cell mass per unit

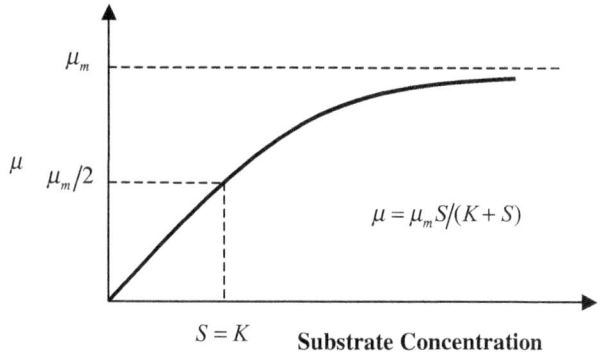

Figure 6.1. Monod form of specific growth rate.

time per gram/liter of cell mass. Therefore, the rate of formation of cell mass per unit culture volume is equal to the product of the specific growth rate and the cell concentration, $r_X = \mu X$, and the total rate of formation of cell mass is $r_X V = \mu X V$. The total growth rate is a term that represents the generation term in the cell mass balance equation (6.4). Thus, for fed-batch cultures,

$$\mu = \frac{1}{XV}\frac{d(VX)}{dt} \tag{6.8}$$

For batch cultures, it is obtained from the batch culture cell balance equation, $\mu^{net} = (dX/dt)/X$. For continuous cultures, it is obtained from its balance equation (Eq. (2.22)) for unsteady state operation or from Eq. (2.23) for steady state operation:

$$(\mu - D)X = \frac{dX}{dt} \Rightarrow \quad \mu = D + \frac{1}{X}\frac{dX}{dt} \tag{2.22}$$

$$\mu = D = \frac{F}{V} \tag{2.23}$$

Various functional forms have been used, including substrate-dependent and substrate-independent growth and inhibited growth by substrates, products, or inhibitors.

6.2.1.1 Substrate Concentration–Dependent Forms, $\mu(S)$
The best known and oldest is the one proposed by Monod for a single cell and is known as the *Monod equation*:[6,7]

$$\mu = \frac{\mu_m S}{K + S} \tag{6.9}$$

where S is the limiting substrate concentration and μ_m and K are constants known as the *maximum specific growth rate* and the *saturation constant*, respectively. Figure 6.1 is a plot of the Monod form of specific growth rate. The rate increases monotonically and approaches asymptotically the maximum rate μ_m. This form is also known as a saturation function and is equivalent to the Langmuir–Hinshelwood kinetics for chemical reactions. The specific growth rate approaches the maximum value asymptotically as the substrate concentration reaches the maximum; that is, when the substrate concentration is substantially larger than the saturation constant, $S \gg K$, $\mu = \mu_m$ (zero-order rate), whereas it is first order, $\mu = (\mu_m/K)S$, when the

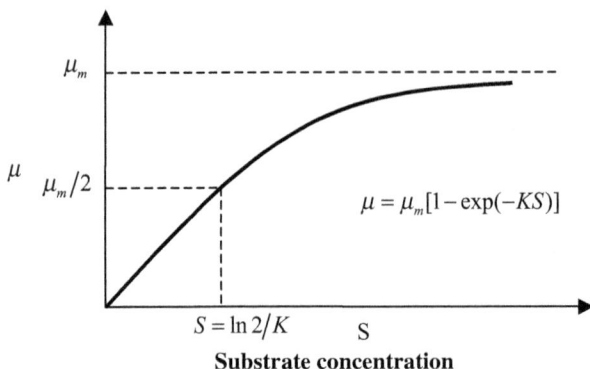

Figure 6.2. Teissier form of specific growth rate.

substrate concentration is substantially smaller than the saturation constant, $S \ll K$. This form is also equivalent to the simplest enzyme kinetics of Briggs–Haldane or Michaelis–Menten. The saturation constant K has physical significance in that when the substrate concentration is equal to the saturation constant, $S = K$, the specific growth rate is one-half of the maximum value, $\mu = \mu_m/2$. At a fixed substrate concentration, the specific growth rate is higher for those organisms with smaller saturation constants. A Monod form can be used when the limiting substrate is a carbon source such as glucose; inorganic salts such as nitrate, phosphate, and sulfate; trace elements; essential amino acids; or vitamins.

Teissier[8] proposed a two-parameter model:

$$\mu = \mu_m[1 - \exp(-KS)] \tag{6.10}$$

where μ_m is the maximum specific growth rate and K is the saturation constant. Figure 6.2 depicts the dependence of the specific growth rate on the limiting substrate concentration. In this form, the specific growth rate is one-half the maximum value when the substrate concentration reaches $\ln 2/K$, that is, $S = \ln 2/K$.

Moser[9] proposed another form of saturation kinetics:

$$\mu = \frac{\mu_m S^n}{K^n + S^n} \tag{6.11}$$

This form of saturation is depicted in Figure 6.3. When $n = 1$, this form reduces to Monod's model, and when $n > 1$, this form is sigmoidal rather than hyperbolic (Monod). This form approaches the asymptotic value more rapidly than the Monod form. All these rates are in the form of saturation functions that depend only on the substrate concentration. For K less than unity, $K < 1$, the Monod equation approaches the maximum value faster than do those of Teissier and Moser, while for $K > 1$, Teissier and Moser equations approach the maximum faster than Monod's. These are shown in Figure 6.4. The Moser form with $n = 2$ has been proposed,[54] and a variation of the Moser form was also proposed,[55] in which the exponents in the numerator and denominator are distinct, $\mu(S) = aS^m/(b + S^n)$.

A specific growth rate in the form of double saturation was proposed by Jost et al.,[10]

$$\mu = \mu_m \frac{S}{K_1 + S} \frac{S}{K_2 + S} \tag{6.12}$$

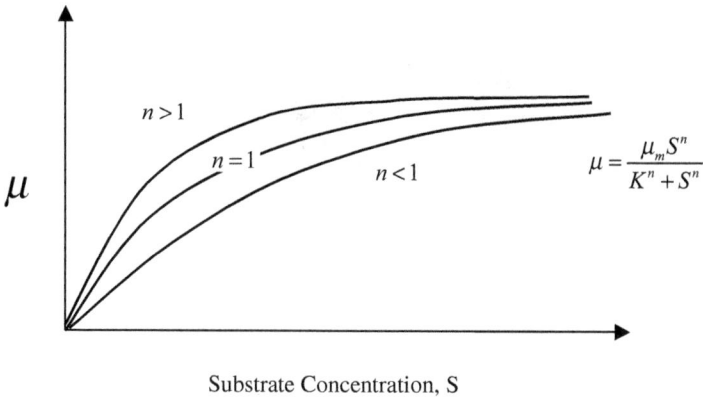

Figure 6.3. Moser form of specific growth rate.

which contains three adjustable parameters: K_1, K_2, and μ_m. As compared to the Monod form of single saturation, this double saturation form approaches more slowly (higher substrate concentration) the asymptotic value of μ_m.

Shehata and Marr[11] proposed a parallel uptake model, which contains four adjustable parameters:

$$\mu = \frac{\alpha S}{K_1 + S} + \frac{\beta S}{K_2 + S} \tag{6.13}$$

When the substrate concentration is substantially larger than both of the saturation constants, $S \gg K_1, K_2$, this model approaches asymptotically the maximum

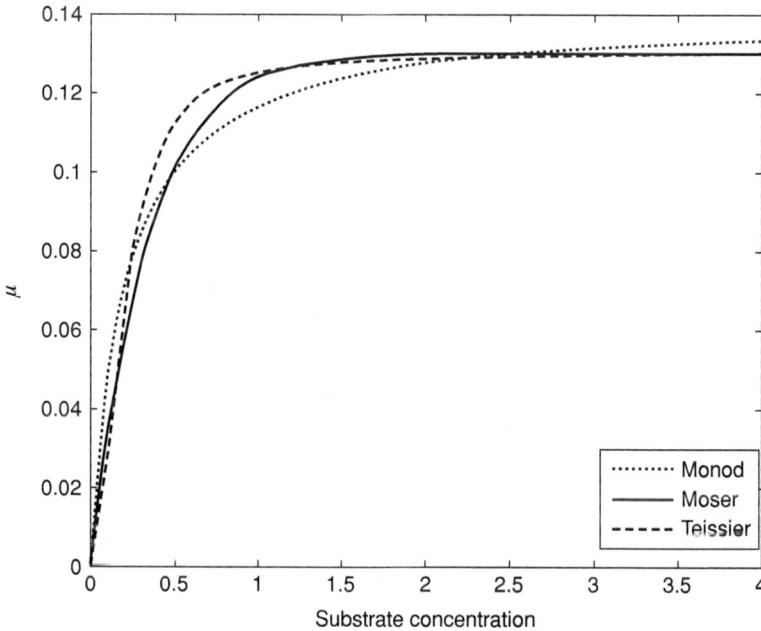

Figure 6.4. Comparison of specific growth rates: Monod $= 0.14S/(0.2 + S)$; Teissier $= 0.13(1- \exp(-3.0S))$; Moser $= 0.13S^2/(0.04 + S^2)$.

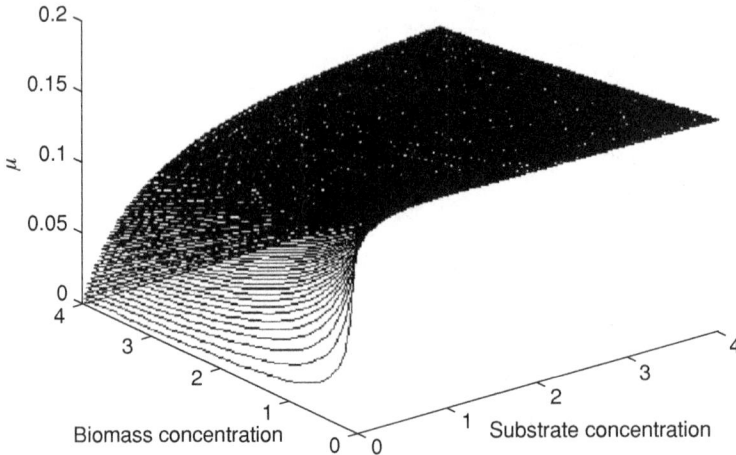

Figure 6.5. Contois specific cell growth rate, $\mu = 0.15S/(0.2X + S)$.

value of $\alpha + \beta$, that is, $\mu_m = \alpha + \beta$. One substrate and oxygen limitation were modeled[12] by

$$\mu = \frac{\mu_m S_1 S_2}{(K_1 + S_1)(K_2 + S_2)} \tag{6.14}$$

in which S_1 represents the limiting substrate concentration and S_2 represents the dissolved oxygen concentration.

6.2.1.2 Substrate and/or Cell Concentration–Dependent Forms, $\mu(S, X), \mu(S, X)$

Observations at high cell concentration that reduced the growth rate led to inclusion of cell concentration. The Contois[13] equation accounts for the reduction in the specific rate at high cell concentration by incorporating the cell concentration into the denominator of the Monod equation:

$$\mu = \frac{\mu_m S}{KX + S} \tag{6.15}$$

where X is the cell concentration. By so doing, the specific rate is made to decrease with high cell concentration. Specific growth rates of filamentous organisms such as molds are known to decrease at high cell concentrations. This may be an attempt to account for the crowding effect due to high cell concentration and/or high concentration that necessitates the incorporation of water activity into the specific growth rate. The dependence of Contois kinetics on the cell mass concentration is shown in Figure 6.5.

Other forms[56,57] proposed include the following:

$$\mu(S, X) = K_1 S - K_2 X \tag{6.16}$$

$$\mu(S, X) = \bar{\mu} + Q_1(X - \bar{X}) + Q_2(S - \bar{S})$$

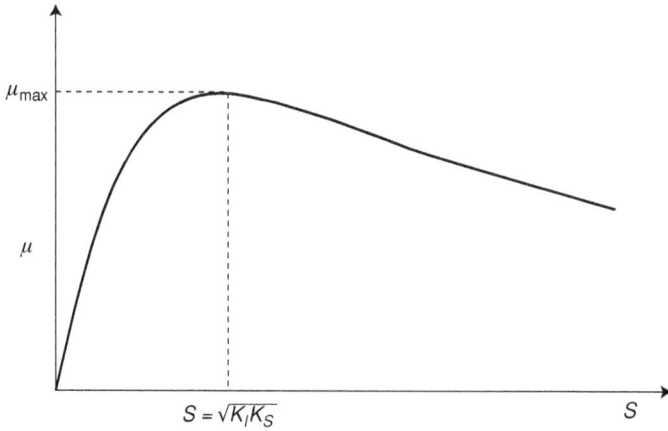

Figure 6.6. Substrate-inhibited specific growth rate, $\mu = \mu_m K_I S/[K_I K_S + (K_I + K_S)S + S^2]$.

where $K_1, K_2, Q_1,$ and Q_2 are adjustable parameters and $\bar{\mu}, \bar{X},$ and $\bar{S}$ are the arithmetic means of $\mu, X,$ and $S,$ respectively. Substrate-independent but a type of population-limited growth model has been proposed by Verhulst,[58]

$$\mu = \alpha(K - X) \tag{6.17}$$

by McKendrick and Pai,[14]

$$\mu = \alpha(X)(\beta - X) \tag{6.18}$$

by Cui and Lawson,[15]

$$\mu = \left(1 - X/X_{max}\right) / \left(1 - X/X_{min}\right) \tag{6.19}$$

and by Fame and Hu,[16]

$$\mu = 1 - \exp\left(-K\left(1 - X_{max}/X\right)\right) \tag{6.20}$$

6.2.1.3 Inhibition Forms, $\mu(S), \mu(S, P), \mu(S, I)$

Substrates, products, intermediates, and inhibitors may inhibit cell growth. Forms used for substrate inhibition are similar to substrate-inhibited enzyme kinetics (noncompetitive and competitive inhibitions) owing to high substrate concentration.

6.2.1.3.1. SUBSTRATE INHIBITION $\mu(S)$. Andrews[17] and Edwards[53] proposed a noncompetitive-type inhibition:

$$\mu = \frac{\mu_m}{\left(1 + \dfrac{K_S}{S}\right)\left(1 + \dfrac{S}{K_I}\right)} = \frac{\mu_m K_I S}{K_I K_S + (K_I + K_S)S + S^2} \tag{6.21}$$

Figure 6.6 shows that this form of specific growth rate has the maximum value at the substrate concentration of $S = \sqrt{K_I K_S}$.

A competitive inhibition can be represented by the following form, which is equivalent in form to the Monod equation:

$$\mu = \frac{\mu_m S}{K_S + (K_S/K_I + 1)S} = \frac{\mu_m S}{a + bS} = \frac{(\mu_m/b)S}{(a/b) + S} \tag{6.22}$$

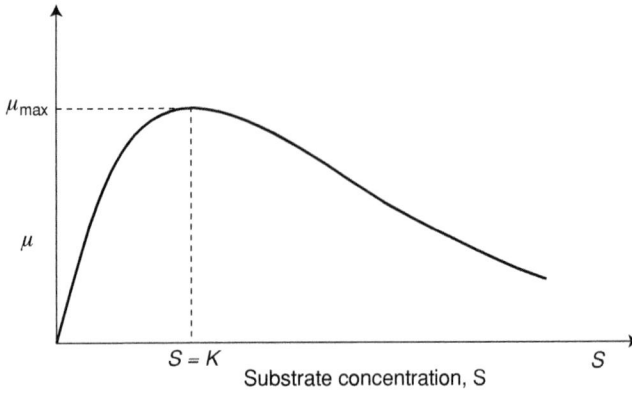

Figure 6.7. Substrate-inhibited specific growth rate.

This Monod-type form does not exhibit a maximum but approaches the asymptotic value of μ_m/b.

A competitive inhibition can be represented by the following form, which is equivalent in form to the Monod equation:

$$\mu = \frac{\mu_m S}{K_S + (K_S/K_I + 1)S} = \frac{\mu_m S}{a + bS} = \frac{(\mu_m/b)S}{(a/b) + S} \tag{6.23}$$

This Monod-type form does not exhibit a maximum but approaches the asymptotic value of μ_m/b.

A one-hump function with only two adjustable parameters represents another form of substrate inhibition:[18]

$$\mu = \mu_m S \exp(-S/K) \tag{6.24}$$

This form is plotted in Figure 6.7, which shows that the maximum rate is $\mu_m K/e$, which occurs at $S = K$.

6.2.1.3.2. PRODUCT INHIBITION $\mu(S, P)$. Products also can inhibit the growth, and a number of specific growth rate expressions have been proposed. As in the case of substrate inhibition, the competitive- and noncompetitive-type inhibitions were used:

$$\mu = \frac{\mu_m}{K_S\left(1 + \dfrac{P}{K_P}\right) + S} = \frac{\mu_m S}{K_S S + S^2 + \dfrac{K_S}{K_P}PS} \tag{6.25}$$

and

$$\mu = \frac{\mu_m}{\left(1 + \dfrac{K_S}{S}\right)\left(1 + \dfrac{P}{K_P}\right)} = \frac{\mu_m K_P S}{(K_S + S)(K_P + P)} \tag{6.26}$$

where P is the product concentration. In Eq. (6.25), the second expression is more appropriate because the specific growth rate must vanish when there is no substrate, $S = 0$. Equation (6.26) was used for the anaerobic glucose fermentation by yeast.[19,20]

The Monod form of specific growth rate expression was modified to include inhibition by the product (alcohol)[19,21] in the form of an exponential decay:

$$\mu = \left(\frac{\mu_m S}{K + S} \right) \exp\left(-\frac{P}{K_P} \right) \tag{6.27}$$

6.2.1.3.3. INHIBITOR INHIBITION $\mu(S, I)$. Inhibitors that may be present in the medium or generated as intermediates may inhibit the growth, such as formaldehyde in methanol fermentation and acetic acid in ethanol fermentation:

$$\mu = \frac{\mu_m S}{(K_S + S) + \dfrac{K_S}{K_I} I} \tag{6.28}$$

$$\mu = \frac{\mu_m}{\left(1 + \dfrac{K_S}{S} \right)\left(1 + \dfrac{I}{K_I} \right)} = \frac{\mu_m S}{(S + K_S)\left(1 + \dfrac{I}{K_I} \right)} \tag{6.29}$$

$$\mu = \frac{\mu_m S}{(S + K_S)\left(1 + \dfrac{SI}{K_I} \right)} \tag{6.30}$$

These forms are equivalent to competitive, noncompetitive, and uncompetitive inhibitions, respectively, in enzyme inhibition kinetics.

All of the preceding expressions cannot account for zero specific growth rates observed at finite concentrations of toxic substrates, products, or inhibitors. The Monod form of specific growth rate expression was modified to include inhibition by the product, as in the following expression:[21]

$$\mu = \left(\frac{\mu_m S}{K + S} \right)\left(1 - \frac{P}{P_m} \right)^n \tag{6.31}$$

where P_m represents the maximum product concentration at which the growth rate ceases and where n is a constant.

6.2.2 Specific Product Formation Rate, π

The yield coefficients $Y_{X/S}$ and $Y_{P/S}$ are similar in concept to selectivity or yield for multiple reactions in chemical kinetics. The cell mass cell yield coefficient $Y_{X/S}$ represents the weight (grams) of cell mass produced per unit weight of substrate consumed to produce new cells, whereas the product yield coefficient $Y_{P/S}$ is the mass of product formed per unit mass of substrate consumed. In chemical kinetics, two types of fractional yields[48] are defined: instantaneous and overall. Likewise, yields can be similarly defined for biological reactions. Instantaneous yields may be defined by

$$Y_{X/S} = \left(\frac{r_X}{-r_S} \right)_{r_p = 0, m = 0} = \left(\frac{\mu X V}{\sigma X V} \right)_{\pi = 0, m = 0} = \left(\frac{\mu}{\sigma} \right)_{\pi = 0, m = 0}$$

$$Y_{P/S} = \left(\frac{r_P}{-r_S} \right)_{r_X = 0, m = 0} = \left(\frac{\pi X V}{\sigma X V} \right)_{\mu = 0, m = 0} = \left(\frac{\pi}{\sigma} \right)_{\mu = 0, m = 0} \tag{6.32}$$

In Eq. (6.32), the subscripts are used to denote the conditions under which the ratios of specific rates are taken. The cell yield $Y_{X/S}$ is defined as the ratio of the specific growth rate to the specific substrate consumption in the absence of product formation $r_p = 0 = \pi$ and any other maintenance $m = 0$. Likewise, the product yield $Y_{P/S}$ is defined as the ratio of the specific product formation to the specific substrate consumption in the absence of cell growth $r_X = 0 = \mu$ and any other maintenance $m = 0$. Because the substrate is consumed to form cell mass and the product and to maintain viability of cells (maintenance energy), the specific substrate consumption rate can be written as

$$\sigma = \frac{\mu}{Y_{X/S}} + \frac{\pi}{Y_{P/S}} + m \tag{6.33}$$

The overall yield coefficient can be defined in a similar manner as the amount of cells (product) produced per amount of substrate consumed in the absence of product (cells) formation and maintenance requirement. Because there is a significant volume change, we must account for the volume change:

$$Y_{X/S} = \left(\frac{X}{-S} \right)_{r_p = 0, m = 0} = \left(\frac{X_f V_f - X_0 V_0}{S_F(V_f - V_0) + S_0 V_0 - S_f V_f} \right)_{r_p = 0, m = 0}$$

$$Y_{P/S} = \left(\frac{P}{-S} \right)_{r_x = 0, m = 0} = \left(\frac{P_f V_f - P_0 V_0}{S_F(V_f - V_0) + S_0 V_0 - S_f V_f} \right)_{r_x = 0, m = 0} \tag{6.34}$$

The overall yield coefficients are difficult to evaluate because of the condition of no product (or no cell) formation and no maintenance requirement. The substrate consumed solely for cell growth (or product formation) is difficult to evaluate in a fed-batch operation where cell growth and product formation may not be separable.

The instantaneous yield coefficients (Eq. (6.32)) are either constants or variables during the operation, depending mostly on the substrate concentration and less on the product concentration and other culture conditions. Apparently, the concept of yield was first used by Raulin to express the nutrient requirements of a fungus, and later, Monod showed that the growth yield is a constant in bacterial cultures when the conditions are maintained constant. Constant-yield coefficients imply that the specific rates of growth, substrate consumption, and specific product formation are functionally identical and differ only in the numerator constants. These overall yield coefficients are the same as the ratios of stoichiometric coefficients in chemical kinetics. Thus, if there is only one limiting reaction throughout the course of fermentation, then the ratios of stoichiometric coefficients remain invariant, and the yield coefficients remain constant, that is, the instantaneous and overall yield coefficients are the same.

6.2.2.1 Constant-Yield Coefficients

When the yield coefficients are constant, the specific substrate consumption rate is simply a weighted sum of the specific rates of cell growth and product formation and the maintenance coefficient (Eq. (6.39)). Therefore, the specification of specific rates $\mu, \pi,$ and m along with constant-yield coefficients $Y_{X/S}$ and $Y_{P/S}$ fixes the specific substrate consumption rate σ. Thus, the functional dependence of σ is limited to those of μ and π.

6.2.2.2 Variable-Yield Coefficients

Product formations are classified into two classes: growth associated and non–growth associated; that is, product formation may be observed while the cell growth takes place, and sometimes the products are formed without cell growth. Leudeking and Piret[22] proposed the following specific product formation rate:

$$\pi = \alpha + \beta\mu \tag{6.35}$$

where α and β are constants. Thus, the total product formation rate $\pi X V = r_p V$ is proportional to the cell concentration as well as the rate of growth:

$$\pi X V = \alpha X V + \beta\mu X V \tag{6.36}$$

The first term represents non–growth-associated product formation, whereas the second term represents growth-associated product formation. For certain fermentation, it is well known that no product is formed unless the cells are allowed to grow, whereas for others, products are formed without the growth of cells. Indeed, according to the classical work of Gaden,[23] fermentation processes can be classified into three types: I, II, and III.

Type I fermentation refers to the fermentation in which there are no separate phases for cell growth and product formation and the products are formed with cell growth so that $\pi = \beta\mu$. Type II fermentation is that in which the cell growth phase is followed by the product formation phase and no product is formed during the growth phase, while during the product formation phase, the product formation is accompanied with cell growth. Type III fermentation is that in which there are separate phases for growth and product formation and during the growth phase, no product is formed, whereas during the product formation phase, a product is formed without cell growth or with negligible cell growth.

It should be pointed out that the specific rates are all related and that the yield coefficients are not necessarily constant and can vary within the phases and from one phase to another. In other words, the yield coefficient in the growth phase may be different from that in the product formation phase, and it can vary even during each phase. When one looks over the various forms of the specific growth rate expressions, it is apparent that the rate expression can be represented by a ratio of polynomials in substrate concentration:

$$\mu = \frac{a_1 S}{1 + b_1 S + b_2 S^2} \tag{6.37}$$

where a_1, b_1, and b_2 are arbitrary constants to be selected appropriately. Appropriate choices of these constants lead to Monod-type growth, $a_1 = \mu_m/K$, $b_1 = 1/K$, and $b_2 = 0$, and to the noncompetitive substrate-inhibited growth rate $a_1 = \mu_m/K$, $b_1 - 1/K_S + 1/K_I$ and $b_2 - 1/K_I K_S$. Thus, it is possible to represent various specific rates by a ratio of a first-order polynomial to a second-order polynomial:

$$\sigma = \frac{c_1 S}{1 + d_1 S + d_2 S^2}, \quad \pi = \frac{e_1 S}{1 + f_1 S + f_2 S^2} \tag{6.38}$$

These expressions can represent substrate inhibition with the maximum values of $c_1/(1 + d_1 + d_2)$ and $e_1/(1 + e_1 + e_2)$, respectively, at the substrate concentrations of $1/d_2$ and $1/f_2$, respectively. These can also represent saturation kinetics with a choice of $d_2 = 0$ or $f_2 = 0$.

6.2.3 Specific Substrate Consumption Rate, σ

The specific substrate consumption rate must account for the generation of cells and product as well as the maintenance of cells. The specific substrate consumption rate is simply a weighted sum of the specific rates of cell growth and product formation and the maintenance coefficient:

$$\sigma = \mu/Y_{X/S} + \pi/Y_{P/S} + m \tag{6.39}$$

Therefore, the specification of the specific growth rate, the specific product formation rate, and the maintenance coefficient, μ, π, and m, along with yield coefficients $Y_{X/S}$ and $Y_{P/S}$, fixes the specific substrate consumption rate σ. When there is no product formation and the maintenance energy is negligible, then the specific substrate consumption rate expressions are usually given in terms of the specific growth rate and yield coefficient, $\sigma = \mu/Y_{X/S}$. Here the yield coefficient can be constant or a function of substrate concentration. It can also include a maintenance term. Thus, it is possible to represent the specific substrate consumption rates σ by a ratio of polynomials:

$$\sigma = \frac{c_1 S}{1 + d_1 S + d_2 S^2} + m \tag{6.40}$$

6.2.4 Net Specific Rates

Sometimes, instead of the usual specific rates, net specific rates for cell growth, substrate consumption, and product formation are denoted by μ^{net}, σ^{net}, and π^{net} and are used to include the cell death rate, maintenance rate, and product decay rate:

$$\mu^{net} XV = \mu XV - k_X XV \tag{6.41}$$

$$\sigma^{net} XV = \mu XV/Y_{X/S} + \pi XV/Y_{P/S} + mXV \tag{6.42}$$

$$\pi^{net} XV = \pi XV - k_p XV \tag{6.43}$$

$$\pi XV = k_1 XV + k_2 \mu XV \tag{6.44}$$

where mXV represents the total maintenance requirement rate, $k_x XV$ and $k_p PV$ represent the total biomass decay rate and metabolite decay rate, respectively, and k_1 and k_2 are constants for non–growth-associated and growth-associated metabolite formation rates, respectively. The maintenance constant, m, and the decay constants, k_x and k_p, are all assumed to be constant.

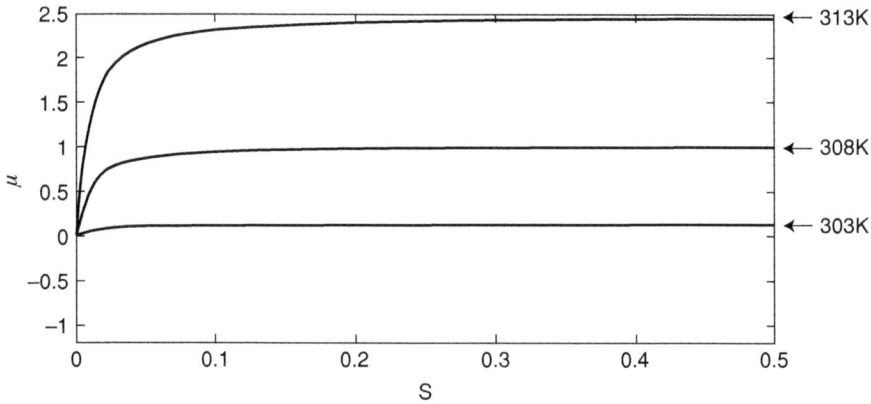

Figure 6.8. Dependence of specific growth rate on temperature, $\mu = [S/0.008 + S]\left[2.45 \cdot 10^{10} e^{-14230/(1.987 \cdot T)} + 10^{23} \cdot e^{-32900/(1.987 \cdot T)} - 1.39\right]$.

For simple cases, it is further assumed that the specific rates μ, σ, and π depend only on the limiting substrate concentration S and are independent of concentrations of cells and product, X and P, and the cell age distribution. The yield coefficients $Y_{X/S}$ and $Y_{P/S}$ are either assumed to be functions of the limiting substrate concentration or are considered constant.

6.2.5 Temperature and pH Effects on Specific Rates

All of the specific rates presented earlier are based on the assumption that pH and temperature are held within a small range by independent control loops. However, if the pH and temperature vary significantly during the operation, their effects must be taken into account. For example, the constants in the specific rates may depend on temperature and pH. Unlike well-defined chemical reactions, in which the rate constants usually follow the Arrhenius form, empirical formulas are use for microbial, animal, and plant cells to describe pH and temperature dependences.

6.2.5.1 Influence of Temperature on Specific Rates
Microbial, animal, and plant cells die at high temperature, and therefore, the specific rates tend to increase with temperature until the critical temperature at which they cease to grow and produce products. Thus, Topiwala and Sinclair[50] used two exponential functions of Arrhenius form to denote the dependence on temperature:

$$\mu(S, T) = \mu(S) f(T)$$
$$= \begin{bmatrix} \mu(S)[a_1 \exp(-E_1/RT) - a_2 \exp(-E_2/RT) + b] & \text{if } T_1 \leq T \leq T_2 \\ 0 & \text{if } T < T_1 \text{ or } T > T_2 \end{bmatrix} \quad (6.45)$$

where a_1, a_2, b, E_1, and E_2 are empirical constants to be determined from experimental data. Thus, it is reasonable to use a similar approach to modify the specific rates of substrate consumption σ and product formation π. This form of dependence of μ on temperature is shown in Figure 6.8.

6.2.5.2 pH Effects on Specific Rates

The dependence of specific rates is also treated empirically. Rozzi[51] proposed a parabolic dependence,

$$\mu(S, \text{pH}) = \mu(S)g(\text{pH}) = \mu(S)[a(\text{pH})^2 + b(\text{pH}) + c] \qquad (6.46)$$

where a, b, and c are constants to be determined from experimental data at various pH values. Conversely, Jackson and Edwards[52] suggested a quadratic fit in terms of hydrogen concentration, taking after the pH effect on enzymatic reactions:

$$\mu(S, \text{pH}) = \mu(S)g(\text{pH}) = \mu(S)\text{H}^+/[K_M + \text{H}^+ + K_I(\text{H}^+)^2] \qquad (6.47)$$

where H^+ represents the hydrogen ion concentration. Similarly, other specific rates σ and π may be modified to reflect the effects of pH.

6.2.6 Maintenance Term

The term *maintenance* refers to the consumption of energy for processes other than the synthesis of new biomass (cells) and products such as the energy required to maintain the chemical potential across the cell membrane, active transport, motility, and macromolecular syntheses. In other words, besides the formation of cells and products that are accounted in mass balance equations, energies are consumed to maintain the cellular identities such as the chemical potential, motility of cells, active transport across cell wall membranes, and synthesis of various macromolecules that are not accounted in the mass balance equations.

Herbert[24] proposed to modify the specific growth rate by considering the oxidation of cell substance:

$$\mu = \frac{\mu_m S}{K + S} - \mu_c \quad \sigma = \frac{\mu}{Y_{X/S}} \qquad (6.48)$$

where μ_c is a constant. This is shown in Figure 6.9.

Conversely, Marr et al.[25] and Pirt[26] proposed to account the maintenance directly into the specific substrate consumption rate:

$$\mu = \frac{\mu_m S}{K + S}, \quad \sigma = \frac{\mu}{Y_{X/S}} + \mu_c' \qquad (6.49)$$

where μ_c' is a constant. Others have combined the approaches of Herbert[24] and Marr et al.[25] and Pirt[26] to account for the maintenance in both the specific growth rate and product formation rate:

$$\mu = \frac{\mu_m S}{K + S} - \mu_c, \quad \sigma = \frac{\mu}{Y_{X/S}} + \mu_c' \qquad (6.50)$$

While all of the preceding expressions adopted constant specific maintenance terms, others proposed maintenance terms that depend on the substrate concentration and also on breaking the cells into viable and nonviable cells. Ramkrishna et al.[27] proposed the following expressions:

$$\mu_{XV} = \frac{\mu_m S}{K + S} - \frac{K'\mu_c}{K' + S}, \quad \mu_{XT} = \frac{\mu_m S}{K + S}, \quad \sigma = \frac{\mu_m S}{K + S}\frac{1}{Y_{X/S}} + \frac{K'\mu_c}{K' + S} \qquad (6.51)$$

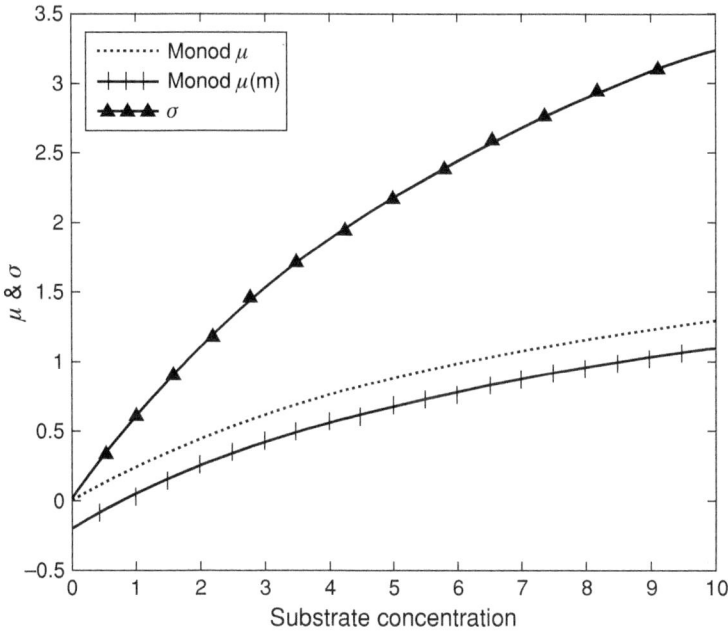

Figure 6.9. Monod specific growth rate containing maintenance term: $\mu = 2.5S/(9.3 + S)$; $\mu(\text{maintenance}) = 2.5S/(9.3 + S) - 0.2$; $\sigma = 2.5S/(9.3 + S)/0.4$.

where the subscripts *XV* and *XT* refer to viable cells and total cells (sum of viable and nonviable cells) and K' and μ_c' are constants.

The maintenance terms proposed for all of the preceding contain a technical difficulty. When there is no more substrate to be consumed, $S = 0$, Eqs. (6.48)–(6.51) indicate that the substrate is consumed for maintenance, an apparent contradiction unless the cells are lysed to provide the necessary energy. This implies that cell lysis terms must be included in the cell mass balance equation. To avoid this type of abnormality, the maintenance term should also depend on the substrate concentration so that when the substrate concentration is zero, the maintenance term also vanishes. For example, one can introduce a functional form of S that vanishes when $S = 0$:

$$\sigma = \frac{\mu}{Y_{X/S}} + \mu_c' \frac{S}{K' + S}, \qquad \sigma = \frac{\mu_m S}{K + S} \frac{1}{Y_{X/S}} + \frac{K' \mu_c S}{K' + S} \qquad (6.52)$$

Conversely, Sinclair and Topiwala[28] introduced the maintenance term in the specific growth rate expression

$$\mu_{XT} = \frac{\mu_m S}{K + S} - (\mu_c + \mu_d), \qquad \mu_{XD} = \mu_d = kX, \qquad \sigma = \frac{\mu_m S}{(K + S)Y_{X/S}} \qquad (6.53)$$

where the subscripts *XT* and *XD* stand for the total cells and dead cells, respectively. In these expressions, the cells go through death and decrease in number owing to maintenance.

6.3 Structured Models

In the unstructured models discussed thus far, the cell behaviors were modeled in terms of extracellular environment. However, in reality, the extracellular environment affects the intracellular environment, to which cells actually respond. Each cell can be viewed as an expanding and dividing complex chemical reactor in which hundreds of enzymatic reactions take place with intimate interactions and internal regulations such as induction, activation, inhibition, and repression. In addition, large cells may encounter internal mass transfer resistance that smaller bacterial cells lack. These reactions can be roughly classified into two categories: those that break up nutrient compounds to derive energy (catabolism) and those that assimilate carbon sources to form cell mass (anabolism). The models that describe the intracellular activities of the organism are called *structured models* and should be developed by selecting properly the parameters that are most relevant for the description of the physiological state of the organism. Structured models take into account the various cellular processes that are important to the cell growth and metabolite production. A highly structured model includes the activities of specific enzymes in the cell and such structures as ribosomes, mitochondria, and macromolecular components such as DNA, RNA, and carbohydrates.

Structured models range from highly sophisticated models of cells, such as *E. coli* and *S. cerevisiae*, which require hundreds of physiological parameters, to relatively simple ones that divide the cell into a number of interacting components rather than a single compartment. An advantage of such models is the potential for modeling the interrelationship of the metabolic processes and predicting the effects of disturbances on the overall microbial process. As stated earlier, the disadvantages of the highly structured model are the difficulty in determining the various kinetic constants in the model that require measurements of activities and amounts of various enzymes and structural components and computational difficulties posed by a large number of equations in optimizing the cellular process.

The concept and details of structured models can best be illustrated by an example of penicillin fermentation by *Penicillium chrysogenum*.

Example 6.E.3: A Structured Model[49]

In this model of penicillin fermentation, the mycelial cell mass is assumed to be made up of three differential states, as illustrated in Figure 6.10: the growing hyphal tips A_0, where mycelial growth is confined to linear extension of hyphal tips; a penicillin-producing fraction A_1, which contains active biomass and is capable of branching to form new tips; and a nonproducing fraction A_2, which is in a degenerate state owing to the loss of cytoplasm within intact cell walls and is therefore nonmetabolic.

The growing tips A_0 are branched from the penicillin-producing fraction A_1 at the rate of $vSA_1V/(K+S)$ and differentiated to A_1 at the rate of $k_1A_0V/(L+S)$. The penicillin-producing fraction is supposed to be right behind the growing tips and grows at the rate of $\mu_mA_0V/(K+S)$, branches to the growing tips at the rate of $vSA_1V/(K+S)$, differentiates to A_1 at the rate of $k_1A_0V/(L+S)$, and decays to the nonviable fraction at the rate of $k_2A_1V/(L+S)$. The substrate, usually glucose, is supplied by the feed stream at the rate of FS_F and is consumed to support the

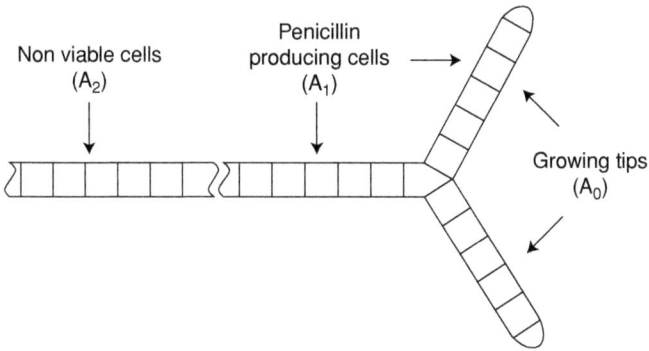

Figure 6.10. Differentiation states in the differentiation state model of Cagney[49] for penicillin production.

growth of penicillin-producing fraction at the rate of $\mu_m A_0 V/(K+S)$, the formation of penicillin at the rate of $k_p S A_1 V/[K_p + S(1+S/K_i)]$, and maintenance of A_1 at the rate of $m_1 S A_1 V/[K_m + S]$. The specific rate of formation of penicillin is substrate inhibited and also subject to loss by hydrolysis at the rate of $-k_h PV$. Thus, the material balance equations can be summarized as follows:

Mass Balance Equations

Growing tips
$$\frac{d(A_0 V)}{dt} = \frac{v S A_1 V}{K+S} - \frac{k_1 A_0 V}{L+S}$$

Penicillin-producing fraction
$$\frac{d(A_1 V)}{dt} = \frac{\mu_m A_0 V}{K+S} - \frac{v S A_1 V}{K+S} + \frac{k_1 A_0 V}{L+S} - \frac{k_2 A_1 V}{L+S}$$

Nonviable fraction
$$\frac{d(A_2 V)}{dt} = \frac{k_2 A_1 V}{L+S}$$

Substrate
$$\frac{d(SV)}{dt} = FS_F - \frac{\mu_m A_0 V}{(K+S)Y_{X/S}} - \left(\frac{k_p S}{[K_p + S(1+S/K_i)]Y_{P/S}} + \frac{m_1 S}{K_m + S}\right)(A_1 V)$$

Penicillin
$$\frac{d(PV)}{dt} = \frac{k_p S A_1 V}{K_p + S(1+S/K_i)} - k_h PV$$

Overall $\quad \dfrac{dV}{dt} = F$

Total cell mass
$$\frac{d(XV)}{dt} = \frac{d(A_0 V + A_1 V + A_2 V)}{dt} = \frac{\mu_m S A_0 V}{K+S}$$

Specific Rates

Growth $\mu = \mu_m/(K+S)$

Penicillin formation $\pi = k_P S / [K_P + S(1 + S/K_I)]$
Cell degradation $k_d = k_2 / (L + S)$
Cell branching $k_{A_0/A_1} = k_1 / (L + S)$
Maintenance $m = m_1 S / (K_m + S)$
Branching $k_b = vS / (K + S)$
Substrate consumption for cell growth $\sigma = \mu / Y_{X/S}$
Substrate consumption for penicillin formation $\eta = \pi / Y_{P/S}$

6.4 Parameter Estimation

The method of maximum likelihood[29–31] is known to be among the most reliable methods of estimating parameters. This is an iterative least squares method that minimizes a criterion for evaluating the goodness of fit between the predicted and measured data. The primary disadvantage of the iterative least squares method is the large computational requirements of a search algorithm for determining the best set of values for model parameters.

The extended Kalman filter (EKF)[32–34] is another method that can simultaneously estimate the values of state variables and the parameters in the model. This method recursively minimizes the sum of the squared error between the predicted and measured values. The recursive nature of this method leads to much more computational efficiency. However, the computational requirements increase with the order of the model (number of equations) owing to the integration of a large gain matrix and inversion of a large matrix. We shall cover these two methods below.

6.4.1 All State Variables Are Measurable

When the state variables are all measurable and therefore available as data, the method of maximum likelihood is the most reliable method for estimating the parameters in the model. The first step is the selection of a criterion for evaluating the goodness of fit between the predicted (estimated) and observed data. This criterion serves as the error model in determining whether a set of parameters with the model can describe sufficiently the dynamic behavior of the system. Any error model can be incorporated into this algorithm, for example, the log likelihood function (LLF) or the sum of the squared error (SSE).

We have a model in the form of ordinary differential equations (e.g., unsteady state mass balance equations of Eqs. (6.3)–(6.7)) in which a number of parameters p are to be determined using experimental data:

$$\frac{d\mathbf{x}}{dt} = \mathbf{f}(\mathbf{x}, \mathbf{p}), \quad \mathbf{f} = (f_1, f_2, f_3, \dots, f_n), \quad \mathbf{x} = (x_1, x_2, x_3, \dots x_n), \tag{6.54}$$

$$\mathbf{p} = (p_1, p_2, p_3, \dots p_m)$$

There are n mass balance equations and m parameters in Eq. (6.54). The parameters appear in specific rates and mass and energy transfer terms. These parameters usually appear nonlinearly and therefore require a nonlinear parameter estimation scheme.

6.4.1.1 Error Criteria

To implement an estimation scheme, one has to select a statistical criterion, for example, least squares. SSE is useful only when the absolute error of the state variables is not a function of its value. Another method is the LLF, which is known to be most versatile and mathematically sound.

The first assumption in LLF is replicated experimental measurements in data that are normally distributed about the true value. The second assumption is that the measured state variables are independent of each other so that the overall probability of obtaining several specific values is the product of the individual probabilities. The joint probability density function, or the *maximum likelihood*, is

$$p(e_1, e_2, \ldots, e_N) = \left(\frac{1}{\sqrt{2\pi}}\right)^N \prod_{i=1}^{N} \left(\frac{1}{s_i}\right) \exp\left(-\sum_{i=1}^{N} \frac{(y_i - x_i)^2}{2s_i^2}\right) \qquad (6.55)$$

where the residual e_i is the difference between the ith measured value y_i and the ith model-predicted value x_i, $e_i = (y_i - x_i)$, N is the number of data points, s_i is the standard deviation for the ith data point, and $p(e_1, e_2, e_3, \ldots, e_N)$ is the probability (likelihood) function, that is, a measure of probability of obtaining precisely these N values of es.

The standard deviations and their dependencies on the measured variables y_i are usually not known a priori, and therefore, it is useful to apply an error model. A simple but useful error model is used in SIMUSOLV,[35,36] a software package designed to handle simulations and parameter estimations for reactions and reactors:

$$s_i^2 = \kappa_i^2 x_i^\gamma \qquad (6.56)$$

where κ_i is a proportionality factor and γ is an adjustable parameter called the *heteroscedasticity factor* and is constrained between 0 and 2. The γ value of 0 implies that the standard deviation is independent of the predicted value of that variable, while the value of 2 implies that the standard deviation is proportional to the predicted value.

It is possible to determine the values of κ_i that maximize the probability function. Differentiating the log of the probability density function with respect to κ_i, setting it to zero, solving it for κ_i, and substituting the resulting expression into Eq. (6.54) yields the following expression:

$$s_i^2 = x_i^\gamma \frac{1}{N} \sum_{i=1}^{N} \frac{(y_i - x_i)^2}{x_i^\gamma} \qquad (6.57)$$

Substituting Eq. (6.57) into Eq. (6.55) and taking the log of the resultant yield, the LLF,

$$\text{LLF} = \log[p(e_1, e_2, \ldots, e_N)] \qquad (6.58)$$

$$= -\frac{N}{2}\left[(\log(2\pi) + 1 + \log\left(\frac{1}{N}\sum_{i=1}^{N}\frac{(y_i - x_i)^2}{x_i^\gamma}\right)\right] - \frac{\gamma}{2}\sum_{i=1}^{N}\log x_i$$

Because there are multiple variables to be estimated, there is an LLF and a γ factor for each of these variables. In addition, the magnitudes of the variables may differ greatly, each of the measured and predicted values must be scaled properly, and the scaled variables must not be zero since the logarithm is taken. The overall LLF to be maximized is the sum of the individual LLFs, and the final expression for the likelihood function can be written in terms of N_y variables,

$$\text{LLF}(\mathbf{p}) = -\frac{N_y N}{2}\left[(\log(2\pi)+1) - \frac{N}{2}\sum_{j=1}^{N_y}\log\left(\frac{1}{N}\sum_{i=1}^{N}\frac{(\bar{y}_{ij}-\bar{x}_{ij})^2}{\bar{x}_{ij}^{\gamma_j}}\right)\right]$$
$$-\sum_{j=1}^{N_j}\frac{\gamma_j}{2}\sum_{i=1}^{N}\log\bar{x}_{ij} \qquad (6.59)$$

where $\bar{y}_{ij}$ and $\bar{x}_{ij}$ are the ith normalized predicted and measured values, respectively, of the jth variable:

$$\bar{y}_{ij} = \frac{y_{ij}+\alpha_j}{\beta_j} \quad and\, \bar{x}_{ij} = \frac{x_{ij}+\alpha_j}{\beta_j} \qquad (6.60)$$

where α_j and β_j are the normalization constants for the jth predicted and measured variables.

A simpler error model is the sum of the squared error,

$$\text{SSE}(\mathbf{p}) = \sum_{i=1}^{N}\sum_{j=1}^{N_y}(\bar{y}_{ij}-\bar{x}_{ij})^2 \qquad (6.61)$$

The weighting factors traditionally used in multivariable error models can be handled indirectly by the normalization factor β_j, and there would be no concern to take a logarithm of a number zero. The SSE is a degenerate form of the likelihood function with $\gamma = 0$. With the selection of an error model, the next step is to select a search procedure to estimate the parameters that minimize the error criterion. There are two search methods for estimating nonlinear parameters. One is a gradient-based method and the other is a direct search method.

6.4.1.2 Estimation Methods
Search procedures for estimating nonlinear parameters can be broadly classified as methods using function values only, methods using first derivatives, Newton's method, and quasi-Newtonian methods. The methods that utilize the values of the function only are direct search methods, whereas the methods that utilize first-order derivatives are gradient-based methods and include the steepest descent (or ascent, if maximization) and conjugate gradient methods.

The *steepest ascent* is based on the gradient that is the vector that gives the local direction of the greatest rate of increase in the function. The search direction is simply the gradient, and the algorithm is called the *steepest ascent for maximization*. For minimization, the search direction is the negative of the gradient, and the algorithm is known as the *steepest descent*:

$$\mathbf{p}^{i+1} = \mathbf{p}^i + \Delta\mathbf{p}^i = \mathbf{p}^i + \alpha^i\nabla f(\mathbf{p}^i) \qquad (6.62)$$

where the vector $\mathbf{p} = (p_1, p_2, p_3, \ldots, p_m)$ is the parameter vector and $\nabla f(\mathbf{p}^i)$ is the gradient and where α^i represents the scalar factor that determines the step length in the direction of the gradient.

The *conjugate gradient method* represents a major improvement over the gradient method by combining the current gradient with the previous gradient. The algorithm begins by evaluating the gradient at the starting point:

$$\mathbf{p}^1 = \mathbf{p}^0 + \alpha^0 \mathbf{s}^0 = \mathbf{p}^0 + \alpha^0 \nabla f(\mathbf{p}^0) \tag{6.63}$$

$$s^1 = s^0 \frac{\nabla^T f(\mathbf{p}^1) \nabla f(\mathbf{p}^1)}{\nabla^T f(\mathbf{p}^0) \nabla f(\mathbf{p}^0)} + \nabla f(\mathbf{p}^1), \mathbf{p}^2 = \mathbf{p}^1 + \alpha^1 s^1 \tag{6.64}$$

$$s^j = s^{j-1} \frac{\nabla^T f(\mathbf{p}^j) \nabla f(\mathbf{p}^j)}{\nabla^T f(\mathbf{p}^{j-1}) \nabla f(\mathbf{p}^{j-1})} + \nabla f(\mathbf{p}^j), \mathbf{p}^{j+1} = \mathbf{p}^j + \alpha^j s^j \tag{6.65}$$

The steepest ascent method is very slow in convergence and may oscillate, whereas the conjugate gradient method is much faster in convergence and more accurate.

6.4.2 Some State Variables Are Not Measurable but Are Observable

In certain situations, some of the state variables are not directly measurable but are observable (they can be calculated from the measurements of other state variables). Therefore, it may be necessary to estimate the state variables that are not directly measurable before carrying out the estimation of parameters in the model. In fact, it is possible to simultaneously predict the unmeasurable state variables and estimation of the parameters.

When a process is linear and a model is available, *Kalman filters*[37] are very powerful tools that can estimate not only the unknown parameters in the model but also the state variables that are not possible to measure on-line. However, this technique requires knowledge of certain stochastic properties of measurement and disturbance noises. For nonlinear systems, this technique is applied to linearized models and is known as the *extended Kalman filter* (EKF). This is a recursive least squares estimator and has been applied to estimate nonlinear bioreactor state variables and kinetic model parameters.

6.4.2.1 Extended Kalman Filter
We begin with a general nonlinear dynamic model with discrete measurements:
System model

$$\dot{\mathbf{x}} = \mathbf{f}(\mathbf{x}(t), t) + \mathbf{G}(\mathbf{x}(t), t)\mathbf{w}(t) \tag{6.66}$$

Measurement model

$$\mathbf{z}(t_i) = \mathbf{h}(\mathbf{x}(t_i), t_i) + \mathbf{v}(t_i) \tag{6.67}$$

where $\mathbf{f}$ is a nonlinear vector function of $\mathbf{x}(t)$ and t, $\mathbf{h}$ is a nonlinear vector function of $\mathbf{x}(t_i)$ and t_i, $\mathbf{G}$ is a matrix function of $\mathbf{x}(t)$ and t, $\mathbf{w}(t)$ is a random disturbance vector, and $\mathbf{v}(t_i)$ is a random error vector in the measurement of $\mathbf{z}(t_i)$. There are assumptions on the random vectors: $\mathbf{w}(t)$ and $\mathbf{v}(t)$ are zero mean white noises with covariance matrices $\mathbf{Q}(t)$ and $\mathbf{R}(t)$, respectively. The noises $\mathbf{w}(t)$ and $\mathbf{v}(t)$ are uncorrelated with each other and also uncorrelated with the initial conditions, $\mathbf{x}(0)$. The mean and covariance of $\mathbf{x}(0)$ are designated as $\mathbf{m}(0)$ and $\mathbf{P_X}(0)$, respectively.

The EKF problem may be stated as follows: given a measurement sequence $\mathbf{Z}(k) = \{\mathbf{z}(t_0), \mathbf{z}(t_1), \mathbf{z}(t_2), \ldots, \mathbf{z}(t_k)\}$ for the model given by Eqs. (6.66) and (6.67), find an estimator to provide unbiased, minimum-variance estimates of $\mathbf{x}(t)$, denoted by $\mathbf{x}(t \,|\, t_k)$, patterned after a linear Kalman filter, and yielding small errors, $\tilde{\mathbf{x}}(t \,|\, t_k) = \mathbf{x}(t) - \hat{\mathbf{x}}(t \,|\, t_k)$. The EKF estimation is described by the following set of filtering equations:

Predictor

$$\dot{\hat{\mathbf{x}}}(t|t_k) = \mathbf{f}(\mathbf{x}(t|t_k), t) \tag{6.68}$$

Error covariance prediction

$$\mathbf{P}(t_k|t_{k-1}) = \Phi(t_k, t_{k-1})\mathbf{P}(t_{k-1}, t_{k-1})\Phi^T(t_k, t_{k-1}) + \mathbf{Q}(t_{k-1}) \tag{6.69}$$

Corrector

$$\hat{\mathbf{x}}(t_k \,|\, t_k) = \hat{\mathbf{x}}(t_k \,|\, t_{k-1}) + \mathbf{K}(t_k)[\mathbf{z}(t_k) - \mathbf{h}(\hat{\mathbf{x}}(t_k \,|\, t_{k-1}, t_k))], \quad \hat{\mathbf{x}}(t_0 \,|\, t_0) = m_0 \tag{6.70}$$

State estimate

$$\mathbf{P}(t_k \,|\, t_k) = [\mathbf{I} - \mathbf{K}(t_k)]\mathbf{P}(t_k \,|\, t_{k-1}), \quad \mathbf{P}(t_0 \,|\, t_0) = \mathbf{P_{x_0}} \tag{6.71}$$

Error covariance matrix

$$\mathbf{K}(t_k) = \mathbf{P}(t_k \,|\, t_{k-1})\,\mathbf{H}^T(t_k)[\mathbf{H}(t_k)\mathbf{P}(t_k \,|\, t_{k-1})\mathbf{H}^T(t_k) + \mathbf{R}(t_k)]^{-1}, \tag{6.72}$$

$$\mathbf{H}(t_k) = [\partial\mathbf{h}/\partial\mathbf{x}]_{\mathbf{x}=\hat{\mathbf{x}}(t_k|t_{k-1})}$$

Kalman gain matrix

$$\frac{\partial \Phi}{\partial t} = \mathbf{F(t)}\Phi(t, t_k), \quad \mathbf{F}(t) = [\partial\mathbf{f}/\partial\mathbf{x}]_{\mathbf{x}=\hat{\mathbf{x}}(t_k|t_{k-1})} \tag{6.73}$$

A schematic implementing the preceding equations to obtain the best estimate is given in Figure 6.11. It is best to illustrate the preceding methods using a simple example so that the methods are clearly understood.

Example 6.E.4 Estimation by EKF

We consider a fed-batch culture of *S. cerevisiae* at low glucose concentrations. Using the experimental data,[59] we shall illustrate the application of an EKF. A Monod model for growth of biomass and substrate consumption is combined with the equation of variable reactor volume:

Process model

$$\frac{dX}{dt} = \frac{\mu_{\max}S}{K_m + S}X - \frac{F}{V}X + u_X \tag{6.E.4.1}$$

$$\frac{dS}{dt} = -\frac{1}{Y_{X/S}}\frac{\mu_{\max}S}{K_m + S}X + \frac{F}{V}(S_0 - S) + u_S \tag{6.E.4.2}$$

$$\frac{d\mu_{\max}}{dt} = u_\mu \tag{6.E.4.3}$$

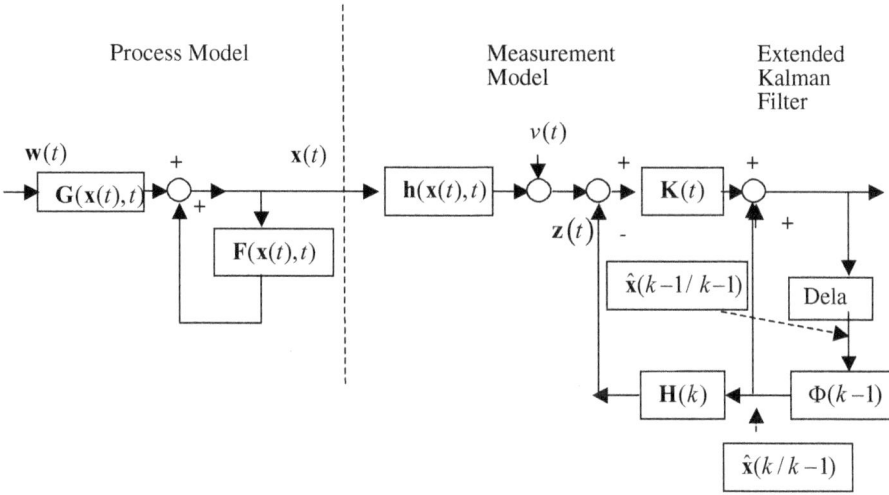

Figure 6.11. System model and extended Kalman filter.

$$\frac{dV}{dt} = F - F_{sam} + u_V \tag{6.E.4.4}$$

where X, S, and V are the cell mass concentration, the substrate concentration, and the reactor volume, respectively; μ_{max}, K_m, $Y_{X/S}$, S_0, F, and F_{sam} are the maximum specific growth rate, the saturation constant, the yield coefficient, the initial substrate concentration, the feeding rate, and the sample stream flow rate, respectively; and u_X, u_S, u_μ, and u_V are the process errors of cell mass, substrate, and maximum specific growth rate and volume, respectively. The spectral density matrix for the process noise was set as follows:

$$Q = \begin{pmatrix} 0.001\frac{g^2}{L^2h} & 0 & 0 & 0 \\ 0 & 0.001\frac{g^2}{L^2h} & 0 & 0 \\ 0 & 0 & 0.05\frac{1}{h^3} & 0 \\ 0 & 0 & 0 & 0 \end{pmatrix} \tag{6.E.4.5}$$

The substrate glucose measurement model is

$$S_{m,t_i} = S(t_i) + s(t_i) \tag{6.E.4.6}$$

where S_{m,t_i}, $S(t_i)$, and $s(t_i)$ are measured and predicted substrate concentrations and the corresponding measurement noise. To maintain the glucose concentration at 0.08 g/L, a feedforward or feedback control was performed. The feedforward flow rate was calculated by the fact that the estimated substrate concentration must not change and was adjusted by the feedback control using a proportional-integral (PI) controller (see later):

$$F(t_i) = \hat{V}(t)\frac{\hat{\mu}_{max}\hat{S}(t_i)\hat{X}(t_i)}{Y_{X/S}[K_m + \hat{S}(t_i)][S_0 - \hat{S}(t_i)]} + F_{PI}(t_i) \tag{6.E.4.7}$$

$$F_{PI}(t_i) = F_{PI}(t_{i-1}) + q_0(S_{set} - \hat{S}(t_i)) + q_1(S_{set} - \hat{S}(t_{i-1})) \tag{6.E.4.8}$$

where S_{set}, q_0, and q_1 are the set point for control and the parameters of the PI controller, which were set to $q_0 = 1.4$ L^2 gh and $q_1 = -1.1$ L^2/gh. Using the EKF

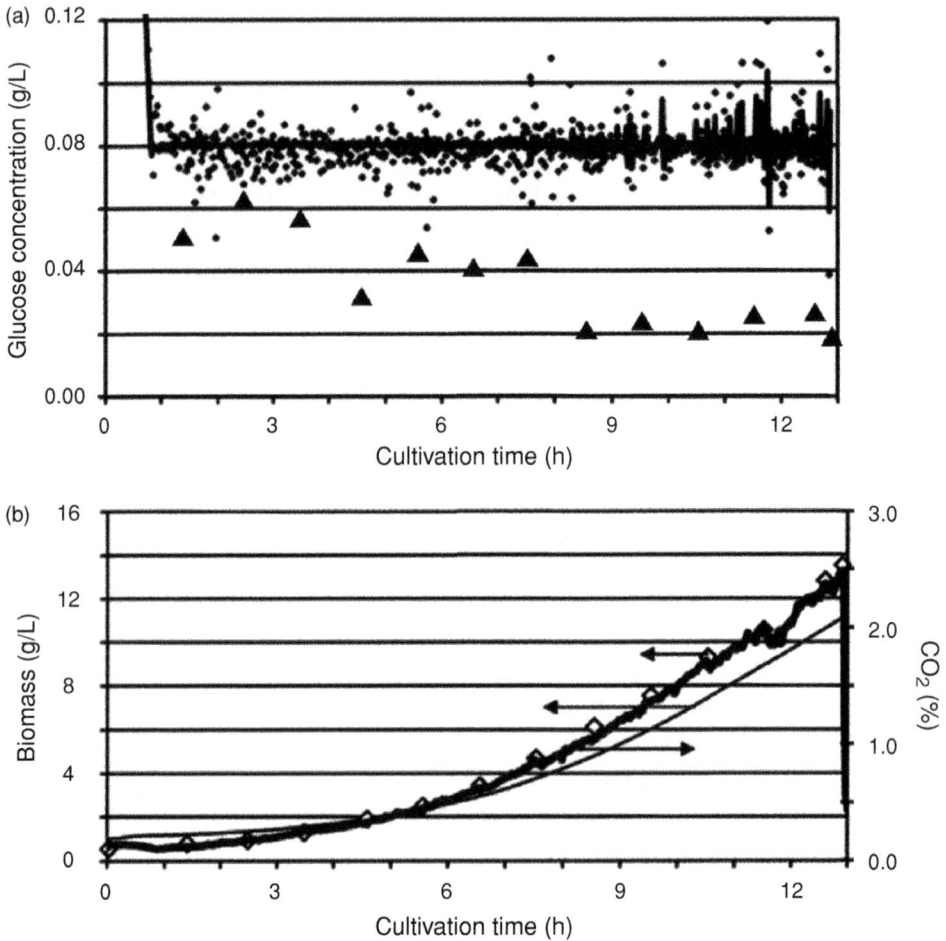

Figure 6.12. Estimation of state variables by extended Kalman filter. (a) On-line glucose measurements (•), predicted glucose concentration (-), and off-line glucose measurements (○). (b) the estimated (-) and off-line measured (◇) biomass concentration and carbon dioxide concentration of the exhaust gas (-). (reproduced from fig. 1(A)b, Ref. 59).

equations, Eqs. (6.66)–(6.73), we estimated simultaneously the cell mass and substrate concentrations and the model parameters. All parameters were changed during simulation runs until the difference between the predicted and set points of glucose became minimal. Kalman estimates of cell mass and substrate (glucose) are shown in Figure 6.12; it is apparent that these state variables are fairly well estimated by the technique.

Although this technique can predict both observable state variables and the unknown parameters in the model, a general drawback of the filer is that it requires for proper filtering a judicious choice of set of the noise covariance matrices $\mathbf{Q}$ and $\mathbf{R}$ and the initial error covariance matrix $\mathbf{P_0}$. This may sometimes be a difficult task.

Appendix: Some Models Proposed in Literature

We list subsequently some models that have been proposed and used for various purposes. We start with the simplest case of cell mass production.

6.A.1 Cell Mass Fermentation[39,40]

Mass Balance Equations

$$\frac{d(XV)}{dt} = \mu XV, \quad \frac{d(SV)}{dt} = FS_F - \frac{\mu XV}{Y_{X/S}}, \quad \frac{dV}{dt} = F$$

Specific Rates

Constant yield[39] $\mu(S) = \dfrac{S}{0.03 + S + 0.5S^2}, Y_{X/S} = 0.5$

Variable yield[40]

$$\mu(S) = \frac{0.504S(1.0 - 0.0204S)}{0.000849 + S + 0.0406S^2} \quad S \leq 4.89\% \,(\mathrm{w/v})$$

$$Y_{X/S}(S) = \frac{0.383(1.0 - 0.0204S)}{1.0 + 2.96S - 0.00501S^2} \quad S \leq 4.89\% \,(\mathrm{w/v})$$

6.A.2 Lysine Fermentation: A Model of Ohno et al.[41]

Mass Balance Equations

$$\frac{d(XV)}{dt} = \mu XV, \quad \frac{d(SV)}{dt} = FS_F - \frac{\mu XV}{Y_{X/S}}, \quad \frac{d(PV)}{dt} = \pi XV, \quad \frac{dV}{dt} = F$$

Specific Rates

$$\mu = 0.125S, \quad \pi = -384\mu^2 + 134\mu, \quad \sigma = \mu/0.135$$

6.A.3 Alcohol Fermentation: A Model of Aiba et al.[42]

Mass Balance Equations

$$\frac{d(XV)}{dt} = \mu XV, \quad \frac{d(SV)}{dt} = FS_F - \frac{\mu XV}{Y_{X/S}}, \quad \frac{d(PV)}{dt} = \pi XV, \quad \frac{dV}{dt} = F$$

Specific Rates

$$\mu = \frac{0.408S}{0.22 + S} \exp(-0.028P), \quad \pi = \frac{S}{0.44 + S} \exp(-0.015P), \quad \sigma = \frac{\mu}{0.1}$$

6.A.4 Penicillin Fermentation: A Model of Bajpai and Reuss[43]

Mass Balance Equations

$$\frac{d(XV)}{dt} = \mu XV, \quad \frac{d(SV)}{dt} = FS_F - \frac{\mu XV}{Y_{X/S}}, \quad \frac{d(PV)}{dt} = (\pi - k)XV, \quad \frac{dV}{dt} = F$$

Specific Rates

$$\mu = \frac{0.11S}{0.006X + S}, \quad \pi = \frac{0.004S}{0.0001 + S + 10S^2}, \quad \sigma = \frac{\mu}{0.47} + \frac{\pi}{1.2} + 0.029, \quad k = 0.01$$

6.A.5 Differential State Model for Penicillin Fermentation by Cagney[49]

Mass Balance Equations

Growing tips $d(A_0 V)/dt = k_b A_1 V - k_{A_0/A_1} A_0 V$

Penicillin-producing fraction $d(A_1 V)/dt = \mu A_0 V - k_b A_1 V + k_{A_0/A_1} A_0 V - k_d A_1 V$

Nonviable fraction $d(A_2 V)/dt = k_d A_1 V$

Substrate $d(SV)/dt = F S_F - (\mu/Y_{X/S})(A_0 V) - \left(\pi/Y_{P/S} + m\right)(A_1 V)$

Penicillin $d(PV)/dt = \pi A_1 V - k_h PV$

Overall $dV/dt = F$

Total cell mass $d(XV)/dt = d(A_0 V)/dt + d(A_1 V)/dt + d(A_2 V)/dt = \mu A_0 V$

Specific Rates

Growth $\mu = \mu_m/(K + S)$

Penicillin formation $\pi = k_P S/[K_P + S(1 + S/K_I)]$

Cell degradation $k_d = k_2/(L + S)$

Cell branching $k_{A_0/A_1} = k_1/(L + S)$

Maintenance $m = m_1 S/(K_m + S)$

Branching $k_b = \nu S/(K + S)$

Substrate consumption for cell growth $\sigma = \mu/Y_{X/S}$

Substrate consumption for penicillin formation $\eta = \pi/Y_{P/S}$

6.A.6 Chittur's Model for Penicillin Fermentation[31]

Mass Balance Equations

Viable cell $d(AV)/dt = (\mu - k_d)AV$

Nonviable fraction $d(A_u V)/dt = k_d AV$

Substrate $d(SV)/dt = F S_F - (\mu/Y_{X/S} + \pi/Y_{P/S} + m)AV$

Penicillin $d(PV)/dt = \pi AV - k_h PV$

Overall $dV/dt = F$

Total cell mass $d(XV)/dt = \mu AV = d(AV)/dt + d(A_u V)/dt$

Specific Rates

Growth $\mu = \mu_m/(K + S)$

Penicillin formation $\pi = k_P S/[K_P + S(1 + S/K_I)]$

Cell degradation and differentiation $k_d = k_2/(L + S)$

Maintenance $m = m_1 S/(K_m + S)$

Branching $k_b = \nu S/(K + S)$

Substrate consumption for cell growth $\sigma = \mu/Y_{X/S}$

Substrate consumption for penicillin formation $\eta = \pi/Y_{P/S}$

6.A.7 Modak–Patkar Model for Invertase Fermentation[60]

Mass Balance Equations

Cell $d(XV)/dt = (\mu_G + \mu_E)XV$

Glucose $d(GV)/dt = F S_F - \sigma XV$

Ethanol $d(EV)/dt = (\pi_E - \eta)XV$
Invertase $d(PXV)/dt = (\pi - k_d P)XV$
Overall $dV/dt = F$

Specific Rates

Growth $\mu = \mu_G + \mu_E$
On glucose $\mu_G = (k_1 G + k_2 G^2)/(k_3 + k_4 G + G^2)$
On ethanol $\mu_E = k_5 E/[(k_6 + k_7\sigma + E)(1 + k_8 E)]$
Fraction of glucose fermented $R = (1 + k_9 G^n)/(k_{10} + k_9 G^n)$
Ethanol production rate $\pi_E = Y_{E/G}^F \sigma R$
Ethanol consumption rate $\eta = \mu_E/Y_{X/E}^R$
Cellular yield $Y_X = (1 - R)Y_{X/G}^R + RY_{X/G}^F$
Invertase formation $\pi = k_P S/[K_P + S(1 + S/K_I)]$
Substrate consumption for cell growth $\sigma = \mu_G/Y_X$

6.A.8 Modified Modak–Patkar Model[44] for Invertase Fermentation: Inhibition by Ethanol of Both Invertase Formation and Cell Growth on Glucose

Mass Balance Equations

Cell $d(XV)/dt = (\mu_G + \mu_E)XV$
Glucose $d(GV)/dt = FS_F - \sigma XV$
Ethanol $d(EV)/dt = (\pi_E - \eta)XV$
Invertase $d(PXV)/dt = (\pi - k_d P)XV$
Overall $dV/dt = F$

Specific Rates

Growth $\mu = \mu_G + \mu_E$
On glucose $\mu_G = \dfrac{(k_1 G + k_2 G^2)}{(k_3 + k_4 G + G^2)}(1 - E/k_E)$
On ethanol $\mu_E = k_5 E/[(k_6 + k_7\sigma + E)(1 + k_8 E)]$
Fraction of glucose fermented $R = (1 + k_9 G^n)/(k_{10} + k_9 G^n)$
Ethanol production rate $\pi_E = Y_{E/G}^F \sigma R$
Ethanol consumption rate $\eta = \mu_E/Y_{X/E}^R$
Cellular yield $Y_X = (1 - R)Y_{X/G}^R + RY_{X/G}^F$
Invertase formation $\pi = k_P S(1 - E/k_E)/[K_P + S(1 + S/K_I)]$
Substrate consumption for cell growth $\sigma = \mu_G/Y_X$

6.A.9 Excreted Protein[45]

Mass Balance Equations

$$\frac{d(XV)}{dt} = \mu XV, \quad \frac{d(SV)}{dt} = FS_F - \frac{\mu XV}{Y_{X/S}}, \quad \frac{d(IV)}{dt} = \pi_I XV, \quad \frac{d(PV)}{dt} = \pi XV$$

Specific Rates

Growth $\mu = \dfrac{21.87S}{(S + 0.4)(S + 62.5)}$
Substrate consumption $\sigma = \frac{\mu}{Y}$

$$\text{Product formation } \pi_{\mathrm{I}} = \frac{S \exp(-5.0S)}{S + 0.1}, \pi = \frac{103.88S(I - P)/X}{0.12S^2 + 29.42S + 3}$$

6.A.10 α-Amylase Fermentation[46]

Mass Balance Equations

$$d(XV)/dt = \mu XV \quad XV(0) = X_0 V_0$$
$$d(S_1 V)/dt = FS_{1F} - \sigma_1 XV \quad S_1 V(0) = S_{10} V_0$$
$$d(S_2 V)/dt = FS_{2F} - \sigma_2 XV \quad S_2 V(0) = S_{20} V_0$$
$$d(PV)/dt = \pi XV \quad PV(0) = P_0 V_0$$
$$dV/dt = F_1 + F_2 \quad V(0) = V_0$$

Specific Rates

$$\text{Growth rate } \mu = \frac{0.86 S_1 S_2}{2.0 + S_1 + S_1^2/33}$$
$$\text{Substrate consumption } \sigma_1 = \frac{\mu}{0.68}, \sigma_2 = \frac{\mu}{1.05}, k = 0.18$$
$$\text{Product formation } \pi = 117.7 \exp(-0.311 S_2), k = 0.18$$

6.A.11 A Poly-β-hydroxybutyric Acid Model[47]

Mass Balance Equations

$$\text{Cell } d(XV)/dt = \mu XV \quad XV(0) = X_0 V_0$$
$$\text{Substrate1 } d(S_1 V)/dt = FS_{1F} - \sigma_1 XV \quad S_1 V(0) = S_{10} V_0$$
$$\text{Substrate2 } d(S_2 V)/dt = FS_{2F} - \sigma_2 XV \quad S_2 V(0) = S_{20} V_0$$
$$\text{Product } d(PV)/dt = \pi XV \quad PV(0) = P_0 V_0$$
$$\text{Total } dV/dt = F_1 + F_2 \quad V(0) = V_0$$

$$0 = F_{1\min} \le F_1 \le F_{1\max}$$
$$0 = F_{2\min} \le F_2 \le F_{2\max}$$
$$V(t_f) = V_{\max}$$

Specific Rates

$$\mu = \frac{0.875 S_1}{5.81 + S_1 + S_1^2/14.5} \frac{S_2}{0.69 + S_2 + S_2^2/0.15}, \quad \sigma_1 = \frac{\mu}{0.45} + \frac{\mu}{0.47} + 0.01$$
$$\pi = \frac{0.402 S_1}{2.09 + S_1 + S_1^2/80} \frac{S_2 + 0.05}{0.05 + S_2 + S_2^2/0.9} \left(1 - \frac{P/X}{0.85}\right), \quad \sigma_2 = \frac{\mu}{2.11}$$

6.A.12 Monoclonal Antibodies by Hybridoma Cells[48]

Mass Balance Equations

$$d(XV)/dt = (\mu - k_d)XV \quad XV(0) = X_0 V_0$$

$$d(S_1V)/dt = FS_{1F} - \sigma_1 XV \quad S_1V(0) = S_{10}V_0$$
$$d(S_2V)/dt = FS_{2F} - \sigma_2 XV \quad S_2V(0) = S_{20}V_0$$
$$d(LV)/dt = \pi_{lac}XV \quad LV(0) = L_0V_0$$
$$d(AV)/dt = \pi_{amm}XV \quad AV(0) = A_0V_0$$
$$d(MV)/dt = \pi_{mab}XV \quad MV(0) = P_0V_0$$
$$dV/dt = F_1 + F_2 \quad V(0) = V_0$$

$X \triangleq$ cellconc., $S_1 \triangleq$ glucoseconc., $S_2 \triangleq$ glycineconc., $V \triangleq$ volume

$A \triangleq$ ammonium, $L \triangleq$ lactate, $M \triangleq$ monoclonolantibody

Specific Rates

$$\mu = \frac{1.09S_1}{1 + S_1}\frac{S_2}{00.3 + S_2}, \quad \sigma_1 = \frac{\mu}{1.09 \times 10^8} + 0.17 \times 10^{-8}\frac{S_1}{19 + S_1},$$

$$\sigma_2 = \frac{\mu}{3.8 \times 10^8}, \quad \pi_{lac} = 1.8\sigma_1, \quad \pi_{amm} = 0.85\sigma_2, \quad \pi_{mab} = \frac{2.56 \times 10^{-8}\mu}{0.02 + \mu} + 0.35$$

REFERENCES

1. Frederickson, A. G., Megee, R. D., III, and Tsuchiya, H. M. 1970. Mathematical models of fermentation processes. *Advances in Applied Microbiology* 13: 419–469.
2. Hatch, R. T. 1982. *Annual Reports on Fermentation Processes*, ed. Tsao, G. T. Academic Press.
3. Kossen, N. W. F. 1982. Computer applications in fermentation technology, in *Third International Conference on Computer Applications in Fermentation Technology*, p. 23. Society of Chemical Industry.
4. Moreira, A. R., Van Dedem, G., and Moo-Young, M. 1979. Process modeling based on biochemical mechanisms of microbial growth. *Biotechnology and Bioengineering Symposium* 9: 179–203.
5. Roels, J. A. 1981. Computer applications in fermentation technology, in *Third International Conference on Computer Applications in Fermentation Technology*, p. 37. Society of Chemical Industry.
6. Monod, J. 1942. *Recherches sur la croissance des cultures bacteriennes.* 2nd ed. Hermann.
7. Monod, J. 1950. La technique de culture continue; theorie et applications. *Annales de L'institut Pasteur* 79: 390–410.
8. Teissier, G. 1942. Croissance des populations bacteriennes et quantite d'alimente disponible. *Revue des Sciences Extrait* 3208: 209–231.
9. Moser, H. 1958. *Dynamics of Bacterial Populations Maintained in the Chemostat.* Carnegie Institution of Washington.
10. Jost, J. L., Drake, J. F., Tsuchia, H. M., and Fredrickson, A. G. 1973. Microbial food chains and food webs. *Journal of Theoretical Biology* 41: 461–484.
11. Shehata, T. E., and Marr, A. G. 1971. Effect of nutrient concentration on the growth of *Escherichia coli. Journal of Bacteriology* 107: 201–216.
12. Ryder, D. N., and Sinclair, C. G. 1972. Model for growth of aerobic microorganisms under oxygen limiting conditions. *Biotechnology and Bioengineering* 14: 787–798.

13. Contois, D. 1959. Relationship between population density and specific growth rate of continuous cultures. *Journal of General Microbiology* 21: 40–50.

14. McKendrick, A. G., and Pai, M. K. 1911. The rate of multiplication of microorganisms: A mathematical study. *Proceedings of the Royal Society, Edinburgh* 31: 649–655.

15. Cui, Q., and Lawson, G. J. 1982. Study on models of single populations: An expansion of the logistic and exponential equation. *Journal of Theoretical Biology* 98: 645–659.

16. Frame, K. K., and Hu, W. S. 1988. A model for density-dependent growth of anchorage-dependent mammalian cells. *Biotechnology and Bioengineering* 32: 1061–1066.

17. Andrews, J. F. 1968. A mathematical model for the continuous culture of microorganisms utilizing inhibitory substrates. *Biotechnology and Bioengineering* 10: 707–723.

18. Agrawal, P., Lee, C., Lim, H. C., and Ramkrishna, D. 1982. Theoretical investigations of dynamic behavior of isothermal continuous stirred tank biological reactors. *Chemical Engineering Science* 37: 453–462.

19. Aiba, S., Shoda, M., and Nagatani, M. 1968. Kinetics of product inhibition in alcohol fermentation. *Biotechnology and Bioengineering* 10: 845–864.

20. Aiba, S., and Shoda, M. 1969. Reassessment of the product inhibition in alcohol fermentation. *Journal of Fermentation Technology* 47: 790–794.

21. Levenspiel, O. 1980. The Monod equation: A visit and a generalization to product inhibition situations. *Biotechnology and Bioengineering* 22: 1671–1687.

22. Leudeking, R., and Piret, E. L. 1959. A kinetic study of the lactic acid fermentation. *Journal of Biochemistry and Microbiological Technology and Engineering* 1: 393–412.

23. Gaden, L., Jr. 1959. Fermentation process kinetics. *Journal of Biochemistry and Microbiology Technology and Engineering* 1: 413–429.

24. Herbert, D. 1959. Continuous culture of microorganisms: Some theoretical aspects, in *Recent Progress in Microbiology, VII International Congress for Microbiology*, ed. Tunevall, G., p. 381. Almquist and Wiksell.

25. Marr A. G., Nilson, E. H., and Clark, D. J. 1963. The maintenance requirement of *Escherichia coli*. *Annals of the New York Academy of Science* 102: 536–548.

26. Pirt, S. J. 1966. The maintenance requirement of bacteria in growing culture. *Proceedings of the Royal Society, B* 163: 224–231.

27. Ramkrishna, D., Frederickson, A. G., and Tsuchiya, H. 1966. Dynamics of microbial propagation: Models considering endogenous metabolism. *Journal of General and Applied Microbiology* 12: 311–327.

28. Sinclaire, G., and Topiwala, H. H. 1970. Model for continuous culture which considers the viability concept. *Biotechnology and Bioengineering* 12: 1069–1079.

29. Astrom, K. J., and Eykoff, P. 1971. System identification – a survey. *Automatica* 7: 123–162.

30. Soderstrom, T. L., and Gustavsson, I. 1978. A theoretical analysis of recursive identification methods. *Automatica* 14: 231–244.

31. Chittur, V. K. 1989. Modeling and optimization of the fed-batch penicillin fermentation. PhD dissertation, Purdue University.

32. Gelb, A. 1974. *Applied Optimal Estimation*. MIT Press.

33. Jazwinski, A. H. 1970. *Stochastic Processes and Filtering Theory*. Academic Press.

34. Sorenson, H. W. 1985. *Kalman Filtering Theory: Theory and Practice*. IEEE Press.

35. Steiner, E. C., Blau, G. E., and Agin, G. L. 1986. *SIMUSOLV – Modeling and Simulation Software*. Mitchell and Gauthier Associates.

36. Burt, C. J. 1989. SIMUSOLV: A new code for modeling and optimization. *CACHE News* 29: 13–16.
37. Kalman, R. E. 1960. A new approach to linear filtering and prediction problems. *Transactions of the ASME – Journal of Basic Engineering* 82: 35–45.
38. Nihtilä, M., and Virkkunen, J. 1977. Practical identifiability of growth and substrate consumption models. *Biotechnology and Bioengineering* 19: 1831–1850.
39. Weigand, W. A., Lim, H. C., Creagan, C., and Mohler, R. 1979. Optimization of a repeated fed-batch reactor for maximum cell productivity. *Biotechnology and Bioengineering Symposium* 9: 335–348.
40. Lim, H. C., Tayeb, I. J., Modak, J. M., and Bonte, P. 1986. Computational algorithms for optimal feed rates for a class of fed-batch fermentation: Numerical results for penicillin and cell mass production. *Biotechnology and Bioengineering* 28: 1408–1420.
41. Ohno, H., Nakanishi, E., and Takamatsu, T. 1976. Optimal control of a semi-batch fermentation. *Biotechnology and Bioengineering* 18: 847–864.
42. Aiba, S., Shoda, M., and Nagatani, M. 1968. Kinetics of product inhibition in alcohol fermentation. *Biotechnology and Bioengineering* 10: 845–864.
43. Bajpai, R. K., and Reuss, M. 1981. Evaluation of feeding strategies in carbon-regulated secondary metabolite production through mathematical modeling. *Biotechnology and Bioengineering* 23: 717–738.
44. Hansen, J. M. 1996. On-line adaptive optimization of fed-batch fermentations. PhD dissertation, University of California, Irvine.
45. Park, S., and Ramirez, W. F. 1988. Optimal production of secreted protein in fed-batch reactors. *American Institute of Chemical Engineering Journal* 34: 1550–1558.
46. Pazlarová, J., Baig, M. A., and Votruba, J. 1984. Kinetics of α-amylase production in a batch and fed-batch culture of *Bacillus subtilis* with caseinate as nitrogen source and starch as carbon source. *Applied Microbiology and Biotechnology* 20: 331–334.
47. Lee, J. H., Lim, H. C., and Hong, J. 1997. Application of non-singular transformation to on-line optimal control of poly-β-hydroxybutyrate fermentation. *Journal of Biotechnology* 55: 135–150.
48. De Tremblay, M., Chavarie, C., and Archambault, J. 1992. Optimization of fed-batch culture of hybridoma cells using dynamic programming: single and multi feed cases. *Bioprocess Engineering* 7: 292–234.
49. Cagney, J. W. 1984. Experimental investigation of a differential state model for fed-batch penicillin fermentation. PhD dissertation, Purdue University.
50. Topiwala, H., and Sinclair, C. C. 1971. Temperature relationship in continuous culture. *Biotechnology and Bioengineering* 13: 795–813.
51. Rozzi, A. 1984. Modeling and control of anaerobic digestion processes. *Transactions of the Institute of Measurements and Control* 6: 153–159.
52. Jackson, J. V., and Edwards, V. H. 1975. Kinetics of substrate inhibition of exponential yeast growth. *Biotechnology and Bioengineering* 17: 943–963.
53. Edwards, V. H. 1970. The influence of high substrate concentrations on microbial kinetics. *Biotechnology and Bioengineering* 12: 679–712.
54. Ming, F., Howell, J. A., and Canovas-Diaz, M. 1988. Mathematical simulation of anaerobic stratified fermentation technology, in *Computer Applications and Fermentation Technology: Modeling and Control of Biotechnological Processes*, p. 69–77. Elsevier.
55. Sokol, W., and Howell, J. A. 1981. Kinetics of phenol oxidation by washed cells. *Biotechnology and Bioengineering* 23: 2039–2049.
56. Staniskis, J., and Levisauskias, D. 1984. Adaptive control algorithm for fed-batch culture. *Biotechnology and Bioengineering* 26: 419–425.

57. Kishimoto, M., Sawano, T., Yoshida, T., and Taguchi, T. 1983. Optimization of a fed-batch culture by statistical data analysis, in *Modelling and Control of Biotechnical Processes*, ed. Halme, A., p. 161. Permagon.
58. Verhulst, P. F. 1838. Notice sur la loi que la population suit dans son accroissement. *Correspondance Mathématique et Physique*. 10: 113–121.
59. Arndt, M., and Hitzmann, B. 2004. Kalman filter based glucose control at small set points during fed-batch cultivation of *Saccharomyces cerevisiae*. *Biotechnology Progress* 20: 377–387.
60. Patkar, A. Y. 1960. Kinetics, modeling and optimization of recombinant yeast fermentations. PhD dissertation, Purdue University.

7 Non–Equation-Based Models

When it is possible to write a complete set of mass and energy balance equations with kinetics and perhaps mass transfer rates, that is, specific rates of cell growth, product formation, formation of intermediates, and substrate consumption and mass transfer coefficients, it is possible to develop a model with a number of process parameters. Then, appropriate experimental data are generated and used to estimate the parameters in the model. However, there are many situations in which it is not possible to write a complete set of mass and energy balance equations due to lack of knowledge and understanding, for example, tissue cultures that are too complex and a lack of knowledge to be able to provide an adequate description with a limited number of balance equations. Effects of medium components on the outcome of fermentation are not well known theoretically, and therefore, empirical approaches have been used. A number of approaches can be adapted for dynamic as well as static relationships. We shall consider these approaches.

In certain situations, it is not possible to measure all state variables that describe the process. Some of the state variables cannot be readily measured or are difficult to measure in the time span necessary. Therefore, we need a method of estimating not only the parameters but also some of the difficult-to-measure state variables.

7.1 Neural Networks

Simplifying mathematical models so that they are manageable for optimization schemes comes at the cost of model accuracy. Worse, if little is known about a process for which obtaining data is very difficult, conventional modeling techniques offer little insight. However, neural networks can work in these difficult situations. Neural networks can model complex systems, particularly those in which few accurate quantitative models exist. Neural networks can be tweaked to approximate any function by readily increasing the number of its parameters with very few changes in its basic structure, even without a priori knowledge about the system. Finding the parameters of a neural network model, particularly a back-propagation model, involves using gradient descent–based techniques to reduce the overall error. Because many optimization schemes also employ techniques that involve calculating gradients, neural network models can be readily used in optimization studies, particularly with fed-batch fermentations described by differential equations.

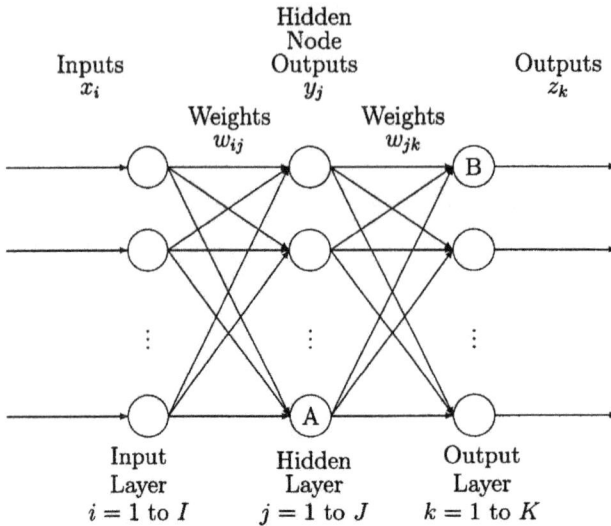

Figure 7.1. Multilayer feedforward neural network architecture.

Neural networks can take on many different forms, each being distinguished by differences in architecture and in how they are applied. Multilayer feedforward neural networks, such as back-propagation and its variants, are commonly used in bioengineering.

7.1.1 Basic Architecture of Neural Networks

Although many kinds of neural networks exist, they have common features in their architecture (Figure 7.1). Multilayer feedforward neural networks generally have nodes arranged in input, hidden, and output layers. Each node within a layer is unilaterally connected to each node in the subsequent layer. Associated with each connection is a weight, a measure of connectivity strength. The weights are analogous to parameters that must be found with an unconstrained optimization procedure.

Inputs are multiplied by their connection weights, then are added together to form the total input to a node. This total input, within the node, is transformed by a transfer function, such as the hyperbolic tangent shown in Figure 7.2, to produce an output signal, which becomes an input to each node in the subsequent layer. Often the sigmoidal transfer function is used, which exhibits similar properties as the hyperbolic tangent, but the outputs are bounded to [0, 1]. This process continues in this mapping phase, finally producing the outputs of the neural network.

The strength of multilayer feedforward neural networks resides in their ability to model nonlinearities through the behavior exhibited by their hidden node's transfer functions. When a multilayer feedforward neural network employs nonlinear transfer functions, such as the sigmoid or hyperbolic tangent, in its hidden nodes, it can model any relationship, thereby functioning as a universal approximator.

The process of adjusting weights is called *training*. During training, the weights are changed according to a learning rule prescribed beforehand. The learning rule regulates how the weights are changed such that the network can learn the presented

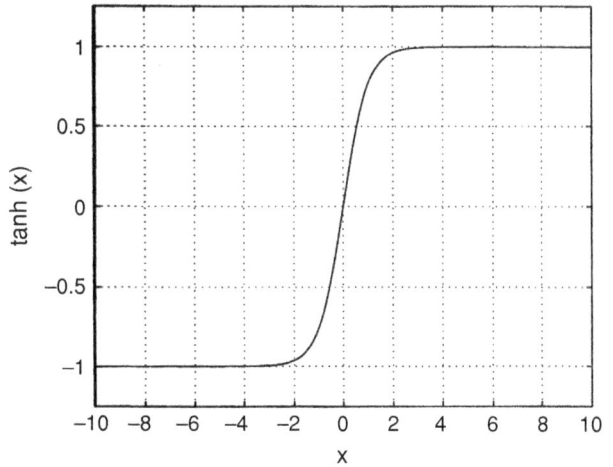

Figure 7.2. Hyperbolic tangent transfer function.

examples (experience). A properly trained neural network, when tested, will produce the correct responses to input examples used in training (recall) and to new ones to demonstrate its ability to interpolate, predict, and extract patterns (generalization).

When the weight changes are based on errors between predictions and actual data values, this process is called *supervised training*. In supervised training, each data set or pattern contains a set of inputs and the corresponding outputs. Several learning rules exist for training multilayer feedforward neural networks. The most popular is called the *generalized delta rule* or *back-propagation*.

7.1.2 Back-Propagation Training Algorithm

Figure 7.3 describes the back-propagation algorithm for adjusting weights. Back-propagation starts with a forward pass through the network, where the outputs to a particular set of inputs are calculated. The network errors are calculated. Then a backward pass begins at the output layer to distribute errors in the form of weight changes:

1. All the weights in the neural network are initialized.
2. A pattern, a set of inputs and outputs, is presented to the neural network.
3. Inputs are mapped to outputs:
   a. Each input, x_i $(i = 1 - I)$ is multiplied by its weight w_{ij}, leading to hidden node j $(j = 1 - J)$.
   b. At hidden node j, all the weighted inputs are summed and added to a unity threshold or bias value T_j, resulting in the total activation input u_j for this hidden node j:

$$u_j = \sum_{i=1}^{I} w_{ij} x_i + T_j \tag{7.1}$$

   c. This total activation or net sum u_j is passed through a transfer function $f(x)$, such as the hyperbolic tangent, resulting in an output signal y_j:

$$f(x) = \tanh x = \frac{e^x - e^{-x}}{e^x + e^{-x}} \tag{7.2}$$

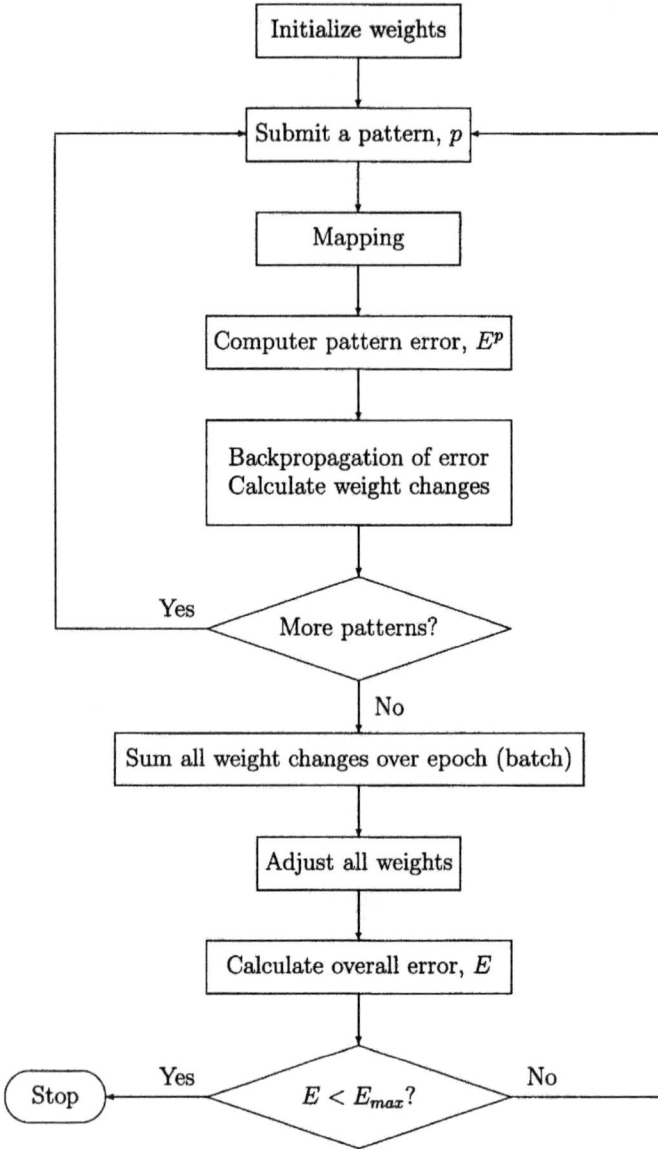

Figure 7.3. Back-propagation algorithm.

$$y_j = f(u_j) = f\left(\sum_{i=1}^{I} w_{ij}x_i + T_j\right) \tag{7.3}$$

d. This output signal from the hidden node j, y_j, becomes the input to each output node k ($k = 1 - K$). All these values from hidden node j are summed to form the net activation input v_k to each output node k to compute the output layer values z_k:

$$v_k = \sum_{j=1}^{J} w_{jk}y_j + T_k \tag{7.4}$$

$$z_k = f(v_k) = f\left(\sum_{j=1}^{J} w_{jk} y_j + T_k\right) \tag{7.5}$$

4. The error per pattern p, E^p, is calculated. Note that the superscript p will be dropped off for t_k and z_k because the same procedure applies to all patterns $p = 1 - P$:

$$E^p = \frac{1}{2} \sum_{k=1}^{K} (t_k^p - z_k^p)^2 \tag{7.6}$$

5. Back propagation of error is applied.
   a. Starting with the output layer, for each node k, $k = 1 - K$, calculate the error signal δ_k for pattern p ($p = 1 - P$) with the mapped output values z_k and actual (target) values t_k:

$$\delta_k^p = (z_k - t_k) f'(v_k) \tag{7.7}$$

   b. Then calculate the error signal δ_j for each hidden node j. Note that δ_k was already calculated with Eq. (7.7):

$$\delta_j^p = \left[\sum_{k=1}^{K} \delta_k^p w_{jk}\right] f'(u_j) \tag{7.8}$$

   c. Calculate the weight changes from mapping pattern p:

$$\Delta w_{jk}^p = \eta_k \delta_k^p y_j^p \tag{7.9}$$

$$\Delta w_{ij}^p = \eta_j \delta_j^p x_i^p \tag{7.10}$$

   d. Present all patterns and repeat mapping (step 3).
   e. Sum all weight changes over all patterns, $p = 1 - P$, for this one batch (epoch) pass n ($n = 1 - N$):

$$\Delta w_{jk}^n = \sum_{p=1}^{P} \Delta w_{jk}^p \tag{7.11}$$

$$\Delta w_{ij}^n = \sum_{p=1}^{P} \Delta w_{ij}^p \tag{7.12}$$

   f. Make weight changes for all weights for the batch pass n where η is a constant called the learning rate, with recommended values $[0, 1]$:

$$\Delta w_{jk}^n = w_{jk}^{n-1} + \Delta w_{jk}^n \tag{7.13}$$

$$\Delta w_{ij}^n = w_{ij}^{n-1} + \Delta w_{ij}^n \tag{7.14}$$

   g. Calculate overall E over all patterns and determine if it is within tolerance, E_{max}:

$$E = \sum_{p=1}^{P} E^p \tag{7.15}$$

h. Begin the next pass of weight changes by presenting all the patterns again at step 3.
i. Continue with $E \leq E_{\max}$.

7.2 Neural Networks in Fed-Batch Fermentation

Neural networks have been applied to fed-batch fermentation in numerous ways. A neural network can be used to model the kinetics of a microorganism, which can then be used in the mass balance equations for control and optimization studies. Neural networks can also be used to model static and dynamic systems, hence serving as state estimators and/or predictors.

Using the Levenberg–Marquardt algorithm, Bulsari et al.[1] trained a feedforward neural network to successfully control *Saccharomyces cerevisiae* fermentation by manipulating the dilution rate to minimize the ethanol concentration. The neural network of Chaudhuri and Modak[2] effectively modeled yeast dynamics using simulated data from the data of Park and Ramirez[3] and experimental data from Patkar and Seo.[4] Chen and Weigand[5] used a neural network to achieve feedback optimization of the same fed-batch bioreactor.

Park and Ramirez[3] maximized the secretion of a heterologous protein (SUC2-s2) from yeast (SEY2102) by formulating the objective function using Pontryagin's maximum principle and a dynamic model developed previously by the authors. For this five-state variable system, they analyzed the behavior of each variable before numerically generating the optimal feed rate policies.

Patkar and Seo[4] studied how different feeding strategies affect yeast (*S. cerevisiae*) cell growth and cloned-gene expression (plasmid containing SUC2 gene for invertase product) to develop kinetic models. They then used a conjugate gradient algorithm with the substrate-inhibition kinetic model to produce the optimal feed policy to successfully maximize invertase productivity.

Syu and Hou[6] used back-propagation neural networks with on-line measurements from their MIMS (membrane introduction mass spectrometer) fermentation system to dynamically model and control the oxygen composition of *Klebsiella oxytoca* fermentation to maximize production of 2,3-butanediol. Their neural network control model directly related products to oxygen composition, allowing the process to be readily controlled from on-line measurements of the main product, 2,3-butanediol.

Multiple neural networks have also been used to model fed-batch fermentations. Ignova et al.[7] successfully combined different pattern recognition neural networks for a supervisory fault detection system for industrial penicillin fed-batch fermentation. Krothapally and Palanki[8] used two neural networks, one to predict switching times and the other for predicting the flow rate during the singular region in a *S. cerevisiae* fed-batch fermentation.

Although feedforward back-propagation neural networks have often been used as predictors for fed-batch fermentation, researchers found that recurrent neural networks modeled better. Karim et al.[9] found that recurrent neural networks were better at modeling fed-batch fermentations of *Zymomonas mobilis*. Syu and Hou[6] showed that their recurrent neural network was better than their back-propagation neural network in modeling a bacterial fermentation *Klebsiella oxytoca* producing 2,3-butanediol. They also successfully used a dynamic recurrent neural network to

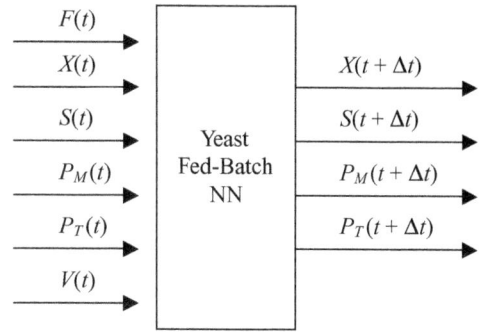

Figure 7.4. Neural network for yeast fed-batch.

control the pH in an *Arthrobacter viscosus* batch fermentation producing penicillin acylase. Saxen and Saxen[10] found that recurrent neural networks were better at modeling kinetics, estimating and identifying states of yeast *S. cerevisiae*. Others have also successfully modeled the kinetics of chemostat fermentations of yeast with recurrent neural networks.

Penicillin fermentation, a bioprocess that exhibits complex kinetic behavior, has been successfully modeled with neural networks. Di Massimo et al.[11] developed biomass and penicillin neural network estimators for on-line application in an industrial fermentation. Thompson and Kramer[12] used simulated data obtained from a modified Bajpai–Reuβ[13] model as process data by adding noise to show that hybrid neural networks performed superior to plain radial basis function and back-propagation neural networks.

If a neural network model trained as a one-step ahead predictor was used to predict multiple time steps ahead, errors accumulated as time progressed. Di Massimo et al.[11] recommended using the back-propagation-in-time training scheme, which minimized the error over a specified range of time or prediction horizon. However, such error propagation was not observed in recombinant yeast fed-batch fermentations modeled and optimized with neural networks developed by Chaudhuri and Modak.[2]

7.2.1 Yeast Fed-Batch Fermentation

This example shows the development of the neural networks for the yeast fermentation, secreting the heterologous protein invertase (SUC2-s2), as shown in Eq. (7.16) and Figure 7.4. The mathematical model exhibits complex substrate inhibition kinetics, which were described by Park and Ramirez[3] as given in Eq. (7.16). To determine the optimal number of hidden nodes for the neural network, several numbers of patterns, time intervals, and hidden nodes were tested, as shown in Table 7.1:

Mass Balance Equations

$$\text{Cell mass(g)} \ d(XV)/dt - C(S)XV$$
$$\text{Glucose(g)} \ d(SV)/dt = FS_F - YC(S)XV$$
$$\text{Level of secreted protein (arbitrary units)} \ d(P_mV)/dt = A(S)(P_t - P_m)V$$
$$\text{Level of total protein (arbitrary units)} \ d(P_tV)/dt = B(S)(XV)$$
$$\text{Culture volume(L)} \ dV/dt = F$$

Table 7.1. *Number of patterns, hidden nodes, and time intervals*

Patterns	Hidden nodes	Time intervals
90	3	15
180	5	30
270	7	45
360	10	60
	15	

Specific Rates

$$\text{Protein secretion rate (hr}^{-1}) \, A(S) = 4.75 C(S)/(0.12 + C(S))$$
$$\text{Protein expression rate (hr}^{-1}) \, B(S) = S e^{-5S}/(0.1 + S) \tag{7.16}$$
$$\text{Cell growth rate (hr}^{-1}) \, C(S) = 21.87 S/[(S + 0.4)(S + 62.5)]$$

Parameters

$$\text{Yield of glucose/cell mass } Y = 7.3$$
$$\text{Feed flow rate (L/hr) } F$$
$$\text{Feed glucose concentration (g/L) } S_F$$

For example, for the fixed final time of 15 hours, data were generated every half hour and yielded a "minimum" set of 180 patterns. Some selected patterns from the minimum set of 180 patterns are used for the fewer numbers of patterns, such as 90, but the minimum set is twice run for the greater numbers of data, such as 360. The transfer function is the hyperbolic tangent, TANSIG in MATLAB. Normalized input and output data are used in the range of $[0, 1]$ and $[-1, 1]$, respectively. The Nguyen–Widrow layer initialization function INITNW from MATLAB generates all initialized weights.

To explore training data issues, several constant flow rates $(0.1; 0.3; 0.5; [1.0, 0];$ $[2.0, 0]; [5.0, 0])$ were integrated with Eq. (7.16) using a Runge–Kutta routine. The initial state variable condition and operating parameters are as follows:

$$[X(0), \, S(0), \, P_m(0), \, P_m(0), \, V(0), \, S_F, \, t_f, \, V_{\max}, \, \Delta t] \tag{7.17}$$
$$= [1, \, 5, \, 0, \, 0, \, 1, \, 20, \, 15, \, 15, \, 0.5]$$

The neural networks were trained using the Levenberg–Marquardt routine TRAINLM (MATLAB). Figure 7.5 shows how the neural networks model is well trained by the training data with a constant flow rate.

7.2.2 Hybrid Neural Networks

In many bioprocesses, complex kinetics are difficult to model, an issue that neural networks resolve well. However, neural networks are expected to rarely predict extrapolated values outside the data ranges, but it is important in process optimization. Incorporating neural networks with mass balances to deal with extrapolation enhances their ability to handle more complex rate expressions, leaving hybrid schemes to model unknown dynamics integrated with conventional models such as

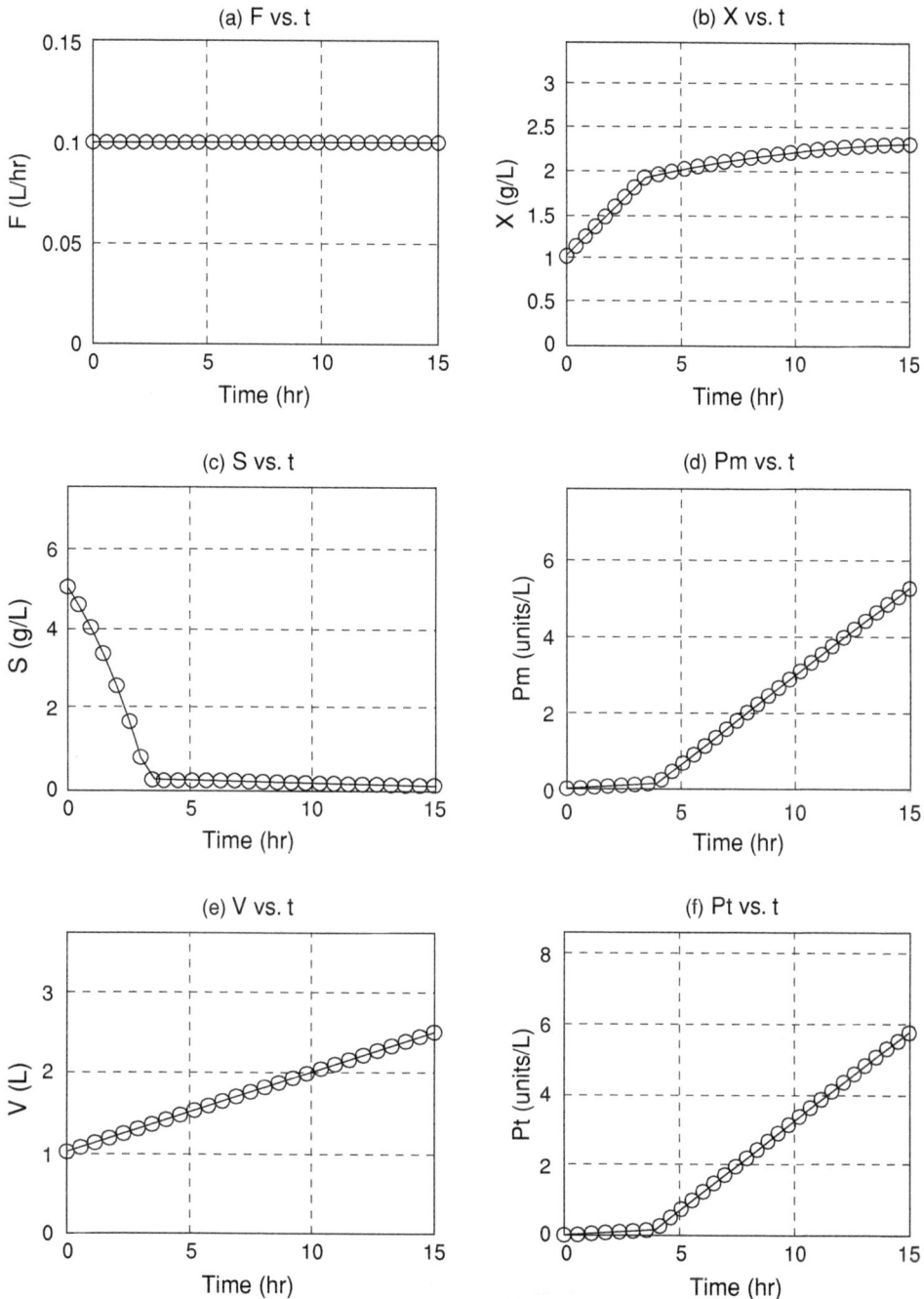

Figure 7.5. Yeast fed-batch fermentation training data with constant flow rate of 1 L/hr.

mass balances. As an example of hybrid neural networks, the problem to maximize the total amount of penicillin produced at a specific time in the fed-batch fermentation with Chittur's model[14] is described in Table 7.2.

Chittur's model of fed-batch penicillin fermentation is represented by mass balance equations (Eq. (7.18)). The hybrid neural networks employed and rearranged Chittur's mass balance equation to model the state variables of substrate, cell (total, viable, and unviable), and penicillin concentrations, as seen in Eq. (7.20):

Table 7.2. *Notation in fed-batch penicillin fermentation model*

Notation	Definition	Units
A	Viable cell concentration	g/L
A_i	Unviable cell concentration	g/L
S	Glucose concentration	g/L
F	Feed flow rate	L/hr
P	Penicillin concentration	g/L
V	Culture volume	L
X	Total cell concentration	g/L
A_f	Viable cell fraction	
A_{if}	Unviable cell fraction	
S_F	Feed stream glucose concentration	g/L
t	Time	hr

Parameter	Definition	Value	Units
i_0	Maximum growth rate	7.72×10^{-2}	hr^{-1}
$\hat{e}_1$	Cell decay rate constant	3.56×10^{-4}	hr^{-1}
K	Saturation constant	4.58×10^{-2}	g/L
L	Decay rate term	0.009	g/L
$\hat{e}_p$	Penicillin synthesis rate constant	1.02×10^{-2}	g/L.hr
K_p	Saturation constant	4.456×10^{-2}	g/L
K_I	Inhibition term	0.075	g/L
$\hat{e}_h$	Penicillin hydrolysis rate constant	0.0093	hr^{-1}
$Y_{X/S}$	Cell yield	0.445	g/g
$Y_{P/S}$	Penicillin yield	1.2	g/g
m_l	Maintenance rate constant	0.0409	hr^{-1}
K_m	Maintenance term	0.0001	g/L

Mass Balance Equations

$$\frac{d(AV)}{dt} = (\mu - k_d)(AV), \quad \frac{d(A_uV)}{dt} = k_d(AV), \quad \frac{d(XV)}{dt} = \mu(AV)$$

$$\frac{d(SV)}{dt} = FS_F - \left(\frac{\mu}{Y_{X/S}} + \frac{\pi}{Y_{P/S}} + m\right)(AV), \quad \frac{d(PV)}{dt} = \pi(AV) - k_d(PV)$$

$$\frac{dV}{dt} = F, \quad A = XA_f, \quad A_u = XA_{uf} \tag{7.18}$$

Specific Rates

$$\mu = \frac{\mu_0 S}{K + S}, \quad \pi = \frac{k_p S}{K_p + S(1 + S/K_I)}, \quad \sigma = \frac{\mu}{Y_{X/S}} \tag{7.19}$$

$$\eta = \frac{\pi}{Y_{P/S}}, \quad m = \frac{m_1 S}{K_m + S}, \quad k_d = \frac{k_1}{L + S}$$

Except penicillin production, the other rate expressions in Eq. (7.20) are generated by neural networks given substrate concentrations (Table 7.3):

$$\frac{d(AV)}{dt} = p_1(S)(AV)$$

Table 7.3. *Rate expressions in hybrid neural networks*

Rate expression	Output	Inputs
Viable cell growth rate	p_1	S
Unviable cell growth rate	p_2	S
Substrate utilization rate	p_3	S
Penicillin production rate	p_4	S, A, and/or P
Cell growth rate	p_5	S

$$\frac{d(A_u V)}{dt} = p_2(S)(AV)$$

$$\frac{d(SV)}{dt} = FS_F - p_3(S)(AV) \qquad (7.20)$$

$$\frac{d(PV)}{dt} = p_4(S, A, P)(AV)$$

$$\frac{d(XV)}{dt} = p_5(S)(AV)$$

Penicillin production rate is more complicated, given with inputs of viable cell and penicillin concentrations. Neural networks to yield rate expressions are depicted in Table 7.4.

The experimental data obtained by Chittur[14] were used to train the hybrid neural networks. Experimental data were measured asynchronously and thus cubic

Table 7.4. *Neural networks for rate expressions*

Rate	Neural networks
Viable cell growth rate	$S(t) \rightarrow$ NN $\rightarrow p_1$
Non-viable cell growth rate	$S(t) \rightarrow$ NN $\rightarrow p_2$
Substrate utilization rate	$S(t) \rightarrow$ NN $\rightarrow p_3$
Penicillin production rate	$S(t) \rightarrow$ NN $\rightarrow p_4$
Biomass growth rate	$S(t) \rightarrow$ NN $\rightarrow p_5$

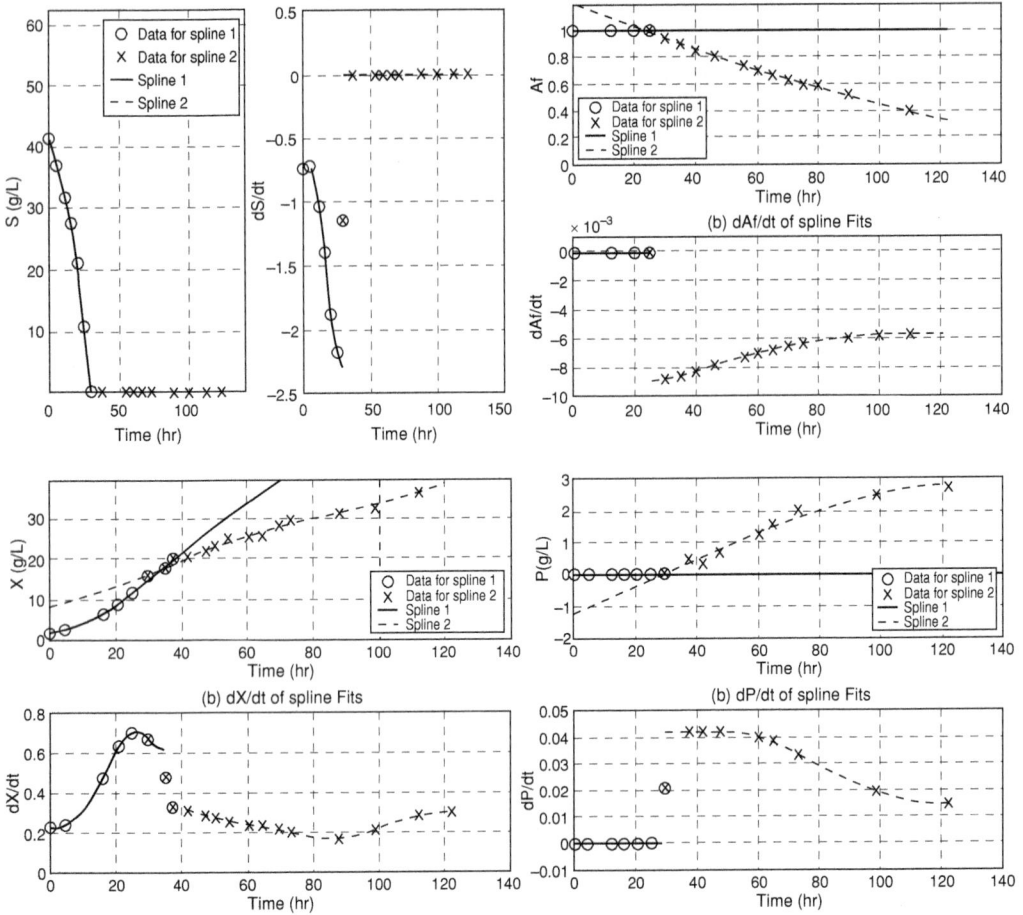

Figure 7.6. Experimental data with spline fits of state variables vs. time.

smoothing splines interpolated missing data before rates could be calculated, as shown in Figure 7.6.

Equation (7.20) is arranged to yield the rates summarized in Eq. (7.21), which can be obtained by analytically calculating derivatives of the cubic spline fits of the experimental data, as shown in Figure 7.6. For simplicity, we modeled the penicillin production rate p_4 as a function of only substrate concentration and set $p_4(S) = \pi$:

$$p_1(S) = \frac{1}{AV}\frac{d(AV)}{dt}$$

$$p_2(S) = \frac{1}{AV}\frac{d(A_uV)}{dt}$$

$$p_3(S) = \frac{1}{AV}\left[FS_F - \frac{d(SV)}{dt}\right] \quad (7.21)$$

$$p_4(S,A,P) = \frac{1}{AV}\frac{d(PV)}{dt}$$

$$p_5(S) = \frac{1}{AV}\frac{d(XV)}{dt}$$

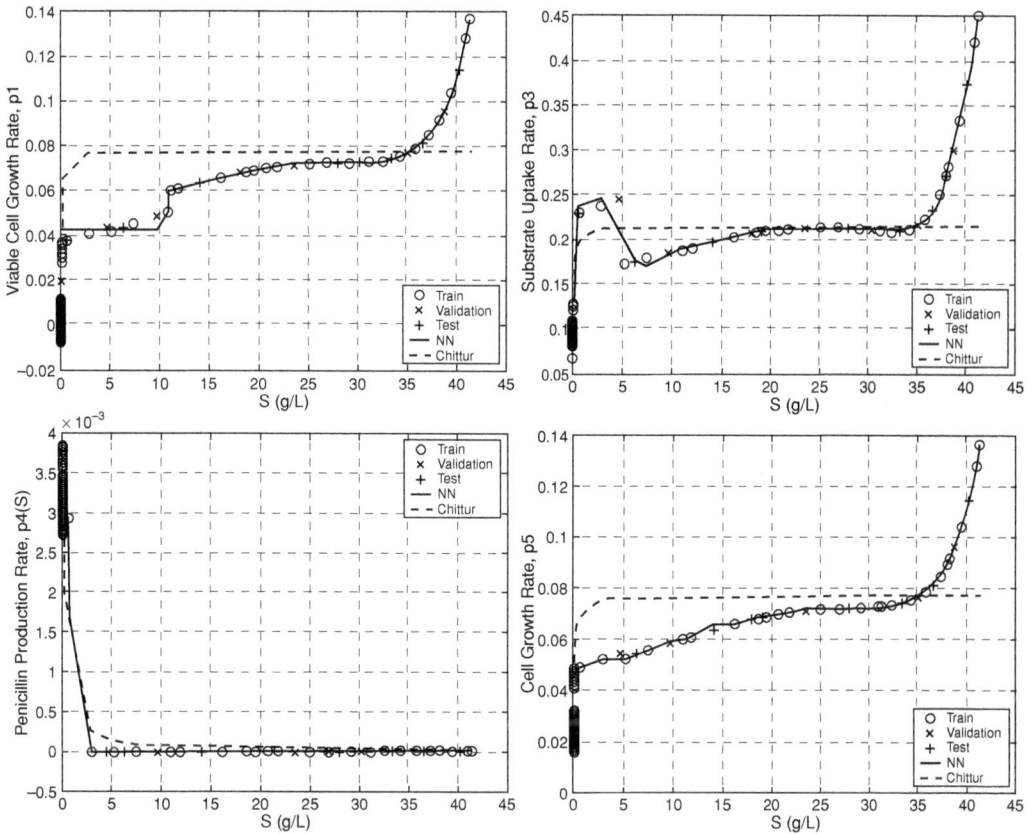

Figure 7.7. Spline fit data for neural network rates.

The calculated rates are summarized in Figure 7.7. Neural networks were trained with the experimental data of the rates (Figure 7.7) versus substrate concentration. Neural networks with the lowest overall mean squared error were chosen, demonstrating the best agreement of the neural network model with the experimental data. Figure 7.7 also shows a good neural network prediction plotted along with the training distributions.

REFERENCES

1. Bulsari, A., Saxen, B., and Saxen, H. 1993. Feedforward neural network for bioreactor control, in *International Workshop on Artificial Neural Networks*, p. 682. Springer.
2. Chaudhuri, B., and Modak, J. 1998. Optimization of fed-batch bioreactor using neural network model. *Bioprocess Engineering* 19: 71–79.
3. Park, S., and Ramirez, W. F. 1998. Optimal production of secreted protein in fed-batch reactors. *American Institute of Chemical Engineers Journal* 34: 1550–1558.
4. Patkar, A., and Seo, J.-H. 1992. Fermentation kinetics of recombinant yeast in batch and fed batch cultures. *Biotechnology and Bioengineering* 40: 103–109.
5. Chen, Q., and Weigand, W. 1992. Feedback optimization of fed-batch bioreactors via neural net learning. *IEEE Conference on Control Applications* 1: 72–77.
6. Syu, M.-J., and Hou, C.-L. 1993. Neural network modeling of batch cell growth pattern. *Biotechnology and Bioengineering* 42: 376–380.

7. Ignova, M., Paul, G., Glassey, J., Ward, A., Montague, G., Thomas, C., and Karim, M. 1996. Towards intelligent process supervision: Industrial penicillin fermentation case study. *Computers in Chemical Engineering* 20: S545–S550.
8. Krothapally, M., and Pulanki, S. 1999. A neural network strategy for end-point optimization of batch processes. *ISA Transactions* 38: 383–396.
9. Karim, M. N., Yoshida, T, Rivera, S. L., Sucedo, V. M., Eikens, B., and Oh, G.-S. 1997. Global and local neural network models in biotechnology: Application to different cultivation processes. *Journal of Fermentation and Bioengineering* 83: 1–11.
10. Saxon, B., and Saxen, H. 1996. A neural-network based model of bioreaction kinetics. *Canadian Journal of Chemical Engineering* 74: 124–131.
11. Di Massimo, C., Montague, G., Willis, M., Tham, M., and Morris, A. 1992. Towards improved penicillin fermentation via artificial neural networks. *Computers in Chemical Engineering* 16: 283–291.
12. Thompson, M. L., and Kramer, M. A. 1993. A hybrid modeling methodology to combine prior knowledge and neural networks. *Proceedings of the 1993 International Joint Conference on Neural Networks* 3: 2987–2990.
13. Bajpai, R., and Reu, M. 1981. Evaluation of feeding strategies in carbon-regulated secondary metabolic production through mathematical modeling. *Biotechnology and Bioengineering* 23: 717–738.
14. Chittur, V. K. 1989. Modeling and optimization of the fed-batch penicillin fermentation. PhD dissertation, Purdue University.

8 Specific Rate Determination

The equation-based models presented in Chapter 6 require specific rates of cell growth, substrate consumption, intermediate formation, and product formation. In this chapter, we will review the traditional methods of determining these specific rates and present a method of determining without taking time derivatives the net specific rates utilizing fed-batch cultures. Traditional methods of generating experimental data involve using primarily shake flasks (batch reactors) and, secondarily, continuous flow reactors. A new method utilizes fed-batch cultures with constant feed rates and does not require differentiation of experimental data.

It is best to generate experimental data using the type of operation that is being contemplated; that is, if a batch culture is the ultimate mode of operation, it is best to generate specific rate data using batch cultures, and if a fed-batch operation is to be used, then it is best to use fed-batch operation to generate specific rate data. Compared to chemical reactions, microbial kinetics are complex and time variant so that frequently, rate data obtained at steady state operations may not apply well to dynamic operations.

8.1 Determination of Specific Rates by Classical Methods

Classical methods utilize shake flasks (batch reactors) and continuous reactors to generate kinetic data. A differential method is the dominant practice, although an integral method can be used to analyze batch data. The former approach takes the time derivative of the concentration profile and is therefore subject to error in numerical differentiation of data that are corrupted with noise, while the integral method assumes a certain form of rate expression with undetermined rate constants, integrates the mass balance equations with the chosen rate expression, and fits the integrated form to the batch data. Use of shake flasks with various initial substrate concentrations and measuring the substrate and cell concentrations during the *exponential growth phase* are very similar to the idea of initial rates in chemical kinetic studies. Cell growth goes through a series of phases: a lag phase, a preexponential phase, an exponential phase, a stationary phase, and a decay phase. Therefore, the data should be taken during the exponential growth phase in a batch reactor or in shake flasks.

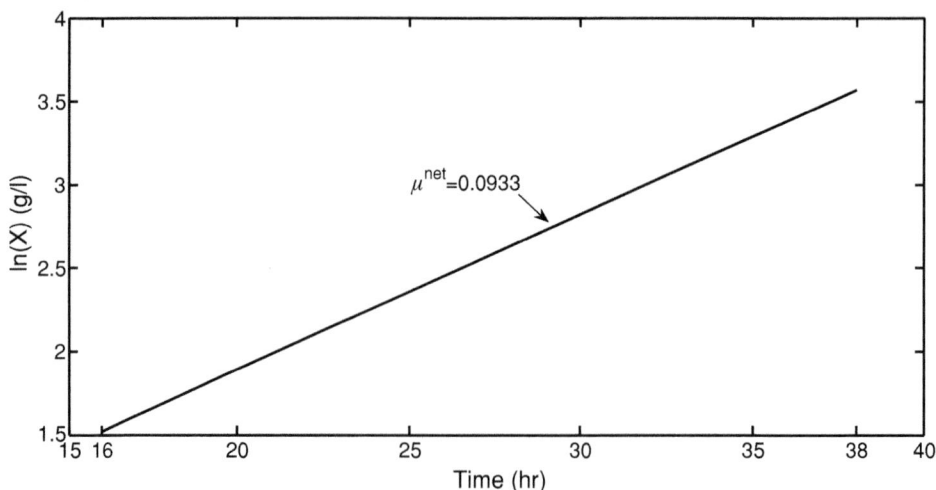

Figure 8.1. A plot of ln X versus time, batch culture.

8.1.1 Specific Rates by Shake Flask Cultures

Growth rates are measured in batch cultures or shake flasks with various initial substrate concentrations. After inoculating with a fixed but small amount of inoculum, the cell, substrate, and product concentrations are monitored at various times. Because neither addition nor withdrawal is made during the course of experiment, shake-flask culture represents a constant-volume batch operation.

8.1.1.1 Specific Cell Growth Rate, μ

The specific growth rate of cells can be determined from the constant-volume batch cell mass balance equation:

$$\frac{dX}{dt} = \mu X \quad \Rightarrow \quad \frac{1}{X}\frac{dX}{dt} = \frac{d\ln X}{dt} = \mu \tag{8.1}$$

according to which the specific growth rate is equal to the slope of the plot of $\ln(\log_e)$ of cell concentration X versus time. Therefore, in the exponential growth phase, the growth is balanced so that μ is constant, and therefore, this equation can be integrated between the time interval $[t_1, t_2]$ to yield

$$\ln X_2 = \ln X_1 + [\mu(S_1)](t_2 - t_1) \tag{8.2}$$

Thus, a plot of ln X versus time (Figure 8.1) should yield a straight line with a slope equal to the specific cell growth rate:

$$[\mu(S_1)] = (\ln X_2 - \ln X_1)/(t_2 - t_1) = \Delta \ln X/\Delta t \tag{8.3}$$

Repetition of runs at various initial substrate concentrations $S_{0i}, i = 1, 2, 3, \ldots, n$ should yield the specific growth rates at various substrate concentrations (Table 8.1), which are then correlated with the substrate concentration to obtain the specific growth rate as a function of substrate concentration, $\mu(S)$, using various functional forms of $\mu(S)$ that were presented in Chapter 6 and picking one that gives the best statistical fit. Conversely, if the specific growth rate is suspected to be a function of both the substrate and product concentrations, $\mu(S, P)$, then one has to record

Table 8.1. *Specific cell growth rate data*

Concentrations of samples taken at various times									
Time	t_1	t_2	t_3	t_4	t_5	.	.	.	t_n
Initial substrate concentration	S_1	S_2	S_3	S_4	S_5	.	.	.	S_n
Product concentration	P_1	P_2	P_3	P_4	P_5	.	.	.	P_n
Specific growth rate	μ_1	μ_2	μ_3	μ_4	μ_5	.	.	.	μ_n

the product concentration as well and correlate the specific cell growth with both concentrations of substrate and product.

8.1.1.2 Specific Substrate Consumption Rate, σ

The specific substrate consumption rate can be determined from experimental data using the substrate balance equation of a constant-volume batch reactor and solving for σ (grams of substrate consumed/unit time/unit gram of cell/liter of cells):

$$dS/dt = \sigma X \Rightarrow \sigma = (dS/dt)(1/X) \tag{8.4}$$

Thus, to obtain σ, it is necessary to differentiate with respect to time the substrate concentration profile, $S(t)$. Because experimental data are available at discrete times, it is necessary to apply an approximation technique to estimate time derivatives of substrate concentrations at discrete times. There are a number of methods for estimating derivatives dS/dt from discrete experimental data: (1) a *graphical* technique of equal-area differentiation of finite differences, $\Delta S/\Delta t$; (2) use of a *numerical* differentiation formula; and (3) *polynomial fit* and analytical differentiation of the fitted polynomial.

8.1.1.2.1. EQUAL-AREA DIFFERENTIATION. In this graphical method, the finite differences $\Delta X / \Delta t$ and $\Delta S / \Delta t$ obtained from discrete experimental data at various times are plotted against time, and the time derivatives dX/dt and dS/dt are obtained using equal-area differentiation. This method is illustrated in the appendix to this chapter.

8.1.1.2.2. NUMERICAL DIFFERENTIATION FORMULA. Numerical differentiation formulas can be applied when experimental data points are equally spaced with respect to time, that is, $t_i - t_{i-1} = t_{i+1} - t_i = \Delta t$; the three-point differentiation formulas for derivatives are as follows:

$$\text{Initial point}\left(\frac{dS}{dt}\right)_{t_0} = \frac{-3S_0 + 4S_1 - S_2}{2\Delta t}$$

$$\text{Interior points}\left(\frac{dS}{dt}\right)_{t_i} = \frac{S_{i+1} - S_{i-1}}{2\Delta t} \tag{8.5}$$

$$\text{End point}\left(\frac{dS}{dt}\right)_{t_n} = \frac{S_{n-2} - 4S_{n-1} + 3S_n}{2\Delta t}$$

Repetition of runs at various initial substrate concentrations yields the specific rates of growth and substrate consumption versus the substrate concentrations. The data

Table 8.2. *Equally spaced experimental data*

Time	t_0	$t_1 = t_0 + \Delta t$	$t_2 = t_0 + 2\Delta t$	$t_3 = t_0 + 3\Delta t$				$t_n = t_0 + n\Delta t$
S conc.	S_0	S_1	S_2	S_3	S_4	.	.	S_n
X conc.	X_0	X_1	X_2	X_3	X_4	.	.	X_n

thus generated, along with the calculated specific growth rate and substrate consumption rates, are summarized in Table 8.2. Then, the rates are correlated with the substrate concentration using various specific rate forms covered in Chapter 6, and the one that gives the best statistical fit is chosen as the rate model.

8.1.1.2.3. POLYNOMIAL FIT AND ANALYTICAL DIFFERENTIATION. The discrete experimental data are fitted to an nth-order polynomial in time by minimizing the error criterion between the experimental values and the predicted values of the polynomial,

$$S(t) = a_0 + a_1 t + a_2 t^2 + a_3 t^3 + \cdots + a_n t^n$$
$$X(t) = b_0 + b_1 t + b_2 t^2 + b_3 t^3 + \cdots + b_n t^n$$
(8.6)

and the polynomial so obtained is analytically differentiated to yield the time derivatives:

$$\frac{dS}{dt} = a_1 + 2a_2 t^1 + 3a_3 t^2 + \cdots + na_n t^{n-1}$$
$$\frac{dX}{dt} = b_1 + 2b_2 t^1 + 3b_3 t^2 + \cdots + nb_n t^{n-1}$$
(8.7)

As in the case of specific growth rate, the specific substrate consumption rate may be a function of concentrations of both substrate and product. In this situation, the specific rate must be correlated with the concentrations of both substrate and product.

8.1.1.3 Specific Product Formation Rate, π

Similarly, the specific product formation rate can be obtained with the aid of the product balance equation:

$$dP/dt = \pi X \quad \Rightarrow \pi = (dP/dt)/X$$
(8.8)

The time derivative of product concentration divided by cell concentration is the specific product formation rate. Thus, differentiation is required for the specific rate. Any one of the three methods discussed in Section 8.1.2 – the graphical, numerical, or analytical method – may be employed to obtain the necessary derivatives.

If the specific product formation rate, $\pi (S, P)$, is a function of not only substrate concentration but also product concentration, then row 3 in Table 8.3 must be correlated with the concentrations of both substrate (row 1) and product (row 2) using the various forms reported in Chapter 6, and the one that gives the best statistical fit can be adopted as the model.

Table 8.3. *Specific product formation rate data*

Time	t_1	t_2	t_3	t_4	·	·	·	t_{n-1}	t_n
Initial substrate concentration	S_1	S_2	S_3	S_4			·	S_{n-1}	S_n
Product concentration	P_1	P_2	P_3	P_4			·	P_{n-1}	P_n
Cell concentration	X_1	X_2	X_3	X_4			·	X_{n-1}	X_n
Specific product formation rate (Eq. (8.6))	π_1	π_2	π_3	π_4			·	π_{n-1}	π_n

8.1.2 Specific Rates by Batch Cultures

Specific rate data may be obtained under controlled conditions using batch reactors. Under a controlled environment of constant pH, dissolved oxygen, and temperature, measurements are more reproducible and reliable than shake-flask data. Ideally, if one can set up a number of small batch reactors under a central digital computer control system equipped with an automatic sampling line connected to an autoanalyzer capable of measuring cell density and concentrations of substrate, intermediates, and product, it would be ideal to obtain kinetic information at various times and expedite data analysis.

The types of experiments needed are very similar to shake-flask experiments, except one makes a fewer number of runs in batch experiments but over a long period of time. Experiments are carried out with different initial conditions (limiting substrate concentration and inoculum size). Estimations of the specific rates can be made by two different methods: differential and integral methods. The differential methods involve taking the derivative of the concentration profiles and then calculating the specific rates using batch mass balance equations, while the integral method relies on choosing a number of suspected forms of the specific rates with yet-to-be determined parameters and integrating the batch mass balance equations with arbitrary parameters and then fitting the integrated forms of mass balance equations to the experimental concentration profiles to obtain the best parameters.

8.1.2.1 Differential Method

This method begins with constant-volume batch mass balance equations:

$$\frac{dX}{dt} = \mu(S, P)X \quad X(0) = X_0 \tag{8.9}$$

$$\frac{dS}{dt} = -\sigma(S, P)X \quad S(0) = S_0 \tag{8.10}$$

$$\frac{dP}{dt} = \pi(S, P)X \quad P(0) = P_0 = 0 \tag{8.11}$$

These equations show that the specific rates of cell growth, substrate consumption, and product formation, μ, σ, and π, are obtained from the time derivatives of concentrations of cells, substrate, and product by dividing by the cell concentration X:

$$\mu = \frac{1}{X}\frac{dX}{dt}, \quad \sigma = \frac{1}{X}\frac{dS}{dt}, \quad \pi = \frac{1}{X}\frac{dP}{dt} \tag{8.12}$$

Table 8.4. *Specific rate data from batch cultures*

Sample times	t_1	t_2	t_3	t_4	t_5	t_n
Cell concentrations	X_1	X_2	X_3	X_4	X_5	X_n
Substrate concentrations	S_1	S_2	S_3	S_4	S_5	S_n
Product concentrations	P_5	P_5	P_5	P_5	P_5	P_n
Finite differences	$(\Delta X/\Delta t)_1$	$(\Delta X/\Delta t)_2$	$(\Delta X/\Delta t)_3$	$(\Delta X/\Delta t)_1$	$(\Delta X/\Delta t)_1$	$(dX/dt)_n$
	$(\Delta S/\Delta t)_1$	$(\Delta S/\Delta t)_2$	$(\Delta S/\Delta t)_3$	$(\Delta S/\Delta t)_4$	$(\Delta S/\Delta t)_5$	$(\Delta S/\Delta t)_n$
	$(\Delta P/\Delta t)_1$	$(\Delta P/\Delta t)_2$	$(\Delta P/\Delta t)_3$	$(\Delta P/\Delta t)_4$	$(\Delta P/\Delta t)_5$	$(\Delta P/\Delta t)_n$

Thus, we need to evaluate the time derivatives at various times and therefore the time profiles of X, S, and P from a set of batch runs. Once the time derivatives are obtained by numerical or graphical means, we can construct a table such as Table 8.4.

To obtain the time derivatives, one has to adopt a method of differentiating time profiles. As discussed in Section 8.1.2, the time derivatives, dX/dt, dS/dt, and dP/dt, of discrete experimental data can be obtained graphically from finite differences $\Delta X/\Delta t$, $\Delta S/\Delta t$, and $\Delta P/\Delta t$ by the equal-area differentiation, numerically using the differentiation formulas (Eq. (8.5)) or analytically fitting the experimental data to an nth-order polynomial and then differentiating it.

Once the derivatives are evaluated and the specific rates are calculated, the remaining task is to correlate the specific rates, μ_i, σ_i, and $\pi_i (i = 1, 2, 3, \ldots, n)$, with S, S, and P or with S, P, and X using the functional forms that were given in Chapter 6 and then to pick the one that gives the best statistical fit.

Example 8.E.1 Estimation of Specific Rates by Differential Methods

Batch cultures were carried out with a microorganism that is known to follow a Monod-type specific cell growth rate, and discrete experimental data of cell and substrate concentrations are given in Table 8.E.1.1. Estimate μ and σ using (1) the equal-area graphical differentiation, (2) numerical differentiation formulas, and (3) polynomial fit followed by analytical differentiation:

1. Equal-area graphical differentiation. The first three columns in Table 8.E.1.1 represent the data to be used in this example. As shown in Table 8.E.1.1, finite differences obtained are used as the estimated derivatives, which are plotted against time as in Figure 8.E.1.1, and visually, curves are drawn for each time increment using the equal-area (above and below the curve) concept. Then, specific rates are determined using Eq. (8.12). The results are given in column 4 for graphical determination of specific growth rate and in column 7 for graphical determination of specific substrate consumption.
2. Numerical differentiation formula. The specific rates were also determined by applying the numerical differentiation formula of Eq. (8.5) to the data of Table 8.E.1.1. The time derivatives are estimated from the data by applying the three-point differentiation formulas of Eq. (8.5), and the specific rates were determined using Eqs. (8.3) and (8.4). The results are listed in Table 8.E.1.1, in column 5 for numerical determination of specific growth rates and in column 8 for numerical determination of specific substrate consumption rates.

Table 8.E.1.1. *Estimation of specific rates of cell growth and substrate consumption by various methods*

t	S	X	Graphical μ	Numerical μ	Polynomial μ	Graphical σ	Numerical σ	Polynomial σ
0	4	0.8	0.43	0.338	0.267	0.67	0.563	0.511
1	3.5	1.1	0.37	0.315	0.327	0.58	0.496	0.509
2	2.9	1.5	0.3	0.263	0.308	0.47	0.400	0.440
3	2.3	1.9	0.29	0.263	0.269	0.38	0.342	0.367
4	1.6	2.5	0.28	0.240	0.201	0.38	0.320	0.269
5	0.7	3.1	0.15	0.145	0.144	0.25	0.210	0.190
6	0.3	3.4	0.08	0.074	0.100	0.11	0.088	0.131
7	0.1	3.6	0.02	0.028	0.050	0.04	0.038	0.067
8	0.03	3.6	0	0.000	0.000	0.03	0.014	0.000
9	0	3.6	0	0.000	0.000	0.01	0.003	0.000

Michaelis–Menten Equation

$$\mu = \frac{\mu_{max}S}{K_s + S}, \quad \sigma = \frac{R_{max}S}{K_s + S}, \quad R_{max} = \frac{\mu_{max}}{Y_{X/S}}$$

μ_{max} or R_{max}		0.62		0.44	0.39	1.16	0.95	0.89
K_s		2.30		1.49	1.08	3.56	3.31	2.94

3. Polynomial fit and analytical differentiation. The data in columns 1–3 were fitted to a polynomial function, such as Eq. (8.6), and then the fitted polynomial functions are differentiated (Eq. (8.7)) to obtain the time derivatives (Eq.(8.7)); the specific rates were evaluated using Eqs. (8.3) and (8.4). The results are listed in column 6 for polynomial determination of specific growth rates and in column 9 for polynomial determination of specific substrate consumption rates.

The estimated specific cell growth rates are plotted in Figure 8.E.1.1, and the specific substrate consumption rates are plotted in Figure 8.E.1.2 against substrate

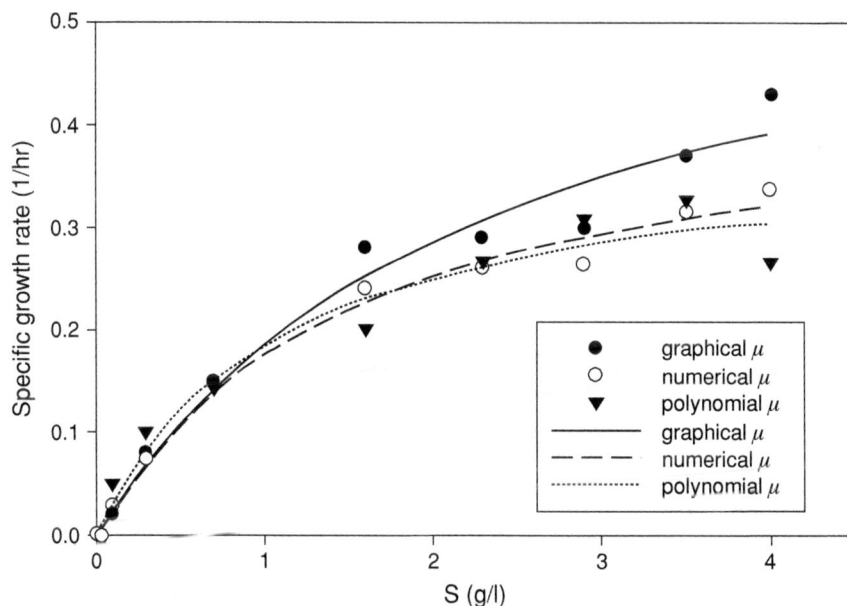

Figure 8.E.1.1. Specific cell growth rates by three different methods.

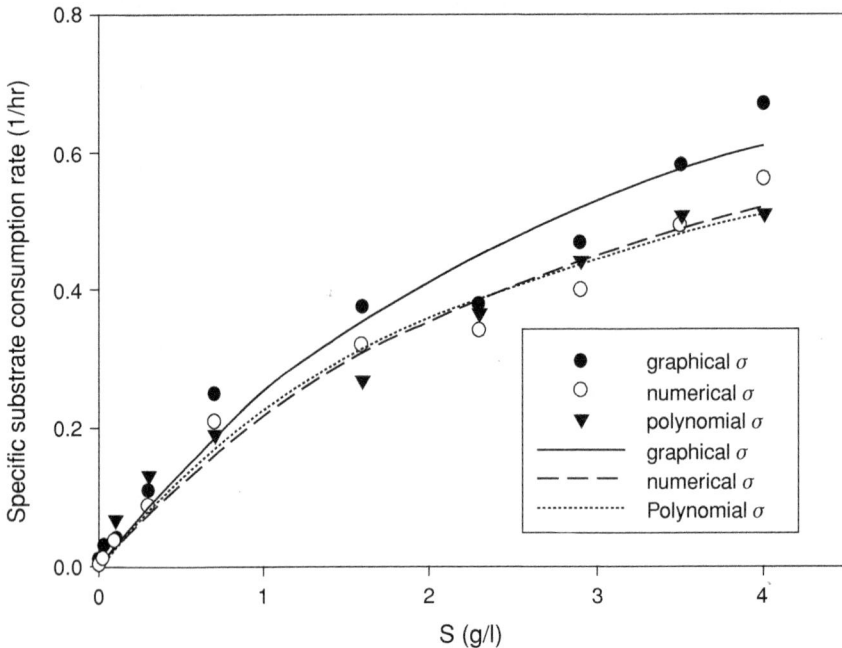

Figure 8.E.1.2. Specific substrate consumption rates by three different methods.

concentrations. The lines (solid, dashed, and dotted) represent the regression fits of the estimated specific rates. It is apparent that the three methods used to estimate the specific rates lead to wide variation. This is expected as the data are differentiated graphically and numerically. The numerical and polynomial estimations are in better agreement than the graphical estimations. The graphical estimation tends to yield higher values of specific rates than the other two. Thus, it appears that the numerical and polynomial fit estimations are superior to the eyeball graphical estimation.

8.1.2.2 Integral Method

As in chemical kinetics, one can use an integral method in which it is not necessary to take time derivatives of various species concentrations, but one has to pick functional forms (models) of specific rates μ, σ, and π, such as those given Chapter 6, with yet-to-be-determined parameters in the model. With a set of initially guessed parameter values in the specific rates, the batch mass balance equations are integrated, and the guessed constants are successively optimized by minimizing a form of error between the experimental data and the integrated equations.

8.1.3 Specific Rates by Continuous Cultures

Specific growth rate can also be determined using a continuous flow reactor such as a chemostat. Because of steady state operations, this mode of generating rate information takes time, especially at low dilution rates. Making use of the continuous reactor treatment in Chapter 2, we begin by writing steady state, constant-volume mass balance equations for cell, substrate, and product:

$$0 - FX + \mu XV = 0 \quad \Rightarrow \mu = F/V = D \tag{8.13}$$

Table 8.5. *Specific rate determination from steady state continuous reactor operation with feed substrate concentration S_F*

	1	2	3	4	5	6	7
Run no.	$D=F/V$	X	S	P	$\mu = D$	$\sigma = D(S_F - S)/X$	$\pi = DP/X$
1	$D_1 = F_1/V$	X_1	S_1	P_1	μ_1	σ_1	π_1
2	D_2	X_2	S_2	P_2	μ_2	σ_2	π_2
3	D_3	X_3	S_3	P_3	μ_3	σ_3	π_3
.	.	.	.	.	.	.	.
.	.	.	.	.	.	.	.
.	.	.	.	.	.	.	.
n	.	X_n	S_n	P_n	μ_n	σ_n	π_n

$$FS_F - FS - \sigma XV = 0 \quad \Rightarrow \sigma = \frac{F(S_F - S)}{XV} \tag{8.14}$$

$$0 - FP + \pi XV = 0 \quad \Rightarrow \pi = DP/X \tag{8.15}$$

According to these algebraic equations, the specific growth rate is simply equal to the dilution rate, and the specific substrate consumption rate is equal to the dilution rate times the difference between the feed and effluent substrate concentrations divided by the cell concentration, while the specific product formation rate is obtained by multiplying the product concentration by the dilution rate and dividing by the cell concentration. Therefore, one can make a series of runs by varying the dilution rate; wait for a steady state; and measure the concentrations of cells, substrate, and product. From these measurements, we can make a table of dilution rates versus concentrations of cells, substrate, and product, as shown in columns 1–4 of Table 8.5.

Once we obtain the entries in columns 1–4, we can calculate the specific rates, $\mu, \sigma,$ and $\pi,$ using Eqs. (8.13)–(8.15), thus completing columns 5, 6, and 7. The remaining task is to correlate these calculated specific rates with the concentrations of various species. For example, if it is known that the specific rate is a function of only substrate concentration, then columns 5, 6, and 7 are correlated with the substrate concentration, column 3 only. If any of the specific rates is suspected to be a function of all three species concentrations, then it must be correlated with columns 1, 2, and 3. In the absence of *prior* knowledge, one can obtain correlations with S alone, with S and X, or with S, X, and P and choose the one with the best correlation coefficient. It would be prudent to make additional runs to ensure that the particular chosen correlation holds up, for example, making runs with different feed concentrations.

It should be noted that although it takes a considerable amount of time to obtain steady state data using a continuous reactor, the rate data are obtained from algebraic equations without differentiation of experimental data. Reiterating what was stated earlier in the introduction, if the use of rate expressions is aimed at improving the operation of continuous reactors, it is best to use the data obtained from continuous reactors. If the intended use is a batch reactor, data should be obtained using a batch reactor.

8.1.4 Specific Rates by Fed-Batch Cultures

In Chapter 4, rate data generation was considered as one of the utilities of fed-batch cultures. Experimental data from fed-batch operation can be used to obtain specific rate data. One way is to utilize directly the mass balance equations. It appears from the mass balance equations given subsequently that differentiation of the time profiles of the total cell mass, substrate, and product concentrations is necessary:

$$\frac{d(XV)}{dt} = \mu XV \quad X(0)V(0) = X_0 V_0 \tag{8.16}$$

$$\frac{d(SV)}{dt} = FS_F - \sigma XV = FS_F - \mu XV/Y_{X/S} \quad S(0)V(0) = S_0 V_0 \tag{8.17}$$

$$\frac{d(PV)}{dt} = \pi XV \quad P(0)V(0) = P_0 V_0 \tag{8.18}$$

$$\frac{dV}{dt} = F \quad V(0) = V_0 \tag{8.19}$$

Solving Eqs. (8.16)–(8.18) for the specific rates, we obtain the following:

$$\mu = \frac{1}{XV}\frac{d(XV)}{dt} = \frac{d(\ln XV)}{dt} \tag{8.20}$$

$$\sigma = -\frac{1}{XV}\frac{d(SV)}{dt} + \frac{FS_F}{XV} \tag{8.21}$$

$$\pi = \frac{1}{XV}\frac{d(PV)}{dt} \tag{8.22}$$

According to these equations, we must obtain the time derivatives of the total cell mass, substrate, and product and then divide them by the total amount of cell mass. The specific rates of cell growth and product formation are similar to those of the variable volume batch, while the specific substrate consumption rate is different from that of the batch because of the substrate feed term in the fed-batch culture. It is therefore obvious that the procedure required to estimate the specific rates is also similar to that which is used for batch culture. Therefore, the procedure is not repeated here. It suffices to state that differentiation of experimental data is a source for large errors.

8.2 A New Method of Determining Specific Rates Using Fed-Batch Cultures

The approach adapted here is an application of the approach proposed by Lee and Yau[1] for a single chemical reaction, which does not require differentiation of experimental discrete data. This method has been shown to yield more accurate rate data for slow and fast reactions than the continuous bioreactor method. In addition, fed-batch cultures with constant feed rates allow determination of rate data over the entire range of conversion.

8.2.1 Constant-Feed Fed-Batch Cultures

The idea behind this approach comes from the mass balance equations for fed-batch cultures, which are

$$\frac{d(XV)}{dt} = \frac{dX}{dt}V + X\frac{dV}{dt} = \frac{dX}{dt}V + XF = \mu XV \quad X(0)V(0) = X_0 V_0 \quad (8.23)$$

$$\frac{d(SV)}{dt} = \frac{dS}{dt}V + S\frac{dV}{dt} = \frac{dS}{dt}V + SF = FS_F - \sigma XV \quad S(0)V(0) = S_0 V_0 \quad (8.24)$$

$$\frac{d(PV)}{dt} = \frac{dP}{dt}V + P\frac{dV}{dt} = \frac{dP}{dt}V + PF = \pi XV \quad P(0)V(0) = P_0 V_0 \quad (8.25)$$

$$\frac{dV}{dt} = F \quad V(0) = V_0 \quad (8.26)$$

We note that in Eqs. (8.23)–(8.26), if measurements are taken at times at which the time derivatives vanish, that is, $dX/dt = 0$, $dS/dt = 0$, and $dP/dt = 0$, we have algebraic mass balance equations:

$$\frac{dX}{dt}V + XF = XF = \mu XV \quad \Rightarrow \mu = \left.\frac{F}{V}\right|_{\frac{dX}{dt}=0} \quad (8.27)$$

$$\frac{dS}{dt}V + SF = FS_F - \sigma XV \quad \Rightarrow \sigma = \left.\frac{F(S_F - S)}{XV}\right|_{\frac{dS}{dt}=0} \quad (8.28)$$

$$\frac{dP}{dt}V + PF = PF = \pi XV \quad \Rightarrow \pi = \left.\frac{PF}{XV}\right|_{\frac{dP}{dt}=0} \quad (8.29)$$

We note that the specific rate expressions for fed-batch culture, Eqs. (8.27)–(8.29), appear identical to those of the steady state continuous culture, Eqs. (8.13)–(8.15). However, there are many differences. First of all, Eqs. (8.27)–(8.29) hold only at the instant of time at which the time derivatives vanish, $dX/dt = 0$, $dS/dt = 0$, and $dP/dt = 0$, whereas those of the continuous culture hold for all times once a steady state is reached. The volume and the feed rate for the fed-batch vary with time, $V(t)$ and $F(t)$, whereas those in the continuous culture remain time invariant.

For a constant feed rate, $F = F_c$ and $V = V_0 + F_c t$, which is substituted into Eqs. (8.27)–(8.29) to obtain

$$\mu = \frac{F}{V} = \frac{F_c}{V_0 + F_c t_1} = \frac{1}{(V_0/F_c) + t_1} \quad (8.30)$$

$$\sigma = \frac{F_c(S_F - S_{t=t_2})}{X_{t=t_2}(V_0 + F_c t_2)} = \frac{(S_F - S_{t=t_2})}{X_{t=t_2}[(V_0/F_c) + t_2]} \quad (8.31)$$

$$\pi = \frac{F_c P_{t=t_3}}{X_{t=t_3}[V_0 + F_c t_3]} = \frac{P_{t=t_3}}{X_{t=t_3}[(V_0/F_c) + t_3]} \quad (8.32)$$

Table 8.6. *Specific rate data from constant-feed fed-batch operations*

	1	2	3		4	5	6	7	8	9	10	11		12	13	14	15	16
Run no.	$D_{0i}V_0/F_C$	t_{1i}	μ_i		S_{1i}	P_{1i}	X_{1i}	t_{2i}	X_{2i}	S_{2i}	P_{2i}	σ_i		t_{3i}	P_{3i}	X_{3i}	S_{3i}	π_i
			Eq. (8.33)									Eq. (8.34)						Eq. (8.35)
1	D_{01}	t_{11}	μ_1		S_{11}	P_{11}	X_{11}	t_{21}	X_{21}	S_{21}	P_{21}	σ_1		t_{31}	P_{31}	X_{31}	S_{31}	π_1
2	D_{02}	t_{12}	μ_2		S_{12}	P_{12}	X_{12}	t_{22}	X_{22}	S_{22}	P_{22}	σ_2		t_{32}	P_{32}	X_{32}	S_{32}	π_2
3	D_{03}	t_{13}	μ_3		S_{13}	P_{13}	X_{13}	t_{23}	X_{23}	S_{23}	P_{23}	σ_3		t_{33}	P_{33}	X_{33}	S_{33}	π_3
n	D_{0n}	t_{1n}	μ_n		S_{1n}	P_{1n}	X_{1n}	t_{2n}	X_{2n}	S_{2n}	P_{2n}	σ_n		t_{3n}	P_{3n}	X_{3n}	S_{3n}	π_n

Equations (8.30)–(8.32) are algebraic equations involving the three times t_1, t_2, and t_3 at which the derivatives vanish, that is, $dX/dt_{t=t_1} = 0$, $dS/dt_{t=t_2} = 0$, and $dP/dt_{t=t_3} = 0$, respectively, the concentrations of X, S, and P, and the ratio of the initial culture volume to the constant feed rate. If we can locate the times at which the concentrations X, S, and P go through extreme points, then the time derivatives would vanish, and the specific rates are obtained from the concentrations of cells, substrates, and product concentrations at their extreme points and the constant feed rate and culture volume. We locate the times at which the concentrations of cell, substrate, and product reach their extreme points and denote them by t_1, t_2, and t_3, respectively, that is, $(dX/dt)_{t_1} = 0$, $(dS/dt)_{t_2} = 0$, and $(dP/dt)_{t_3} = 0$. The corresponding concentrations of cell, substrate, and product are noted as $X_1 = X(t_1)$, $S_2 = S(t_2)$, and $P_3 = P(t_3)$, respectively. Then, the specific rates are expressed as

$$\mu|_{X=X(t_1)} = \mu_1 = \frac{1}{(V_0/F_c) + t_1} \tag{8.33}$$

$$\sigma|_{S=S(t_2)} = \sigma_2 = \frac{(S_F - S_2)}{X_2[(V_0/F_c) + t_2]} \tag{8.34}$$

$$\pi|_{P=P(t_3)} = \pi_3 = \frac{P_3}{X_3[(V_0/F_c) + t_3]} \tag{8.35}$$

It is apparent that if we can locate the times t_1, t_2, and t_3 and the corresponding values of concentrations, then the net specific rates, μ, σ, and π, can be calculated algebraically using Eqs. (8.33)–(8.35). No differentiation of experimental data is necessary. The experimental data needed are illustrated in Table 8.6. It is apparent that these times, t_1, t_2, and t_3, need not be distinct, that is, the zero derivatives, some of which can occur at the same time. We also note from Eq. (8.33) that the peak or valley $(dX/dt = 0)$ must be unique, that is, there should be only one, whereas Eqs. (8.34) and (8.35) suggest that the peaks or valleys in S and P are not unique. There can be a number of peaks and/or valleys.

Example 8.E.2 Monod Model with Constant Cell Yield Coefficient

For the purpose of illustrating the preceding new method of determining specific rates from constant feed rate fed-batch reactor (FBR) data, a Monod model $\mu = 0.15S/(0.2 + S)$, a yield coefficient of 0.8, a feed substrate concentration of 10 g/L, a reactor volume of 1 L, and the initial conditions $[XV, SV, V]_{t=0} = [0.5, 0.05, 0.5]$ are used. The simulated data are generated by integrating the mass balance equations

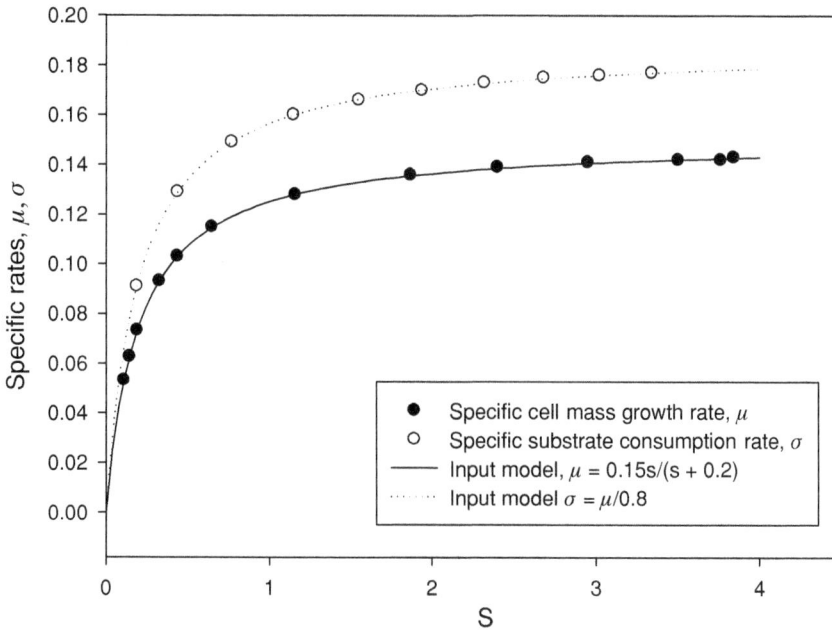

Figure 8.E.2.1. Estimated specific rates μ *and* σ from constant-feed FBR: Monod model with constant cell yield coefficient.

(8.23), (8.24), and (8.26). No experimental errors were added. The feed rate was varied from 6 mL/hr to 121 mL/hr. The times (t_{1i} and t_{2i}) of the peaks and valleys of cell concentrations ($dX/dt = 0$) and substrate concentrations ($dS/dt = 0$) are identified and recorded in columns 3 and 7, respectively, of Table 8.E.2.1. The substrate (S_{1i}) and cell (X_{1i}) concentrations at the peak and valley times of cell concentration are recorded in columns 4 and 5, respectively. Likewise, the substrate (S_{2i}) and cell (X_{2i}) concentrations at the peak and valley times of substrate concentrations are recorded in columns 8 and 9, respectively. Then, the specific rates of growth rate (column 6) and substrate consumption (column 10) are calculated using Eqs. (8.33) and (8.34), respectively.

Plots of specific rates, μ and σ, against substrate concentration are shown in Figure 8.E.2.1. The figure shows that the peaks and valleys at which the derivatives vanish, $(dX/dt) = 0$ and $(dS/dt) = 0$, appear at different times or at a same time at a constant feed rate. An automated device that determines the peak and valley would be very useful in this case.

Example 8.E.3 Substrate-Inhibited Model with Constant Cell Yield Coefficient

A substrate inhibition model $\mu = S/(0.5S^2 + S + 0.03)$, $Y_{X/S} = 0.3$, is used to generate the necessary data. The feed substrate concentration is $S_F = 10$ g/L, the reactor volume is 1 L, and the initial conditions are $[XV, SV, V]_{t=0} = [0.1\text{ g}, 0.01\text{ g}, 0.1\text{ L}]$:

$$\mu = 0.7S/(0.2S^2 + S + 0.5); \quad Y_{X/S} = 0.3; \ S_F = 10\text{g/L}; \ t_f = 5\,\text{hr};$$

$$F = 0.005\text{L/hr incrementing } 0.005 \text{ up to } F = 0.1\text{ L/hr};$$

$$D = V_0/F = 0.1/F\,\text{hr}, t \text{ in hr}$$

Table 8.E.2.1. *Estimation of specific rates from constant-feed FBR: Monod model with constant-yield coefficient*

Run	1 F hr^{-1}	2 $D_{0i} = V_0/F_c$ hr	3 $t_{1i} = (dX/dt)_{t_{1i}}$ $= 0$ hr	4 $S_{1i} = S(t_{1i})$ g/L	5 $X_{1i} = X(t_{1i})$ g/L	6 μ_i hr^{-1} (Eq. (8.33))	7 $t_{2i} = (dS/dt)_{t_{2i}}$ $= 0$ hr	8 $S_{2i} = S(t_{2i})$ g/L	9 $X_{2i} = X(t_{2i})$ g/L	10 σ_i hr^{-1} (Eq. (8.34))
1	0.006	83.3					4.10	0.188	1.25	0.076
2	0.011	45.5					5.52	0.438	1.48	0.126
3	0.016	31.3					6.47	0.770	1.65	0.148
4	0.021	23.8					7.34	1.15	1.79	0.159
5	0.026	19.2	0.022	0.110	1.00	0.0520	7.98	1.55	1.87	0.166
6	0.031	16.1	0.081	0.144	1.00	0.0618	8.56	1.94	1.93	0.169
7	0.036	13.9	0.150	0.191	1.00	0.0712	9.20	2.32	1.98	0.168
8	0.041	12.2	0.330	0.327	0.997	0.0798	9.57	2.68	1.97	0.171
9	0.046	10.9	0.290	0.328	0.996	0.0814	9.73	3.02	1.93	0.175
10	0.051	9.80	0.390	0.437	0.990	0.0981	10	3.34	1.90	0.177
11	0.061	8.20	0.540	0.649	0.990	0.114				
12	0.071	7.04	0.950	1.16	0.977	0.125				
13	0.081	6.17	1.53	1.87	0.962	0.130				
14	0.091	5.50	1.91	2.40	0.942	0.135				
15	0.101	4.95	2.31	2.95	0.918	0.138				
16	0.111	4.50	2.73	3.50	0.892	0.138				
17	0.116	4.31	2.94	3.76	0.879	0.138				
18	0.121	4.13	2.91	3.84	0.865	0.142				

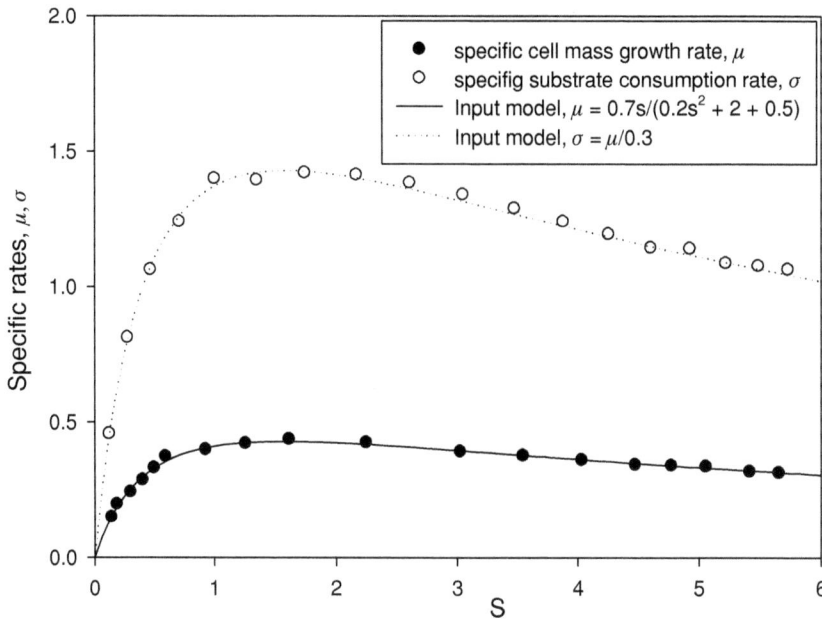

Figure 8.E.3.1. Estimated specific rates μ and σ from constant-feed FBR: substrate inhibition model with constant cell yield coefficient.

As in the case of Example 8.E.2, the data are generated by integrating the mass balance equations (8.2), (8.3), and (8.5) using the preceding information. The feed rate is varied from 5 to 100 mL/hr. The results are tabulated in Table 8.E.3.1. The times (t_{1i}) at which the cell concentration goes through the peaks and valleys $(dX/dt = 0)$ and the corresponding substrate concentrations (S_{1i}) and cell mass (X_{1i}) are recorded in columns 3, 4, and 5, respectively. The specific growth rates are calculated using Eq. (8.33) and recorded in column 6. Likewise, the times (t_{2i}) at which the substrate concentration goes through the peaks and valleys $(dS/dt = 0)$ and the corresponding substrate concentrations (S_{2i}) and cell mass (X_{2i}) are recorded in columns 7, 8, and 9, respectively. Then, the substrate consumption rates are calculated using Eq. (8.34) and tabulated in column 10.

The specific rates of growth (column 6, Table 8.E.3.1) and substrate consumption (column 10) are plotted against the corresponding substrate concentrations (column 4, Table 8.E.3.1) in Figure 8.E.3.1. It is clear that both the specific growth rates and specific substrate consumption rate determined from the constant feed rate fed-batch operation agree well with the original input data generated from the substrate-inhibited model. Because the data generated from the model are used without the addition of experimental error, the agreement is extremely good. In actual experimental evaluation of specific rates, it is anticipated that the agreement will be within the experimental error.

Example 8.E.4 Substrate Inhibition Model with Variable Cell Yield

The simulated data are generated using a substrate inhibition model with a variable-yield coefficient,

$$\mu = 0.7S/(0.2S^2 + S + 0.5); \quad Y = \frac{0.35(0.3S^2 + 1.5S + 0.7)}{(0.2S^2 + S + 0.5)}; \quad S_F = 10 \, \text{g/L}; \quad t_f = 5 \, \text{hr};$$

Table 8.E.3.1. *Estimation of specific rates from constant-feed FBR: Substrate inhibition model with constant-yield coefficient*

Run	1 $F\,\mathrm{hr}^{-1}$	2 $D_{0i} = V_0/F_c$ hr	3 $t_{1i} = (dX/dt)_{t_{1i}}$ $= 0$ hr	4 $S_{1i} = S(t_{1i})$ g/L	5 $X_{1i} = X(t_{1i})$ g/L	6 $\mu_i\,\mathrm{hr}^{-1}$ (Eq. (8.33))	7 $t_{2i} = (dS/dt)_{t_{2i}}$ $= 0$ hr	8 $S_{2i} = S(t_{2i})$ g/L	9 $X_{2i} = X(t_{2i})$ g/L	10 $\sigma_i\,\mathrm{hr}^{-1}$ (Eq. (8.34))
1	0.005	20					0.595	0.123	1.05	0.457
2	0.01	10					0.849	0.272	1.10	0.813
3	0.015	6.67	0.0406	0.142	0.999	0.149	1.06	0.464	1.16	1.06
4	0.02	5	0.0592	0.186	0.997	0.198	1.22	0.706	1.20	1.24
5	0.025	4	0.113	0.299	0.994	0.243	1.30	1.00	1.21	1.40
6	0.03	3.33	0.142	0.400	0.991	0.288	1.59	1.35	1.26	1.39
7	0.035	2.86	0.160	0.495	0.986	0.331	1.76	1.74	1.26	1.42
8	0.04	2.5	0.172	0.589	0.980	0.374	1.95	2.17	1.24	1.41
9	0.045	2.22	0.281	0.927	0.973	0.400	2.16	2.61	1.22	1.38
10	0.05	2	0.374	1.259	0.963	0.421	2.37	3.05	1.18	1.34
11	0.055	1.82	0.470	1.619	0.949	0.437	2.59	3.48	1.15	1.29
12	0.06	1.67	0.685	2.249	0.929	0.425	2.81	3.88	1.10	1.24
13	0.065	1.54	1.01	3.023	0.904	0.392	3.01	4.26	1.06	1.19
14	0.07	1.43	1.22	3.55	0.874	0.377	3.22	4.61	1.01	1.15
15	0.075	1.33	1.44	4.03	0.842	0.361	3.31	4.93	0.956	1.14
16	0.08	1.25	1.66	4.48	0.809	0.344	3.53	5.22	0.919	1.09
17	0.085	1.18	1.75	4.78	0.776	0.341	3.63	5.49	0.872	1.08
18	0.09	1.11	1.85	5.06	0.744	0.338	3.73	5.73	0.829	1.07
19	0.095	1.05	2.08	5.41	0.713	0.319	3.95	5.95	0.801	1.01
20	0.1	1	2.18	5.66	0.68	0.314	4.06	6.15	0.765	0.995

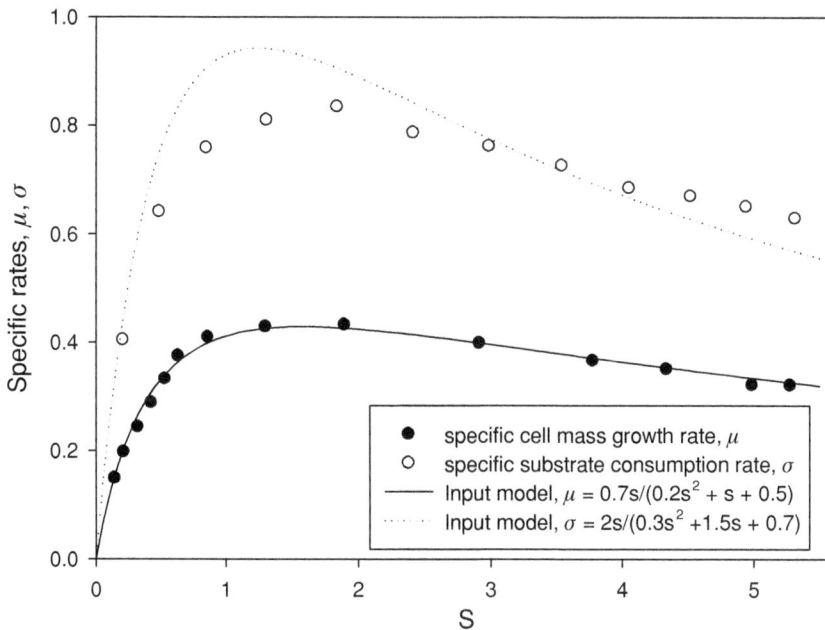

Figure 8.E.4.1. Estimated specific rates μ and σ from constant-feed FBR: substrate inhibition model with variable cell yield coefficient.

$F = 0.005\,\text{L/hr}$ incrementing 0.005 up to $F = 0.1\,\text{L/hr}$;

$D = V_0/F = 0.1/F\,\text{hr}, t$ in hr

using the feed substrate concentration of $S_F = 10$ g/L, the reactor volume of 1 L, and the initial conditions $[XV, SV, V]_{t=0} = [0.1\,\text{g}, 0.01\,\text{g}, 0.1\,\text{L}]$. The feed rate is varied from 5 to 100 mL/hr. As in Example 8.E.3, the data are generated by integrating the mass balance equations (8.23), (8.24), and (8.26) using the preceding information. The results are tabulated in Table 8.E.4.1. The times (t_{1i}) at which the cell concentration goes through the peaks and valleys $(dX/dt = 0)$ and the corresponding concentrations of substrate (S_{1i}) and cell mass (X_{1i}) are recorded in columns 3, 4, and 5, respectively. The specific growth rates are calculated using Eq.(8.33) and recorded in column 6. Likewise, the times (t_{2i}) at which the substrate concentration goes through the peaks and valleys $(dS/dt = 0)$ and the corresponding substrate concentrations (S_{2i}) and cell mass (X_{2i}) are recorded in columns 7, 8, and 9, respectively. Then, the substrate consumption rates are calculated using Eq. (8.34) and tabulated in column 10.

The specific rates of growth (column 6) and substrate consumption (column 10) are plotted against the corresponding substrate concentrations (column 4) in Figure 8.E.4.1. It is clear that both the specific growth rates and specific substrate consumption rate determined from the constant feed rate fed-batch operation agree well with the original input data generated from the substrate-inhibited model. Because the data generated from the model are used without the addition of experimental error, the agreement is extremely good.

Figure 8.E.4.1 shows the plot of μ and σ against the substrate concentration. As compared with the model, the specific rates of cell growth and substrate consumption are in good agreement.

Table 8.E.4.1. *Estimation of specific rates from constant-feed FBR: Substrate inhibition model with variable-yield coefficient*

	1	2	3	4	5	6	7	8	9	10
Run	F hr^{-1}	$D_{0i} = V_0/F_c$ hr	$t_{1i}(dX/dt)_{t_{1i}} = 0$ hr	$S_{1i} = S(t_{1i})$	$X_{1i} = X(t_{1i})$	μ_i hr^{-1} (Eq. (8.33))	$t_{2i}(dS/dt)_{t_{2i}} = 0$ hr	$S_{2i} = S(t_{2i})$	$X_{2i} = X(t_{2i})$	σ_i hr^{-1} (Eq. (8.34))
1	0.005	20					1.08	0.205	1.15	0.404
2	0.01	10					1.41	0.481	1.30	0.641
3	0.015	6.67	0.036	0.143	1.00	0.149	1.70	0.850	1.44	0.759
4	0.02	5	0.068	0.211	0.998	0.197	1.98	1.31	1.54	0.810
5	0.025	4	0.109	0.319	0.995	0.243	2.22	1.84	1.57	0.834
6	0.03	3.33	0.134	0.423	0.992	0.288	2.62	2.41	1.62	0.787
7	0.035	2.86	0.152	0.526	0.987	0.332	2.92	2.99	1.59	0.762
8	0.04	2.5	0.165	0.628	0.982	0.375	3.23	3.54	1.55	0.726
9	0.045	2.22	0.219	0.860	0.975	0.410	3.56	4.06	1.50	0.685
10	0.05	2	0.332	1.30	0.965	0.429	3.77	4.52	1.42	0.670
11	0.055	1.82	0.493	1.89	0.951	0.433	3.99	4.94	1.34	0.651
12	0.06	1.67	0.842	2.91	0.929	0.399	4.22	5.31	1.27	0.629
13	0.065	1.54	1.19	3.78	0.900	0.366				
14	0.07	1.43	1.43	4.34	0.866	0.350				
15	0.075	1.33	1.78	4.99	0.832	0.321				
16	0.08	1.25	1.87	5.27	0.797	0.320				

The proposed method relies on the occurrence of maxima or minima in the concentrations at various constant feed rates. Let us investigate whether these extreme values occur. Inspection of Eq. (8.23) shows that

$$\frac{dX}{dt} \begin{cases} < 0 \\ = 0 \text{ if } \\ > 0 \end{cases} \begin{cases} \mu < \dfrac{1}{\tau + t} \\ \mu = \dfrac{1}{\tau + t} \quad \tau = \dfrac{V_0}{F_c} \\ \mu > \dfrac{1}{\tau + t} \end{cases} \tag{8.36}$$

Because the right-hand side of the inequalities, $1/(\tau + t)$, decreases with time, one should pick a large feed rate F_c (a small τ) so that at near $t = 0$, $\mu < F_c/V_0$ and $(dX/dt) < 0$. Then, the cell concentration first decreases with time, goes through a minimum, and gradually increases with time. For constant feed rates that are too low, the time profiles may not exhibit minima, and therefore, the feed rate must be increased to obtain the minimum. In other words, if no minimum is observed with a chosen constant feed rate, then it should be increased. Likewise, for the substrate, we observe from Eq. (8.24) that

$$\frac{dS}{dt} \begin{cases} > 0 \\ = 0 \text{ if } \\ < 0 \end{cases} \begin{cases} (S_F - S) > \sigma X(\tau + t) \\ (S_F - S) = \sigma X(\tau + t) \quad \tau = \dfrac{V_0}{F_c} \\ (S_F - S) < \sigma X(\tau + t) \end{cases} \tag{8.37}$$

Because the right-hand side of the inequality in Eq. (8.37) increases with time, a proper choice of low value of $\tau = V_0/F_c$ would lead to $dS/dt > 0$ and increasing substrate concentration near the initial time. As time increases, the right-hand side increases without a bound so that $dS/dt < 0$, and the substrate concentration should decrease with time. Therefore, the substrate concentration should go through a maximum, making it possible to calculate the specific substrate consumption rate using Eq. (8.34). For a specific product formation rate, we observe from Eq. (8.25) that

$$\frac{dP}{dt} \begin{cases} < 0 \\ = 0 \text{ if } \\ > 0 \end{cases} \begin{cases} \pi X < P/(\tau + t) \\ \pi X = P/(\tau + t) \quad \tau = \dfrac{V_0}{F_c} \\ \pi X > P/(\tau + t) \end{cases} \tag{8.38}$$

Once again, with proper choices, we should be able to observe the product concentration to increase first, go through a maximum, and decrease with time.

Typical time profiles of concentration profiles of cell mass, substrate, and product are shown in Figure 8.2. The extreme times, t_1, t_2, and t_3, are the times at which X, S, and P go through the extreme values, respectively. It is not necessary for these extreme values to take place at distinct times. In other words, more than one extreme value may take place simultaneously.

The peaks and valleys for cell mass, substrate, and product concentration are identified, and the corresponding concentrations are identified; these concentrations and the time at which they occur are recorded in Table 8.6. Having done this once, we repeat the process with a number of different constant feed rates to generate a data set of specific rates with corresponding concentrations of cells, substrates,

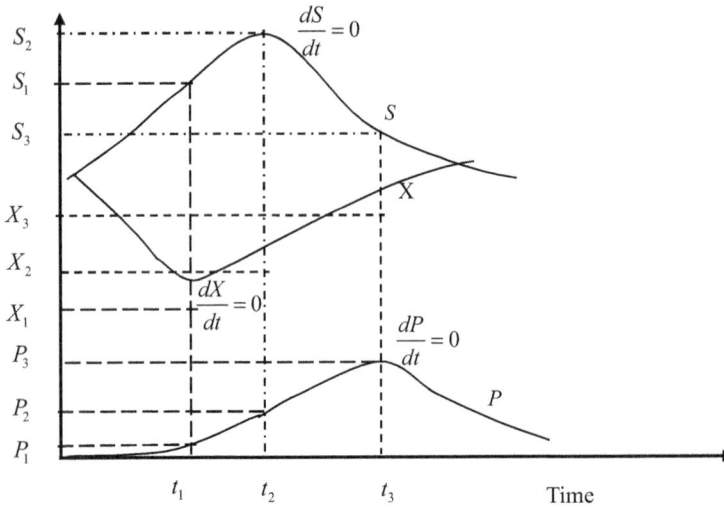

Figure 8.2. Time profiles of S, X, and P of constant-feed fed-batch operations.

and products. The constant feed rates must be chosen to cover a wide range of substrate and product concentrations because the objective is to correlate the specific rates as functions of substrate concentrations, both substrate and product concentrations, or (but rarely) substrate, product, and cell concentrations. The results are then a table of calculated specific rates of cell growth, substrate consumption, and product formation μ, σ, and π versus concentrations of cells, substrate, and product, X, S, and P.

The remaining task is the same as any method of rate data analysis, that is, to correlate the net specific rates with S, with S and P, or with S, P, and X. The method outlined previously avoids the need to take derivatives of rate data. Rate data can be obtained using a steady state continuous-stirred tank reactor (CSTR), also requiring no derivatives of experimental data. However, for biological reactors, the rate data obtained at steady states do not necessarily hold up well for unsteady state operations such as batch and fed-batch. Even if the steady state rate data are applicable to unsteady state operations, it takes a long time for CSTRs to reach a steady state, especially when the dilution rate is small (rate is low). When the dilution rate is high (rates are high), the outlet substrate concentration approaches that of the feed, and the required difference between these two is subject to a higher degree of error. In fact, a fed-batch reactor with a constant feed rate has been reported to yield more accurate rate data for slow reactions than those obtained from the CSTR.[1] Determination of rates over the entire conversion range is made easier for fast reactions with FBR.

In Table 8.6, the net specific growth rates μ of column 3 are calculated using Eq. (8.33) from columns 1 and 2, which are obtained from experimental data, as illustrated in Figure 8.2. The net specific substrate consumption rates σ of column 10 are obtained from columns 6–8 using Eq. (8.34). Finally, the net specific product formation rates π (column 12) are calculated from Eq. (8.35) using the data in columns 12–14.

The remaining task is to correlate the specific growth rates, column 3, with column 4 if the net specific growth rate is a function of substrate concentration alone, $\mu(S)$. If it is a function of both concentrations of substrate and product, $\mu(S, P)$, column 3 is correlated with columns 4 and 5. Conversely, if it is a function of concentrations of substrate, product, and cells $\mu(S, P, X)$; column 3 is correlated with columns 4–6. The specific substrate consumption rates, column 11, should be correlated with column 9 if they depend on S only, $\sigma(S)$, and with columns 9 and 10 if $\sigma(S, P)$, or with columns 8–10 if $\sigma(S, P, X)$. For the specific product formation rate, we do the same: column 16 with column 15, with columns 13 and 15, or with columns 13–15.

Because most common forms of specific rates are in the form of the ratio of polynomials in substrate concentration, one could curve fit the specific rates to one of the following forms:

$$\frac{aS}{1 + bS + cS^2}, \quad \frac{aS}{1 + bS + cS^2}\left(1 - \frac{P}{d}\right), \quad \frac{aS}{1 + bS + cS^2}\exp(-eP), \quad \frac{aS}{X + bS} \quad (8.39)$$

where a, b, c, d, and e are constants. These forms include Monod form ($c = 0$), inhibitions due to substrate S and product P, and contour kinetics. Using a method of parameter optimization, the parameters a–e are estimated. If a particular functional form is known to hold or is preferred, the specific rates may be correlated to that form.

8.2.2 Utilization of Quasi Steady State

It is possible to make use of quasi steady states to obtain kinetic data. *Quasi steady state* was first defined by Pirt[2] and refers to conditions in which approximate steady states are achieved in terms of constant cell and substrate concentrations, $dX/dt = 0$ and $dS/dt = 0$. It is obvious that to maintain these two variables X and S constant, generally, two inputs are needed. These are the feed rate and feed substrate concentration, $F(t)$ and $S_F(t)$. However, it should be noted that if a product is formed in addition to cell mass, then another input is necessary to maintain a quasi steady state, making it difficult to realize a quasi steady state with respect to three concentrations, S, X, and P.

The concept is based on the manipulation of feed rate ($F = \mu V$) and feed substrate concentration ($S_F = S + X/Y_{X/S}$) to force the fed-batch to show constant concentrations of cells ($dX/dt = 0$) and substrate ($dS/dt = 0$):

$$\frac{d(XV)}{dt} = \frac{dX}{dt}V + XF = \mu XV \Rightarrow \frac{dX}{dt} = \left(\mu - \frac{F}{V}\right)X \Rightarrow \frac{dX}{dt} \simeq 0 \text{ if } F = \mu V$$

$$\Rightarrow \frac{dV}{dt} = F = \mu V \Rightarrow V = V_0 \exp\left(\int_0^t \mu(S)d\tau\right) \Rightarrow F = V_0 \mu \exp\left(\int_0^t \mu(S)d\tau\right)$$

$$(8.40)$$

$$\frac{d(SV)}{dt} = \frac{dS}{dt}V + SF = FS_F - \sigma XV \Rightarrow \frac{dS}{dt} = \frac{F}{V}(S_F - S) - \sigma X$$

$$\Rightarrow \frac{dS}{dt} = 0 \text{ if } S_F = S + \frac{\sigma XV}{F} = S + \frac{\sigma XV}{\mu V} = S + \frac{\sigma X}{\mu} = S + \frac{X}{Y_{X/S}}$$

$$(8.41)$$

Table 8.A.1. *Equal-area graphical differentiation*

Time	Substrate concentration			Finite difference of slope	Estimated derivative
t_1	S_1	ΔS	Δt	$\Delta S/\Delta t$	$(dS/dt)_{t_1}$
		$S_2 - S_1$	$t_2 - t_1$	$(\Delta S/\Delta t)_2 = (S_2 - S_1)/(t_2 - t_1)$	
t_2	S_2				$(dS/dt)_{t_2}$
		$S_3 - S_2$	$t_3 - t_2$	$(\Delta S/\Delta t)_3 = (S_3 - S_2)/(t_3 - t_2)$	
t_3	S_3				$(dS/dt)_{t_3}$
		$S_4 - S_3$	$t_4 - t_3$	$(\Delta S/\Delta t)_4 = (S_4 - S_3)/(t_4 - t_3)$	
t_4	S_4				$(dS/dt)_{t_4}$
		$S_5 - S_4$	$t_5 - t_4$	$(\Delta S/\Delta t)_5 = (S_5 - S_4)/(t_5 - t_4)$	
t_5	S_5				$(dS/dt)_{t_5}$
.	.				
.	.				
.	.			$(\Delta S/\Delta t)_n = (S_n - S_{n-1})/(t_n - t_{n-1})$	
		$S_n - S_{n-1}$	$t_n - t_{n-1}$		
t_n	S_n				$(dS/dt)_{t_n}$

Because the substrate concentration is constant owing to the manipulation of $S_F(t)$ according to Eq. (8.17), the feed rate simplifies to

$$F(t) = V_0 \mu \exp(\mu t) \tag{8.42}$$

and

$$S_F(t) = S + \frac{X}{Y_{X/S}} \tag{8.43}$$

Therefore, by manipulating the feed rate and feed substrate concentration in accordance with Eqs. (8.42) and (8.43), it is theoretically possible to achieve quasi steady states for cell and substrate concentrations. However, in practice, it would be difficult, if not impossible, to manipulate the feed substrate concentration as indicated by Eq. (8.43).

Once the rate data are obtained, it is necessary to fit the rate data to a model involving concentrations of substrate, product, intermediate(s), and cells. A model is chosen from careful theoretical analysis and experimental data, and the parameters in the model are determined by fitting the experimental rate data to the chosen model.

Appendix: Equal-Area Graphical Differentiation of Discrete Experimental Data

Let us consider a set of discrete data:

1. Tabulate the discrete experimental data in tabular form, as shown in Table 8.A.1.
2. For each time interval, calculate Δt and ΔS.
3. Calculate the finite difference $(\Delta S/\Delta t)_i = (S_i - S_{i-1})/(t_i - t_{i-1})$ as the estimate of the average slope in an interval between t_i and t_{i-1}.

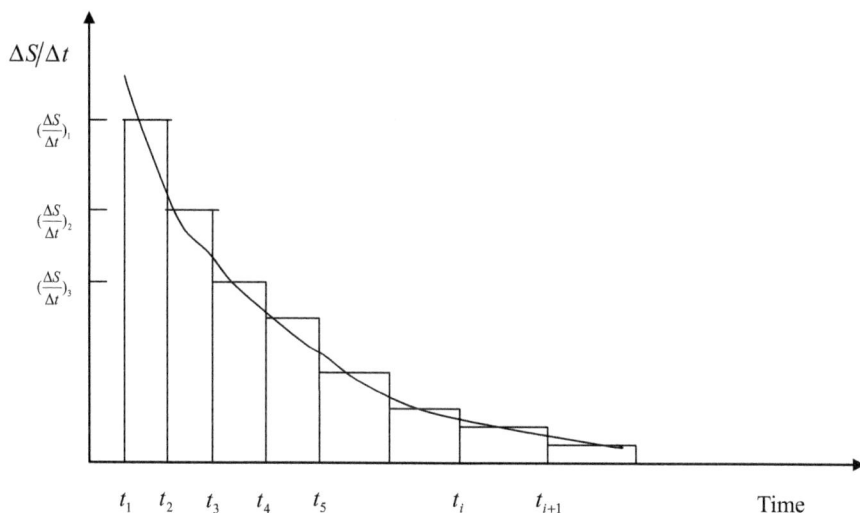

Figure 8.A.1. Equal-area differentiation.

4. Construct a histogram of the finite differences $(\Delta S/\Delta t)_i$ versus time t_i (see Figure 8.A.1).
5. Draw in the smooth curve through the histogram so that for each or multiple intervals, the area above the curve is equal to the area under the curve.
6. Read estimates of derivative dS/dt from the curve at the data points $t_1, t_2, \ldots, t_n$.

REFERENCES

1. Lee, H. H., and Yau, B. O. 1981. An experimental reactor for kinetic studies: Continuously fed-batch reactor. *Chemical Engineering Science* 36: 483–488.
2. Pirt, S. J. 1974. The theory of fed batch culture with reference to the penicillin fermentation. *Journal of Applied Chemistry and Biotechnology* 24: 415–424.

9 Optimization by Pontryagin's Maximum Principle

A number of items need to be considered before getting into a formal optimization. We must decide what to optimize. Here we shall call the objective function the performance index we wish to maximize, and we will maximize the performance index. When the objective is to minimize rather than maximize, such as the minimum time to achieve a desired concentration, one would change the sign of the performance index (from positive to negative) and maximize it instead. Next is the question of what would be the best initial conditions and whether the final conditions should be left unspecified or specified. There is the question of what input (control) variables we should manipulate to maximize the chosen performance index. Finally, there are questions regarding physical constraints on the manipulated variable as well as the state variables. We shall look at these questions.

One of the primary objectives of reaction engineering is to maximize the rate of formation of the product (productivity) and/or the yield (selectivity or relative yield). For an existing plant, a faster product formation rate implies a higher productivity or a corresponding reduction in plant operating time and operating cost. The increased rate implies a smaller reactor and therefore a lower capital investment cost for new plants to be built. Likewise, an improved yield implies lower raw material costs for existing and new plants. One may wish to maximize the total amount of product produced per cycle of fed-batch operation or maximize the productivity, the amount of product produced per cycle per unit operating time.

9.1 Impulse and Parameter Optimizations

There are two types of optimization problems for dynamic processes such as batch and semi-batch bioreactors that involve inputs that are either functions of time or constant. The first type is impulse optimization, and as such, it deals with the determination of best values of manipulated variables as functions of time during the entire operation, the *time profile* (trajectory) *optimization*. Time profile optimization problems include determination of optimal feed rate profiles for such variables as the substrate, phosphate, inducers, nitrogen sources, temperature, pH, and medium composition. In these problems, we are seeking the optimal functions, and therefore, the tools to be used are the variational calculus or some versions of it. The other type is determination of the best constant values such as the initial bioreactor volume; the inoculum size; the initial medium composition, including the substrate; the optimum isothermal temperature; the optimum constant pH; and the fermentation time. These problems are classified as parameter optimizations. For these problems, the ordinary calculus, or some variations of it, are used.

The medium optimization is in general a parameter optimization problem in the sense that traditionally, *the best initial values* are determined by some statistical means owing to a lack of mathematical models that can quantitatively predict the effect of medium composition on the outcome of bioreactor operation. However, the current thought is to look at it as a time profile optimization problem in the sense that the limiting component may change during the course of fermentation. The conditions that are optimal for cell growth may differ from those for product formation. Therefore, the composition ought to be optimized at all times, not just initially. Thus, it may be advantageous to add medium components and nutrients during the course of fed-batch operation.

Unsteady state mass balance equations and energy balance equations are used to model fed-batch cultures. Normally, fed-batch cultures are performed isothermally, and therefore, the energy balance is decoupled from the mass balance equations. Then, the energy balance is used to calculate the heat removal or addition rate. The unsteady state mass and energy balances are a set of ordinary differential equations. In this chapter, we consider the optimization of fed-batch cultures that are modeled by a set of ordinary differential equations. However, there are situations in which not enough information is available to complete some mass balances, for example, the medium compositions as they affect the growth and product formation rates, and therefore, it is difficult, if not impossible, to develop a model in the form of differential equations. Sometimes our lack of knowledge prevents us from writing any mass balance equation. In these situations, a statistical model may be developed instead, one that relates the effects of the medium components for which we cannot write a mass balance on cell and product concentrations and intermediates. A neural network model may be developed that can reflect the effects at various fermentation times. In other cases, the kinetic parameters in the differential equations are allowed to vary with time rather than assuming constant values throughout the course of the culture. In this way, the model is updated at various times or time intervals to obtain better matches between the model prediction and actual experimental data. Thus, a method dealing with models not based on equations must be used.

9.2 Optimization Criteria

Having decided on the manipulated variables, the next decision to make regards the performance index. Various performance indices that have been used for optimization of fed-batch fermentation are listed in Table 9.1. In addition, the process operation time (final time) can be specified (fixed) a priori or left free to be optimized as well. The initial conditions may be fixed or left free to be optimally determined. Some or all of the final conditions may be fixed to meet the specified conversion, residual concentration of substrate, or product concentration. There are certain constraints that must be met, such as that which specifies the full use of bioreactor volume, the upper and lower limits on the feed rate, and others, such as the maximum oxygen transfer when enriched oxygen or pure oxygen is not used.

Table 9.1. *Various performance indices used for optimization of fed-batch cultures*

Product	Objective	Performance index	Reference
Cell mass	Maximum productivity	$P = [X(t_f)V(t_f) - X(0)V(0)]/t_f$ free	13
	Maximum amount in	$P = [X(t_f)V(t_f) - X(0)V(0)]$	14, 30
	fixed time	t_f fixed	15–20
	Minimum time to	$P = t_f$	
	obtain a fixed amount	$X(t_f), V(t_f), X(0), V(0)$ fixed	
Metabolite	Maximum productivity	$P = [P(t_f)V(t_f) - P(0)V(0)]/t_f$ free	21–24
	Maximum amount in	$P = [P(t_f)V(t_f) - P(0)V(0)] t_f$ fixed	24–28
	fixed time	$P = t_f$	31–32
	Minimum time to	$P(t_f), V(t_f), P(0), V(0)$ fixed	
	obtain a fixed amount	$P = \{p_c[X(t_f)V(t_f) - X(0)V(0)]$	24
	Maximum profit rate	$\quad + p_m[P(t_f)V(t_f) - P(0)V(0)]\}/t_f$	
Cell mass and	Maximum profit rate	$P = \{p_c[X(t_f)V(t_f) - X(0)V(0)]$	29
metabolite	Maximum profit	$\quad + p_m[P(t_f)V(t_f) - P(0)V(0)]\}/t_f$	
	Minimum time to	$P = p_c[X(t_f)V(t_f) - X(0)V(0)]$	
	obtain a fixed amount	$\quad + p_m[P(t_f)V(t_f) - P(0)V(0)]t_f$ fixed	
		$P = t_f P(t_f), X(t_f), V(t_f),$	
		$P(0), X(0), V(0)$ fixed	

9.2.1 Performance Indices

A general performance index reflecting the profit may consist of the value of cell mass plus the value of product minus the cost of substrate and medium minus the operating cost and the amortization cost:

$$P[\mathbf{x}(t_f)] = p_c[V(t_f)X(t_f) - V(0)X(0)] + p_p[V(t_f)P(t_f) - V(0)P(0)] \quad (9.1)$$
$$-\{(p_s S_F + p_m)[V(t_f) - V(0)] - p_{op}t_f - p_{am}V(t_f)\}$$

where t_f is the final time, $V(t_f) = V_{\max}$ is the final (maximum) volume, $V(0)$ is the initial volume, and p_c, p_m, p_p, and p_s are the price of unit cell mass, the medium cost per unit volume, the unit product price, and the unit cost of limiting substrate, respectively. The operating cost per unit time is denoted by p_{op}; the total operating cost is assumed to be proportional to the operating time, t_f; and the amortization cost is assumed to be proportional to the reactor volume and represented by $p_{am}V(t_f)$. The first and second terms in Eq. (9.1) represent the values associated with the final cell mass, if any, and the value of metabolites produced, respectively; the third term represents the costs associated with the substrate and medium charged into the culture, which are fixed since all terms are fixed. Because amortization cost and medium cost are independent of the fed-batch operation, we may drop them from the performance index. We note that the initial culture volume $V(0)$ and the inoculum concentration $X(0)$ are parameters that can appear in the performance index P and therefore can also be optimized. If the final time t_f is chosen a priori, then the operating cost is fixed, and if the initial volume is fixed, the substrate cost is also fixed so that the performance index may be simplified to

$$P[\mathbf{x}(t_f)] = p_c V(t_f)X(t_f) + p_p V(t_f)P(t_f) \quad (9.2)$$

where it is assumed that no product is present initially and the initial amount of cells $V(0)X(0)$ is negligible in comparison to the final amount $V(t_f)X(t_f)$.

If cell mass does not contribute to the profit, p_c is set to zero, and if the cell mass is the product, then p_p should be set to zero. One can also specify the amount of metabolite (and/or cell mass) at the final time and minimize the time to obtain the specified (fixed) amount of product, that is, the *minimum time problem*:

$$P[\mathbf{x}(t_f)] = -t_f, \quad P(t_f) = P_f \tag{9.3}$$

This minimum time problem may be used to solve the optimal productivity problem by repeatedly solving the minimum time problems with varying final amounts of metabolite and picking the one with the best productivity, $P(t_f)/t_f = P_f/t_f$. Thus, the performance index could be the profit rate, the amount of profit, or the final time, as shown in Table 9.1[12] for the production of cell mass, metabolites, and cell mass plus metabolites.

Sometimes it is desirable to *maximize the profit rate*, especially when the process volume is large. Therefore, one can modify Eq. (9.2) to incorporate the rate:

$$P[\mathbf{x}(t_f)] = [p_c V(t_f)X(t_f) + p_p V(t_f)P(t_f)]/t_f \tag{9.4}$$

Various performance indices have been used such as the profit rate, the amount of profit, or the final time, as shown in Table 9.1.

9.2.2 Free and Fixed Final Times

The final time t_f, that is, the time to complete the operation of cultures, can be either left unspecified (to be chosen optimally) or specified a priori. The final time may be specified indirectly by other constraints such as a productivity requirement $P_{\text{req}} = P(t_f)V(t_f)/t_f$ or a constraint on the final concentration of substrate, $S(t_f) \leq S_{\text{min}}$. Unspecified final time may be included in the performance index in the form of productivity or operational costs, which can be assumed approximately proportional to the final time, $P[\mathbf{x}(t_f)] = p_p[V(t_f)P(t_f)] - p_{op}t_f$. When the final time is free to be chosen, it is necessary, as shown earlier, to augment the state variables by introducing an additional state variable, $\dot{x}_{n+1} = 1$, $x_{n+1}(0) = 0$, so that the final time is converted to the augmented state variable at the final time, $t_f = x_{n+1}(t_f)$. By so doing, the performance index is put into a standard form, that is, a function of the final state variables.

9.2.3 Free and Fixed Initial and Final States

In the preceding treatment, we have restricted the initial conditions on state variables to be all fixed (given). However, we should be able to select optimally some initial conditions such as the initial volume $V(0)$, the initial substrate concentration $S(0)$, and the inoculum size $X(0)$. The larger the inoculum size, the better is the result, and it may be restricted by a practical consideration, but the initial volume and substrate concentrations must be selected optimally. Parameter optimization can be applied to determine the best values of initial volume and substrate concentration. Sometimes it may be desirable to specify the final substrate concentration because excess residual concentration may lead to additional separation costs. Sometimes one may wish to obtain a desired product concentration in a minimum time. Thus, one or more of the final states may be fixed. These situations lead to different boundary conditions, as we see in Chapter 12.

9.2.4 Various Constraints

There are also practical situations in which one wishes to impose one or more constraints. It is obvious that the culture volume must be constrained so that the fed-batch operation makes full use of maximum working volume, $V(t_f) = V_{max}$. Otherwise, one may end up getting a result that calls for an infinite culture volume. It may be necessary to put a low enough constraint on the final substrate concentration so that it does not hinder a separation process. In other situations, the oxygen demand may be too excessive for aeration and may have to impose a limit, unless enriched air or pure oxygen can be used.

Let us consider maximizing a performance index defined by the final outcome for a process described by a set of ordinary differential equations resulting from unsteady state mass and energy balances. The initial conditions, such as the initial volume and substrate concentration, should also be optimized. However, these are parameter optimization problems and not profile optimization problems.

9.3 Choice of Manipulated Variables

In addition to the substrate feed rate profile $F(t)$ that appears to be the most direct and logical choice for manipulated variables, there are a number of other choices for the manipulated variables such as the culture volume profile proposed by Modak and Lim,[1] $V(t)$, the variable feed substrate concentration profile proposed by Guthke and Knorre[2] and Alvarez and Alvarez,[3] $S_F(t)$, the substrate concentration profile proposed by Guthke and Knorre and San and Stephanopoulos,[4] $S(t)$, the mass flow rate of substrate proposed by Fishman and Biryukov,[5] $S_F F(t)$, or the specific growth rate proposed by Yamane et al.,[6] μ.

The selection of the *feed flow rate* $F(t)$ is rather obvious as it is practical and its effect is most direct. However, the choice of the volumetric feed rate as the manipulated variable leads to a *singular control problem* that is difficult to solve. It is possible to formulate the problem through a transformation of variables so that the feed flow rate does not appear in the transformed equations and thus one can avoid a singular control problem and proceed to obtain the *optimal substrate concentration profiles*, $S(t)$ (sometimes ignoring the substrate and overall balance equations), by reducing to a computationally simpler nonsingular problem. However, to realize the optimal profile $S(t)$, one has to determine the corresponding feed flow rate profile $F(t)$. If the calculated optimal profile violates the physical constraints (through the substrate and overall mass balances), the minimum and/or maximum flow rates, this approach needs to be modified to include an iterative solution to meet the constraints, leading to a very difficult problem. Hence, there is no distinct advantage to using the substrate concentration profile as the control variable. Similar to this approach is a nonsingular transformation method proposed by Yamane et al.,[6] Menawat et al.,[7] and Jayant and Pushpavanam,[8] in which the feed rate is made to appear nonlinearly and thus avoid the singular problem. However, as pointed out by Shin and Lim,[9] there remains a numerical convergence problem and/or appropriate constraints[9] that need to be checked at each time interval.

Choice of the *feed substrate concentration*, $S_F(t)$, as the manipulated variable retains all mass balance equations. Optimizing the performance index using as the control variable the feed substrate concentration $S_F(t)$, while keeping the feed flow rate constant, may appear to make the problem a bit easier to solve, but the physical implementation of the feed substrate concentration as a function of time poses another difficulty that may be hard to overcome.

Another choice is the *mass feed rate of substrate*, $S_F F(t)$, to optimize the performance index. This approach actually ignores the overall mass balance and can lead to a large error, as pointed out by Lee et al.[10] It has been shown by Modak and Lim[1] and Modak[11] that the optimality conditions obtained for either the feed rate $F(t)$ or the *reactor volume* $V(t)$ as the manipulated variable are identical.

Modak[11] has shown that the choice of the substrate concentration $S(t)$ in the reactor, the feed substrate concentration S_F, and the substrate mass flow rate FS_F leads to suboptimal operating policies as compared to the choice of the *substrate feed rate* $F(t)$ and that choosing them as the manipulated variables instead of the feed rate does not take into account possible dependence of the specific rates on cell and/or product concentrations and the substrate balance equations. Thus, the substrate feed rate profile or multiple feed rate profiles of substrates, precursors, and other nutrients are the logical choice for the manipulated (control) variables.

Finally, a new transformation technique has been proposed by Lee et al.,[10,36,37] in which the substrate concentration profiles are used as the manipulated variables. Unlike the previous approach, this incorporates the substrate balance equations.

9.4 Feed Rate Problem Formulation and Solution

The process is described by a set of unsteady state balance equations of mass and energy,

$$\frac{dx_i(t)}{dt} = f_i[x_1(t), x_2(t), x_3(t), \ldots, x_n; u_1(t), u_2(t), u_3(t), \ldots, u_m(t)]$$

$$i = 1, 2, 3, \ldots, n$$

(9.5)

where $x_i (i = 1, 2, 3, \ldots, n)$ are called the state variables representing species masses, the reactor volume, and temperature and $u_j (j = 1, 2, 3, \ldots, m)$ are the manipulated (control) variables, such as the feed flow rates, that are used to influence the state variables. In vector notation, we define the state and control vectors as follows:

$$\mathbf{x}(t) = [x_1 x_2 x_3 \ldots x_n]^T \quad \mathbf{u}(t) = [u_1 u_2 u_3 \ldots u_m]^T$$

(9.6)

Then, the dynamic equation in vector form is

$$\frac{d\mathbf{x}}{dt} = \mathbf{f}[\mathbf{x}(t), \mathbf{u}(t)] \quad \mathbf{x}(0) = \mathbf{x_0}$$

(9.7)

The initial conditions are assumed specified (given) for the time being. The right-hand side (RHS) of Eq. (9.7) does not depend explicitly on time t, and therefore, the processes described by Eq. (9.7) are called *autonomous processes*. In some cases, the

RHS may depend explicitly on time, for example, some parameters depend explicitly on time. These situations will be handled by introducing an additional auxiliary state variable. It is assumed that all initial conditions are fixed (given) here; cases in which certain initial conditions are free and to be chosen optimally will be treated later. The objective is to maximize the performance index that depends on the outcome at the final time,

$$\underset{u_1,u_2,u_3,\dots u_m}{\text{Max}} P[x_1(t_f), x_2(t_f), x_3(t_f), \dots, x_n(t_f)] = \underset{\mathbf{u}(t)}{\text{Max}} P[\mathbf{x}(t_f)] \tag{9.8}$$

by manipulating the control vector, $\mathbf{u}(t)$. The constraints on the manipulated variables are as follows:

$$u_{i_{\min}} \le u_i(t) \le u_{i_{\max}}, \quad i = 1, 2, 3, \dots m \tag{9.9}$$

According to Pontryagin's maximum principle (PMP),[33] instead of maximizing the original performance index, one maximizes the Hamiltonian, which is defined as a scalar product of the adjoint vector λ and the RHS of the differential equation $\mathbf{f}$. In other words, each adjoint variable λ_i is multiplied by the corresponding RHS of the mass balance equation f_i and summed for all components:

$$\underset{u_1,u_2,u_3,\dots u_m}{\text{Max}} \left(H = \lambda^{\mathbf{T}}\mathbf{f} = \sum_{i=1}^{n} \lambda_i f_i \right) \tag{9.10}$$

where the adjoint vector satisfies the following differential equations:

$$\frac{d\lambda}{dt} = -\frac{\partial H}{\partial \mathbf{x}} = -\frac{\partial (\mathbf{f}^T \lambda)}{\partial \mathbf{x}} = -\left(\frac{\partial \mathbf{f}}{\partial \mathbf{x}} \right)^T \lambda \tag{9.11}$$

subject to certain boundary conditions that are detailed later. We demonstrate (not derive) PMP.

9.4.1 Pontryagin's Maximum Principle

Consider the optimal control function $u^*(t)$ and a comparison (nonoptimal) function $u(t, \varepsilon)$, where ε is a small parameter such that when $\varepsilon = 0$, $u(t, \varepsilon = 0) = u^*(t)$. Now, we make a Taylor series expansion of the comparison function around the optimal function,

$$u(t, \varepsilon) = u(t, \varepsilon = 0) + \left(\frac{\partial u}{\partial \varepsilon} \right)_{\varepsilon=0} \varepsilon + \frac{1}{2} \left(\frac{\partial^2 u}{\partial \varepsilon^2} \right)_{\varepsilon=0} \varepsilon^2 + \cdots \overset{\Delta}{=} u^*(t)$$

$$+\delta u + \delta^2 u + \cdots \tag{9.12}$$

where the first variation of u, δu, is defined as

$$\delta u \overset{\Delta}{=} \left(\frac{\partial u}{\partial \varepsilon} \right)_{\varepsilon=0} \varepsilon \tag{9.13}$$

and the second variation of u, $\delta^2 u$, is defined as

$$\delta^2 u \overset{\Delta}{=} \frac{1}{2} \left(\frac{\partial^2 u}{\partial \varepsilon^2} \right)_{\varepsilon=0} \varepsilon^2 \tag{9.14}$$

In terms of these variations, the comparison function can be written as

$$u(t, \varepsilon) = u^*(t) + \delta u + \delta^2 u + \cdots \tag{9.15}$$

or

$$\delta u = u(t, \varepsilon) - u^*(t) + o(\varepsilon), \quad o(\varepsilon) \triangleq \lim_{\varepsilon \to 0} \frac{o(\varepsilon)}{\varepsilon} = 0 \tag{9.16}$$

Therefore,

$$\delta u = u(t, \varepsilon) - u^*(t) \tag{9.17}$$

The first variation is the difference between the comparison (nonoptimal) function $u(t, \varepsilon)$ and the optimal function $u^*(t) \triangleq u(t, \varepsilon = 0)$.

Let the first variations be the differences between the nonoptimal and optimal functions:

$$\begin{aligned}
\delta \mathbf{x} &\triangleq \mathbf{x}(t, \varepsilon) - \mathbf{x} * (t) \\
\delta \mathbf{u} &\triangleq \mathbf{u}(t, \varepsilon) - \mathbf{u} * (t) \\
\delta \mathbf{f} &\triangleq \mathbf{f}(\mathbf{x}, \mathbf{u}) - \mathbf{f}^*(\mathbf{x}^*, \mathbf{u}^*) = \left(\frac{\partial \mathbf{f}}{\partial \mathbf{x}}\right)^* \delta \mathbf{x} + \left(\frac{\partial \mathbf{f}}{\partial \mathbf{u}}\right)^* \delta \mathbf{u}
\end{aligned} \tag{9.18}$$

where the asterisk refers to the optimum. We take the first variation of Eq. (9.7):

$$\delta \left(\frac{d\mathbf{x}}{dt}\right) = \delta \mathbf{f}(\mathbf{x}, \mathbf{u}) = \left(\frac{\partial \mathbf{f}}{\partial \mathbf{x}}\right) \delta \mathbf{x} + \left(\frac{\partial \mathbf{f}}{\partial \mathbf{u}}\right) \delta \mathbf{u} \tag{9.19}$$

Solving for $(\partial \mathbf{f}/\partial \mathbf{u})\delta \mathbf{u}$ and recognizing that the d/dt operator and δ operator commute so that the order can be interchanged in Eq. (9.19), we obtain

$$\left(\frac{\partial \mathbf{f}}{\partial \mathbf{u}}\right) \delta \mathbf{u} = \delta \left(\frac{d\mathbf{x}}{dt}\right) - \left(\frac{\partial \mathbf{f}}{\partial \mathbf{x}}\right) \delta \mathbf{x} = \frac{d}{dt}(\delta \mathbf{x}) - \left(\frac{\partial \mathbf{f}}{\partial \mathbf{x}}\right) \delta \mathbf{x} \tag{9.20}$$

This is a linear ordinary differential equation in $\delta \mathbf{x}$. We form a scalar product with the adjoint vector λ (introducing one degree of freedom),

$$\lambda^T \left[\frac{d}{dt}(\delta \mathbf{x}) - \left(\frac{\partial \mathbf{f}}{\partial \mathbf{x}}\right) \delta \mathbf{x}\right] = \lambda^T \left(\frac{\partial \mathbf{f}}{\partial \mathbf{u}}\right) \delta \mathbf{u} \tag{9.21}$$

and force the left-hand side (LHS) of Eq. (9.21) to be the total derivative of a scalar product (using up to one degree of freedom, a procedure similar to the concept of the integrating factor) so that it can be integrated readily:

$$\lambda^T \left[\frac{d}{dt}(\delta \mathbf{x}) - \left(\frac{\partial \mathbf{f}}{\partial \mathbf{x}}\right) \delta \mathbf{x}\right] \equiv \frac{d}{dt}(\lambda^T \delta \mathbf{x}) = \lambda^T \left(\frac{\partial \mathbf{f}}{\partial \mathbf{u}}\right) \delta \mathbf{u} \tag{9.22}$$

This is done by matching the LHS with the middle term, which is expanded, and canceling a common term on both sides:

$$\begin{aligned}
\lambda^T \left[\frac{d}{\cancel{dt}}\cancel{(\delta x)} - \left(\frac{\partial \mathbf{f}}{\partial \mathbf{x}}\right) \delta \mathbf{x}\right] &= \frac{d}{dt}(\lambda^T \delta \mathbf{x}) = \frac{d\lambda^T}{dt}(\delta \mathbf{x}) + \lambda^T \frac{d}{\cancel{dt}}\cancel{(\delta x)} \quad \Rightarrow \\
-\lambda^T \left(\frac{\partial \mathbf{f}}{\partial \mathbf{x}}\right) \delta \mathbf{x} &= \frac{d\lambda^T}{dt}(\delta \mathbf{x})
\end{aligned} \tag{9.23}$$

or

$$\frac{d\boldsymbol{\lambda}}{dt} = -\left(\frac{\partial \mathbf{f}}{\partial \mathbf{x}}\right)^T \boldsymbol{\lambda} \tag{9.24}$$

Equation (9.24) defines the adjoint vector. This set of differential equations was provided as Eq. (9.11). Now, the scalar product, Eq. (9.22), is integrated to obtain

$$\int_0^{t_f} \boldsymbol{\lambda}^T \left(\frac{\partial \mathbf{f}}{\partial \mathbf{u}}\right) \delta \mathbf{u}\, dt = \int_0^{t_f} \frac{d}{dt} (\boldsymbol{\lambda}^T \delta \mathbf{x})\, dt = \boldsymbol{\lambda}^T(t_f)\delta \mathbf{x}(t_f) - \boldsymbol{\lambda}^T(0)\delta \mathbf{x}(0) \tag{9.25}$$

Rewriting Eq. (9.25) as

$$\boldsymbol{\lambda}^T(t_f)\delta \mathbf{x}(t_f) - \boldsymbol{\lambda}^T(0)\delta \mathbf{x}(0) = \int_0^{t_f} \boldsymbol{\lambda}^T \left(\frac{\partial \mathbf{f}}{\partial \mathbf{u}}\right) \delta \mathbf{u}\, dt \tag{9.26}$$

the next step is to take the variation of the performance index, δP, and match it with the LHS of Eq. (9.25). The variation of the performance index is

$$\delta P = \left[\frac{\partial P}{\partial \mathbf{x}(t_f)}\right]^T \delta \mathbf{x}(t_f) \tag{9.27}$$

We force δP of Eq. (9.27) to match the LHS of Eq. (9.26):

$$\delta P = \left[\frac{\partial P}{\partial \mathbf{x}(t_f)}\right]^T \delta \mathbf{x}(t_f) = \boldsymbol{\lambda}^T(t_f)\delta \mathbf{x}(t_f) - \boldsymbol{\lambda}^T(0)\delta \mathbf{x}(0) = \int_0^{t_f} \boldsymbol{\lambda}^T \left(\frac{\partial \mathbf{f}}{\partial \mathbf{u}}\right) \delta \mathbf{u}\, dt \tag{9.28}$$

Because the initial conditions are assumed to be all given (fixed), $\delta \mathbf{x}(0) = \mathbf{0}$, and therefore,

$$\frac{\partial P}{\partial \mathbf{x}(t_f)} = \boldsymbol{\lambda}(t_f) \tag{9.29}$$

Equation (9.29) provides the final conditions on the adjoint variables.

Now, for convenience, we introduce a Hamiltonian as the scalar product of the adjoint vector $\boldsymbol{\lambda}(t)$ with the RHS of mass and energy balance equation $\mathbf{f}(\mathbf{x}, \mathbf{u})$:

$$H = \boldsymbol{\lambda}^{\mathbf{T}}\mathbf{f} = \mathbf{f}^{\mathbf{T}}\boldsymbol{\lambda} = \sum_{i=1}^{n} \lambda_i f_i \tag{9.30}$$

Then, the integrand in Eq. (9.28) can be expressed in terms of the Hamiltonian:

$$\boldsymbol{\lambda}^T \left(\frac{\partial \mathbf{f}}{\partial \mathbf{u}}\right) \delta \mathbf{u} = \left[\left(\frac{\partial \mathbf{f}}{\partial \mathbf{u}}\right)^T \boldsymbol{\lambda}\right]^T \delta \mathbf{u} = \left(\frac{\partial H}{\partial \mathbf{u}}\right)^T \delta \mathbf{u} \tag{9.31}$$

Substitution of Eq. (9.31) into Eq. (9.28) yields

$$\delta P = \int_0^{t_f} \left(\frac{\partial H}{\partial \mathbf{u}}\right)^T \delta \mathbf{u}\, dt \tag{9.32}$$

Because all initial conditions on the state variables are assumed to be fixed,

$$\boldsymbol{\lambda}^T(0)\delta \mathbf{x}(0) = \lambda_1(0)\delta x_1(0) + \lambda_2(0)\delta x_2(0) + \cdots + \lambda_k(0)\delta x_k(0)$$
$$+ \lambda_{k+1}(0)\delta x_{k+1}(0) + \cdots + \lambda_n(0)\delta x_n(0) = 0 \tag{9.33}$$

When one or more of the initial states is not specified, say, $x_k(0) \neq 0$ and $x_{k+1}(0) \neq 0$, then we set the corresponding initial value of the adjoint variables to zero, $\lambda_k(0) = 0$, and

$$\lambda_{k+1}(0) = 0 \text{ to force } \boldsymbol{\lambda}^T(0)\delta\mathbf{x}(0) = 0$$

To force $\delta P = 0$, the integrand in Eq. (9.32) is made to vanish by picking

$$\delta\mathbf{u} = 0 \ (\mathbf{u} \text{ is on the constrained boundary}) \tag{9.34}$$

$$\delta\mathbf{u} \neq 0 \ (\mathbf{u} \text{ is not on the constrained boundary}), \quad \frac{\partial H}{\partial \mathbf{u}} = 0$$

Equation (9.34) implies[33] that the Hamiltonian must be maximized by picking $\mathbf{u}$.

9.4.2 Boundary Conditions on Adjoint Variables

The boundary conditions on the adjoint variables are obtained first from Eq. (9.28) by matching terms on the LHS with those on the RHS, the *transversality conditions*:

$$\delta P = \left[\frac{\partial P}{\partial \mathbf{x}(t_f)} \right]^T \delta\mathbf{x}(t_f) = \boldsymbol{\lambda}^T(t_f)\delta\mathbf{x}(t_f) - \boldsymbol{\lambda}^T(0)\delta\mathbf{x}(0) \tag{9.28}$$

We consider some specific examples to obtain the proper boundary conditions on the adjoint variables:

1. If all initial conditions on the state variables are given (fixed), then their variations are zero, $\delta\mathbf{x}(0) = \mathbf{0}$, and thus, the final conditions on the adjoint variables are all specified because of Eq. (9.28):

$$\lambda(t_f) = \left[\frac{\partial P}{\partial \mathbf{x}(t_f)} \right] \tag{9.29}$$

2. If some of the initial conditions are not fixed but free, say, $x_j(0)$ is free ($\delta x_j(0) \neq 0$ and $\delta x_i(0) = 0, i \neq j$), then to force $\lambda_j(0)\delta x_j(0) = 0$ in Eq. (9.28), the initial condition on the corresponding adjoint variable is set to zero:

$$\lambda_j(0) = 0 \tag{9.35}$$

3. If there is a recycle of some species, there would be periodic (cyclic) boundary conditions, such as $x_k(0) = x_k(t_f)$, and therefore, $\delta x_k(0) = \delta x_k(t_f)$, and Eq. (9.28) is used to obtain the boundary condition:

$$\left[\frac{\partial P}{\partial x_k(t_f)} \right] \delta x_k(t_f) = \lambda_k(t_f)\delta x_k(t_f) - \lambda_k(0)\delta x_k(0)$$

$$= [\lambda_k(t_f) - \lambda_k(0)]\delta x_k(t_f) \tag{9.36}$$

By matching the LHS with the RHS, we obtain the corresponding boundary condition on the adjoint variable, a cyclic condition on the corresponding adjoint variable:

$$\left[\frac{\partial P}{\partial x_k(t_f)} \right] = [\lambda_k(t_f) - \lambda_k(0)] \tag{9.37}$$

4. If a final condition is specified on the state variable, say, $x_p(t_f) = x_{pf}$, then

$$\left[\frac{\partial P}{\partial x_p(t_f)} \right] \delta x_p(t_f) = \lambda_p(t_f) \underbrace{\delta x_p(t_f)}_{=0} - \lambda_p(0) \underbrace{\delta x_p(t_f)}_{=0} \tag{9.38}$$

so that the final condition on the corresponding adjoint variable is unspecified, that is,

$$\lambda_p(t_f) \text{ is unknown} \tag{9.39}$$

With the necessary conditions for optimum and proper boundary conditions on the adjoint variables, we can summarize the maximum (minimum) principle of Pontryagin.

9.4.2.1 A Summary of Pontryagin's Maximum Principle

PMP can be summarized as follows: maximization of the performance index $P[\mathbf{x}(t_f)]$ for processes described by a set of ordinary differential equations, $d\mathbf{x}\,dt = \mathbf{f}[\mathbf{x}(t), \mathbf{u}(t)]$, with appropriate initial and final conditions is equivalent to maximizing the Hamiltonian function, $H = \boldsymbol{\lambda}^T \mathbf{f}$, where the adjoint vector satisfies the differential equations, $d\boldsymbol{\lambda}/dt = -(\partial \mathbf{f}/\partial \mathbf{x})^T \boldsymbol{\lambda} = -\partial H/\partial \mathbf{x}$, with appropriate boundary conditions, as specified previously.

9.4.3 Hamiltonian

To gain the properties of the Hamiltonian function, we begin by time differentiating it and making use of Eqs. (9.7) and (9.11):

$$\frac{dH}{dt} = \left(\frac{\partial H}{\partial \mathbf{x}} \right)^T \frac{d\mathbf{x}}{dt} + \left(\frac{\partial H}{\partial \mathbf{u}} \right)^T \frac{d\mathbf{u}}{dt} + \left(\frac{\partial H}{\partial \boldsymbol{\lambda}} \right)^T \frac{d\boldsymbol{\lambda}}{dt}$$

$$= -\left(\frac{d\boldsymbol{\lambda}}{dt} \right)^T \mathbf{f} + \left(\frac{\partial H}{\partial \mathbf{u}} \right)^T \frac{d\mathbf{u}}{dt} + \left(\frac{d\boldsymbol{\lambda}}{dt} \right)^T \mathbf{f} = 0 \tag{9.40}$$

$$\left\{ \begin{array}{l} \text{on the boundry } d\mathbf{u}/dt = \mathbf{0} \\ \text{oP boundary } \partial H/\partial \mathbf{u} = \mathbf{0} \end{array} \right\}$$

Therefore, *the Hamiltonian is a constant on the optimal trajectory*,

$$H(\mathbf{x}^*, \mathbf{u}^*, \boldsymbol{\lambda}^*) = \text{a constant} \tag{9.41}$$

When the final time is free, we have

$$\delta P = \left[\frac{\partial P}{\partial \mathbf{x}(t_f)} \right]^T \delta \mathbf{x}(t_f) = \left[\frac{\partial P}{\partial \mathbf{x}(t_f)} \right]^T \left(\frac{d\mathbf{x}}{dt} \right)_{t=t_f} \delta t_f = [\boldsymbol{\lambda}^T(t_f)\mathbf{f}(t_f)]\delta t_f \tag{9.42}$$

$$= H(t_f)\delta t_f = 0 \Rightarrow H(t_f) = 0$$

According to Eq. (9.42), the Hamiltonian at the final time is zero, and because the Hamiltonian is constant according to Eq. (9.41), Eq. (9.42) implies that the Hamiltonian is identically zero over the entire time period:

$$H^*(t) = 0 \qquad (9.43)$$

Thus, when the final time is free (to be chosen), the Hamiltonian is identically zero, which provides an additional equation, Eq. (9.43). Maximization of the Hamiltonian function by the manipulated variables is now carried out. There are two different cases: the Hamiltonian is (1) nonlinear or (2) linear in the manipulated variables, u_i. The nonlinear case is considered first, followed by the linear case.

9.4.3.1 Nonlinear in Manipulated Variables
In this case, the usual maximization is carried out:

$$\frac{\partial H}{\partial \mathbf{u}} = 0 \quad \text{or} \quad \frac{\partial H}{\partial u_i} = 0, \quad i = 1, 2, 3, \ldots, m \qquad (9.44)$$

Equation (9.44) states that the Hamiltonian is maximized by each one of the manipulated variables. Because the manipulated variables are constrained within a range of magnitude, the optimum may lie on the constraint boundary, where $\partial H/\partial u_i \neq 0$. Conversely, if the Hamiltonian is linear in u, then the optimal u_i depends on the sign of the coefficient associated with the manipulated variable.

9.4.3.2 Linear in Manipulated Variables
If the coefficients associated with the manipulated variables in the Hamiltonian function are all positive, the maximum values of the manipulated variables maximize the Hamiltonian. Conversely, if the coefficients are all negative, the minimum values of the manipulated variables maximize the Hamiltonian. To illustrate this point, we rearrange the RHS of the state equation into two terms, one that is free of manipulated variables and another that depends on the manipulated variables:

$$\frac{d\mathbf{x}}{dt} = \mathbf{f}(\mathbf{x}, \mathbf{u}) = \mathbf{f}_1(\mathbf{x}) + f_2(\mathbf{x})\mathbf{u} \qquad (9.45)$$

Then, the corresponding Hamiltonian is

$$H = \boldsymbol{\lambda}^T[\mathbf{f}_1(\mathbf{x}) + f_2\mathbf{u}] = \sum_{i=1}^{n} \lambda_i f_{1i}(\mathbf{x}) + \sum_{i=1}^{n} f_2(\mathbf{x})\lambda_i u_i = \sum_{i=1}^{n} \lambda_i f_{1i}(\mathbf{x}) + \sum_{i=1}^{n} \phi_i u_i \qquad (9.46)$$

Note that the Hamiltonian is linear in the manipulated variables and the coefficients, that multiply the manipulated variables are $\phi_i = f_2 \lambda_i$ so that the choice of u_i depends on the sign of ϕ_i, the *switching function*:

$$u_i = \begin{pmatrix} u_{i\,\text{max}} & \text{if} \phi_i > 0 \\ u_{i\,\text{min}} & \text{if} \phi_i < 0 \\ u_{i\,\text{sin}} = ? \text{ if } \phi_i \equiv 0 \text{ over finite interval(s)} \end{pmatrix} \qquad (9.47)$$

where the subscripts max, min, and sin refer to the maximum, minimum, and singular values, respectively, of the manipulated variables. As the switching function changes in sign, say, from plus to minus (or minus to plus), the manipulated variable takes on the maximum value and switches to the minimum value (and vice versa). Therefore, the coefficients ϕ_i are appropriately known as the *switching functions*.

We note that the maximum principle does not provide any solution when the coefficients vanish over a finite time interval, not at a point. This does not imply that any value of the manipulated variables would maximize the Hamiltonian. The time interval in which the coefficients vanish identically over a finite time interval is known as the *singular interval* and the control (manipulated variables) during the time interval as the *singular control*. This phenomenon appears frequently in semi-batch chemical reactors, catalyst distribution problems, and semi-batch polymerization reactions and most frequently in fed-batch operations. The singular feed rate must lie between the minimum and maximum values given by Eq. (9.9) and is determined subsequently.

9.4.3.3 Singular Control

Singular control problems arise when the Hamiltonian is linear in manipulated variables. Because the Hamiltonian is a scalar product of the adjoint variables with the RHSs of the ordinary differential equations of unsteady state mass and energy balances, the Hamiltonian is linear when the manipulated variables appear linearly in the mass and/or energy balances. This happens if one manipulates the feed flow rate of chemical, biological, and polymerization reactors, particularly semi-batch operations and separation units, and in binary catalyst distribution problems. The switching functions $\phi_i(t)$ dictate the control function, as shown in Eq. (9.47).

Let us look at a graphical illustration of Eq. (9.47) in Figure 9.1, where some feasible switching functions and the corresponding sequence of feed rate profiles are presented. This discussion is restricted to a segment having one singular period but readily carries over to multiple singular intervals. The switching function can have a sequence of positives or negatives, or a period of zero value followed by a period of positive (negative) values, all of which are not illustrated in Figure 9.1. The switching function can start with a positive value, go through an interval of zero value, and take on a negative value (case A). The feed rate sequence according to Eq. (9.47) is maximum → singular (intermediate values) → minimum. Conversely, case B, which is the direct opposite of case A, leads to the feed rate sequence of minimum → singular → maximum. The switching function can begin with a singular interval, which is then followed by negative or positive values (cases C and D), respectively. The feed rate sequence is singular → minimum for case C, and for case D, it is singular → maximum.

As shown in Chapters 12–15, the initial conditions, in particular, the initial substrate concentration, dictate the initial sign of the switching function. If the initial substrate concentration is higher than a critical value, the switching function takes on positive values, and therefore, the initial feed rate is the maximum value, whereas if the initial substrate concentration is lower than the critical value, the switching function takes on negative values, and therefore, the initial feed rate is the minimum value (no feeding, batch).

General dynamic process equations, the unsteady state mass balance equations, showing explicitly the linear appearance of manipulated variables, can be written as

$$\frac{d\mathbf{x}}{dt} = \mathbf{g}(\mathbf{x}) + \mathbf{H}(\mathbf{x})\mathbf{u} \qquad (9.48)$$

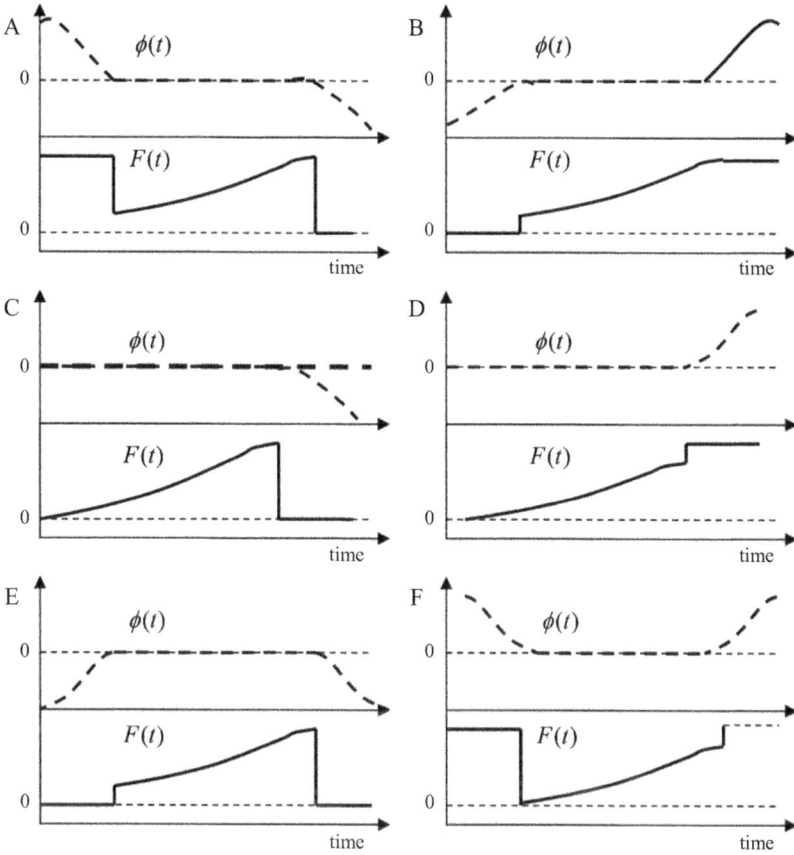

Figure 9.1. Various forms of switching functions $\phi(t)$ and the corresponding feed rate profiles $F(t)$.

The RHS, $\mathbf{g}(\mathbf{x})$, is an n-component vector function that depends on the state vector only, and $\mathbf{H}(\mathbf{x})$ is an $n \times m$ matrix that may depend on the state vector. However, most frequently, it is a constant matrix. In many fed-batch operations, there is only one feed rate so that there is only one manipulated variable, and Eq. (9.48) can be reduced to

$$\frac{d\mathbf{x}}{dt} = \mathbf{f}_1(\mathbf{x}) + \mathbf{f}_2(\mathbf{x})u \tag{9.49}$$

Furthermore, the vector function $\mathbf{f}_2(\mathbf{x})$ is a constant vector so that Eq. (9.49) can be further reduced to

$$\frac{d\mathbf{x}}{dt} = \mathbf{f}_1(\mathbf{x}) + \mathbf{f}_2 u \tag{9.50}$$

where $\mathbf{f}_2$ is a constant vector. Because there is only one control variable, there is also a single switching function. The Hamiltonian associated with the state equation, Eq. (9.50), is

$$H = \lambda^{\mathrm{T}}[\mathbf{f}_1(\mathbf{x}) + \mathbf{f}_2 u] = \sum_{i=1}^{n} \lambda_i f_{1i}(\mathbf{x}) + u \sum_{i=1}^{n} \lambda_i f_{2i}(\mathbf{x}) = \sum_{i=1}^{n} \lambda_i f_{1i}(\mathbf{x}) + \phi u \tag{9.51}$$

where the single switching function is $\phi = \sum_{i=1}^{n} \lambda_i f_{2i}$.

9.5 Handling of Problems in Nonstandard Forms

In some cases, either or both of the state equations and the performance index may not be in the standard forms used in the preceding section. In other words, the state equations may depend explicitly on the clock time, that is, they may be nonautonomous. The performance index may not depend on only the final state vector, for example, it may depend on the final state as well as some initial state variables. We handle subsequently these nonstandard forms.

9.5.1 Nonautonomous Processes

If the time t appears explicitly in the state equation, $d\mathbf{x}/dt = \mathbf{f}[\mathbf{x}(t), \mathbf{u}(t), t]$, it is necessary to introduce an augmented state variable, say, $x_{n+1} = t$, through an additional state equation:

$$\frac{dx_{n+1}}{dt} = 1, \quad x_{n+1}(0) = 0 \tag{9.52}$$

This state equation is added to the original state equation so that we now have an augmented state equation, which appears autonomous:

$$\frac{d\mathbf{x_a}}{dt} = \frac{d}{dt}\begin{pmatrix} \mathbf{x} \\ x_{n+1} \end{pmatrix} = \begin{pmatrix} \mathbf{f}(\mathbf{x}, x_{n+1}) \\ 1 \end{pmatrix} = \mathbf{f_a} \tag{9.53}$$

With the introduction of the augmented state variable, the explicit time dependence disappears in the state equations.

9.5.2 Performance Indices Depend Explicitly on the Final Time

If the performance index depends explicitly on the final time in addition to the final state variables, $P[\mathbf{x}(t_f), t_f]$, we introduce, as earlier, an augmented state variable (Eq. (9.52)) so that $t_f = x_{n+1}(t_f)$ and

$$P[\mathbf{x}(t_f), t_f] = P[\mathbf{x}(t_f), x_{n+1}(t_f)] = P[\mathbf{x_a}(t_f)] \tag{9.54}$$

With the introduction of an augmented state vector, $\mathbf{x_a}(t)$, the explicit time dependence disappears from the performance index.

The performance may also depend on one or more initial state variables, for example, $P[\mathbf{x}(t_f), x_r(0)]$. Because the performance index depends both on the final state and on the initial state, the matching conditions according to Eq. (9.28) are

$$\delta P = \left[\frac{\partial P}{\partial \mathbf{x}(t_f)} \right]^T \delta \mathbf{x}(t_f) + \left[\frac{\partial P}{\partial \mathbf{x}(0)} \right]^T \delta \mathbf{x}(0) = \boldsymbol{\lambda}^T(t_f)\delta\mathbf{x}(t_f) - \boldsymbol{\lambda}^T(0)\delta\mathbf{x}(0) \tag{9.55}$$

so that for the rth component,

$$\left[\frac{\partial P}{\partial x_r(t_f)} \right] \delta x_r(t_f) + \left[\frac{\partial P}{\partial x_r(0)} \right] \delta x_r(0) = \lambda_r(t_f)\delta x_r(t_f) - \lambda_r(0)\delta x_r(0) \tag{9.56}$$

Thus, the boundary conditions are

$$\lambda_r(t_f) = \frac{\partial P}{\partial x_r(t_f)}, \quad \lambda_r(0) = -\frac{\partial P}{\partial x_r(0)} \tag{9.57}$$

Thus, the appearance of a free initial state variable in the performance index imposes an initial condition on the corresponding adjoint variable.

9.5.3 Other Forms of Performance Index

Sometimes the performance index may contain an integral:

$$P = P[\mathbf{x}(t_f)] + \int_0^{t_f} g(\mathbf{x}, \mathbf{u})dt \tag{9.58}$$

The integral can be replaced by introducing an additional state variable, say, x_{n+2}, through an additional state equation:

$$\frac{dx_{n+2}}{dt} = g(\mathbf{x}, \mathbf{u}) \quad x_{n+2}(0) = 0 \tag{9.59}$$

Then, the integral in Eq. (9.58) is replaced by

$$\int_0^{t_f} g(\mathbf{x}, \mathbf{u})dt = x_{n+2}(t_f) \tag{9.60}$$

and Eq. (9.58) becomes

$$P = P[\mathbf{x}(t_f)] + x_{n+2}(t_f) \tag{9.61}$$

Thus, the performance index now becomes a function only of the final states.

9.5.4 Constraints on State Variables

Sometimes it may be necessary to impose a constraint involving state variables. In fed-batch operations, there is always a volume constraint, and at times, it may be desirable to impose a constraint on the cell concentration, the reactor volume, or the specific growth rate to remain free of oxygen demand limitations. Let us consider a scalar inequality constraint on the state first:

$$h[\mathbf{x}(t)] \leq 0 \tag{9.62}$$

In the presence of the inequality constraint, the Hamiltonian must be modified[34] to

$$H[\mathbf{x}(t), \boldsymbol{\lambda}(t), \mathbf{u}(t)] = \boldsymbol{\lambda}^T \mathbf{f} + \eta h(\mathbf{x}) \tag{9.63}$$

where the Lagrangian $\eta(t) \geq 0$ is a scalar function that satisfies $\eta(t)h(\mathbf{x}) \equiv 0$ so that

$$\eta(t) \begin{bmatrix} = 0 \text{ when } h(\mathbf{x}) < 0 \\ < 0 \text{ when } h(\mathbf{x}) = 0 \end{bmatrix} \tag{9.64}$$

When the feed rate $F(t)$ is chosen as the manipulated variable, we note that the state variable constraint, $V_0 \leq V \leq V_{max}$, is reduced to a terminal state constraint because $dV/dt = F \geq 0$ implies that $V(t_f) = V_{max}$ and that $V(t) \leq V_{max}$:

$$V(t_f) = x_k(t_f) = V_{max} \qquad (9.65)$$

This makes unknown the terminal condition on the corresponding adjoint variable, $\lambda_k(t_f) = $ unknown.

9.5.5 Constraints on Control Variables

There are situations in which the control variable or combinations with state variables are constrained. We consider the general case of a single control variable with multiple state variables given analytically by

$$C[\mathbf{x}(t), u(t), t] \leq 0 \qquad (9.66)$$

We take a variation of Eq. (9.66) on the control boundary $C(\mathbf{x}, u, t) = 0$:

$$\sum_{i=1}^{n} \frac{\partial C}{\partial x_i} \delta x_i + \frac{\partial C}{\partial u} \delta u = 0 \qquad (9.67)$$

Therefore, on the control boundary, it is seen that

$$\delta u = \frac{-\sum_{i=1}^{n} (\partial C / \partial x_i) \delta x_i}{\partial C / \partial u} \qquad (9.68)$$

9.6 Optimization of Initial Conditions

As seen in the preceding example problems, the performance index and the sequence of optimal feed rates (profiles) depend on the given initial conditions. Therefore, it is important to choose the best initial conditions, that is, the initial inoculum size, the initial reactor volume, and the initial substrate concentration. Concerning the inoculum size, in general, the bigger the inoculum size is, the better the performance will be. In other words, the bigger the initial cell concentration is, the higher the amount or the rate of generation of cell mass and products will be. However, the initial inoculum size is limited by practical constraints and traditional practices. Therefore, we take whatever inoculum size can be handled, and this leaves the initial conditions to be optimally selected as the reactor volume and the substrate concentration. As we show in Chapters 12–14, the initial substrate concentration should be selected to eliminate the initial sequence of feed rates of maximum or maximum–minimum (minimum or minimum–maximum) to place the state onto a hypersurface so that the initial rate is indeed the singular feed rate. The optimum initial reactor volume must be determined by carrying a parameter optimization on it.

9.7 Generalized Legendre–Clebsch Condition

Another important condition is the *generalized Legendre–Clebsch condtion:*[35]

$$\frac{\partial}{\partial F}\frac{d^p\phi}{dt^p} = 0, \quad p \text{ is an odd integer}$$

$$(-)^q\frac{\partial}{\partial F}\frac{d^{2q}\phi}{dt^{2q}} \leq 0$$

(9.69)

where q is an order of singularity, $2q$ being the number of nonvanishing derivatives of the Hamiltonian when a control variable appears explicitly. In general, the singular control is a function of both state and variables. Because a feedback control is preferred, it is necessary to eliminate the adjoint variables using $\phi = 0, d\phi/dt = 0$, and $H = 0$ (free t_f).

Through PMP and singular control theory, the optimization problem is reduced to solving a two-point boundary value, with the singular feed rate deduced from the time derivatives of the switching function. However, there is not enough information to deduce the singular feed rate despite knowing the relationship between the feed rate and the switching function given by Eq. (9.47). Thus, it is necessary to investigate thoroughly all the conditions imposed and implied by the theory.

9.8 Transformation to Nonsingular Problem[36–38]

We have so far considered fed-batch optimization through the manipulation of feed rates using the maximum principle. In general, the objective of fed-batch fermentation is maximization of performance indices, such as the profit, productivity, or final product concentration, by manipulating the manipulated variables such as pH, temperature, or the feed rate(s) of substrates, inducers, and precursors and optimally selecting the initial conditions such as the substrate concentration, reactor volume, and inoculum sizes. Whereas the determination of optimum initial conditions is a problem in ordinary calculus, parameter optimization, the determination of optimal time profiles, is a variational calculus problem. The temperature and pH usually appear nonlinearly in the fed-batch mass balance and energy balance equations and therefore also appear nonlinearly in the Hamiltonian; thus, determining their optimal profiles is a nonsingular problem and is readily made using the maximum principle of Pontryagin and a standard numerical gradient method. Conversely, the feed rates appear linearly in the mass balance equation and therefore in the Hamiltonian. Thus, the problem becomes a singular problem that is numerically difficult to solve, especially for high-order processes.

As discussed in Section 9.3, although the substrate feed rate profile $F(t)$ appears to be a most direct and logical choice for manipulated variables, there are a number of other choices for the manipulated variables such as the culture volume profile[1] $V(t)$, variable feed substrate concentration profile[2,3] $S_F(t)$, substrate concentration profile[2,4] $S(t)$, mass flow rate of substrate[5] $S_FF(t)$, or specific growth rate[6] μ.

It is possible to formulate the problem through a transformation of variables so that the feed flow rate does not appear in the transformed equations, or appears nonlinearly, and thus one can avoid a singular control problem. When the feed rate is eliminated from the mass balance equations and the substrate concentration is then chosen as the manipulated variable, the optimal feed rate $F(t)$ must be determined from the optimal substrate concentration $S(t)$,[2,4] which was obtained by ignoring the substrate and overall balance equations. If the calculated optimal profile violates the physical constraints (through the substrate and overall mass balances), the minimum and/or maximum flow rates, this approach needs to be modified to include an iterative solution to meet the constraints, leading to a very difficult problem. Similar to this approach is a nonsingular transformation method[6–9] by which the feed rate is made to appear nonlinearly and thus avoid a singular problem. However, there remains a numerical convergence problem and/or appropriate constraints that need to be checked at each time interval. Optimizing the performance index using as the control variable the feed substrate concentration $S_F(t)$[2,4], while keeping the feed flow rate constant, may appear to make the problem a bit easier to solve, but the physical implementation of the feed substrate concentration as a function of time poses another difficulty. A general method of transforming singular problems into nonsingular problems is introduced subsequently.

9.8.1 Transformation of Singular Problems into Nonsingular Problems

A general transformation based on the amounts of substrate consumed in place of the substrate balances can be used to convert the feed rate optimization problem into the problem of optimal substrate concentration profiles with constraints on state and control variables. This transformation avoids the difficulty associated with dealing with singular control problems, as presented earlier. This means we can avoid the numerical difficulty of optimization of the singular control problem, especially for high-order processes. It will be shown that the transformation involves alternate forms of mass balance equations for the limiting substrate, and therefore, it can be used for processes with multiple feed rates and high-order processes.

We shall introduce a transformation method in which the feed rate is eliminated from the mass balance equations and the substrate concentration is manipulated to optimize the performance index. Let us start with the problem of maximizing cell mass. The fed-batch mass balance equations for cell mass, the limiting substrate, and the overall mass are given in Chapter 6 and repeated here:

$$\frac{dx_1}{dt} = \mu x_1 \tag{9.70}$$

$$\frac{dx_2}{dt} = FS_F - \sigma x_1 \tag{9.71}$$

$$\frac{dV}{dt} = F \tag{9.72}$$

where $x_1 = XV$ is the total amount of cell mass, $x_2 = SV$ is the total amount of glucose, and $x_3 = V$ is the culture volume. The performance index is

$$\underset{F(t)}{\text{Max}}[P = x_1(t_f)], \quad t_f \text{ fixed} \tag{9.73}$$

There is a volume constraint:

$$V_0 \leq V(t) \leq V_{\max} \qquad (9.74)$$

The flow rates are also constrained by

$$0 = F_{\min} \leq F \leq F_{\max} \qquad (9.75)$$

To eliminate F from the process model, we introduce a new state variable that represents the total amounts of substrate consumed $x_4(t)$, which is equal to the amount present initially $(S_0 V_0)$ plus the amount added $(S_F(V - V_0))$ minus the amount remaining (VS) currently:

$$x_4 \overset{\Delta}{=} S_0 V_0 + S_F(V - V_0) - VS = x_{20} + S_F(x_3 - V_0) - x_2 \qquad (9.76)$$

Then, the time derivative of Eq. (9.76), dx_4/dt, should represent the consumption rate of substrate, $\sigma XV = \sigma x_1$:

$$\sigma XV = \sigma x_1 = \frac{dx_4}{dt} = S_F \frac{dx_3}{dt} - \frac{dx_2}{dt} = S_F F - \frac{dx_2}{dt} \qquad (9.77)$$

It is apparent that Eq. (9.77) contains the substrate balance equation and represents another form of substrate balance equation (9.71). Thus, we can rearrange Eq. (9.77) as

$$\frac{dx_4}{dt} = \sigma x_1 \qquad (9.78)$$

Equation (9.78) is another form of substrate balance equation written explicitly without the feed rate. The initial condition on x_4, the amount consumed initially, is obviously zero, as verified from Eq. (9.76):

$$x_4(0) = S_0 V_0 + S_F(V_0 - V_0) - (SV)_{t=0} = S_0 V_0 + S_F(V_0 - V_0) - S_0 V_0 = 0 \quad (9.79)$$

Therefore, we can replace Eq. (9.71) with Eq. (9.78). It is also clear that the overall balance (Eq. (9.72)) need not be included because it does not appear explicitly in three-component balance equations,

$$\frac{dx_1}{dt} = \mu(S)x_1 \; x_1(0) = X_0 V_0 \qquad (9.80)$$

$$\frac{dx_4}{dt} = \sigma(S)x_1 \; x_4(0) = 0 \qquad (9.81)$$

and the performance index is

$$\underset{S(t)}{\text{Max}}[P = x_1(t_f)] \qquad (9.82)$$

The problem posed by Eqs. (9.80)–(9.82) is nonsingular because the substrate concentration $S(t)$ appears nonlinearly, and therefore, a numerical solution can readily be obtained, say, using the steepest ascend. At this point, it should be noted that this approach is limited to processes in which the specific rates are functions only of substrate concentration. If the specific rate is a function of concentrations of cell mass and/or product, that is, $\mu(S, X, P)$, then this approach cannot be used. If the process is to make cell mass, or the cell mass is the product, and the cell mass yield coefficient is a constant, then the cell mass concentration becomes a function only of substrate concentration, and therefore, the problem reduces effectively to that in which the specific rates are functions only of substrate concentration.

The volume constraint is obtained from Eq. (9.74) using Eq. (9.76):

$$V - V_{max} = \frac{x_4 + V_0(S_F - S_0)}{(S_F - S)} - V_{max} = g(x_4, S) - V_{max} \leq 0 \qquad (9.83)$$

Equation (9.83) is a constraint on the state and manipulated variables. Once we obtain the optimum substrate concentration profile $S^*(t)$ and therefore dS^*/dt, the corresponding optimum feed rate $F^*(t)$ is obtained from Eq. (9.71):

$$F^*(t) = [\sigma^*(XV)^* + V^* dS^*/dt]/(S_F - S^*) \qquad (9.84)$$

The reactor volume is obtained from Eq. (9.83):

$$V^* = [x_4^* + V_0(S_F - S_0)]/(S_F - S^*) \qquad (9.85)$$

Therefore, the optimal feed rate is obtained by substituting Eq. (9.85) into Eq. (9.84):

$$F^*(t) = \{\sigma^* x_1^* + [x_4^* + V_0(S_F - S_0)](dS/dt)/(S_F - S^*)\}/(S_F - S^*) \qquad (9.86)$$

According to Eq. (9.86), the optimal feed rate profile is obtained from the time profiles of substrate concentration, the time derivative of substrate concentration, the profile of the total amount of cell mass, and the amount of substrate consumed. The original volume constraint is now a constraint in state variable $x_4(t)$ and control variabl $S(t)$

$$V_0 \leq V = [x_4 + V_0(S_F - S_0)]/(S_F - S) \leq V_{max} \qquad (9.87)$$

The constraint on the manipulated variable is

$$S(0) = S_0, \quad S_{min} \leq S(t) \leq S_{max} \qquad (9.88)$$

where S_{min} and S_{max} are to be determined from the constraints on the magnitude constraints on the feed rate:

$$0 = F_{min} \leq F \leq F_{max} \qquad (9.89)$$

As illustrated, we can replace the substrate balance equation (9.71) with the substrate consumption rate equation (9.77) and use the substrate concentration as the manipulated variable instead of the feed rate, thus avoiding a singular problem.

9.8.2 Substrate Concentration as Single Manipulated Variable

The solution to the problem posed by Eqs. (9.70), (9.73), (9.78), and (9.83) is obtained via PMP:

$$\begin{aligned} \underset{S(t)}{Max}\{H &= \lambda_1 \mu x_1 + \lambda_4 \sigma x_1 + \eta[g(x_4, S) - V_{max}] \\ &= [\lambda_1 \mu(S) + \lambda_4 \sigma(S)]x_1 + \eta(t)g(x_4, S)\} \end{aligned} \qquad (9.90)$$

where $\eta(t)$ is a Lagrangian multiplier. We note that $S(t)$ appears nonlinearly. Thus, the singular problem with the feed rate $F(t)$ as the manipulated variable is avoided, and a nonsingular problem is formulated with the substrate concentration $S(t)$ as the manipulated variable. The Hamiltonian is maximized:

$$\underset{S(t)}{\text{Max}} H; \quad \frac{\partial[(\lambda_1\mu + \lambda_4\sigma)x_1 + \eta g]}{\partial S} = 0 \tag{9.91}$$

where

$$\eta \begin{cases} = 0 & \text{when } g < 0, \text{ off equality control boundary} \\ \leq 0 & \text{when } g = 0, \text{ on equality control boundary} \end{cases} \tag{9.92}$$

Therefore, off the equality control boundary, the adjoint variables and control function must satisfy

$$\frac{d}{dt}\begin{pmatrix} \lambda_1 \\ \lambda_4 \end{pmatrix} = -\begin{pmatrix} \partial H/\partial x_1 \\ \partial H/\partial x_4 \end{pmatrix} = -\begin{pmatrix} \lambda_1\mu + \lambda_4\sigma \\ 0 \end{pmatrix} \tag{9.93}$$

$$\partial H/\partial S = 0 = (\lambda_1\partial\mu/\partial S + \lambda_4\partial\sigma/\partial S) \tag{9.94}$$

On the control constraint boundary, the corresponding adjoint equations and control function are given by

$$\frac{d}{dt}\begin{pmatrix} \lambda_1 \\ \lambda_4 \end{pmatrix} = -\begin{pmatrix} \partial H/\partial x_1 \\ \partial H/\partial x_4 \end{pmatrix} = -\begin{pmatrix} \lambda_1\mu + \lambda_4\sigma \\ 0 \end{pmatrix} \quad \begin{pmatrix} \lambda_1(t_f) \\ \lambda_4(t_f) \end{pmatrix} = \begin{pmatrix} 1 \\ 0 \end{pmatrix} \tag{9.95}$$

Because $d\lambda_4/dt = 0$ and $\lambda_4(t_f) = 0$, we conclude that $\lambda_4(t) = 0$. Therefore, the Hamiltonian and adjoint equation reduce to

$$\underset{S(t)}{\text{Max}}[H = \lambda_1\mu x_1 + \lambda_4\sigma x_1 = \lambda_1\mu x_1] \tag{9.96}$$

$$\frac{d\lambda_1}{dt} = (\partial H/\partial x_1) = -\lambda_1\mu \tag{9.97}$$

Maximizing the Hamiltonian, we obtain

$$\underset{S(t)}{\text{Max}}[H = \lambda_1\mu x_1] \quad \Rightarrow \partial H/\partial S = 0 = \partial\mu/\partial S \tag{9.98}$$

According to Eq. (9.98), the specific growth rate must be maximized to maximize the amount of cells at the final time. Therefore, the substrate concentration must be maintained at the value $S = S_m$, corresponding to the maximum specific growth rate, $\mu_{max} = \mu(S = S_m)$. The feed rate to maintain the specific growth rate at its maximum value is obtained from Eq. (9.86) with $dS/dt = 0$:

$$F^*(t) = \frac{\sigma(S_m)x_1}{(S_F - S_m)} = \frac{\sigma(S_m)XV}{(S_F - S_m)} \tag{9.99}$$

The total amount of cells XV is obtained from the cell balance equation (9.70), if the feed rate of Eq. (9.99) is applied from $t = 0$ so that $S_0 = S_m$:

$$\frac{dx_1}{dt} = \frac{d(XV)}{dt} = \mu x_1 = \mu(S_m)XV = \mu_m XV \Rightarrow XV = (XV)_0\exp(\mu_m t) \tag{9.100}$$

Equation (9.100) is then substituted into Eq. (9.99) to obtain the feed rate that is exponential:

$$F^*(t) = \frac{\sigma(S_m)x_1}{(S_F - S_m)} = \frac{\sigma(S_m)XV}{(S_F - S_m)} = \frac{(XV)_0 \exp(\mu_m t)}{(S_F - S_m)}$$

$$= \frac{(XV)_0}{(S_F - S_m)} \exp(\mu_m t) = \alpha \exp(\mu_m t) \qquad (9.101)$$

9.8.3 Singular Problems with Multiple Manipulated Variables

Consider now the problem of manipulating two feed streams, for example, the fed-batch culture of poly-β-hydroxyl butyric acid (PHB[6]) without the inhibition of product formation rate by the product, $\pi(S_1, S_2)$. The specific rates of cell growth, substrate consumption, and product formation are functions of the two substrates, glucose and ammonium chloride. Normally, the specific product formation rate is a function of both substrates as well as the product, PHB. Here we deal with a situation in which the product inhibition is negligible:

$$\frac{dx_1}{dt} = \frac{d(XV)}{dt} = \mu(S_1, S_2)XV = \mu(S_1, S_2)x_1 \qquad (9.102)$$

$$\frac{dx_2}{dt} = \frac{d(S_1 V)}{dt} = S_{1F}F_1 - \sigma_1(S_1, S_2)XV = S_{1F}F_1 - \sigma_1(S_1, S_2)x_1 \qquad (9.103)$$

$$\frac{dx_3}{dt} = \frac{d(S_2 V)}{dt} = S_{2F}F_2 - \sigma_2(S_1, S_2)XV = S_{2F}F_2 - \sigma_2(S_1, S_2)x_1 \qquad (9.104)$$

$$\frac{dx_4}{dt} = \frac{d(PV)}{dt} = \pi(S_1, S_2)XV = \pi(S_1, S_2)x_1 \qquad (9.105)$$

$$\frac{dx_5\rho}{dt} = \frac{d(V\rho)}{dt} = F_1\rho_1 + F_2\rho_2 \qquad (9.106)$$

In this PHB model, S_1 and S_2 represent the concentrations of glucose and ammonium chloride, respectively, P is the concentration of PHB, X is the concentration of active cell mass (cell concentration − PHB concentration), V is the fermentor volume, S_{1F} and S_{2F} are the feed concentrations of glucose and ammonia, respectively, F_1 and F_2 are the glucose and ammonium chloride feed rates, and ρ, ρ_1, and ρ_2 are the densities of the culture, the glucose feed, and the ammonium chloride feed, respectively. It is assumed in general that the densities are the same, $\rho = \rho_1 = \rho_2$, so that Eq. (9.106) reduces to

$$\frac{dx_5}{dt} = \frac{dV}{dt} = F_1 + F_2 \qquad (9.107)$$

Because there are two feed streams, glucose and ammonium chloride, we introduce the amounts of glucose and ammonium chloride consumed, x_6 and x_7, the amount present initially plus the amount added minus the amount remaining:

$$x_6(t) \stackrel{\Delta}{=} S_{10}V_0 + S_{1F} \int_0^t F_1 d\tau - S_1 V(t) = x_2(0) - x_2(t) + S_{1F} \int_0^t F_1 d\tau \qquad (9.108)$$

$$x_7(t) \overset{\Delta}{=} S_{20}V_0 + S_{2F} \int_0^t F_2 d\tau - S_2 V(t) = x_3(0) - x_3(t) + S_{2F} \int_0^t F_2 d\tau \quad (9.109)$$

Then, the time derivatives of x_6 and x_7 are equal to the consumption of substrates S_1 and S_2, respectively:

$$\frac{dx_6}{dt} = \sigma_1(S_1, S_2)x_1 \quad (9.110)$$

$$\frac{dx_7}{dt} = \sigma_2(S_1, S_2)x_1 \quad (9.111)$$

Conversely, differentiation of Eqs. (9.108) and (9.109) yields

$$\frac{dx_6}{dt} = S_{1F}F_1 - \frac{dx_2}{dt} = S_{1F}F_1 - (S_{1F}F_1 - \sigma_1 x_1) = \sigma_1(S_1, S_2)x_1$$
$$\frac{dx_7}{dt} = S_{2F}F_2 - \frac{dx_3}{dt} = S_{2F}F_2 - (S_{2F}F_2 - \sigma_2 x_1) = \sigma_2(S_1, S_2)x_1 \quad (9.112)$$

Equation (9.112) confirms that by introducing the amounts of substrates consumed (Eqs. (9.108) and (9.109)), we can replace the two substrate balance equations (Eqs. (9.103) and (9.104)), yielding a new set of balance equations:

$$\frac{dx_1}{dt} = \mu(S_1, S_2)x_1 \quad x_1(0) = x_{10} \quad (9.113)$$

$$\frac{dx_4}{dt} = \pi(S_1, S_2)x_1 \quad x_4(0) = x_{40} \quad (9.114)$$

$$\frac{dx_6}{dt} = \sigma_1(S_1, S_2)x_1 \quad x_6(0) = 0 \quad (9.115)$$

$$\frac{dx_7}{dt} = \sigma_2(S_1, S_2)x_1 \quad x_7(0) = 0 \quad (9.116)$$

For this problem, both $S_1(t)$ and $S_2(t)$ appear nonlinearly in the specific rates, and therefore, this problem is nonsingular with respect to both $S_1(t)$ and $S_2(t)$, which are the only inputs to the differential equations.

The performance index is maximized using the concentration profiles of $S_1(t)$ and $S_2(t)$:

$$\underset{S_1(t),S_2(t)}{\text{Max}} P[x_4(t_f)] \quad (9.117)$$

The Hamiltonian to be maximized is

$$\underset{S_1(t),S_2(t)}{\text{Max}} \{H = \lambda_1 \mu x_1 + \lambda_6 \sigma_1 x_1 + \lambda_7 \sigma_2 x_1 + \lambda_4 \pi x_1 + \eta[g(x_4, S) - V_{max}] \quad (9.118)$$
$$= [\lambda_1 \mu + \lambda_6 \sigma_1 + \lambda_7 \sigma_2 + \lambda_4 \pi]x_1 + \eta(t)h(x_4, S)\}$$

$$\frac{\partial H}{\partial S_1} = 0 \quad \text{and} \quad \frac{\partial H}{\partial S_2} = 0 \quad (9.119)$$

The optimal numerical solutions, $S_1^*(t)$ and $S_2^*(t)$, can be obtained using the steepest ascent method, as described in Chapter 10. With the optimal substrate concentrations, one can construct a control strategy[40] in which the measured substrate concentration profiles are forced to follow the optimal paths, $S_1^*(t)$ and $S_2^*(t)$, by manipulating the feed rates, $F_1(t)$ and $F_2(t)$.

If it is desired to obtain the optimal feed rates, F_1^* and F_2^*, from the preceding optimal $S_1^*(t)$ and $S_2^*(t)$, we first collect the data: $x_1^*(t) = (XV)^*$, $x_6^*(t)$, $x_7^*(t)$, and $x_4^*(t)$. Then, expressions for F_1^* and F_2^* are obtained from the substrate balances (Eqs. (9.103) and (9.104)):

$$\frac{d(S_1 V)}{dt} = \frac{dS_1}{dt}V + S_1\frac{dV}{dt} = \frac{dS_1}{dt}V + S_1(F_1 + F_2) = S_{1F}F_1 - \sigma_1 x_1$$

or

$$-(S_{1F} - S_1)F_1 + S_1 F_2 = -\frac{dS_1}{dt}V - \sigma_1 x_1 \tag{9.120}$$

Likewise, from Eq. (9.104),

$$S_2 F_1 - (S_{2F} - S_2)F_2 = -\frac{dS_2}{dt}V - \sigma_2 x_1 \tag{9.121}$$

Solving Eqs. (9.120) and (9.121) for F_1 and F_2,

$$F_1 = \frac{[(S_{2F} - S_2)\dot{S}_1 + S_1\dot{S}_2]V + [(S_{2F} - S_2)\sigma_1 + S_1\sigma_2]x_1}{(S_{1F} - S_1)(S_{2F} - S_2) - S_1 S_2} \triangleq \frac{N_{11}V + N_{12}}{D} \tag{9.122}$$

and

$$F_2 = \frac{[(S_{1F} - S_1)\dot{S}_2 + S_2\dot{S}_1]V + [(S_{1F} - S_1)\sigma_2 + S_2\sigma_1]x_1}{(S_{1F} - S_1)(S_{2F} - S_2) - S_1 S_2} \triangleq \frac{N_{21}V + N_{22}}{D} \tag{9.123}$$

where

$$N_{11} \triangleq [(S_{2F} - S_2)\dot{S}_1 + S_1\dot{S}_2], \quad N_{12} \triangleq [(S_{2F} - S_2)\sigma_1 + S_1\sigma_2]x_1$$
$$N_{21} \triangleq [(S_{1F} - S_1)\dot{S}_2 + S_2\dot{S}_1], \quad N_{22} \triangleq [(S_{1F} - S_1)\sigma_2 + S_2\sigma_1]x_1 \tag{9.124}$$

These two feed rates are the functions of the culture volume, $V(t)$, which is also the function of $F_1(t)$ and $F_2(t)$, $V(t) = \int_0^t (F_1 + F_2)d\tau$. Hence, we have two coupled integral equations. We can solve Eqs. (9.122) and (9.123) for V, and by equating them, we obtain the relationship between F_1 and F_2:

$$F_1 = \frac{N_{11}}{N_{21}}F_2 + \frac{N_{12}N_{21} - N_{11}N_{22}}{DN_{21}} \triangleq \alpha F_2 + \beta \tag{9.125}$$

Therefore,

$$F_1 + F_2 = \alpha F_2 + \beta + F_2 = (1 + \alpha)F_2 + \beta \tag{9.126}$$

Substitution of Eq. (9.126) into Eq. (9.107) yields

$$F_2 = \frac{N_{11}}{D\alpha}\int_0^t [(1 + \alpha)F_2 + \beta]d\tau + \frac{N_{12}}{D\alpha} - \frac{\beta}{\alpha} \tag{9.127}$$

which is a special form of Volterra equation of the second kind:

$$F_2(t) = f(t) + \int\limits_0^t K(\tau)F_2(\tau)d\tau \qquad (9.128)$$

Once we solve this equation numerically, we obtain the numerical values of $F_2(t)$. Obviously, then, $F_1(t)$ is obtained using Eq. (9.125). Numerical schemes to solve the Volterra equation of the second kind are available elsewhere[39] and are not covered here.

REFERENCES

1. Modak, J. M., and Lim, H. C. 1989. A simple nonsingular control approach to fed-batch fermentation optimization. *Biotechnology and Bioengineering* 33: 1–15.
2. Guthke, R., and Knorre, W. A. 1981. Optimal substrate profile for antibiotic fermentation. *Biotechnology and Bioengineering* 23: 2771–2777.
3. Alvarez, J., and Alvarez, J. 1988. Analysis and control of fermentation processes by optimal and geometric methods, in *Proceedings of the 1988 Automatic Control Conference*, p. 1112. American Automatic Control Council.
4. San, K.-Y., and Stephanopoulos, G. 1989. Optimization of fed-batch penicillin fermentation: A case of singular optimal control with state constraints. *Biotechnology and Bioengineering* 34: 72–78.
5. Fishman, V. M., and Biryukov, V. V. 1974. Kinetic model of secondary metabolite production and its use in the computation of optimal conditions. *Biotechnology and Bioengineering Symposium* 4: 647–662.
6. Yamane, T., Kume, T., Sada, E., and Takamatsu, T. 1997. A simple optimization technique for fed-batch culture. *Journal of Fermentation Technology* 55: 587–598.
7. Menawat, A., Mutharasan, R., and Coughanowr, D. R. 1987. Singular optimal control strategy for a fed-batch bioreactor: Numerical approach. *American Institute of Chemical Engineers Journal* 33: 776–783.
8. Jayant, A., and Pushpavanam, S. 1998. Optimization of biochemical fed-batch reactor – transition from a nonsingular to a singular problem. *Industrial Engineering Chemistry Research* 37: 4314–4321.
9. Shin, H. S., and Lim, H. C. 2006. Comment on "Optimization of biochemical fed-batch reactor – transition from a nonsingular to a singular problem." *Industrial Engineering Chemistry Research* 45: 4851–4854.
10. Lee, J. H., Lim, H. C., and Hong, J. 1997. Application of nonsingular transformation to on-line optimal control of fed-batch fermentation. *Journal of Biotechnology* 55: 135–150.
11. Modak, J. 1993. Choice of control variable for optimization of fed-batch fermentation. *Chemical Engineering Journal* 32: B59–B69.
12. Parelukar, S. J., and Lim, H. C. 1985. Modeling, optimization and control of semi- batch bioreactors. *Advances in Biochemical Engineering/Biotechnology* 32: 207–258.
13. Peringer, P., and Blachere, H. T. 1979. Modeling and optimal control of bakers' yeast production in repeated fed-batch culture. *Biotechnology and Bioengineering Symposium* 9: 205–213.
14. Mori, H., Yamane, T., Kobayashi, T., and Shimizu, S. 1983. Comparison of cell productivities among fed-batch, repeated fed-batch and continuous cultures at high cell concentration. *Journal of Fermentation Technology* 61: 391–401.
15. Heninger, P. J. 1983. Computer control and optimization of a repeated fed-batch bioreactor. PhD dissertation, Purdue University.

16. Weigand, W. A., Lim, H. C., Creagan, C. C., and Mohler, R. D. 1978. *Second International Conference on Computer Applications in Fermentation Technology*. Philadelphia, Wiley.
17. Weigand, W. A., Lim, H. C., Creagan, C. C., and Mohler, R. D. 1979. Optimization of a repeated fed-batch reactor for maximum cell productivity. *Biotechnology and Bioengineering Symposium* 9: 335–348.
18. Weigand, W. A. 1981. Maximum cell productivity by repeated fed-batch culture: Constant yield case. *Biotechnology and Bioengineering* 23: 249–266.
19. Dairaku, K., Yamasaki, Y., Morikawa, H., Shioya, S., and Takamatsu, T. 1982. Experimental study of time-optimal control of fed-batch culture of baker's yeast. *Journal of Fermentation Technology* 60: 67–75.
20. Yamane, T., Sada, E., and Takamatsu, T. 1979. Start-up of chemostat: Application of fed-batch culture. *Biotechnology and Bioengineering* 21: 111–129.
21. Fishman, V. M., and Biryukov, V. V. 1974. Kinetic model of secondary metabolite production and its use in computation of optimal conditions. *Biotechnology and Bioengineering Symposium* 4: 647–662.
22. Ohno, H., Nakanishi, E., and Takamatsu, T. 1976. Optimal control of a semi-batch fermentation. *Biotechnology and Bioengineering* 18: 847–864.
23. Andeyeva, L. N., and Biryukov, V. V. 1973. Analysis of mathematical models of the effect of pH on fermentation processes and their use for calculating optimal fermentation conditions. *Biotechnology and Bioengineering Symposium* 4: 61–76.
24. Yamane, T., Kume, T., Sada, E., and Takamatsu, T. 1977. A simple optimization technique for fed batch culture. *Journal of Fermentation Technology* 55: 587–598.
25. Kishimoto, M., Yoshida, T., and Taguchi, H. 1980. Optimization of fed batch culture by dynamic programming and regression analysis. *Biotechnology Letters* 2: 403–406.
26. Kishimoto, M., Yoshida, T., and Taguchi, H. 1981. On-line optimal control of fed-batch culture of glutamic acid production. *Journal of Fermentation Technology* 59: 125–129.
27. Aiba, S. 1979. Review of process control and optimization in fermentation. *Biotechnology and Bioengineering Symposium* 9: 269–281.
28. Ohno, H., Nakanishi, E., and Takamatsu, T. 1978. Optimum operating mode for a class of fermentation. *Biotechnology and Bioengineering* 20: 625–636.
29. Choi, C. Y., and Park, S. Y. 1981. The parametric sensitivity of the optimal fed-batch fermentation policy. *Journal of Fermentation Technology* 59: 65–71.
30. Shin, S. H., and Lim, H. C. 2007. Cell-mass maximization in fed-batch culture: Sufficient conditions for singular arc and optimal feed rate profiles. *Bioprocess and Biosystems Engineering* 29: 335–347.
31. Shin, S. H., and Lim, H. C. 2007. Optimization of metabolite production in fed-batch culture: Use of sufficiency and characteristics of singular arc and properties of adjoint vector in numerical computation. *Industrial Engineering Chemistry Research* 46: 2526–2534.
32. Shin, H., and Lim, H. C. 2007. Maximization of metabolite in fed-batch cultures: Sufficient conditions for singular arc and optimal feed rate profiles. *Biochemical Engineering Journal* 37: 62–74.
33. Pontryagin, L. S., Boltynskii, V. G., Gamkrelidze, R. V., and Mishchenko, E. F. 1962. *The Mathematical Theory of Optimal Processes*. Interscience.
34. Maurer, H. 1977. On optimal control problems with bounded state variable and control appearing linearly. *Society of Industrial and Applied Mathematics Journal, Control* 15: 345–362.
35. Kelly, H. J. 1965. A transformation approach to singular subarcs in optimal trajectory and control problems. *Society of Industrial and Applied Mathematics Journal, Control* 2: 234–241.

36. Lee, J. H., Lim, H. C., Yoo, Y. H., and Park, Y. H. 1999. Optimization of feed rate profile for the monoclonal antibody production. *Bioprocess Engineering* 20: 137–146.
37. Lee, J. H., Lim, H. C., and Kim, S. I. 2001. A nonsingular optimization approach to the feed rare profile optimization of fed-batch cultures. *Bioprocess and Biosystems Engineering* 24: 115–125.
38. Linz, P. 1985. *Analytical and Numerical Methods for Volterra Equations.* Society of Industrial and Applied Mathematics.
39. Merriam, C. W., III. 1964. *Optimization Theory and the Design of Feedback Control Systems.* McGraw-Hill.

10 Computational Techniques

There are two different types of methods for profile optimization. When a process is modeled by a set of differential mass and energy balance equations, we can apply variational methods such as Pontryagin's maximum principle (PMP), as described in Chapter 9. This approach usually leads to a split two-point boundary value problem in which the initial values of the state variables and the final values of the adjoint variables are normally known and the problem reduces to maximizing the Hamiltonian. However, there are situations in which it is not possible to write one or more balance equations or no equations at all. Thus, there may be no equation or an inadequate number of equations to describe the process. These situations require a statistical means to model the process and apply optimization to optimize the process operations. In this chapter, we consider numerical techniques to solve the split boundary value problems that result from the application of PMP and statistical optimization techniques.

10.1 Computational Techniques for Processes with Known Mathematical Models

The optimization problem is formulated as in Chapter 9. The objective is to maximize the performance index, which depends on the final values of the state variables,

$$\underset{\mathbf{u}(t)}{\mathrm{Max}}\, P[\mathbf{x}(t_f)] \tag{10.1}$$

for the processes described by a set of ordinary differential mass and energy balance equations,

$$\frac{d\mathbf{x}}{dt} = \mathbf{f}[\mathbf{x}(t), \mathbf{u}(t)] \tag{10.2}$$

where $\mathbf{x}$ is an n-component state variable, $\mathbf{f}$ is an n-component vector function, and $\mathbf{u}$ is an m-component manipulated-variable vector. A set of specified initial values is

$$\mathbf{x}(0) = \mathbf{x_0} \tag{10.3}$$

The manipulated variables are magnitude constrained:

$$u_{i_{min}} \leq u_i(t) \leq u_{i_{max}} \tag{10.4}$$

The necessary conditions for optimality for this standard problem are obtained from PMP from Chapter 9 in the form of a two-point split boundary value problem, maximizing the Hamiltonian by choosing the manipulated variables properly,

$$\underset{u_1, u_2, u_3, \ldots u_m}{\text{Max}} \left(H = \boldsymbol{\lambda}^{\mathrm{T}} \mathbf{f} = \sum_{i=1}^{n} \lambda_i f_i \right) \tag{10.5}$$

where the adjoint variables satisfy the following differential equations:

$$\frac{d\boldsymbol{\lambda}}{dt} = -\frac{\partial H}{\partial \mathbf{x}} = -\frac{\partial (\mathbf{f}^T \boldsymbol{\lambda})}{\partial \mathbf{x}} = -\left(\frac{\partial \mathbf{f}}{\partial \mathbf{x}} \right)^T \boldsymbol{\lambda} \tag{10.6}$$

Subject to the final conditions,

$$\boldsymbol{\lambda}(t_f) = \left[\frac{\partial P}{\partial \mathbf{x}(t_f)} \right] \tag{10.7}$$

This split boundary value problem, that is, the state vector with the known *initial* values (Eq. (10.3)) and the adjoint vector with the known *final* values (Eq. (10.7)), can be solved by two different iteration methods: (1) the boundary condition iteration and (2) the control variables iteration. The boundary condition iteration begins by assuming initially the missing boundary conditions (either the final conditions of the state variables or the initial conditions of the adjoint variables) and integrating the state and adjoint variables simultaneously to check if the guessed boundary conditions agree with the given boundary conditions. If not, the boundary conditions are improved and the process is repeated until the computed boundary conditions agree with the given boundary conditions. The control vector iteration begins by first assuming the control variables, integrating the state equations forward and the adjoint equations backward, evaluating the Hamiltonian and maximizing it to obtain a new set of manipulated variables, and repeating the process until the Hamiltonian does not change appreciably from one iteration to the next. We now present in detail each of these iteration methods.

10.1.1 Boundary Condition Iterations (Simple Shooting Method)[1]

In this scheme, the missing boundary conditions, either the missing initial conditions on the adjoint vector $\boldsymbol{\lambda}(0)$ or the final conditions on the state variables $\mathbf{x}(t_f)$, are first guessed so that we have either the initial conditions or final conditions on both the state and adjoint variables. First, we consider guessing the initial conditions on the adjoint variables and iterating on them: the *initial value problem*.

10.1.1.1 Iterations on Guessed Initial Values of Adjoint Variables (a Simple Shooting Method)

With the guessed initial conditions on the adjoint vector, $\boldsymbol{\lambda}^0(0) = \boldsymbol{\lambda}_0^0$, set $i = 1$:

1. Using the guessed initial values of the adjoint vector, integrate forward from 0 to t_f the state and adjoint differential equations using the feed rate $F^i(t)$ that

maximizes the Hamiltonian, $\partial H/\partial F^i = \partial(\lambda^T \mathbf{f})/\partial F^i = 0$, to obtain the final condition on the adjoint vector, $\lambda^i(t_f)$ $d\mathbf{x}/dt = \mathbf{f}(\mathbf{x}, \mathbf{u}), \mathbf{x}(0) = \mathbf{x}_0; \dot{\lambda} = -(\partial \mathbf{f}/\partial \mathbf{x})^T \lambda, \lambda^i(0) = \lambda_0^{i-1}$ and compare it with the final condition given by PMP: $\lambda(t_f) = \partial P/\partial \mathbf{x}(t_f)$. If the calculated values agree with the given values, the guessed values are correct, and the resulting $F^i(t)$ is the optimal feed rate. If there is no agreement, then one must iterate on the initial values of the adjoint vector.

2. Vary the initial value of the guessed adjoint variables by small amounts, (ε) $\lambda^{i+1}(0) = \lambda^i(0) + \varepsilon^i$, and carry out the integration as in step 1 to obtain $\lambda^{i+1}(t_f)$.

3. Using the two final values of the adjoint variables, calculate the gradient, $\Delta^{i+1} = \{[(\lambda^{i+1}(t_f) - \partial P/\partial \mathbf{x}(t_f)] - [(\lambda^i(t_f) - \partial P/\partial \mathbf{x}(t_f)]\}/[\lambda^{i+1}(0) - \lambda^i(0)]$ to obtain a new initial adjoint vector, $\lambda^{i+2}(0) = \lambda^{i+1}(0) + \varepsilon \Delta^{i+1}$, and integrate as in step 1 to see if the calculated final adjoint vector $\lambda^{i+2}(t_f)$ agrees with that of PMP, $\lambda^{i+2}(t_f) = \partial P/\partial \mathbf{x}(t_f)$. If it does, the iterated initial adjoint vector is correct, and the calculated $F^{i+2}(t)$ is the optimum. If it does not, set $i = i + 1$.

4. Iterate on the initial values of the adjoint vector until the calculated final value of the adjoint vector by integration agrees with the final conditions given by PMP: $\lambda(t_f) = \partial P/\partial \mathbf{x}(t_f)$.

It is noted here that it is difficult to guess reasonable initial values of the adjoint variables as no physical insight is available on the adjoint variables. The other alternative is to guess the missing final conditions on the state variables and iterate on them. For state variables, we have some insight as to the approximate range based on physical grounds. We can approximate reasonable ranges for the state variables because the state variables are species concentrations and bioreactor volume.

10.1.1.2 Iterations on Guessed Final Values of State Variables (a Simple Shooting Method)

With a guessed set of final values of the state vector, $\mathbf{x}^0(t_f) = \mathbf{x}^0$, set $i = 1$:

1. Using the initially guessed final values of the state variables, integrate backward in time from t_f to 0 both the state and adjoint differential equations using the feed rate $F^i(t)$ that maximizes the Hamiltonian, $\partial H/\partial F^i = \partial(\lambda^T \mathbf{f})/\partial F^i = 0$, to obtain the initial values of the state vector, $\mathbf{x}^i(t_f), d\mathbf{x}/dt = \mathbf{f}(\mathbf{x}, \mathbf{u}), \mathbf{x}^0(t_f) = \mathbf{x}^0; \dot{\lambda} = -(\partial \mathbf{f}/\partial \mathbf{x})^T \lambda, \lambda(t_f) = \partial P/\partial \mathbf{x}(t_f)$ and compare it with the given initial values of the state variables, $\mathbf{x}(0) = \mathbf{x}_0$. If the calculated values agree with the given values, the guessed final values are correct, and the resulting $F^i(t)$ is the optimal feed rate. If there is no agreement, then one must begin iteration on the final values of the state vector.

2. Vary the guessed final value of the state variables by small amounts $(\varepsilon) \mathbf{x}^{i+1}(t_f) = \mathbf{x}^i(t_f) + \varepsilon^i$, and carry out the integration as in step 1 to obtain $\mathbf{x}^{i+1}(0)$.

3. Using the two sets of initial values of the state variables, calculate the gradient, $\Delta^{i+1} = \{[\mathbf{x}^{i+1}(0) - \mathbf{x}_0] - [\mathbf{x}^i(0) - \mathbf{x}_0]\}/\varepsilon$, and follow it by a small amount to obtain new final values of the state vector, $\mathbf{x}^{i+2}(t_f) = \mathbf{x}^{i+1}(t_f) + \alpha \Delta^{i+1}$, and integrate as in step 1 to see if the calculated initial values of the state vector $\mathbf{x}^{i+2}(0)$ agree with the given initial value, $\mathbf{x}_0$. If they do, the iterated final values of the state

vector are correct, and the calculated $F^{i+2}(t)$ is the optimum. If they do not, set $i = i + 1$.

4. Iterate on the final values of the state vector until the iterated initial values of the state vector agree with the given initial values, $\mathbf{x}_0$.

10.1.2 Multiple Shooting Method

An extension to the simple shooting method is the *multiple shooting method*.[2] This method involves breaking up a trajectory into a number of subintervals and formulating an initial value problem for each subinterval. The differential equations are integrated over each of the subintervals, $[t_i, t_{i+1}]$. The initial conditions at t_i are forced to match with the final conditions obtained from the integration over the previous subinterval $[t_{i-1}, t_i]$ to meet the continuity conditions by adjusting the initial conditions as free parameters for each subinterval to achieve continuity equations. Thus, if we have m subintervals for each of n state and adjoint variables, there are $2n \times m$ initial conditions to guess and $2n \times m$ continuity equations. Although this appears to be formidable, essentially, we are solving $2n \times m$ algebraic equations. We now cover briefly this multiple shooting method.

Break up the total time $(0, t_f)$ into $N - 1$ points, where the initial conditions, $\boldsymbol{\varphi}_i (i = 1, 2, 3, \dots N - 1)$, are allocated at $t = t_i$. Let us denote by $\mathbf{y}^{k+1}(t)$ the solutions of the differential equations over the time interval $[t_k, t_{k+1}]$ with the initial conditions $\mathbf{y}_k$. The continuity conditions demand that $\mathbf{y}^{k+1} = \boldsymbol{\varphi}_{k+1}$. In addition, the boundary condition must be met at $t = 0$ and $t = t_f$. Thus, this is a problem of $n \times m$ equations and $n \times m$ parameters.

A potential disadvantage of this boundary value iteration is apparent if one looks at the set of state and adjoint differential equations. The state equation in forward time integration is usually stable (negative eigenvalues), and therefore, the backward integration is unstable (positive eigenvalues). The adjoint equation in backward time integration is stable, while the forward integration is unstable. Thus, the iteration on the final values of state variables has a distinct advantage of realistically guessing the final values but can cause a numerical difficulty.

The next approach is called the control vector iteration and begins with a guessed profile for the control vector, the feed rate $F^0(t)$, and iterates on the control vector.

10.1.3 Control Vector Iterations[3]

This method begins with guessed time profiles for the control vector over the entire time interval, $0 \le t \le t_f$. With the guessed time profile, $\mathbf{u}^1(t)$, which can be a constant vector, do the following:

1. Integrate the state equations forward from 0 to t_f with the given initial values $d\mathbf{x}/dt = \mathbf{f}(\mathbf{x}, \mathbf{u}), \mathbf{x}(0) = \mathbf{x}_0$ to obtain $\mathbf{x}^1(t)$ and evaluate $\partial \mathbf{f}/\partial \mathbf{x}$ so that the adjoint equations can be integrated backward from t_f to 0 using the given final values to obtain $\boldsymbol{\lambda}^1(t), \dot{\boldsymbol{\lambda}} = -(\partial \mathbf{f}/\partial \mathbf{x})^T \boldsymbol{\lambda}, \boldsymbol{\lambda}(t_f) = \partial P/\partial \mathbf{x}(t_f)$. Evaluate the Hamiltonian, $H^1 = (\boldsymbol{\lambda}^1)^T \mathbf{f}^1$, and its gradient, $\partial H^1/\partial \mathbf{u}^1$.
2. Improve the control vector by the following steepest ascent algorithm:

$$\mathbf{u}^{i+1} = \mathbf{u}^i + W(t) \partial H^i/\partial \mathbf{u}^i$$

Here $W(t)$ is a positive definite weighting matrix to be specified based on the steepest ascent method. In case of a single control variable, it is constant. Using the improved control vector, $\mathbf{u}^{i+1}(t)$, repeat step 1 to obtain $\mathbf{x}^{i+1}(t)$, $\lambda^{i+1}(t)$, and $H^{i+1}(t)$. Check if the values of the new Hamiltonian and performance index are appreciably larger than the old, $[(H^{i+1} - H^i)/H^i] \le \varepsilon_1$ and $\{P[\mathbf{x}^{i+1}(t_f)] - P[\mathbf{x}^i(t_f)]\}/P[\mathbf{x}^i(t_f)] \le \varepsilon_2$. If they are, continue with the iteration of step 1. If they are not, the solution is on hand; stop the iteration.

EXAMPLE 10.E.1: OPTIMIZATION OF CELL MASS PRODUCTION It is instructive to give an example of a process to be optimized. Consider the simplest fed-batch process, in which a substrate is converted to cell mass. The unsteady state cell mass, substrate, and total mass balance are as follows:

$$\frac{d(XV)}{dt} = \mu XV \qquad X(0) = X_0$$
$$\frac{d(SV)}{dt} = S_F F - \sigma XV \quad S(0) = S_0 \qquad (10.E.1.1)$$
$$\frac{dV}{dt} = F \qquad V(0) = V_0$$

where X and S stand for the concentrations of cells and substrate, respectively, V is the culture volume, S_F is the constant feed substrate concentration, and F is the volumetric feed flow rate. The usual assumption of constant density is made for the feed and the bioreactor contents. The specific growth rate μ and the cell mass yield coefficient $Y_{X/S}$ are assumed, for simplicity, to be functions only of the limiting substrate concentration S.

It is convenient to define as state variables the total amount of cells and substrate and culture volume, XV, SV, V, and the feed rate as the manipulated variable $F(t)$:

$$\frac{d\mathbf{x}}{dt} = \mathbf{f}_1(\mathbf{x}) + \mathbf{f}_2 u, \quad \frac{d\mathbf{x}}{dt} = \frac{d}{dt}\begin{pmatrix} x_1 \\ x_2 \\ x_3 \end{pmatrix} = \frac{d}{dt}\begin{pmatrix} XV \\ SV \\ V \end{pmatrix} = \begin{pmatrix} \mu(\mathbf{x})x_1 \\ -\sigma(\mathbf{x})x_1 \\ 0 \end{pmatrix} + \begin{pmatrix} 0 \\ S_F \\ 1 \end{pmatrix} F$$

$$(10.E.1.2)$$

$$\mathbf{f}_1(\mathbf{x}) = \begin{pmatrix} \mu(\mathbf{x})x_1 \\ -\sigma(\mathbf{x})x_1 \\ 0 \end{pmatrix}, \quad \mathbf{f}_2 = \begin{pmatrix} 0 \\ S_F \\ 1 \end{pmatrix} \qquad (10.E.1.3)$$

The specific growth and substrate consumption rates are functions of S and therefore of x_2/x_3:

$$\mu(S) = \mu(SV/V) = \mu(x_2/x_3), \quad \sigma(S) = \sigma(SV/V) = \sigma(x_2/x_3) \quad (10.E.1.4)$$

The objective is to maximize the profit function that depends on the outcome at the final time,

$$\underset{\mathbf{u}(t)=F(t)}{\text{Max}} \; P[\mathbf{x}(t_f)] \qquad (10.E.1.5)$$

by manipulating the manipulated variables, $u = F$. According to PMP, one maximizes the Hamiltonian, which is a scalar product of adjoint variables, with the right-hand side of the differential equations that define the process,

$$H = \lambda^T[\mathbf{f_1}(\mathbf{x}) + \mathbf{f_2}u] = (\lambda_1\mu - \lambda_2\sigma)x_1 + (\lambda_2 S_F + \lambda_3)F = (\lambda_1\mu - \lambda_2\sigma)x_1 + \phi F$$

$$(10.E.1.6)$$

where the adjoint variables λ_i satisfy the following differential equations:

$$\frac{d\lambda}{dt} = -\frac{\partial H}{\partial \mathbf{x}} = -\left[\left(\frac{\partial \mathbf{f_1}}{\partial \mathbf{x}}\right)^T + \left(\frac{\partial \mathbf{f_2}}{\partial \mathbf{x}}\right)^T u\right]\lambda = -\left[\left(\frac{\partial \mathbf{f_1}}{\partial \mathbf{x}}\right)^T\right]\lambda \quad \lambda(t_f) = \frac{\partial P}{\partial \mathbf{x}(t_f)}$$

$$\frac{d}{dt}\begin{pmatrix}\lambda_1\\\lambda_2\\\lambda_3\end{pmatrix} = -\begin{pmatrix}\dfrac{\partial H}{\partial x_1}\\[4pt]\dfrac{\partial H}{\partial x_2}\\[4pt]\dfrac{\partial H}{\partial x_3}\end{pmatrix} = -\begin{pmatrix}\lambda_1\mu - \lambda_2\sigma\\(\lambda_1\mu' - \lambda_2\sigma')x_1/x_3\\-(\lambda_1\mu' - \lambda_2\sigma')x_1x_2/x_3^2\end{pmatrix}, \quad \begin{pmatrix}\lambda_1(t_f)\\\lambda_2(t_f)\\\lambda_3(t_f)\end{pmatrix} = \begin{pmatrix}\dfrac{\partial P}{\partial x_1(t_f)}\\[4pt]\dfrac{\partial P}{\partial x_2(t_f)}\\[4pt]\dfrac{\partial P}{\partial x_3(t_f)}\end{pmatrix}$$

$$(10.E.1.7)$$

where the primes denote partial differentiation with respect to S. These are subject to certain boundary conditions, as detailed earlier.

Because the Hamiltonian, Eq. (10.E1.6), is linear in F, the optimum F depends on the sign of the switching function, $\phi(t) = S_F\lambda_2 + \lambda_3$:

$$F = \begin{bmatrix}F_{max} \text{ when } \phi > 0\\ F_{min} \text{ when } \phi < 0\\ F_{sin} \text{ when } \phi = 0 \text{ over finite time interval(s)}\end{bmatrix}$$

$$(10.E.1.8)$$

We note that PMP does not provide any solution (singular) when the coefficient $\phi(t)$ vanishes over a finite time interval. This does not imply that any value of the manipulated variables would serve. We need to investigate.

During the singular interval, the switching function, $\phi(t)$, is identically zero over a finite time period, and therefore, their higher-order derivative must also vanish:

$$\phi = 0, \quad \frac{d\phi}{dt} = 0, \quad \frac{d^k\phi}{dt^k} = 0, \quad k = 1, 2, 3, \ldots n \qquad (10.E.1.9)$$

Almost all cases involving only one manipulated variable require only up to the second derivative, in which the manipulated variable appears explicitly. This type of problem is classified as singular control of the first order. During the singular interval, we have

$$\phi = S_F\lambda_2 + \lambda_3 = 0 \qquad (10.E.1.10)$$

$$d\phi/dt = -S_F(\lambda_1\mu' - \lambda_2\sigma')x_1/x_3 + (\lambda_1\mu' - \lambda_2\sigma')x_1x_2/x_3^2$$
$$= (\lambda_1\mu' - \lambda_2\sigma')(-S_F + x_2/x_3^{\neq 0}) = 0 \Rightarrow \lambda_1\mu' - \lambda_2\sigma' = 0 \quad (10.E.1.11)$$

$$d^2\phi/dt^2 = 0 \Rightarrow d(\lambda_1\mu' - \lambda_2\sigma')/dt = 0 \Rightarrow \frac{dS}{dt} = \frac{(\lambda_1\mu - \lambda_2\sigma)\mu'}{\lambda_1\mu'' - \lambda_2\sigma''} \quad (10.E.1.12)$$

The singular feed rate is obtained by substituting Eq. (10.E.1.12) into the substrate balance equation (10.E.1.2):

$$\frac{d(SV)}{dt} = S\frac{dV}{dt} + V\frac{dS}{dt} = SF_{sin} + V\frac{(\lambda_1\mu - \lambda_2\sigma)\mu'}{\lambda_1\mu'' - \lambda_2\sigma''} = -\sigma x_1 + S_F F_{sin} \quad (10.E.1.13)$$

or

$$F_{\sin} = \frac{\sigma x_1}{S_F - S} + \frac{V[\mu - (\lambda_2/\lambda_1)\sigma]\mu'}{(S_F - S)[\mu'' - (\lambda_2/\lambda_1)\sigma'']} \qquad (10.E.1.14)$$

The singular feed rate expression given in Eq. (10.E.1.14) contains unknown adjoint variables λ_1 and λ_2. To obtain a feedback solution in terms of only state variables, we must eliminate the adjoint variables. In this example, only the ratio λ_2/λ_1 must be eliminated. This can be done by substituting Eq. (10.E.1.11) into Eq. (10.E.1.14):

$$F_{\sin} = \frac{\sigma XV}{S_F - S} + \frac{V(\mu - \sigma\mu'/\sigma')\mu'}{(S_F - S)(\mu'' - \sigma''\mu'/\sigma')} = \frac{\sigma XV}{S_F - S} + \frac{V(Y_{X/S})'\mu'}{(S_F - S)(\mu'/\sigma')'} \qquad (10.E.1.15)$$

Equation (10.E.1.15) is the feedback solution in that the measurements of state variables, $x_1, x_2,$ and x_3 ($XV, SV,$ and V), and calculation of specific rates and their derivatives, $\mu, \sigma, \mu', \sigma', \mu'',$ and σ'', are needed to implement the closed-loop feedback control of the feed rate.

It is also interesting to note that if the yield coefficient is constant, not a function of substrate concentration, then the second term in Eq. (10.E.1.15) vanishes owing to $(Y_{X/S})' = 0$ so that

$$F_{\sin} = \frac{\sigma XV}{S_F - S} \qquad (10.E.1.16)$$

which, when substituted into the substrate balance equation (10.E.1.1), yields

$$\frac{d(SV)}{dt} = S\frac{dV}{dt} + V\frac{dS}{dt} = SF_{\sin} + V\frac{dS}{dt} = S_F F_{\sin} - \sigma XV \Rightarrow$$

$$V\frac{dS}{dt} = (S_F - S)F_{\sin} - \sigma XV = (S_F - S)\frac{(S_F - S)}{(S_F - S)} - (S_F - S) = 0 \Rightarrow \frac{dS}{dt} = 0 \qquad (10.E.1.17)$$

Equation (10.E.1.17) implies that the substrate concentration is held constant during the singular interval at the value corresponding to the maximum value of the specific growth rate, $S = S_m, \max \mu = \mu(S = S_m)$.

In general, for a third-order process, there are three adjoint variables, $\lambda_1, \lambda_2,$ and λ_3, and therefore, three equations involving the adjoint variables are required. Two equations are $\phi = 0$ of Eq. (10.E.1.10) and $\dot{\phi} = 0$ of Eq. (10.E.1.11). Equation (10.E.1.12) leads to the singular feed rate expression. Therefore, in general, one more equation is needed. The third equation is available if the final time t_f is free so that the Hamiltonian is zero:

$$S_F\lambda_2 + \lambda_3 = 0 \qquad (10.E.1.10)$$

$$\mu'\lambda_1 - \sigma'\lambda_2 = 0 \qquad (10.E.1.11)$$

$$H = (\lambda_1\mu - \lambda_2\sigma)x_1 + \phi F = (\lambda_1\mu - \lambda_2\sigma)x_1 = 0 \Rightarrow \mu\lambda_1 - \sigma\lambda_2 = 0 \qquad (10.E.1.6)$$

Solving Eqs. (10.E.1.11) and (10.E.1.6) simultaneously for the adjoint variables λ_1 and λ_2,

$$\lambda_1/\lambda_2 = \sigma'/\mu' = \sigma/\mu \qquad (10.E.1.18)$$

Rearrangement of Eq. (10.E.1.18) yields

$$\mu\sigma' - \sigma\mu' = 0 \Rightarrow (\sigma\mu' - \mu\sigma')/\sigma^2 = 0 \Rightarrow (\mu/\sigma)' = 0 \Rightarrow (Y_{X/S})' = 0 \quad (10.E.1.19)$$

This equation implies that when the final time is free, the cell mass yield should be maximized. When the final time is free, there is an infinite time available, and therefore, the optimal policy should not be concerned with the rate of reaction but with maximizing the yield to obtain the maximum amount of cell mass by maintaining the substrate concentration constant at the value corresponding to the maximum cell yield coefficient. Conversely, if the yield coefficient is constant, the optimal policy does not exist. Any feed rate profile would be due to the infinite time and would yield the same amount of cell mass.

Conversely, if the final time is fixed, the Hamiltonian is an unknown constant, and therefore, the optimal feed rate profile does not maximize the cell mass yield. Instead, the optimal feed rate profile compromises between the cell growth rate and cell mass yield. In general, the singular control is a function of both state and adjoint variables. Because a feedback control based on the state variables only is preferred, it is necessary to eliminate the adjoint variables using $\phi = 0$, $d\phi/dt = 0$ $H = 0$ (free t_f). It should be noted that we have three homogeneous equations that are linear in the adjoint variables. Therefore, *it is possible to obtain the singular control in feedback mode for third-order processes with free final time.* For higher-order processes and fixed final time problems, it may not be possible to obtain in general the singular feed rate in feedback mode, with the exception of limiting cases.

The methods described here are based on solving the necessary conditions of PMP. There are other approaches that may be called *direct methods*. The basic idea is to transform the optimization problem into a problem in nonlinear programming, without using the optimality conditions.

10.1.4 Nonlinear Programming

Nonlinear programming (NLP) is an algorithm for solving a system of equality or inequality constraints with an objective function to be maximized. Nonlinearities are there in at least one of the constraints and the objective function. The problem is to maximize an objective function of n variables subject to m equality and r inequality constraints:

$$
\begin{aligned}
\text{Maximize:} \quad & f(\mathbf{x}) & \mathbf{x} = [x_1 \, x_2 \ldots, x_n]^T \\
\text{Subject to:} \quad & \mathbf{g}(\mathbf{x}) \geq \mathbf{0} & \mathbf{g} = [g_1 \, g_2 \ldots, g_r]^T \\
& \mathbf{h}(\mathbf{x}) = \mathbf{0} & \mathbf{h} = [h_1 \, h_2 \ldots, h_m]^T
\end{aligned}
\quad (10.8)
$$

10.1.4.1 Penalty Function Method[4]

The penalty function approach is based on augmenting the objective function with the constraints as a penalty function and then solving the resulting unconstrained optimization problem repeatedly. The first is the exterior penalty function method. The objective function is augmented with a penalty function to transform the constrained optimization problem into an unconstrained problem by forming the

Lagrangian

$$\text{Max } L(\mathbf{x}, r_h, r_g) = f(\mathbf{x}) + P(\mathbf{x}, r_h, r_g) = f(\mathbf{x}) + r_h \sum_{i=1}^{m} h_i(\mathbf{x})^2 + r_g \sum_{j=1}^{r} (\max\{0, g_j(\mathbf{x})\})^2$$

(10.9)

where $P(\mathbf{x}, r_h, r_g)$ is the penalty function and r_h and r_g are negative penalty constants. If the constraints are not satisfied, the values are squared and added to the objective function. The original constrained problem is solved by repeatedly solving the unconstrained problem with increasing (infinity) penalty constants.

To apply the NLP method of computation, the dynamic optimization problem described by differential equations must be converted into static optimization described by algebraic equations. Let the time interval $t_0 = 0$ to t_f be discretized into N number of nodes, $t_0 < t_1 < t_2 < \ldots < t_{(N-1)} = t_f$. At each node, the control variable $F(t)$ is parameterized as a function of time such as a polynomial or piecewise constant, for example, constant control parameters $\widetilde{F(t)} = [F_1, F_2, \ldots, F_{N-1}]$. Now, we can obtain the state vector trajectory $\underline{x}(\tilde{F}, t)$ by integrating the differential equations with the given initial conditions and the control parameters at each time interval $[t_i, t_{i+1}]$, where $i = 0, N - 1$. The objective function is in the form of $L(\tilde{F})$ and can be calculated from the state vector trajectory $\underline{X}(\tilde{F}, t)$. Therefore, the problem is transformed into a static optimization problem of maximizing $L(\tilde{F})$. Without equality or inequality constraints, the steepest descent or conjugate gradient methods are directly used with gradient generation methods described previously. However, if the original problem contains constraints, we must transform the problem into a constrained static optimization problem.

The computational algorithm is as follows:

1. Determine initial candidate of solution, initial penalty constants, the scaling value for multipliers, the number of unconstrained optimization iterations, and the number of penalty function iterations.
2. Solve the transformed unconstrained optimization problem
3. Check stopping criteria. First, check the satisfaction of constraints. If feasible, check the stopping criteria; if they are not satisfied, update and go to step 2.

10.1.4.2 Square Quadratic Programming[5]
A quadratic programming problem is an optimization problem in which a quadratic objective function of n variables is minimized subject to m linear equality or inequality constraints:

$$\begin{aligned}
\text{Minimize:} \quad & f(\mathbf{x}) = \mathbf{c}^T \mathbf{x} + \frac{1}{2} \mathbf{x}^T \mathbf{Q} \mathbf{x} \\
\text{Subject to:} \quad & \mathbf{h} = \mathbf{A} \mathbf{x} - \mathbf{b} = \mathbf{0} \\
& \mathbf{g} = \mathbf{x} \geq \mathbf{0}
\end{aligned}$$

(10.10)

where $\mathbf{c}$ is an n-component constant vector, $\mathbf{A}$ is an $(m \times n)$ matrix of constants, and $\mathbf{Q}$ is an $(n \times n)$ symmetric matrix.

First, the Lagrangian function is formed by adjoining the objective function with the equality and inequality constraints,

$$L(\mathbf{x}, \lambda, \eta) = f + \lambda^T \mathbf{g} + \eta^T \mathbf{h}$$

(10.11)

where λ and η are the Lagrange multiplier vectors for the equality and inequality constraints, respectively. According to optimality criteria (Kuhn–Tucker conditions), the optimum values of λ^* and η^* satisfy the following condition:

$$\nabla_x L = \begin{bmatrix} f_\eta + \mathbf{g}_\eta^T \lambda \\ \mathbf{g} \end{bmatrix}_{(\eta^*, \lambda^*)} = 0 \qquad (10.12)$$

where $\eta^*, \lambda^*, f_\eta = \partial f/\partial \eta$ are $(m \times 1), (r \times 1), (m \times 1)$ vectors, respectively, and $\mathbf{g}_\eta^T = (\partial \mathbf{g}/\partial \eta)^T$ is a $(n \times m)(m \times n)$ matrix.

The square quadratic programming (SQP) method is to iteratively find out the optimum solution (η^*, λ^*). With the initial estimate (η_0, λ_0), the second iterations are represented by $\eta_1 = \eta_0 + d\eta$ and $\lambda_1 = \lambda_0 + d\lambda_0$. Taking a Taylor series expansion around the initial estimate (η_0, λ_0) and trimming higher-order terms,

$$\begin{bmatrix} L_{\eta\eta} & \mathbf{g}_\eta^T \\ \mathbf{g}_\eta & 0 \end{bmatrix} \begin{bmatrix} d\eta \\ \lambda_1 \end{bmatrix} = - \begin{bmatrix} f_\eta \\ \mathbf{g} \end{bmatrix} \qquad (10.13)$$

where $L_{\eta\eta}$ is the Hessian of L, $\partial^2 L/\partial \eta^2$. The derivatives $d\eta$ and $d\lambda$ can be directly used to improve the next iteration. However, to ensure the global convergence of optimum values, the iteration can be modified like $\eta_1 = \eta_0 + \alpha d\eta$, where α is a parameter guaranteeing the improved penalty function, $P(\eta_1) < P(\eta_0)$.

10.1.4.3 Other Methods

The augmented Lagrange multiplier method is the strongest penalty function method. It overcomes many difficulties originating from the penalty function. The most peculiar point is its ability to get the Lagrange multiplier as well as penalty multipliers. In addition, sequential linear programming, the generalized reduced gradient method, and the sequential gradient restoration algorithm are popular constrained optimization algorithms.

The basic idea is to transform the optimal control problem into an NLP, without using the optimality condition. To use the NLP techniques described previously, the optimization problem must first be transformed into the appropriate form for NLP. Two methods have been used: sequential and simultaneous approaches. We will show how to formulate the transformed NLP problem in this chapter. Once this is done, all optimization techniques, including constrained methods, are applicable.

To enforce the constraints at each time node, $t_0, t_1, t_2, \ldots, t_{N-1}$, we can approximate the constraints. The numbers of equality and inequality constraints are $N^*n_e + n_f$ and N^*n_i, respectively. As a result, the constrained optimal control problem is transformed into a constrained static optimization problem, as follows:

$$\begin{aligned} \underset{\tilde{F}}{\text{Max}} \ & L(\tilde{F}) \\ g(\tilde{F}) &= 0, \ h(\tilde{F}) \leq 0 \end{aligned} \qquad (10.14)$$

We already covered the gradient-based optimization methods for this form of problem. The penalty function method or SQP is usually used.

EXAMPLE 10.E.2: MAXIMIZATION OF LYSINE PRODUCTION USING SQP   The following is an example of fed-batch fermentation for maximizing lysine at the final time using SQP, as illustrated by Pushpavanam et al.[5] The unsteady state cell mass, substrate, product, and total mass balances are

$$\frac{dX}{dt} = \mu X - \frac{FX}{V}, \quad \frac{dS}{dt} = -\sigma X + \frac{F}{V}(S_F - S)$$

$$\frac{dP}{dt} = \pi X - \frac{FP}{V}, \quad \frac{dV}{dt} = F \tag{10.E.2.1}$$

where X, S, and P stand for the concentrations of cells, substrate, and product, respectively, V is the culture volume, S_F is the constant feed substrate concentration, and F is the volumetric feed flow rate, which maximizes the product concentration at the final process time, that is,

$$\underset{F(t)}{\mathrm{Max}} \, [P(t_f)] \tag{10.E.2.2}$$

As stated earlier, the continuous time is divided into N equal subintervals. The fed-batch operation is represented by a batch operation with impulse-type feeding of substrate at the start of each subinterval, resulting in the following equations of the batch reactor:

$$\dot{X} = \mu X$$
$$\dot{S} = -\sigma X \tag{10.E.2.3}$$
$$\dot{P} = \pi X$$

V_i is the reactor volume in the interval (t_{i-1}, t_i) and ΔV_i is the pulse volume added at t_{i-1}. The resultant objective is to find the optimal set of ΔV_is at t_is for $i = 0, 1, \ldots, N-1$ so that the product concentration is maximized at the final time t_f. The initial conditions before adding the pulse volume at t_0 are

$$[X, \, S, \, P, \, V](t_0^-) = [X_0, \, S_0, \, P_0, \, V_0] \tag{10.E.2.4}$$

When the pulses ΔV_i are added, the mass balance at the time instants t_{i-1} gives rise to

$$X(t_{i-1}^+) = (V_{i-1}/V_i)X(t_{i-1}^-)$$
$$S(t_{i-1}^+) = (V_{i-1}/V_i)S(t_{i-1}^-) + [(V_i - V_{i-1})/V_i]S_F \tag{10.E.2.5}$$
$$P(t_{i-1}^+) = [V_{i-1}/V_i]P(t_{i-1}^-)$$

To calculate sensitivity with respect to ΔV_is, the mass balance equations were differentiated with respect to ΔV_js for $j = 1, 2, \ldots, N$:

$$\frac{\partial \dot{X}}{\partial \Delta V_j} = \frac{\partial}{\partial X}(\mu X)\frac{\partial X}{\partial \Delta V_j} + \frac{\partial}{\partial S}(\mu X)\frac{\partial S}{\partial \Delta V_j} + (\mu X)\frac{\partial P}{\partial \Delta V_j}$$

$$\frac{\partial \dot{S}}{\partial \Delta V_j} = \frac{\partial}{\partial X}(-\sigma X)\frac{\partial X}{\partial \Delta V_j} + \frac{\partial}{\partial S}(-\sigma X)\frac{\partial S}{\partial \Delta V_j} + \frac{\partial}{\partial P}(-\sigma X)\frac{\partial P}{\partial \Delta V_j} \tag{10.E.2.6}$$

$$\frac{\partial \dot{P}}{\partial \Delta V_j} = \frac{\partial}{\partial X}(\pi X)\frac{\partial X}{\partial \Delta V_j} + \frac{\partial}{\partial S}(\pi X)\frac{\partial S}{\partial \Delta V_j} + \frac{\partial}{\partial P}(\pi X)\frac{\partial P}{\partial \Delta V_j}$$

The preceding sensitivity differential equations require the set of initial conditions with respect to the jth pulse ΔV_j at t_{j-1}, which are obtained by differentiating Eq. (10.E.2.5):

$$\frac{\partial X}{\partial \Delta V_j}(t_{j-1}^+) = -X(t_{j-1}^-)\frac{V_{j-1}}{V_j^2}$$

$$\frac{\partial S}{\partial \Delta V_j}(t_{j-1}^+) = (S_F - S(t_{j-1}^-))\frac{V_{j-1}}{V_j^2}$$

$$\frac{\partial P}{\partial \Delta V_j}(t_{j-1}^+) = -P(t_{j-1}^-)\frac{V_{j-1}}{V_j^2} \qquad (10.E.2.7)$$

The following set comprises updating conditions for the sensitivity at the junction points t_{i-1} for $i = j+1, \ldots, N$:

$$\frac{\partial X}{\partial \Delta V_j}(t_{i-1}^+) = X(t_{i-1}^-)\frac{V_i - V_{i-1}}{V_i^2} + \frac{V_{i-1}}{V_i^2}\frac{\partial X}{\partial \Delta V_j}(t_{i-1}^-)$$

$$\frac{\partial S}{\partial \Delta V_j}(t_{i-1}^+) = \frac{[S(t_{i-1}^-) - S_F](V_i - V_{i-1})}{V_i^2} + \frac{V_{i-1}}{V_i}\frac{\partial S}{\partial \Delta V_j}(t_{i-1}^-) \qquad (10.E.2.8)$$

$$\frac{\partial P}{\partial \Delta V_j}(t_{i-1}^+) = P(t_{i-1}^-)\frac{V_i - V_{i-1}}{V_i^2} + \frac{V_{i-1}}{V_i^2}\frac{\partial P}{\partial \Delta V_j}(t_{i-1}^-)$$

Along with these updating conditions, integrating Eqs. (10.E.2.3) and (10.E.2.6) from t_j^+ to t_N gives rise to the sensitivity of the state variables with respect to ΔV_j, for example, $\partial X(t_f)/\partial \Delta V_j$.

As an example, alcohol fermentation is considered. The rate expressions are given by Modak and Lim:[6]

$$\mu = \frac{0.408S}{(0.22 + S)}e^{-0.028P}, \quad \pi = \frac{S}{(0.44 + S)}e^{-0.015P}, \quad \sigma = \mu/0.1 \qquad (10.E.2.9)$$

Initial biomass concentration X, substrate concentration S, and reactor volume V were chosen as 0.2 g/L, 100 g/L, and 5 L, respectively. The feed substrate concentration, the final reactor volume, and the total process time were chosen as $S_F = 100$ g/L, $V_f = 20$ L, and $t_f = 20$ hr, respectively. Under these conditions, Modak and Lim,[6] via PMP, showed that the optimal feed strategy is $F_{min} = 0$, a batch period (11.25 hrs), F_{sin} (8 hrs), and $F_b = 0$, a batch (0.75 hrs). Table 10.E.2.1 shows the optimal pulsed feed rate strategies with respect to subinterval numbers of 1, 3, 5, 7, 10, and 20, respectively. When the time interval is 20 so that the impulse feeding is made every 1 hr, the optimal impulse feeding strategy calls for 10 hrs of no feeding ($F_{min} = 0$) followed by 10 hrs of pulse feeding (F_{sin}) to obtain 32.99 g/L of alcohol. It is expected that there would be a short batch period after the singular feeding and that the performance index would approach even closer to that (34.52 g/L) obtained by PMP, with additional subintervals, perhaps 100 or more. This clearly demonstrates the shortcomings of dynamic programming, which requires extensive numerical computation.

The optimal biomass and alcohol concentration profiles are shown in Figure 10.E.2.1. As observed, the results of nonsingular and singular approaches differ in the initial and final portions. The performance index is comparable to that optimized

Table 10.E.2.1. *Optimal pulsed feed rates and performance indices by SQP with several subinterval numbers*

N	1	3	5	7	10	20
Charges	15	0	0	0	0	0
		3.682	0	0	0	0
		11.32	1.372	0	0	0
			5.066	0	0	0
			8.562	3.369	0	0
				5.010	0.310	0
				6.621	1.397	0
					1.587	0
					4.206	0
					0	0
						0
						0.5587
						1.2322
						1.3164
						1.5852
						1.8333
						2.2030
						2.3521
						3.9191
						0
$P(t_f)$ g/l alcohol	29.08	30.67	31.62	32.13	32.63	32.99

by PMP, and the discontinuities in the state variables due to the discredited time intervals caused the performance index of the nonsingular approach to be inferior to that obtained by the singular approach. This is in part due to an insufficient number of subintervals used in the optimization study.

10.1.5 A Special Transformation to Convert Singular to Nonsingular Problems[7]

In a fed-batch reactor system described by mass balance differential equations, the singular feed rate problem occurs as the feed rate appears linearly. Therefore, to transform the singular feed rate problem into a nonsingular problem, an appropriate control should appear in a nonlinear form. This approach was presented in detail in Chapter 9. It should be noted that this approach is limited to processes in which *the specific rates are functions only of substrate concentration*. It is best to illustrate this approach with an example in Chapter 9.

EXAMPLE 10.E.3: MAXIMIZATION OF CELL MASS PRODUCTION FOR A PROCESS WITH A CONSTANT CELL MASS YIELD COEFFICIENT[7] Let us consider the cell mass maximization problem in fed-batch fermentation with a constant cell mass yield coefficient. The specific cell mass growth rate and yield coefficient are as follows:[8]

$$\mu(S) = \frac{S}{0.03 + S + 0.5S^2}, \quad \sigma = \mu/Y, \quad Y = 0.5 \qquad (10.E.3.1)$$

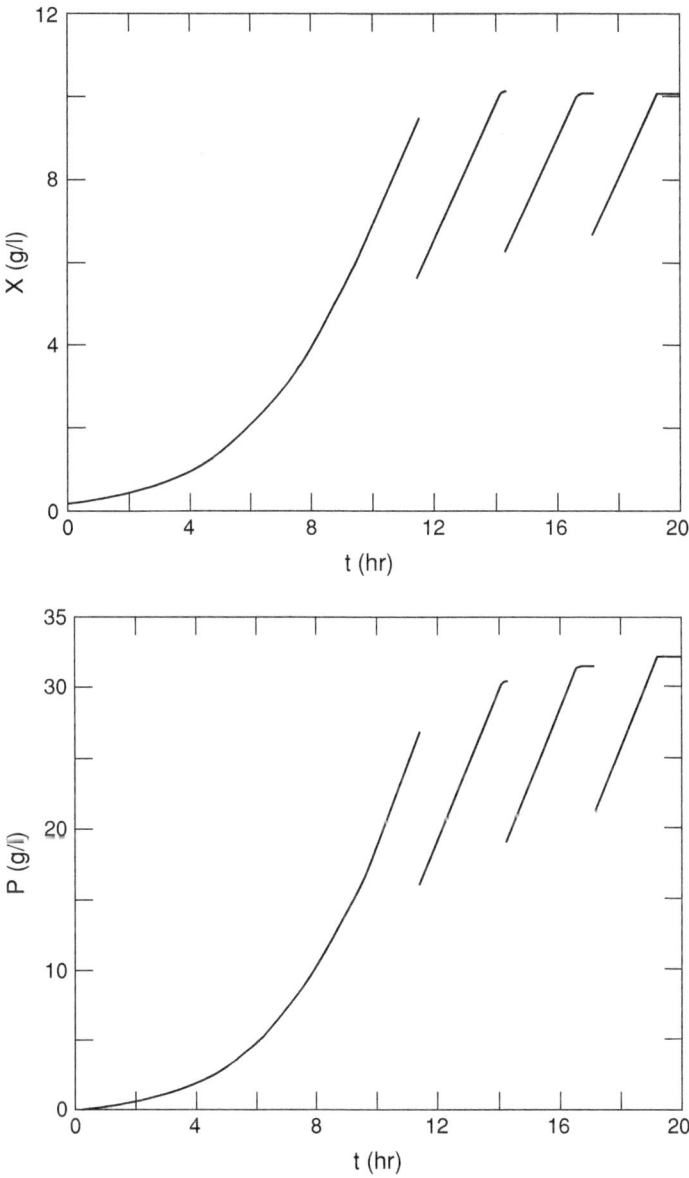

Figure 10.E.2.1. Optimal concentrations of biomass and alcohol ($N = 7$) adapted from reference 5.

The usual mass balance equations are

$$\frac{dx_1}{dt} = \mu x_1 = \frac{(x_2/x_3)x_1}{0.03 + (x_2/x_3) + 0.5(x_2/x_3)^2} \tag{10.E.3.2}$$

$$\frac{dx_2}{dt} = FS_F - \sigma x_1 = FS_F - \frac{2(x_2/x_3)x_1}{0.03 + (x_2/x_3) + 0.5(x_2/x_3)^2} \tag{10.E.3.3}$$

$$\frac{dx_3}{dt} = F \tag{10.E.3.4}$$

The objective is to maximize the total amount of cells at the final time:

$$\underset{F(t)}{\text{Max}}[P = x_1(t_f)], \quad t_f \text{ fixed} \tag{10.E.3.5}$$

$$V_0 \le V(t) \le V_{max} \tag{10.E.3.6}$$

$$0 = F_{min} \le F \le F_{max} \tag{10.E.3.7}$$

To eliminate F from the process model, we introduce, as in Chapter 9, a new state variable that represents the total amounts of substrate consumed, $x_4(t)$:

$$x_4 \overset{\Delta}{=} S_0V_0 + S_F(V - V_0) - VS = x_{20} + S_F(x_3 - V_0) - x_2 \tag{10.E.3.8}$$

Then, the time derivative of Eq. (10.E.3.8), dx_4/dt, should represent the consumption rate of substrate, $\sigma XV = \sigma x_1$:

$$\frac{dx_4}{dt} = S_F \frac{dx_3}{dt} - \frac{dx_2}{dt} = \sigma XV = \sigma x_1 \tag{10.E.3.9}$$

With the introduction of $x_4(t)$, the process can be represented by the following:

$$\frac{dx_1}{dt} = \mu(S)x_1 = \frac{Sx_1}{0.03 + S + 0.5S^2} \quad x_1(0) = X_0V_0 \tag{10.E.3.10}$$

$$\frac{dx_4}{dt} = \sigma(S)x_1 = \frac{Sx_1}{0.03 + S + 0.5S^2} \quad x_4(0) = 0 \tag{10.E.3.11}$$

The process presented here is represented by two state variables, x_1 and x_4, and we can consider the substrate concentration, $S(t)$, which appears nonlinearly, as the manipulated variable. Therefore, a numerical solution can be obtained readily, say, using the steepest ascent. As shown in Chapter 9, the solution to the problem is obtained via PMP:

$$\underset{S(t)}{\text{Max}}\{H = \lambda_1\mu x_1 + \lambda_4\sigma x_1 + \eta[g(x_4, S) - V_{max}]\} \tag{10.E.3.12}$$

The Hamiltonian is maximized to obtain

$$\underset{S(t)}{\text{Max}}[H = \lambda_1\mu x_1] \Rightarrow \partial H/\partial S = 0 = \partial\mu/\partial S \tag{9.98}$$

According to Eq. (9.98), the specific growth rate must be maximized to maximize the amount of cell mass at the final time. Therefore, the substrate concentration must be maintained at the value $S = S_m = 0.173$ g/L, which maximizes the specific growth rate, $\mu_{max} = \mu(S = 0.173) = 0.794$. The feed rate to maintain the specific growth rate at its maximum value is obtained from the substrate balance equation by setting $dS/dt = 0$:

$$F^*(t) = \frac{\sigma(S_m)x_1}{(S_F - S_m)} = \frac{1.59XV}{(10 - 0.173)} = 0.162XV \tag{10.E.3.13}$$

The following initial state variables and parameters were used to carry out a steepest ascent on the Hamiltonian using the substrate concentration as the manipulated variable:

$$[XV(t_0), SV(t_0), V(t_0), V_{max}, F_{max}, S_F, t_f]$$
$$= [1 \text{ g}, 0 \text{ g}, 1 \text{ L}, 5 \text{ L}, 4 \text{ L/hr}, 10 \text{ g/L}, 3.8 \text{ hr}] \tag{10.E.3.14}$$

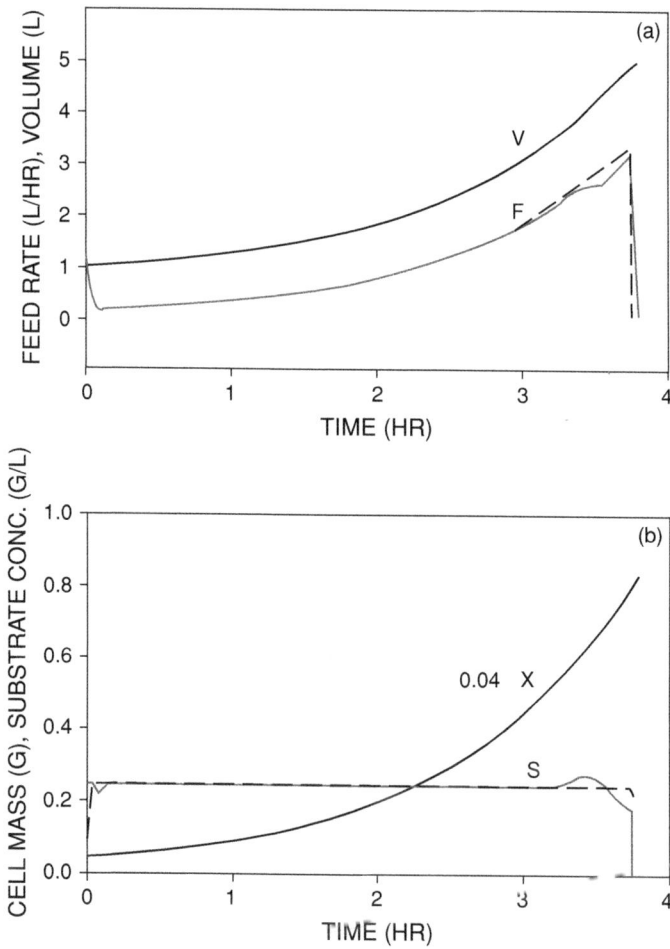

Figure 10.E.3.1. Optimal profiles of production with a constant yield by nonsingular approach. Solid lines: non-singular approach, dotted lines: singular approach.

As seen in Figure 10.E.3.1, the optimal feed rate profile nearly follows that which was calculated by singular optimal control scheme (a bang followed by a singular period), except the final feed rate region.

10.2 Numerical Techniques for Processes without Mathematical Models

10.2.1 Neural Network[9]

The neural network calculation scheme and procedure are illustrated in Chapter 7. The mathematical model of yeast fermentation for the production of heterologous protein invertase (SUC2-s2), as shown by Eq. (7.16), was used to generate the neural network data and is used in Example 10.E.4.

Table 10.E.4.1. *Performance index for best neural networks*

NN F_{opt} in Math Model	NN F_{opt} in NN	F_{opt}, Reported
26.7	41.2	32.4

EXAMPLE 10.E.4: MAXIMIZATION OF SECRETED PROTEIN USING NEURAL NETWORK MODEL[9] On the basis of the neural network model, the neural networks were optimized to maximize the amount of secreted protein at the final time:

$$\max_{\substack{0\leq F(t)\leq10\,l/hr \\ V_{max}=14.35\,L}} P = (P_m V)_{t_f=15\,hr} \qquad (10.E.4.1)$$

The initial state variable conditions and operating parameters are as follows:

$$[X(0), S(0), P_m(0), P_t(0), V(0), S_F, t_f, V_{max}, \Delta t] = [1, 5, 0, 0, 1, 20, 15, 15, 0.5] \qquad (10.E.4.2)$$

The optimal feed rate calculated from the neural network optimization was implemented into the mass balance equation (Eq. (7.16)), which was then integrated to produce process data. The process data were compared to the predicted data, which were calculated by mapping the neural networks with the same feed rate. The process data, the predicted data, and the reported data from Park and Ramirez[14] were, respectively, labeled "NN F_{opt} in math model," "NN F_{opt} in NN," and "F_{opt}, reported." Figure 10.E.4.1 shows how the neural networks performed better in optimization.

The performance index from the optimized neural networks was compared with the references in Table 10.E.4.1. The amount of secreted protein obtained at the final time by applying the optimal feed rate obtained from neural network optimization to the mass balance model (NN F_{opt} in math model) resulted in the lowest performance index of 26.7, which was much inferior to 32.4, as reported (F_{opt}, reported). Obviously, the neural network approach resulted in a false result of 41.2, indicating an inferior result that can be obtained with the neural network approach. When some information is available, the neural network approach is inferior as its premise is that one does not know anything about the process.

10.2.2 Genetic Algorithm[10]

A genetic algorithm (GA) is a stochastic search method in analogy to the natural principles of genetics and survival. Biological evolution itself is an optimization process because the reproductive elements are optimized by mutation, crossover, and selection mechanisms. GA deals with a set of potential solution populations and probabilistic rules for evolution, and these are the reasons why GA drives us to find the global optimum solution.[11] Moreover, GA is a relatively robust algorithm as it is not restricted to several constraint conditions such as convexity, sensitivity, continuity, and nonlinearity of the optimization problem. GA has been adopted to optimize fed-batch cultures.[12]

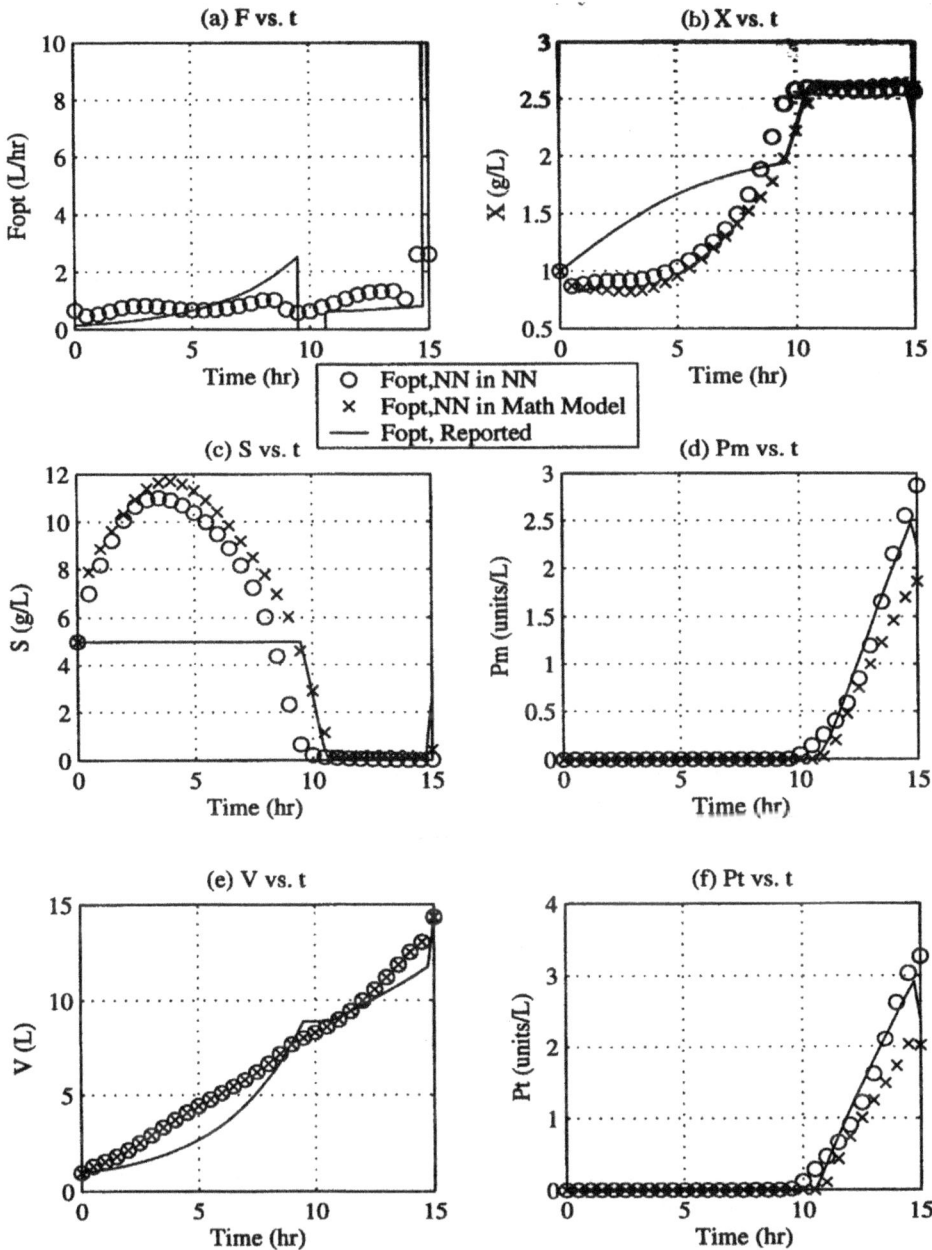

Figure 10.E.4.1. Comparison of F_{opt} in neural network (adapted from reference 9).

GA is performed with artificial chromosomes composed of binary strings. Each string represents information of the parameter to be optimized, for example, in the case of fed-batch optimization, the feed flow rate at certain times and switching times. GA operation comprises reproduction, crossover, and mutation. Some chromosomes are reproduced when their objective functions are relatively better than those of others. Crossover generates next offspring chromosomes so that the objective function can approach the optimal region. Mutation is useful to preserve some important information at particular regions or to prevent the solution from

Table 10.E.5.1. *An unstructured model for aerobic yeast growth*[10]

$$\mu = \frac{0.0747G + 0.4G^2}{0.0011 + 0.351G + G^2} + \frac{0.114E/0.005 + 0.59\sigma + E}{1 + 1.43\sigma}$$

$$\sigma = \frac{\dfrac{0.0747G + 0.4G^2}{0.0011 + 0.351G + G^2}}{0.5435 - 0.3659\dfrac{1 + 20712.9G^2}{100 + 20712.9G^2}}$$

$$\pi = 0.459\sigma\frac{1 + 20712.9G^2}{100 + 20712.9G^2}$$

$$\eta = \frac{0.114E/0.005 + 0.59\sigma + E}{0.625(1 + 1.43\sigma)}$$

Source: Modified from Table of Na et al.[10]

falling in a local optimum. A GA-based optimization routine is as follows. Variables or parameters to be optimized are coded as binary integers, which constitute a chromosome. Populations of chromosomes are randomly generated for initial calculation. Chromosomes are selected that yield the best performance index and are copied. Crossover proceeds with the selected and copied chromosomes, which look for mates, followed by mutation. With increasing generation, it is expected that the chromosomes with performance indices toward the optimal value will dominate.

EXAMPLE 10.E.5: FED-BATCH OPTIMIZATION OF YEAST PRODUCTION USING GA   Let us introduce an example of fed-batch optimization of yeast using GA.[10] The problem is to maximize yeast cell mass at the final process time t_f:

$$\max_{F(t)}[\mathrm{P}(t_f) = (XV)_f] \tag{10.E.5.1}$$

The performance index is subject to the following mass balance equations:

$$\frac{d}{dt}\begin{bmatrix} XV \\ GV \\ EV \\ V \end{bmatrix} = \begin{bmatrix} \mu \\ -\sigma \\ \pi - \eta \\ 0 \end{bmatrix} XV + \begin{bmatrix} 0 \\ G_F \\ 0 \\ 1 \end{bmatrix} F \tag{10.E.5.2}$$

where X, G, and E are the concentrations of cell mass, glucose, and ethanol, respectively, V is reactor volume, G_F is the feed glucose concentration, and F is the feed rate. The specific rates μ, σ, π, and η are for cell mass growth, glucose consumption, ethanol production, and ethanol consumption, respectively. As shown in Table 10.E.5.1, μ and η are functions of G and E, while σ and π are functions of G only.

The volume and feed rate are constrained as follows:

$$V_0 \leq V(t) \leq V_{\max} = V(t_f), \quad 0 = F_{\min} \leq F(t) \leq F_{\max} \tag{10.E.5.3}$$

The total process time of 18 hrs is evenly divided into 36 subintervals, where the set of optimal F_is are to be determined. The efficiency of GA operation is highly affected by the probabilities of crossover (P_c) and mutation (P_m) and population size. In this example, P_c, P_m and the population size were 0.003, 0.6, and 100, respectively. The

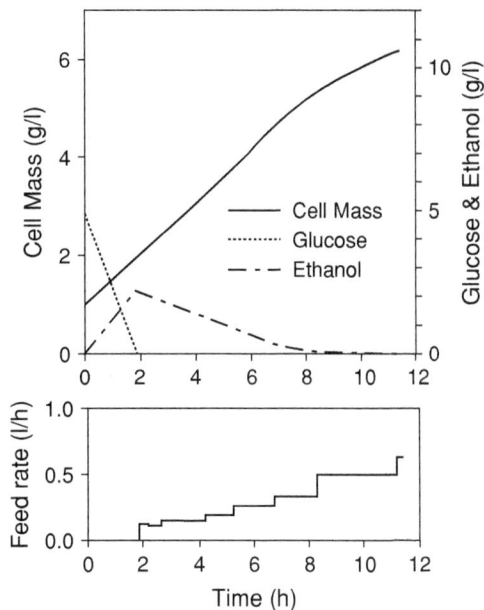

Figure 10.E.5.1. Profiles of optimal feed rate and state variables by genetic algorithm (adapted from reference 10).

initial conditions for state variables are as follows:

$$[X, G, V, E, V_{max}, F_{max}](t_0) = [1, 5, 2, 0, 5, 1]$$

$$\text{Total amount of glucose} = 45\,\text{g}, G_F = 35/3$$

(10.E.5.4)

The calculated performance index, the yeast cell mass concentration at the final time, was 24.60 g/L. The optimal feed rate and state variables are shown in Figure 10.E.5.1. The feed rate profile suggests a batch period of 1.87 hrs followed by a singular period of 9.62 hrs. The feed flow begins when the initially present glucose is almost exhausted, and ethanol, which formed during the batch period, is gradually consumed during the first 2/3 of the period of singular feed rate.

REFERENCES

1. Jaspan, R. K., and Coull, J. 1971. Trajectory optimization techniques in chemical reaction engineering. *AIChE Journal* 17: 111–115.
2. Morrison, D. D., Riley, J. D., and Zancanaro, J. F. 1962. Multiple shooting method for two-point boundary value problem. *Communications of the ACM* 5: 613–614.
3. Fine, F. A., and Bankoff, S. G. 1967. Control vector iteration in chemical plant optimization. *I&EC Fundamentals* 6: 288–293.
4. Fletcher, R. 1975. An ideal penalty function for constrained optimization. *IMC Journal of Applied Mathematics* 15: 319–342.
5. Pushpavanam, S., Rao, S., and Khan, I. 1999. Optimization of a biochemical fed batch reactor using sequential quadratic programming. *Industrial and Engineering Chemistry Research* 38: 1998–2004.
6. Modak, J., and Lim, H. C. 1987. Feedback optimization of fed-batch fermentation. *Biotechnology and Bioengineering* 30: 528–540.
7. Modak, J. M., and Lim, H. C. 1989. Simple nonsingular control approach to fed-batch fermentation optimization. *Biotechnology and Bioengineering* 33: 11–15.

8. Weigand, W. A., Lim, H. C., Creagan, C. C., and Mohler, R. D. 1979. Optimization of a repeated fed batch reactor for maximum cell productivity. *Biotechnology Bioengineering Symposium* 9: 335–348.
9. Chan, T. Y.-L. 2005. Application of neural networks to fed-batch fermentation. PhD dissertation, University of California, Irvine.
10. Na, J. G., Chang, Y. K., Chung, B. H., and Lim, H. C. 2002. Adaptive optimization of fed-batch culture of yeast by using genetic algorithms. *Bioprocess and Biosystems Engineering* 24: 299–308.
11. Goldberg, D. E. 1989. *Genetic Algorithms in Search, Optimization and Machine Learning: Reading.* Addison-Wesley.
12. Roubos, J. A., van Straten, G., and van Boxtel, A. J. B. 1999. An evolutionary strategy for fed-batch bioreactor optimisation: Concepts and performance. *Journal of Biotechnology* 67: 173–178.
13. Modak, J. M. 1988. A theoretical and experimental optimization of fed-batch fermentation processes. PhD dissertation, Purdue University.
14. Park, S., and Ramirez, W. F. 1988. Optimal production of secreted protein in fed-batch reactors. *AIChE Journal* 34: 1550–1558.

11 Optimization of Single and Multiple Reactions

In previous chapters, we looked at the basic concept of fed-batch operation as a means of manipulating the feed rate or substrate concentration to maximize the cell growth and product formation rates and harvest only at the end of the operation, not during the operation. Let us first maximize rigorously the conversion of a single reaction of an arbitrary rate expression. Then, we consider the question of the timing of the withdrawal. In the conventional fed-batch operation, the withdrawal is made only at the end of operation and not during the course of operation. Should we withdraw the reaction mixture only at the end of the run? Should we withdraw a part of the culture intermittently or continuously throughout the course of the operation? If the product is harvested all at once only at the end of the run, then the operation is the traditional fed-batch or repeated fed-batch (if a portion of the final reactor content is retained for the next cycle). However, if the product stream is withdrawn during the course in some fashion, then the operation resembles dynamic (variable volume) operation of a continuous-stirred tank reactor (CSTR). By allowing impulse feeding and withdrawal, one can also theoretically mimic temporally the spatial operation of a plug-flow reactor (PFR). Impulse feeding or withdrawal refers to adding or withdrawing a fixed amount instantaneously in the form of an impulse function as one would dump a bucketful of feed in an infinitesimally small time interval. Intuitively, a withdrawal during the course of operation does not change the reaction composition, and therefore, it appears that a withdrawal during the course of operation would not help. To answer this important question, we consider in the following section a rigorous solution to this fundamental question of when to withdraw.

11.1 Single Reactions with a Single Feed Rate

Consider the optimization of isothermal operation of a single reaction $(A \rightarrow B)$ of arbitrary rate expression $r_A(C_A)$ by manipulating the feed flow rate. A schematic diagram is shown in Figure 11.1.

The objective is to maximize the conversion of species A in a given time t_f by feeding the reactant A as an arbitrary function of time:

$$\operatorname*{Max}_{F(t)} P = [C_{AF} - C_A(t_f)]/C_{AF} \tag{11.1}$$

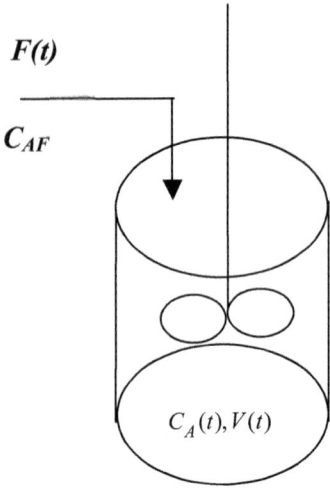

$F(t)$

C_{AF}

$C_A(t), V(t)$

Figure 11.1. A fed-batch operation with a feed rate manipulation.

The species A balance is

$$\frac{d(C_A V)}{dt} = C_{AF}F(t) + r(C_A)V(t) \quad C_A(0) = C_{A0} \tag{11.2}$$

and the overall balance, assuming equal density for the feed and reactor content, is

$$\frac{dV}{dt} = F(t) \quad V(0) = V_0 \quad 0 \le V \le V_{max} \tag{11.3}$$

Applying Pontryagin's maximum principle (PMP),[1] given in Chapter 9, we let $x_1 = C_A V, x_2 = V,$ and $u = F$ and rewrite Eqs. (11.1)–(11.3):

$$\underset{u(t)}{\text{Max}}\, P = [C_{AF} - x_1(t_f)]/C_{AF} \tag{11.4}$$

$$\frac{dx_1}{dt} = C_{AF}u(t) + r(x_1/x_2)x_2 \quad x_1(0) = C_{A0}V_0 \tag{11.5}$$

$$\frac{dx_2}{dt} = u \quad x_2(0) = V_0 \tag{11.6}$$

where the reaction rate is a function of reactant concentration, $r(x_1/x_2)$. The Hamiltonian to be maximized is

$$\underset{u}{\text{Max}}[H = \lambda_1(C_{AF}u + rx_2) + \lambda_2 u = \lambda_1 rx_2 + (C_{AF}\lambda_1 + \lambda_2)u \overset{\Delta}{=} H_1 + \phi u] \tag{11.7}$$

where the switching function is defined as

$$\phi = C_{AF}\lambda_1 + \lambda_2 \tag{11.8}$$

The adjoint variables λ must satisfy the following differential equations:

$$\frac{d}{dt}\begin{pmatrix} \lambda_1 \\ \lambda_2 \end{pmatrix} = -\begin{pmatrix} \partial H/\partial x_1 \\ \partial H/\partial x_2 \end{pmatrix} = -\begin{pmatrix} \lambda_1 x_2\,(\partial r/\partial x_1) \\ \lambda_1 x_2(\partial r/\partial x_2) + \lambda_1 r \end{pmatrix}$$

$$= -\begin{pmatrix} \lambda_1 r' \\ -\lambda_1 r' x_1/x_2 + \lambda_1 r \end{pmatrix} \tag{11.9}$$

where $r' = dr/dC_A$. The boundary conditions are

$$\begin{pmatrix} \lambda_1(t_f) \\ \lambda_2(t_f) \end{pmatrix} = \begin{pmatrix} \partial P/\partial x_1(t_f) \\ \partial P/\partial x_2(t_f) \end{pmatrix} = \begin{pmatrix} -1/C_{AF} \\ 0 \end{pmatrix} \qquad (11.10)$$

Because the Hamiltonian is linear in u, the optimal u depends on the sign of the *switching function* $\phi(t)$; if it is positive, the Hamiltonian is maximized by taking the maximum value of u, u_{max}, whereas if it is negative, we take the minimum value u_{min}, and if it is identically zero over a finite time interval, then PMP does not provide the solution:

$$u^* = \begin{cases} u_{max} & \text{if } \phi > 0 \\ u_{min} & \text{if } \phi < 0 \\ u_{sin} & \text{if } \phi \equiv 0 \text{ over finite time interval(s)} \end{cases} \qquad (11.11)$$

The time interval in which the switching function ϕ is identically zero is called the *singular interval*, and the manipulated variable is the *singular control*. The solution to the singular problem is obtained by recognizing that if the switching function is identically zero over finite time intervals, then its time derivatives must also be zero. This problem of singular feed rate was fully covered in Chapter 9.

To obtain the singular feed rate, we differentiate successively the switching function

$$\phi = (C_{AF}\lambda_1 + \lambda_2) = 0 \qquad (11.12)$$

$$\dot{\phi} = (C_{AF}\dot{\lambda}_1 + \dot{\lambda}_2) = -C_{AF}\lambda_1 r' + \lambda_1 r' r_1/r_2 - \lambda_1 r = \lambda_1[r'(C_A - C_{AF}) - r] = 0$$
$$\Rightarrow -r = -r'(C_A - C_{AF}) \qquad (11.13)$$

Equation (11.13) defines the reactant concentration in the singular interval. Equation (11.13) represents the intersection of the reaction rate curve with a tangent to the rate curve drawn from the feed concentration. Thus, the reactant concentration must be held constant at the intersection, $C_A = C_s$, during the singular interval. A graphical representation is given in Figure 11.2, in which it is also clear that the reaction rate must be a nonmonotonic function of the reactant concentration and that the feed concentration must be greater than C_{0m}, the intercept of the tangent at its inflection point with the abscissa.

The second derivative of the switching function yields

$$\ddot{\phi} = \dot{\lambda}_1[r'(C_A - C_{AF}) - r] + \lambda_1 d[r'(C_A - C_{AF}) - r]/dt = 0$$
$$\Rightarrow [r''(C_A - C_{AF}) - r' - r'](dC_A/dt) \quad \Rightarrow (dC_A/dt) = 0 \qquad (11.14)$$

Equation (11.14) states that the reactant concentration is a constant value. Substitution of this fact into the reactant balance equation (11.2) and use of Eq. (11.3) yield

$$\frac{d(C_A V)}{dt} = C_S \frac{dV}{dt} = C_S F(t) = C_{AF} F(t) + r(C_S) V(t) \qquad (11.15)$$

Or, solving for the feed rate, we obtain

$$F(t) = \frac{-r(C_S)}{C_{AF} - C_S} V(t) \overset{\Delta}{=} \alpha V(t) \qquad (11.16)$$

NECESSARY CONDITION FOR SINGULAR FEED RATE

$$-r = \frac{d(-r)}{dC_A}(C_A - C_{AF})$$

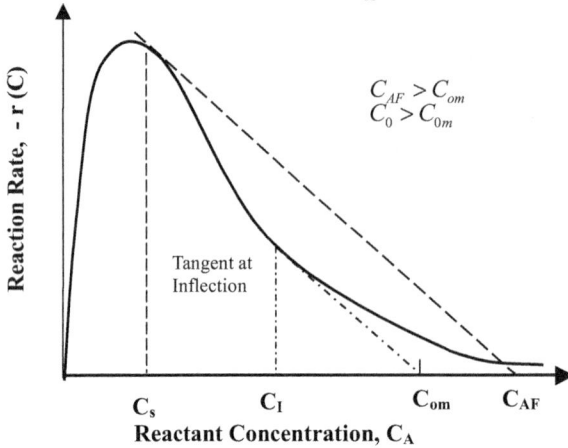

Figure 11.2. General characteristics of reaction rate curves for singular control.[3]

Equation (11.16) states that the feed rate is proportional to the reactor volume. Substitution of Eq. (11.16) into Eq. (11.3) yields

$$\frac{dV}{dt} = F = \alpha V \tag{11.17}$$

or

$$V(t) = V_0 \exp(\alpha t) \tag{11.18}$$

According to Eq. (11.18), the reactor volume increases exponentially with time. Finally, the feed rate is obtained by substituting Eq. (11.18) into Eq. (11.16):

$$F(t) = \alpha V_0 \exp(\alpha t) = [-r(C_S)/(C_{AF} - C_S)]V_0 \exp\{[-r(C_S)/(C_{AF} - C_S)]t\} \tag{11.19}$$

Thus, the feed rate and the reactor volume increase exponentially during the singular period.

11.1.1 Optimal Feed Rate Profile

Because the singular feed rate maintains the reactant concentration constant at C_S, it is obvious that the initial reactant concentration should be also C_S. If it is less than C_S, it should be brought to C_S as soon as possible by applying the maximum feed rate, F_{max}, and if it is greater than C_S, it should be reduced to C_S as quickly as possible by applying $F_{min} = 0$; that is, run the reactor in a batch mode. Then, the singular feed rate F_{sin} should be applied to hold the reactant concentration constant at C_S until the reactor is full, which is followed by a period of batch until the desired conversion is met.

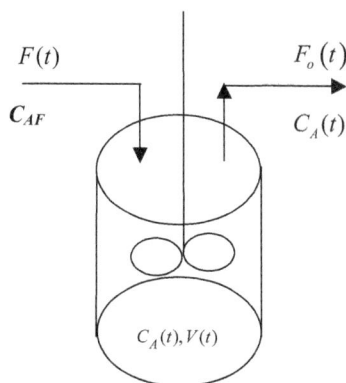

Figure 11.3. A generalized reactor operation with feed and withdrawal rate manipulations.

11.2 Single Reactions with Both Feed and Withdrawal Rates

For the purpose of determining how and when the withdrawal should be made, we shall consider the optimization of isothermal operation of a single reaction ($A \rightarrow B$) by manipulating both the inlet and outlet flow rates. A generalized reactor operation scheme is given in Figure 11.3. With this scheme, it is obvious that $F = F_o = 0$ (with proper initial conditions) represents the batch operation, while the steady state CSTR operation is represented by $F = F_o > 0$. PFR operation is represented by an impulse function, $F = F_o = \delta(t)$, and the semi-batch (fed-batch) operation by $F > 0$, $F_o = 0$ until the end, a delayed step function starting at the final time, $F_0 S(t - t_f)$. By choosing to manipulate not only the feed rate but also the withdrawal rate as functions of time, it may be possible to discover a new mode of operation that has not been observed yet, or it may confirm the traditional fed-batch operation as the optimum.

For simplicity, we assume that the density is constant and identical for the feed and withdrawal streams and the reactor content. The overall and the reactant mass balance equations are

$$\frac{dV}{dt} = F(t) - F_o(t) \quad V(0) = V(t_f) \quad 0 \leq V \leq V_{\max} \tag{11.20}$$

and

$$\frac{d(C_A V)}{dt} = C_{AF} F(t) - C_A(t) F_o(t) + r(C_A) V(t) \quad C_A(0) = C_A(t_f) \tag{11.21}$$

where $C_A(t)$ stands for the reactant concentration, $V(t)$ is the reactor volume at any time, $V_{\max}$ is the maximum reactor volume, $F(t)$ is the inlet flow rate, $F_o(t)$ is the outlet flow rate, $r(C_A)$ is the reaction rate as a function of C_A, C_{AF} is the feed reactant concentration, and t_f is the final time. The boundary conditions depict a cyclic process in which a single cycle is repeated presumably without a down time between cycles.

The objective is to determine the optimum feed and withdrawal rates, F and F_o, that maximize the productivity,

$$\underset{F, F_o}{\text{Max}} \left[P = \int_0^{t_f} (C_{AF} - C_A) F_o \, dt / V_{\max} C_{AF} t_f \right] \tag{11.22}$$

subject to a conversion constraint

$$\int_0^{t_f} (1 - C_A/C_{AF})F_o dt / \int_0^{t_f} F_o dt = 1 - C_{Af}/C_{AF} \qquad (11.23)$$

where C_{Af} is the final reactant concentration.

We recast the preceding problem into a standard form to be solved by PMP (Chapter 9) by introducing the following state variables:

$$x_1(t) = V(t)/V_{\max}, \quad x_2(t) = 1 - C_A(t)/C_{AF}, \quad x_3(t) = t, \quad x_4(t) = \int_0^t u_2 d\tau,$$
$$(11.24)$$
$$x_5(t) = \int_0^t x_2 u_2 d\tau, \quad \theta = t_f, \quad u_1(t) = F(t)/V_{\max}, \quad u_2(t) = F_o(t)/V_{\max}$$

The state variable $x_1(t)$ represents the dimensionless reactor volume, $x_2(t)$ the dimensionless conversion, $x_3(t)$ the running time, $x_4(t)$ the accumulative withdrawal volume, $x_5(t)$ the accumulative product of conversion and withdrawal rate, $u_1(t)$ the feed rate, and $u_2(t)$ the withdrawal rate. In terms of these variables, the preceding problem is recast as

$$dx_1/dt = f_1 = u_1 - u_2 \quad x_1(0) = x_1(\theta) \qquad (11.25)$$

$$dx_2/dt = f_2 = -r_A/C_{AF} - u_1 x_2/x_1 \quad x_2(0) = x_2(\theta) \qquad (11.26)$$

$$dx_3/dt = f_3 = 1 \quad x_3(0) = 0 \qquad (11.27)$$

$$dx_4/dt = f_4 = u_2 \quad x_4(0) = 0 \qquad (11.28)$$

$$dx_5/dt = f_5 = x_2 u_2 \quad x_5(0) = 0 \qquad (11.29)$$

$$x_5(\theta) = x_{2f} x_4(\theta), \quad x_{2f} = 1 - C_{Af}/C_{AF} \qquad (11.30)$$

$$\underset{\substack{u_1, u_2 \geq 0 \\ \theta > 0}}{\text{Max}} [P = x_5(\theta)/x_3(\theta)] \qquad (11.31)$$

The solution to this problem is obtained using PMP.

11.2.1 Solution via Pontryagin's Maximum Principle[2,3]

The necessary conditions for optimality are obtained using PMP, which was covered in detail in Chapter 9. A step-by-step recipe-type application is given subsequently. According to PMP, the maximization of a performance index P is equivalent to maximizing the Hamiltonian, which is the inner product of the adjoint vector λ with the right-hand side of the state equation vector $\mathbf{f}$,

$$H = \lambda^{\mathbf{T}} \mathbf{f} = \sum_{i=1}^{i=5} \lambda_i f_i = \lambda_1(u_1 - u_2) + \lambda_2(-r_A/C_{AF} - u_1 x_2/x_1) + \lambda_3 + \lambda_4 u_2 + \lambda_5 x_2 u_2$$
$$= [\lambda_1 - (x_2/x_1)\lambda_2]u_1 + (-\lambda_1 + \lambda_4 + x_2\lambda_5)u_2 + (-r_A/C_{AF})\lambda_2 + \lambda_3 \qquad (11.32)$$

where the adjoint variables λ_i must satisfy the following differential equations $(d\lambda_i/dt = -\partial H/\partial x_i)$ and boundary conditions $(\lambda_i(\theta) = \partial P/\partial x_i(\theta))$:

$$d\lambda_1/dt = -\lambda_2\left(x_2/x_1^2\right)u_1; \qquad\qquad \lambda_1(0) = \lambda_1(\theta) \qquad (11.33)$$

$$d\lambda_2/dt = [u_1/x_1 + (dr_A/dx_2)/C_{AF}] - \lambda_5 u_2; \quad \lambda_2(0) = \lambda_2(\theta) \qquad (11.34)$$

$$\dot{\lambda}_3 = 0 \quad \Rightarrow \lambda_3(t) = \lambda_3(\theta) = -x_5(\theta)/x_3^2(\theta) \qquad (11.35)$$

$$\lambda_4(t) = \rho x_{2f} \qquad (11.36)$$

and

$$\lambda_5 = \lambda_5(\theta) = \rho + 1/x_3(\theta) \qquad (8.37)$$

The stopping conditions are given by

$$H[\mathbf{x}^*(t), \boldsymbol{\lambda}^*(t), \mathbf{u}^*(t)] = H[\mathbf{x}^*(\theta), \boldsymbol{\lambda}^*(\theta)] = 0 \qquad (11.38)$$

Equation (11.38) states that the Hamiltonian H evaluated with the optimum flow rate profile $\mathbf{u}^*(t) = [u_1(t), u_2(t)]^T$, the state profile $\mathbf{x}^*(t) = [x_1(t), x_2(t), \ldots x_5(t)]^T$, and adjoint profile $\boldsymbol{\lambda}^*(t) = [\lambda_1(t), \lambda_2(t), \ldots \lambda_5(t)]^T$ is identically zero. Therefore, with the optimal flow rate profiles, the Hamiltonian is zero over the entire time interval, $(0, t_f)$, and this is the stopping condition in a numerical scheme.

11.2.2 Switching Space Analyses[3]

The Hamiltonian, Eq. (11.32), is rewritten as

$$H = \phi_1 u_1 + \phi_2 u_2 + \lambda_3 - \lambda_2 r_A/C_{AF} \qquad (11.39)$$

where the switching functions ϕ_i are the coefficients associated with u_i:

$$\phi_1 = \lambda_1 - \lambda_2 x_2/x_1 \qquad (11.40)$$

and

$$\phi_2 = \lambda_4 - \lambda_1 - x_2\lambda_5 \qquad (11.41)$$

Because the Hamiltonian is linear in u_1 and u_2, its maximization depends on the signs of ϕ_1 and ϕ_2. For example, if $\phi_i < 0$, then u_i should take on the minimum value, and conversely, if $\phi_i > 0$, then u_i should take on the maximum value. Thus, ϕ_1 and ϕ_2 are known as the switching functions for u_1 and u_2, respectively. However, when the switching function is identically zero over a finite time intervals $[t_1, t_2]$, the maximum principle does not provide a well-defined control, and this control is known as the singular control. Therefore, the Hamiltonian is maximized by choosing u_i as follows:

$$u_i = \begin{bmatrix} u_{i,\max} \text{ if } \phi_i > 0 & \text{maximum control} \\ u_{i,\min} \text{ if } \phi_i < 0 & \text{minimum control} \\ u_{i,\sin} \text{ if } \phi_i = 0 \text{ over finite interval(s)} & \text{singular (intermediate) control} \\ u_{i,\min} \le u_{i,\sin} \le u_{i,\max} & \end{bmatrix} \qquad (11.42)$$

(a) Low Initial Reactant Concentration

(b) High Initial Reactant Concentration

(c) Appropriate Initial Reactant Concentration

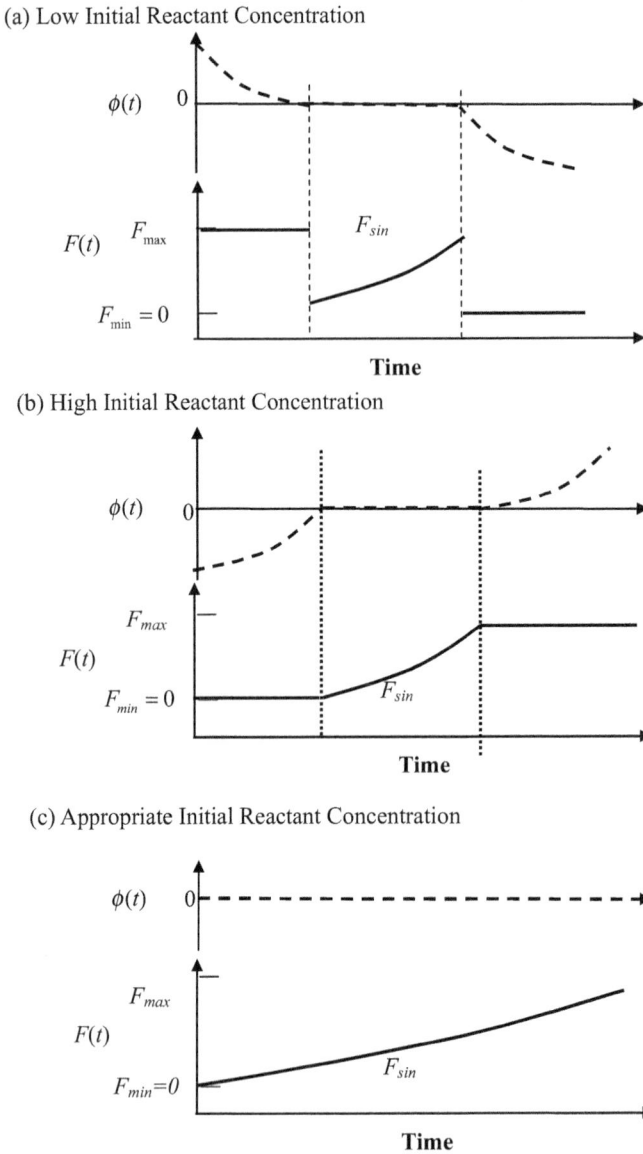

Figure 11.4. Switching functions and optimal feed rate profiles. Adapted from reference 3.

The fact that the switching functions are identically zero over a finite time interval requires that all orders of their time derivatives also vanish. The type of problem in which flow rates appear linearly in the mass balance equations is known as singular control of first order[3] so that only the first two derivatives, $d\phi_i/dt = 0$ and $d^2\phi_i/dt^2 = 0$, are needed to determine the singular control. Thus, the problem reduces to knowing when the switching functions, ϕ_1 and ϕ_2, are positive, negative, or identically zero over certain time intervals and, accordingly, determining the singular control variables. Various structural forms of the optimum flow rate are depicted in Figure 11.4. Intuitively, it is clear that when the initial reactant concentration is low, the initial feed rate should be the maximum (bang), whereas when the initial concentration is high, the initial feed rate should be the minimum (bang, no feed, so that it is a batch operation). When the initial conditions are chosen appropriately to be on the singular arc, the singular feed rate is applied for the entire interval.

11.2.3 Modal Analyses

The analysis is made systematic through modal analysis.[3] Mode is an ordered tuple variable in which each element corresponds to the switching stage of its respective control variable (Eq. (11.42)). As shown by Eq. (11.42), there are three possibilities for each control variable. The maximum (an impulse or the upper bang if the maximum is finite) is given a numeral 1, while the singular control is given numeral 2, and the minimum (the lower bang) is given numeral 3. The first digit refers to the stage of the first control u_1, while the second digit refers to the stage of the second control u_2. Thus, for example, mode 12 represents the maximum value (the upper bang control) for u_1 and the singular control for u_2, while mode 23 represents u_1 in the singular control and u_2 in the minimum (the lower bang). The set of all possible modes makes up the switching space control. The state and adjoint variables change with time in any mode, leading to changes in the switching functions. This change leads to "transition" to another mode. When the boundary conditions are periodic, as is the case here, the optimal policy is represented by a stationary mode (mode 22) or a closed loop in the switching space.

Although this approach appears to be formidable due to the horrendous number of feasible loops and the requirement of analytical solutions for the system equations, which are, in general, nonlinear, this is not the case as the examination of the system equations in each mode renders only a few modes feasible. Additionally, only a few specific transitions are permissible from a given mode, and inspection of the feasible modes eliminates most closed loops, leaving only a few candidates for the optimal policy.

11.2.4 Feasible Modes

1. **Mode 12(21).** This mode must occur instantaneously, and the control variable $u_1(u_2)$ would be proportional to the Dirac delta function (impulse function) owing to an unbounded control vector and bounded x_1. Now, if the other control variable $u_2(u_1)$ is singular, this mode must occur over a finite time interval for only then would the derivatives of ϕ_2 be identically zero. This apparent contradiction simply means that mode 12(21) cannot occur in the optimal policy, that is, the policy of maximum for u_1 and singular for u_2 (singular for u_1 and maximum for u_2) is not feasible.

2. **Mode 22.** Substitution of $\phi_1 = 0$, $\phi_2 = 0$ and Eqs. (11.26) and (11.33) into $d\phi_2/dt = 0$ yields

$$\lambda_4 u_1/x_1 - \lambda_5 r_A/C_{AF} = 0 \qquad (11.43)$$

Substitution of dx_2/dt into Eq. (11.43) and rearrangement yield

$$(1 + x_2 C_{AF}/r_A)/dx_2/dt = -\lambda_5/\lambda_4 = \text{constant} > 0 \qquad (11.44)$$

This can be satisfied either by a stationary value or a nonstationary function of time. Let us consider the stationary case first, $dx_2/dt = 0$. With this control scheme, Eqs. (11.26), (11.34), (11.35), and (11.38) require that $1/\theta = 0$. An infinite time interval implies that the system remains stationary at a point in both the state and the switching spaces. The periodic nature of the problem makes transitions to this mode obviously forbidden. Conversely, for the case of a nonstationary function of time, Eqs. (11.43) and (11.23) imply that

$$\lambda_5/\lambda_4 = -C_{AF}u_1/x_1r_A = [r_A + C_{AF}(dx_2/dt)]/r_Ax_2 > 0 \qquad (11.45)$$

A careful examination of the switching functions, Eqs. (11.40) and (11.41), shows that a transition can occur due to a sign change only in ϕ_1 by saturation of $x_1 (0 \le x_1 \le 1)$. Therefore, a transition to mode 32 may be possible. However, as shown later, $\lambda_5 = 0$ for mode 32. With Eq. (11.43) and $u_1 \ne 0$, it is obvious that $1/\theta = 0$. Therefore, this mode is necessarily stationary.

Another possibility is oscillatory control, in which the state variables oscillate without a transition. Clearly such a cycle is physically possible since $u_1 \ne 0$ and $u_2 \ne 0$. However, rearrangement of Eq. (11.44) yields

$$dx_2/dt = [(-\lambda_5/\lambda_4)x_2 + 1](r_A/C_{AF}) < 0 \qquad (11.46)$$

Because the derivative must be negative for all times, no oscillatory control is optimal.

3. **Mode 32.** Because $u_1 = 0$, $\phi_2 = 0$ and $d\phi_2/dt = 0$, $\lambda_5 = 0$ so that $\phi_2 = -\lambda_1 + \lambda_4$. Because $d\lambda_1/dt = 0$ from Eq. (11.33) and $d\lambda_4/dt = 0$, ϕ_2 remains time invariant. Therefore, transition from this mode can occur only to modes 12 and 22. However, transitions to both of these modes are not allowed as mode 12 does not exist and mode 22 is stationary. Consequently, mode 32 cannot occur in an optimal loop; that is, the minimum for u_1 and singular for u_2 is not possible.

4. **Mode 23.** Because $u_2 = 0$, $\phi_1 = 0$, $d\phi_1/dt = 0$ and $u_2 = 0$,

$$\lambda_2[(dr_A/dC_A)x_2 + r_A/C_{AF})]/x_1 = 0 \qquad (11.47)$$

From Eqs. (11.37) and (11.38), $\lambda_2(\theta)r_A(\theta)/C_{AF} = P/\theta \ne 0$. Therefore,

$$(dr_A/dC_A)x_2 + r_A/C_{AF} = 0 \qquad (11.48)$$

Because $x_2 = 1 - C_A/C_{AF} = -(C_A - C_{AF})/C_{AF}$, rearrangement yields

$$-r_A = \frac{d(-r_A)}{dC_A}(C_A - C_{AF}) \qquad (11.49)$$

The right-hand side of Eq. (11.49) is a straight line passing through C_{AF} with a slope of $d(-r_A)/dC_A$ (tangent to the rate curve), while the left-hand side is the rate curve. This result was previously obtained (Eq. (11.13)). Therefore, the plot of Eq. (11.49) is the same as in Figure 11.2, where it is clear that the tangent drawn to the rate curve, $r_A(C_A)$, from the feed concentration, C_{AF}, represents the right-hand side, while the left-hand side is the rate curve itself, $-r_A(C_A)$. As indicated, $C_A = C_s$ is the concentration at which the tangent line meets the rate curve and is the concentration that remains constant during the singular feed rate.

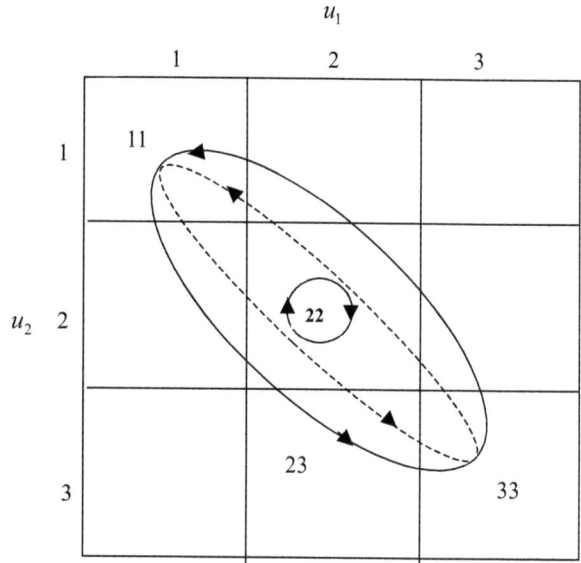

Figure 11.5. Feasible optimal policies.[3]

Assuming that the reaction rate $-r_A(C_A)$ is finite for all concentrations, this suggests that the rate curve must go through a maximum, has an inflection point, and approaches zero asymptotically. *Thus, the necessary condition for singular control suggests that the rate curve must be nonmonotonic.* In other words, singular control is not the optimum for a single reaction that obeys a monotonic rate expression such as the Monod form, power laws, and saturation kinetics. It is apparent that the feed concentration C_{AF} must be greater than the value C_{om} at which the tangent at the inflection point C_I intersects the C_A axis. Because the concentration is maintained at a constant value, C_s, during the period, $dx_{2s}/dt = 0$ is substituted into Eqs. (11.25) and (11.26) along with $u_2 = 0$ and is integrated to obtain

$$x_1 = x_{1s} \exp[-r_A(x_{2s})t/C_{AF}x_{2s}] \tag{11.50}$$

and

$$u_1 = (x_{1s}/x_{2s})[-r_A(x_{2s})/C_{AF}] \exp[-r_A(x_{2s})t/C_{AF}x_{2s}] \tag{11.51}$$

where $x_{1s} = V_s/V_m$ and $x_{2s} = 1 - C_{As}/C_{AF}$ are the fractional volume and the conversion, respectively, at the beginning of the singular control. According to Eqs. (11.50) and (11.51), the feed flow rate and the reactor volume increase exponentially during this mode, while the concentration remains constant.

The transition from this mode occurs due to saturation in x_1 (the reactor is filled up). When $\phi_2 < 0$, it can be shown that $\phi_1 < 0$ when $x_1 = 1 + 0$, and therefore, only mode 33 can be the exit mode.

11.2.5 Optimal Policies

From the preceding, we come to the following observations. Two closed-loop modes (11-33) and (11-23-33) and one stationary mode (22) are feasible, as shown in Figure 11.5. To fill the reactor after discharge, mode 13 should follow mode 31 so that they merge into mode 11. Thus, mode 11 can represent various combinations of modes

13 and 31 by appropriate selection of the constant of proportionality for the Dirac delta function for u_1 and u_2. It should be noted also that in general, transition may occur to 13 from 31 and then to 23 so that 11-23-33 appears to be the most general case of singular control. These optimal policies will be examined in more detail subsequently.

1. Stationary operation (mode 22). The basic equation for this policy is $dx_2/dt = 0$. Thus, the reactor operates as a steady state CSTR, and there are two possibilities, depending on whether the maximum rate occurs at a lower or higher concentration than the desired conversion.

   a. $-r_A(x_{2f}) \geq -r_A(x_2)$ for all $x_2 > x_{2f}$. This is the case in which the rate at the desired (final) conversion $-r_A(x_{2f})$ is higher or equal to the rates at conversions that are higher than the desired conversion $-r_A(x_2)$ for all $x_2 > x_{2f}$. In this case, it can be readily shown that the reactor should be run at the desired conversion, $x_2 = x_{2f}$, and that the steady state flow rate should be obtained from Eqs. (11.25) and (11.26),

   $$u_1 = u_2 = -r_A(C_{Af})/C_{AF}x_{2f} \tag{11.52}$$

   and the productivity is

   $$P_{ss} = -r_A(x_{2f})/C_{AF} \tag{11.53}$$

   The other possibility is as follows.

   b. $-r_A(x_{2m}) > -r_A(x_{2f})$ for some $x_{2m} > x_{2f}$. This is the case in which the maximum rate occurs at a conversion higher than the desired conversion, that is, the maximum rate occurs at a reactant concentration that is lower than the desired final concentration. It is obvious that the reactor should be operated at the concentration corresponding to the maximum rate and a part of the feed should be bypassed and mixed with the reactor output stream to meet the final desired concentration. Thus, the operational is characterized by

   $$u_1 = u_2 = -r_{Am}/C_{AF}x_{2f} \tag{11.54}$$

   $$u_{bp} = [-r_{Am}/C_{AF}](1/x_{2f} - 1/x_{2m}) \tag{11.55}$$

   where u_{bp} is the bypass flow rate. The productivity is

   $$P_{ss} = -r_{Am}/C_{AF} \tag{11.56}$$

2. Maximum–minimum (bang-bang, if finite) operation (modes 11–33). This operation consists of filling the reactor from its initial volume, V_0, to the maximum value, V_m, as fast as possible (instantaneously, using an impulse) with the feed concentration C_{AF}, then allowing the reaction to take place batchwise until the desired conversion, x_{2f}, is reached, and then harvesting as fast as possible (instantaneously, using an impulse), retaining volume V_0 for the next cycle. The spatial dual of this operation is a PFR with a recycle ratio of V_0/V_m. The productivity of this maximum–minimum (bang-bang) operation is

   $$P_{BB} = (1 - V_0/V_m)V_m)x_{2f} \bigg/ C_{AF} \int_{x_{2b}}^{x_{2f}} [1/-r_A(x_2)]dx_2 \tag{11.57}$$

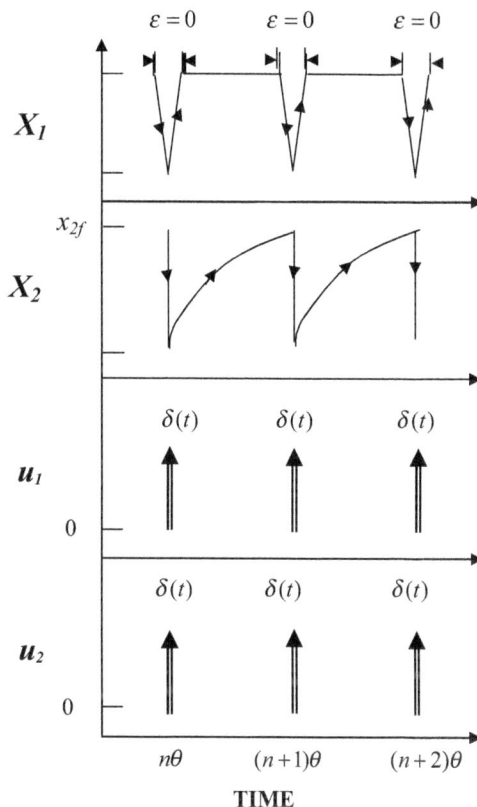

Figure 11.6. Typical maximum–minimum (bang-bang) profiles.[3]

where x_{2b} is the conversion at the start of the batch period, and it is related to the initial volume V_0 by the mass balance

$$V_0 x_{2f} = V_m x_{2b} \qquad (11.58)$$

The maximum productivity is obtained from Eq. (11.57) by differentiating with respect to x_{2b} and setting the resultant to zero:

$$(x_{2f} - x_{2b})/-r_A(x_{2b}) = \int_{x_{2b}}^{x_{2f}} (-1/r_A)dx_2 \qquad (11.59)$$

Equation (11.59) contains x_{2b} implicitly and therefore can be solved either analytically or graphically. In terms of x_{2b}, the optimal recycled volume and the optimal productivity are

$$\alpha = V_0/V_m = x_{2b}/x_{2f} \qquad (11.60)$$

and

$$P_{BB} = r(x_{2b})/C_{AF} \qquad (11.61)$$

For the case in which $-r_A(x_2) \geq -r_A(x_{2f})$ for $x_2 > x_{2f}$, it can be shown that $V_0 = V_m$ so that the operation reduces to the stationary case with bypass (case 1b). Typical profiles of volume, conversion, and flow rates are given in Figure 11.6.

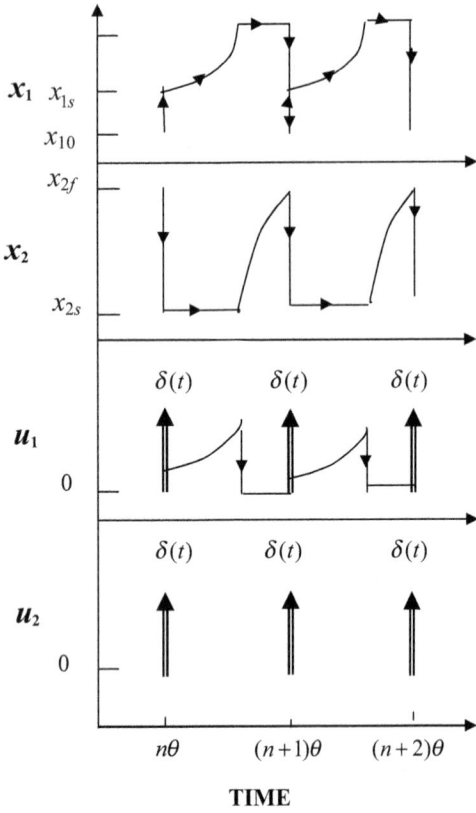

Figure 11.7. Typical singular feed rate profiles.[3]

3. Singular operation (modes 11-23-33). A general version of this operation consists of retention of a fraction of the reactor volume (αV_m) during the harvest (instantaneous dumping of $(1 - \alpha)V_m$), an instantaneous addition of fresh feed to volume, V_s, a singular addition of the feed to fill the reactor volume, V_m, and a batch operation until the desired conversion is achieved, x_{2f}. Then, the reactor content is again instantaneously discharged to a volume of αV_m, and the operations described earlier are repeated. During the period of singular feed rate, the feed rate and the reactor volume increase exponentially (Eqs. (11.50) and (11.51)), and the reactant concentration is maintained constant at C_s. Typical profiles are shown in Figure 11.7.

A theoretical spatial dual of this is a PFR with recycle and distributed feed, as shown in Figure 11.8. The front portion is tapered, and the feed is distributed to give constant flow rate throughout the reactor; in the tapered section, the reactant concentration remains unchanged.

The productivity is

$$P_{SI} = x_{2f}(1 - \alpha) / \left[t_s + C_{AF} \int_{x_{2s}}^{x_{2f}} dx_2/r \right] \qquad (11.62)$$

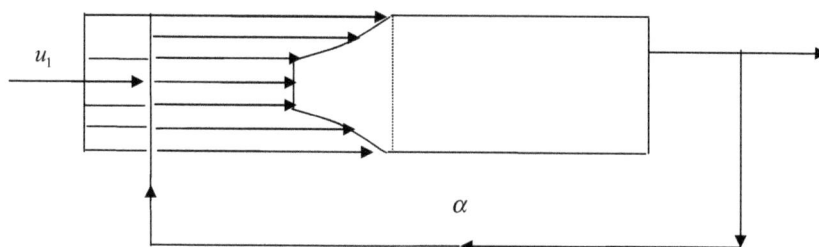

Figure 11.8. PFR with recycle and distributed feed, a spatial dual of singular[3] operation.

Here t_s is the time at which the fractional reactor volume changes from x_{1s} to 1 and x_{2s} is the conversion at the start of the mode 23. The switching time t_s is obtained from Eq. (11.51) as

$$t_s = x_{2s}/r(x_{2s}/C_{AF}) \ln(1/x_{1s}) \qquad (11.63)$$

The mass balance equation relates x_{1s} to x_{2s}:

$$\alpha x_{2f} = x_{1s} x_{2s} \qquad (11.64)$$

The performance index, P_{SI}, is a function of the fractional recycle (retention) volume α, and maximization of it with respect to α yields an implicit relationship for α:

$$(1 - \alpha)/\alpha + \ln \alpha = \ln(x_{2s}/x_{2f}) + [r(x_{2s})C_{AF}x_{2s}] \int_{x_{2s}}^{x_{2f}} dx_2/r(x_2) \overset{\Delta}{=} z \qquad (11.65)$$

The productivity is then given by

$$P_{SI} = \alpha x_{2f}[r(x_{2s})/_{AF}] \quad x_{2f}/x_{2s} \qquad (11.66)$$

The feasibility conditions for singular control are as follows:

1. x_{2s} exists
2. $z > 0$
3. $\alpha x_{2f}/x_{2s} < 1$ (or $V_s < V_m$)

In general, there will be two values of x_{2s}, as one can observe from Figure 11.2, that will satisfy the condition given by Eq. (11.49) for a given C_{AF}. Of the two, it can be shown that the larger value is x_{2s}.

If the feasibility conditions are satisfied, the singular control is optimal. Frequently, the optimal policy would be bang-bang type. If the conversion at which the rate is maximum is greater than the desired conversion, that is, $x_{2m} > x_{2f}$ ($C_{Am} < C_{Af}$), then the stationary policy is optimal. In other words, the reactant concentration at which the rate is maximum is smaller than the desired reactant concentration, and the stationary policy (CSTR) is optimal.

As illustrated in Figure 11.2, singular control is feasible when the rate expression $r(C)$ has an inflection point with negative slope and the feed concentration C_{AF} is greater than the intersection of the C axis by the tangent line drawn at the inflection point C_{om}. Thus, it is sufficient to consider cases in which the rate goes through a

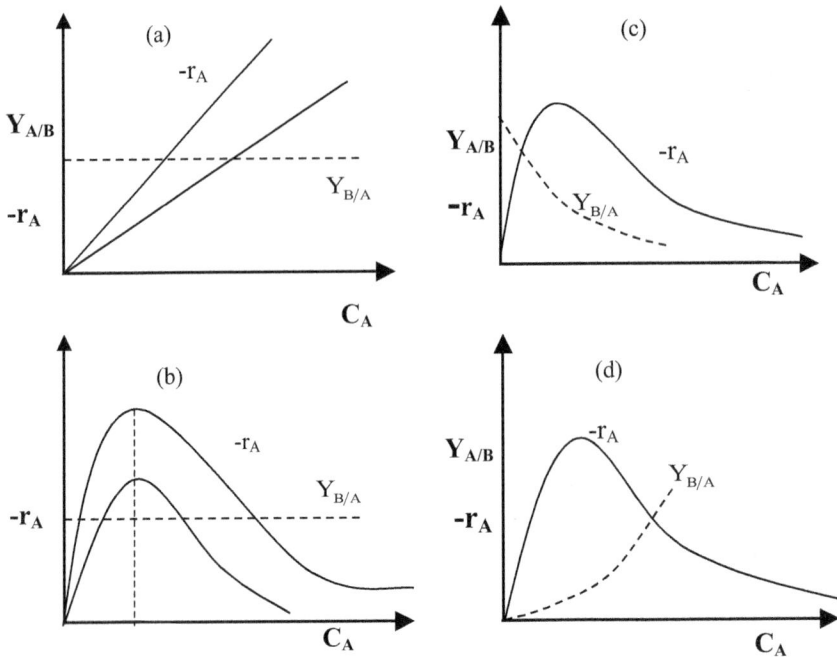

Figure 11.9. Various forms of yield coefficient as a function of reactant concentration.

maximum and decreases asymptotically with further increases in the reactant con-centration. Therefore, autocatalytic, adiabatic exothermic, Lagmuir–Hinshelwood-type catalytic and substrate-inhibited enzyme reactions are potential cases in which singular controls may outperform a CSTR, a PFR, and a PFR with recycle.

From the preceding analyses, it is clear that the withdrawal rate u_2 is never singular and is applied only at the end of the run, not during the course of operation. Thus, the semi-batch operation is optimized by manipulating the feed rate u_1, and the withdrawal u_2 is applied only at the end. Therefore, henceforth, the optimization is carried out with u_1 alone.

11.3 Optimization of Multiple Reactions with a Feed Rate

Consider that a simple multiple reaction scheme consists of two parallel reactions where the desired product is B and the side reaction yields the undesired product C. Both rates of formation, r_B and r_C, are functions of the reactant concentration C_A only. Figure 11.9 depicts simple examples of constant and variable yields. If the order of reaction or the rate expressions were the same for the desired and undesired products, then the yield would be constant, independent of the reactant concentration. Conversely, if the order of the desired reaction were higher than that

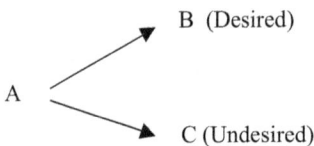

of the undesired reaction, the yield would increase with the reactant concentration. For example, if the order of the desired reaction were second and the order of the undesired reaction were first, then the yield would increase linearly with the reactant concentration. Conversely, if the order of the desired reaction were lower than that of the undesired reaction, the yield would decrease with the reactant concentration.

The yield may also depend on other environmental factors such as temperature and pH. The objective is to maximize the total rate of formation of species B, $r_B(C_A)V$, by controlling the reactant concentration C_A in the reactor. The instantaneous yield for B as a function of the reactant concentration is defined as

$$Y_{B/A}(C_A) = \frac{r_B(C_A)}{-r_A(C_A)} = \frac{r_B(C_A)}{r_B(C_A) + r_C(C_A)} = \frac{1}{1 + r_C/r_B} \qquad (11.67)$$

so that the total rate of formation of the desired product is

$$r_B(C_A)V = [-r_A(C_A)V][Y_{B/A}(C_A)] \qquad (11.68)$$

Thus, maximization of the total rate of formation of B is equivalent to maximizing the right-hand side of Eq. (11.68), that is, maximizing the product of the rate of disappearance of A and the yield coefficient. This process is considered subsequently for the general cases of constant and variable yields.

11.3.1 Constant Yields

When the reaction rate expressions or the reaction order are the same but differ in the numerator constants, the yield coefficient is constant (Figure 11.9a), that is, independent of the reactant concentration. When the yield coefficient is constant, the reactant concentration does not affect the yield, and the total rate is maximized by maximizing the rate of disappearance of species A. Thus, when the rate of disappearance of species A increases monotonically and the yield coefficient is constant, then the rate of formation of the desired species B is maximized by operating the reactor at the maximum reactant concentration, that is, a batch reactor. Conversely, when both reaction rates are nonmonotonic but peak at the same reactant concentration so that the yield coefficient is constant (see Figure 11.9b), then the rate of formation of the desired species B is maximized by maximizing the rate for species A, reducing to the case of a single reaction. Thus, the singular feed rate is applied to maintain the reactant concentration constant at the value corresponding to Eq. (11.16), that is, a semi-batch reactor operation.

When the rate for species A is nonmonotonic and the yield coefficient is not a constant but a function of reactant concentration, as depicted in Figures 11.9c and 11.9d, then the situation is much more complex, and it is no longer optimum to maintain the reactant concentration constant. It must be varied during the course of the reaction to maximize the product of the total rate of disappearance of A and the yield coefficient.

For the case of nonmonotonic rates with constant-yield coefficients, as depicted in Figure 11.9b, the optimal policy to maximize the formation of species B is to maintain the reactant concentration constant at the value that corresponds to C_S in Eq. (11.16). Alternatively to using a fed-batch reactor with the feed rate manipulated to keep the reactant concentration at C_S, a conventional steady state CSTR may be used to maintain the reactant concentration constant at C_S. However, the reactant concentration C_S may be higher than the desired final concentration and may require an additional reactor to achieve the desired conversion. The fermentation industry is very reluctant to use a continuous reactor, let alone two reactors in series, owing to potential contaminations and mutations and the very poor results they have experienced with continuous fermentation for reasons yet to be known and contrary to theoretical predictions. With a batch process, the volume remains at the maximum, but the reactant concentration, which affects the rate and/or yield, cannot be regulated. A fed-batch process provides the means to manipulate the rate and/or yield.

11.3.2 Variable Yields

When the yield depends on the reactant concentration, the right-hand side of Eq. (11.68) must be maximized. We formulate the simplest problem of maximizing the amount of the desired product B produced at a fixed final time by manipulating the feed rate of reactant A,

$$\underset{F(t)}{\text{Max}}\,[\text{P} = C_B(t_f)V(t_f)] \tag{11.69}$$

for the parallel reactions in which species B is the desired product, which is described by the following mass balance equations:

$$\frac{d(C_A V)}{dt} = C_{AF}F + r_A(C_A)V \tag{11.70}$$

$$\frac{d(C_B V)}{dt} = r_B(C_A)V \tag{11.71}$$

$$\frac{dV}{dt} = F, \quad 0 = F_{\min} \leq F(t) \leq F_{\max}, \quad V_0 \leq V(t) \leq V_{\max} \tag{11.72}$$

11.3.2.1 Formulation and Solution

Putting this problem into a standard form, we set $C_A V = x_1$, $C_B V = x_2$, $V = x_3$, and $F = u$, and the objective function and the mass balance equations are

$$\underset{F(t)}{\text{Max}}\,[\text{P} = x_2(t_f)] \tag{11.73}$$

$$\frac{dx}{dt} = C_{AF}u + r_A(x_1/x_3)x_3 \tag{11.74}$$

$$\frac{dx_2}{dt} = r_B(x_1/x_3)x_3 \tag{11.75}$$

$$\frac{dx_3}{dt} = u \tag{11.76}$$

The Hamiltonian is

$$H = \lambda_1(C_{AF}u + r_A x_3) + \lambda_2 r_B x_3 + \lambda_3 u = (\lambda_1 C_{AF} + \lambda_3)u + (\lambda_1 r_A + \lambda_2 r_B)x_3 \tag{11.77}$$

Thus, the switching function is

$$\phi = C_{AF}\lambda_1 + \lambda_3 \tag{11.78}$$

The adjoint variables must satisfy the following differential equations:

$$\dot{\lambda}_1 = -\frac{\partial H}{\partial x_1} = -(\lambda_1 r_A' + r_B'), \quad \lambda_1(t_f) = \frac{\partial P}{\partial x_1(t_f)} = 0 \tag{11.79}$$

$$\dot{\lambda}_2 = -\frac{\partial H}{\partial x_2} = 0 \quad \lambda_2(t_f) = \frac{\partial P}{\partial x_2(t_f)} = 1 \quad \Rightarrow \lambda_2 = 1 \tag{11.80}$$

$$\dot{\lambda}_3 = -\frac{\partial H}{\partial x_3} = (\lambda_1 r_A' + r_B')(x_1/x_3) - (\lambda_1 r_A + r_B), \quad \lambda_3(t_f) = \frac{\partial P}{\partial x_3(t_f)} = 0 \tag{11.81}$$

Because the Hamiltonian is linear in u, the optimal policy depends on the sign of the switching function ϕ,

$$u = \begin{cases} u_{\max} & \text{if } \phi > 0 \\ u_{\min} & \text{if } \phi < 0 \\ u_{\sin} & \text{if } \phi \equiv 0 \text{ over finite time interval(s)} \end{cases} \tag{11.82}$$

where $u_{\max}$, $u_{\min}$, and $u_{\sin}$ refer to the maximum feed rate, the minimum feed rate, and the singular feed rate, respectively. The singular feed rate $u_{\sin}$ is applied when the Hamiltonian is identically zero over a finite time interval. Thus, we need to investigate the switching function ϕ:

$$\phi = C_{AF}\lambda_1 + \lambda_3 = 0 \tag{11.83}$$

Differentiating ϕ once with respect to time, we obtain

$$\dot{\phi} = C_{AF}\dot{\lambda}_1 + \dot{\lambda}_3 = (\lambda_1 r_A' + r_B')(x_1/x_3 - C_{AF}) - (\lambda_1 r_A + r_B) = 0 \tag{11.84}$$

or

$$\lambda_1 = [r_B - r_B'(x_1/x_3 - C_{AF})]/[r_A'(x_1/x_3 - C_{AF}) - r_A] \tag{11.85}$$

Differentiating once more, we obtain

$$\ddot{\phi} = (\lambda_1 r_A' + r_B')r_A'(x_1/x_3 - C_{AF}) - 2(\lambda_1 r_A' + r_B')r_A \\ + (\lambda_1 r_A'' + r_B'')(x_1/x_3 - C_{AF})\dot{C}_A = 0 \tag{11.86}$$

or

$$\dot{C}_A = \frac{(\lambda_1 r_A' + r_B')[(C_A - C_{AF})r_A' - 2r_A]}{(\lambda_1 r_A'' + r_B'')[(C_A - C_{AF})]} \tag{11.87}$$

Substitution of Eq. (11.85) into Eq. (11.87) yields

$$\dot{C}_A = \frac{\{r_A' r_B - r_A r_B'\}[r_A' - 2r_A/(x_1/x_3 - C_{AF})]}{\{r_A'' r_B - r_A r_B'' + (r_A' r_B'' - r_A'' r_B')(x_1/x_3 \quad C_{AF})\}} \tag{11.88}$$

The singular feed rate $F_{\sin}$ required to force $\dot{C}_A$ to satisfy Eq. (11.88) is obtained from the mass balance on species A (Eq. (11.70)):

$$F_{\sin} = \frac{(\dot{C}_A - r_A)V}{C_{AF} - C_A} \tag{11.89}$$

Finally, substituting Eq. (11.88) into Eq. (11.89) yields the singular feed rate,

$$F_{\text{sin}} = \frac{\{r'_A r_B - r_A r'_B\}\left[r'_A - \frac{2r_A}{(C_A - C_{AF})}\right] - r_A\{r''_A r_B - r_A r''_B + (r'_A r''_B - r''_A r'_B)(C_A - C_{AF})\}}{\{r''_A r_B - r_A r''_B + (r'_A r''_B - r''_A r'_B)(C_A - C_{AF})\}(C_A - C_{AF})}$$

(11.90)

The singular feed rate expression is too complex to make general remarks. The real question to be asked at this point is, when is a semi-batch mode of operation feasible for the two parallel reactions?

The optimal feed rate sequences depend on the initial condition C_{A0}. If it is low, then the maximum feed rate should be applied to bring C_A to C_{AS}, at which the singular feed rate F_{sing} begins until the reactor is full, $V = V_{\text{max}}$, and finally followed by a batch ($F = 0$) period until a desired conversion is achieved. Unlike the case of a single reaction, C_S is not defined and therefore must be obtained numerically. Conversely, if it is too high, a minimum feed rate ($F = 0$, a batch) is applied to reduce the concentration to C_{As} before a singular feed rate is applied. Finally, if the initial condition is proper, the singular feed rate is applied from the beginning:

$$\begin{cases} F_{\text{max}} & \rightarrow F_{\text{sin}} \rightarrow F = 0 & \text{initial condition } C_{A0} \text{ too low} \\ F_{\text{min}} = 0 \rightarrow F_{\text{sin}} \rightarrow F = 0 & \text{initial condition } C_{A0} \text{ too high} \\ F_{\text{sin}} & \rightarrow F = 0 & \text{initial condition } C_{A0} \text{ just right} \end{cases}$$

(11.91)

For the best result, it is obvious that the initial condition C_{A0} should be picked just right so that the singular feed rate is applied from the beginning. The remaining parameter for optimization is the initial volume V_0. Thus, there are two parameters that need to be optimized. This is an ordinary calculus problem and therefore could be searched numerically.

REFERENCES

1. Pontryagin, L. S., Boltyanskii, Y. G., Gamkrelidze, R. V., and Mischenko, E. F. 1962. *The Mathematical Theory of Optimal Processes*. Wiley-Interscience.
2. Bryson, A. E., and Ho, Y.-C. 1975. *Applied Optimal Control*. John Wiley.
3. Wagmare, R. S., and Lim, H. C. 1981. Optimal operation of isothermal reactors. *Industrial and Engineering Chemistry, Fundamentals* 20: 361–336.

12 Optimization for Cell Mass Production

Optimization of bioprocesses is very important because these processes require capital-intensive plants, yield products that are low in concentration, and sometimes use expensive raw materials. The objective of bioprocess optimization is to maximize the profit of the process. More specifically, it frequently involves the maximization of the volumetric productivity, metabolite concentration, conversion, or yield and minimization of capital and operating costs or time to achieve a desired conversion.

Vital to the success of optimization methods is the development of mathematical models that describe adequately the behavior of the process under various conditions and therefore provide quantitative relationships between the outcome (outputs) and the manipulated variables (inputs) of the process. Some models contain a number of process parameters that need to be determined directly or indirectly from experimental data. The optimal control is aimed at achieving process optimization. A model being the foundation for process optimization and control, any increase in complexity of the model is justified only if it results in a significant improvement in the process performance. In light of the complex nature of the microbial and cellular processes, improved on-line measurement and data acquisition methods are of central importance to optimization of these bioprocesses. In this chapter, we deal with the optimization of bioreactors that are modeled by ordinary differential equations, which results from the material balances of species involved: the cells, substrates, products, intermediates, nutrients, various enzymes, and other chemical entities, as we have seen in previous chapters. In this chapter, we consider an impulse optimization through the application of PMP, which is most well suited for processes described by a set of ordinary differential equations.

12.1 Optimization by Pontryagin's Maximum Principle

Ordinary calculus or equivalent numerical schemes are used to solve parameter optimization problems. Best constant values are determined through the necessary conditions resulting from the ordinary calculus. Various numerical techniques are used to search for the solution. However, when it comes to determination of time profiles, that is, the best function of time, which is equivalent to finding the best constant value at each and every time increment, variational calculus provides the necessary conditions. Variational calculus is used to solve the problem of finding the best function, as compared to the ordinary calculus, which allows determination of the best constant values. Function optimization based on Pontryagin's maximum principle[1] (PMP) is particularly useful when the manipulated variables and/or state variables are constrained. Consequently, the maximum principle has been used extensively to determine the optimal time profiles for fed-batch bioreactors.[2-9] A brief introduction to and a summary of PMP are given in Chapter 9. Some familiarity of Chapter 9 would be essential in understanding the application of this optimization technique.

For second-order processes described by two ordinary differential equations (ODEs) with both terminal states fixed (the initial and final conditions must be specified), the optimization based on Green's theorem[10] is easier to apply than the maximum principle, which leads to a two-point boundary problem that is numerically difficult to solve. Application of Miele's technique[11] avoids the two-point boundary value problem. However, industrial problems are much more complex and require more than two ODEs to describe them. Thus, these techniques cannot be applied to higher-order bioprocesses.

Various performance indices have been considered for the optimization of bioreactors. For any reactor optimization, in general, two criteria need to be considered: the total rate (productivity) and the selectivity (yield). Intuitively, it is appealing to maximize the rate of reaction if the raw material is relatively inexpensive, while the yield should be maximized if the raw material cost is relatively high. Excess of residual substrate may make purification difficult and contribute significantly to the separation cost. Thus, the conversion may be included in the performance criterion to reduce the substrate concentration to an acceptable value. Performance indices of various kinds have received attention in the past for optimization of fed-batch cultures, including the maximum amount of product at the final time, the maximum productivity (the total amount divided by the time of operation), the maximum profit, the minimum operating costs, and the minimum time to achieve a desired conversion. In addition to these, other empirical performance indices were used, such as maximizing the specific growth rate during the course of bioreactor operation. Such an approach would be optimal for the simple case of biomass production if the yield coefficient were constant. Various criteria are covered in Chapter 9.

12.2 Maximization of Cell Mass at Fixed and Free Final Times

First, we consider the simplest case of producing cell mass as the final product and only one feed stream containing the limiting substrate – a case of a scalar control variable. This type of situation arises when the cell mass is the product, such as single-cell proteins, yeasts, and products that are intracellular and whose fractions in the cell remain approximately constant. In this situation, maximization of cell mass is equivalent to maximization of the product.

12.2.1 Problem Formulation

We begin by writing the mass balance equations for cell mass and substrate and the overall mass balance, as was done in Chapter 3. The simplest and idealized mass balances are as follows:

$$\frac{d(XV)}{dt} = \mu XV \quad X(0)V(0) = X_0 V_0 \tag{12.1}$$

$$\frac{d(SV)}{dt} = S_F F - \sigma XV = S_F F - \mu XV / Y_{X/S} \quad S(0)V(0) = S_0 V_0 \tag{12.2}$$

$$\frac{dV}{dt} = F \quad V(0) = V_0 \tag{12.3}$$

where X and S stand for the concentrations of cells and substrate, respectively, V is the culture volume, S_F is the feed substrate concentration, F is the volumetric feed flow rate, μ is the specific growth rate of cells, σ is the specific consumption rate of substrate, and $Y_{X/S}$ is the cell mass yield coefficient. The usual assumption of constant and equal density for the feed and bioreactor content is made to arrive at Eq. (12.3). The specific growth rate μ and the cell mass yield coefficient $Y_{X/S}$ are assumed, for simplicity, to be functions only of the limiting substrate concentration S. Various forms of specific rates were presented in Chapter 6. The objective is to determine the optimal feed rate profile, $F(t)^*$, that maximizes the total amount of cells produced, $X(t_f)V(t_f)$, at a final time, which may be fixed or free (to be chosen),

$$\underset{F(t)^*}{\text{Max}} [P = X(t_f)V(t_f)] \tag{12.4}$$

subject to the feed rate (manipulated variable) constraint

$$0 = F_{\min} \leq F(t) \leq F_{\max} \tag{12.5}$$

and the volume constraint

$$V(t) \leq V_{\max} \tag{12.6}$$

which states that the reactor operating volume should be fully utilized.

It is convenient to define as state variables the total amount of cells and amount of substrate and the culture volume, $x_1 = XV$, $x_2 = SV$ and $x_3 = V$, as well as the feed rate as the manipulated variable, $F(t)$. Then, the preceding mass balance equations (12.1)–(12.3) can be rewritten as

$$\frac{d}{dt} \begin{pmatrix} x_1 \\ x_2 \\ x_3 \end{pmatrix} = \frac{d}{dt} \begin{pmatrix} XV \\ SV \\ V \end{pmatrix} = \begin{pmatrix} \mu x_1 \\ -\sigma x \\ 0 \end{pmatrix} + \begin{pmatrix} 0 \\ S_F \\ 1 \end{pmatrix} F, \quad \begin{pmatrix} x_1(0) = X_0 V_0 \\ x_2(0) = S_0 V_0 \\ x_3(0) = V_0 \end{pmatrix} \tag{12.7}$$

or in standard form,

$$\frac{d}{dt} \begin{bmatrix} x_1 \\ x_2 \\ x_3 \end{bmatrix} = \begin{bmatrix} a_1 = \mu x_1 \\ a_2 = -\sigma x_1 \\ a_3 = 0 \end{bmatrix} + \begin{bmatrix} b_1 - 0 \\ b_2 = S_F \\ b_3 = 1 \end{bmatrix} F, \quad \dot{\mathbf{x}} = \mathbf{a}(\mathbf{x}) + \mathbf{b}F = \mathbf{f}(\mathbf{x}, F) \tag{12.8}$$

In terms of the state variables, the performance index is

$$P[\mathbf{x}(t_f)] = x_1(t_f) \tag{12.9}$$

and the volume constraint is

$$h(\mathbf{x}) = x_3(t) - V_{\max} \leq 0 \tag{12.10}$$

However, this state constraint can be converted into a final state constraint by recognizing that the reactor volume cannot decrease, $dV/dt = F \geq 0$, so that imposing the volume constraint at the final time would automatically satisfy Eq. (12.10). Therefore, Eq. (12.10) is equivalent to a final state constraint:

$$x_3(t_f) = V_{\max} \tag{12.11}$$

With this notation, the cell and substrate concentrations are

$$X = \frac{VX}{V} = \frac{x_1}{x_3}, \quad S = \frac{VS}{V} = \frac{x_2}{x_3} \tag{12.12}$$

so that the specific rates are functions of x_2/x_3, that is,

$$\mu(S) = \mu(x_2/x_3), \quad \sigma(S) = \sigma(x_2/x_3), \quad Y_{X/S}(S) = Y_{X/S}(x_2/x_3) \tag{12.13}$$

12.2.2 Solution by Pontryagin's Maximum Principle

According to PMP (Chapter 9), we form the Hamiltonian, which is a scalar product of adjoint variables λ_i with the corresponding right-hand sides of the mass balance equations (Eq. (12.7)), and maximize it instead of the original performance index of Eq. (12.9) (see Chapter 9):

$$\underset{F*(t)}{\text{Max}} H = \boldsymbol{\lambda}^{\mathbf{T}}[\mathbf{a}(\mathbf{x}) + \mathbf{b}F] = \boldsymbol{\lambda}^{\mathbf{T}}\mathbf{f} = \sum_{i=1}^{3} \lambda_i f_i = \lambda_1 \mu x_1 + \lambda_2(S_F F - \sigma x_1) + \lambda_3 F$$

$$= (\lambda_1 \mu - \lambda_2 \sigma)x_1 + (S_F \lambda_2 + \lambda_3)F \overset{\Delta}{=} H_1 + \phi F \tag{12.14}$$

where

$$H_1 = (\lambda_1 \mu - \lambda_2 \sigma)x_1, \quad \phi = (S_F \lambda_2 + \lambda_3) \tag{12.15}$$

The adjoint variables must satisfy the following differential equations:

$$\frac{d\boldsymbol{\lambda}}{dt} = \begin{bmatrix} d\lambda_1/dt \\ d\lambda_2/dt \\ d\lambda_3/dt \end{bmatrix} = -\begin{bmatrix} \partial H/\partial x_1 \\ \partial H/\partial x_2 \\ \partial H/\partial x_3 \end{bmatrix} = -\begin{bmatrix} \lambda_1 \mu - \lambda_2 \sigma \\ (\lambda_1 \mu' - \lambda_2 \sigma')x_1/x_3 \\ -(\lambda_1 \mu' - \lambda_2 \sigma')x_1 x_2/x_3^2 \end{bmatrix} \tag{12.16}$$

where the primes denote the differentiation with respect to the substrate concentration, that is, $\mu' = \partial \mu/\partial S$ and $\sigma' = \partial \sigma/\partial S$. The final conditions on the adjoint variables are obtained from the *transversality conditions* (Eq. (9.27)) by matching coefficients:

$$\delta P[\mathbf{x}(t_f)] = \sum_{i=1}^{n} \left[\frac{\partial P}{\partial x_i(t_f)}\right] \delta x_i(t_f) = \sum_{i=1}^{n} \lambda_i(t_f)\delta x_i(t_f) - \sum_{i=1}^{n} \lambda_i(0)\delta x_i(0) \tag{9.27}$$

Specifically, for this problem, we have $\partial P/\partial x_1(t_f) = 1$, $\partial P/\partial x_2(t_f) = \partial P/\partial x_3(t_f) = 0$, and $\delta x_2(t_f) = 0$. Because $x_3(t_f) = V_{\max}$ is fixed and all the initial conditions on the state variables are also fixed so that their variations are all zero, $\delta x_i(0) = 0$, ($i = 1, 2, 3$) so that

$$\delta P = \frac{\partial P}{\partial x_1(t_f)}^{=1} \delta x_1(t_f) + \frac{\partial P}{\partial x_2(t_f)}^{=0} \delta x_2(t_f) + \frac{\partial P}{\partial x_3(t_f)}^{=0} \delta x_3(t_f)$$

$$= \lambda_1(t_f)\delta x_1(t_f) + \lambda_2(t_f)\delta x_2(t_f) + \lambda_3(t_f)\delta x_3(t_f)^{=0} - \sum_{i=1}^{3} \lambda_i(0)\delta x_i(0)^{=0} \quad (12.17)$$

Therefore,

$$\boldsymbol{\lambda}(t_f) = \begin{bmatrix} \lambda_1(t_f) \\ \lambda_2(t_f) \\ \lambda_3(t_f) \end{bmatrix} = \begin{bmatrix} \partial P/\partial x_1(t_f) \\ \partial P/\partial x_2(t_f) \\ \text{free} \end{bmatrix} = \begin{bmatrix} 1 \\ 0 \\ \text{free} \end{bmatrix} \quad (12.18)$$

12.2.2.1 Optimal Feed Rate Profile

Because the feed rate, $F(t)$, appears linearly in Eq. (12.14), the Hamiltonian is maximized by picking the feed rate according to the sign of the coefficient of F, the *switching function*, $\phi = (\lambda_2 + S_F\lambda_3)$. Thus, to maximize the Hamiltonian, the feed rate should take on the maximum value when the switching function is positive, $\phi > 0$, the minimum value (zero) when the switching function is negative, $\phi < 0$, and zero at the final time when the reactor volume is full. But when $\phi \equiv 0$ over a finite time interval or intervals, PMP fails to provide the solution. This interval is called the *singular interval*, the flow rate the *singular feed rate*, and the trajectory the *singular arc*. Thus, the feed rate profile is

$$F = \begin{cases} F_{\max} & \text{when } \phi = (S_F\lambda_2 + \lambda_3) > 0 \\ F_{\min} = 0 & \text{when } \phi = (S_F\lambda_2 + \lambda_3) < 0 \\ F_{\text{sin}} & \text{when } \phi = (\lambda S_{F_2} + \lambda_3) \equiv 0 \text{ over } t_q < t < t_{q+1} \\ F_b = 0 & \text{when } x_3(t) = V_{\max} \end{cases} \quad (12.19)$$

The remaining task is to determine the singular feed rate and the sequence and time intervals of the maximum, minimum, singular, and zero feed rates. Thus, the sequence of feed rates is determined from the time behavior of the switching function $\phi(t)$, and there are many possibilities. For example, it can start with a positive (or negative) value, go through a period of zero value, and then become negative (or positive); it can start with a period of zero value and change to a positive (or negative) value. Some of these combinations are depicted, and the corresponding sequences of feed rates are sketched, in Figure 12.1.

12.2.2.2 Factors Influencing Optimal Feed Rate Profile

It is clear from earlier that the optimal feed rate profile $F^*(t)$ would be influenced by the initial state $\mathbf{x}(0)$, the length of final time t_f, the limits on the feed rate $F_{\min}$ and $F_{\max}$, and the functional forms of the specific rates of cell growth and substrate consumption μ and σ. We begin with the simplest case, the case of constant cell mass yield. The insights gained from this case can provide a foundation for comprehending the more complex case of variable yields.

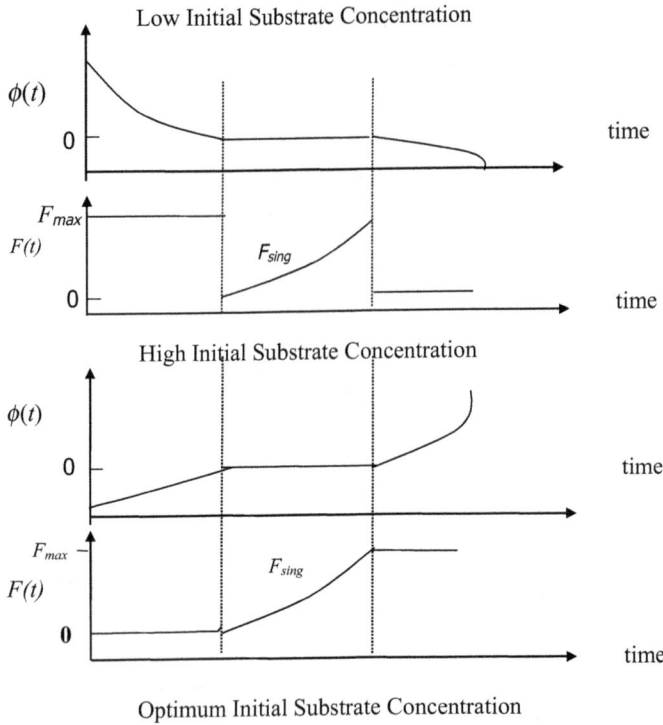

Figure 12.1. Switching functions and corresponding feed rate sequences.

12.2.3 Constant-Yield Coefficients

This is a degenerate case of the general biomass problem because the constant-yield coefficient $Y_{X/S}$ reduces the earlier seemingly third-order (three differential equations) problem to a second-order problem (two differential equations and an algebraic equation). Consider the mass balance equations (12.1)–(12.3). Equations (12.1) and (12.3) are substituted into Eq. (12.2) to obtain

$$\frac{d(SV)}{dt} = S_F F - \frac{\mu X V}{Y_{X/S}} = S_F F - \frac{1}{Y_{X/S}}\frac{d(XV)}{dt} = S_F \frac{dV}{dt} - \frac{1}{Y_{X/S}}\frac{d(XV)}{dt} \qquad (12.20)$$

Because the yield coefficient is constant, we can integrate Eq. (12.20) term by term from 0 to t, making use of Eq. (12.3), to obtain

$$SV - S_0 V_0 = S_F(V - V_0) - (XV - X_0 V_0)/Y_{X/S} \qquad (12.21)$$

or in terms of state variables,

$$x_2 = x_{20} + S_F(x_3 - x_{30}) - (x_1 - x_{10})/Y_{X/S} \overset{\Delta}{=} g(x_1, x_3) \qquad (12.22)$$

According to Eq. (12.22), x_2 is not an independent variable; that is, it can be replaced by the remaining two state variables, x_1 and x_3. Therefore, we can represent the case of constant-yield coefficient by two state variables x_1 and x_3 with two differential equations,

$$\frac{d\mathbf{x}}{dt} = \frac{d}{dt}\begin{pmatrix} x_1 \\ x_3 \end{pmatrix} = \frac{d}{dt}\begin{pmatrix} XV \\ V \end{pmatrix} = \begin{pmatrix} \mu x_1 \\ 0 \end{pmatrix} + \begin{pmatrix} 0 \\ 1 \end{pmatrix} F \quad (12.23)$$

where the specific growth rate, which is a function of $S = SV/V$, is now a function of two state variables:

$$\mu(S) = \mu(SV/V) = \mu(x_2/x_3) = \mu[g(x_1, x_3)/x_3] \quad (12.24)$$

The Hamiltonian (Eq. (12.14)) reduces to

$$\underset{F^*(t)}{\text{Max}} H = \boldsymbol{\lambda}^{\mathrm{T}}[\mathbf{a}(\mathbf{x}) + \mathbf{b}F] = \lambda_1 \mu x_1 + \lambda_3 F \overset{\Delta}{=} H_1 + \phi F \quad (12.25)$$

where

$$H_1 = \lambda_1 \mu x_1, \quad \phi = \lambda_3 \quad (12.26)$$

The differential equations for the adjoint variables (Eq. (12.16)) reduce to

$$\frac{d\boldsymbol{\lambda}}{dt} = \frac{d}{dt}\begin{bmatrix} \lambda_1 \\ \lambda_3 \end{bmatrix} = -\begin{bmatrix} \dfrac{\partial H}{\partial x_1} \\ \dfrac{\partial H}{\partial x_3} \end{bmatrix} = -\begin{pmatrix} \lambda_1 \mu - \lambda_1 x_1 \dfrac{\partial \mu}{\partial x_1} \\ \lambda_1 x_1 \dfrac{\partial \mu}{\partial x_3} \end{pmatrix} = -\begin{pmatrix} \lambda_1 \mu + \dfrac{\lambda_1 x_1 \mu'}{x_3 Y_{X/S}} \\ \lambda_1 x_1 \mu'(S_F - g/x_3)/x_3 \end{pmatrix} \quad (12.27)$$

and the boundary conditions are obtained from Eq. (12.18) by matching the coefficients:

$$\boldsymbol{\lambda}(t_f) = \begin{bmatrix} \lambda_1(t_f) \\ \lambda_3(t_f) \end{bmatrix} = \frac{\partial P}{\partial \mathbf{x}}(t_f) = \begin{bmatrix} 1 \\ \text{free} \end{bmatrix} \quad (12.28)$$

We proceed to consider the final process time, which can be either fixed a priori or free to be chosen. A particular choice of the final time can lead to different results.

12.2.3.1 Free Final Time, t_f

If the final time is free (also to be optimized), then the Hamiltonian must vanish everywhere according to PMP (Eq. (9.43)):

$$H^* = \lambda_1 \mu x_1 + \lambda_3 F \overset{\Delta}{=} H_1 + \phi F = 0 \quad (12.29)$$

At the final time, the feed rate is zero, $F = 0$, because the volume is full, $V(t_f) = V_{\max}$. Thus, the last term in Eq. (12.29) vanishes, leaving the first term only. Because $\lambda_1(t_f) = 1$ according to Eq. (12.28), the Hamiltonian at the free final time is

$$H^*(t_f) = H_1^*(t_f) = \lambda_1(t_f)\mu(t_f)x_1(t_f) = \mu(t_f)x_1(t_f) = 0 \quad (12.30)$$

Because $x_1(t_f) = XV \neq 0$, the specific growth rate at the final time must be zero,

$$\mu(t_f) = 0 \quad (12.31)$$

For the specific growth rate to vanish, the substrate concentration must approach zero for monotonic μ, while it must approach either zero or infinity for nonmonotonic μ.

However, the substrate concentration cannot exceed the feed concentration, $S \leq S_F$. Therefore, it must approach zero so that

$$\mu(t_f) = 0 \Rightarrow \mu(S \to 0) \Rightarrow \mu(t_f \to \infty) \tag{12.31a}$$

Equation (12.31a) states that substrate concentration at the final time must approach zero, and because the substrate concentration approaches zero asymptotically, *the final time must also approach infinity*, $t_f \to \infty$. A rigorous proof is given elsewhere.[6] Because the performance index does not depend on the final time, one should be able to pick any value of time without affecting the performance index. That the final time is infinity is not surprising. Later we will consider finite final times.

12.2.3.2 On Interior Singular Arc

During the period of singular feed rate on an interior region, the switching function is identically zero, $\phi = \lambda_3 \equiv 0$. Therefore, the last term in the Hamiltonian, Eq. (12.29), vanishes, leaving only the first term:

$$H^* = \lambda_1 \mu x_1 = 0 \tag{12.32}$$

Because $x_1 = XV \neq 0$ and $\mu \neq 0$ during the interval, Eq. (12.32) implies the adjoint variable $\lambda_1 = 0$ on the entire singular arc. Thus, both λ_1 and λ_3 are zero during the singular interval, trivially satisfying the adjoint equation (12.27). This means that when the final time is free, any mode of feed, not necessarily the singular feed rate, satisfies Eq. (12.27). In fact, it is simple to show that any mode of operation yields the same amount of cells at the final time, as long as the final time is free (infinity) and the yield coefficient is constant.

It is clear from the preceding result that with a particular choice of performance index and specific rate expressions, maximizing the cell mass has to be done at a fixed (finite) final time. If one allows the final time to be free, it calls for an infinite time operation so that the entire amount of substrate charged would be completely converted into the cell mass. Because the yield is constant, the total amount of cells obtained from the amount of substrate charged would be independent of the substrate concentration in the bioreactor and hence independent of the feeding profile. Indeed, any feeding mode yields the same theoretical number of cells, as given by Eq. (12.21), by setting $t = t_f$ and $S(t_f) = 0$:

$$XV(t_f) = X_0 V_0 + S_0 V_0 Y_{X/S} + [V(t_f) - V_0] S_F Y_{X/S} \tag{12.33}$$

12.2.3.3 Optimal Feed Rate Profile

There is no singular arc, and any mode of feed rate leads to the same amount of cell mass at the final time, which approaches infinity. A number of purely numerical approaches to solve the preceding free final time problem have been reported, which all lead to misleading results[12,13] at a finite time, clearly pointing out the need for more careful analysis to gain insight prior to resorting to a numerical effort. The prior conclusion is only true if the cell mass is not subject to lyses or decay or the final time does not appear in the performance index. It is instructive to investigate what factors influence the final time if the final time is not specified. These situations are examined subsequently.

12.2.3.4 Cell Lyses or Decay

If the cell mass is subject to lyses or decay, the cell balance must account for these phenomena by incorporating a death term, $k_d XV$:

$$\frac{d(XV)}{dt} = \mu XV - k_d XV = (\mu - k_d)XV$$
$$\frac{dx_1}{dt} = \mu x_1 - k_d x_1 = (\mu - k_d)x_1 \tag{12.34}$$

Then, the Hamiltonian is

$$H = \lambda_1(\mu - k_d)x_1 + \lambda_3 F = 0 \tag{12.35}$$

Because $F = 0$ at the final time, the Hamiltonian reduces to

$$H^*(t_f) = \lambda_1(t_f)[\mu(t_f) - k_d]x_1(t_f) = [\mu(t_f) - k_d]x_1(t_f) = 0 \tag{12.36}$$

According to Eq. (12.36),

$$\mu(t_f) = k_d \tag{12.37}$$

Equation (12.37) implies that the specific growth rate must equal the decay constant at the final time. In other words, the fed-batch operation ceases at the final time at which the specific growth rate equals the specific decay rate. Any operation beyond this point leads to less cell mass as the rate of decay is then greater than the rate of growth. Thus, the final time is not infinite but finite.

12.2.3.5 Performance Indices and Optimization

If the performance index contains explicitly the final time, obviously, the final time must be free (to be determined optimally). For example, if the cell mass productivity is to be maximized, $P = XV(t_f)/t_f = x_1(t_f)/t_f$, or the minimum time to obtain a fixed amount of cell mass is the objective, $P = -t_f$, then one has to introduce an additional state variable, $x_4(t) = t$, $(dx_4/dt = 1$ and $x_4(0) = 0)$, to force the performance index to be a function of the state at the final time only:

$$P[\mathbf{x}(t_f)] = x_1(t_f)/t_f = x_1(t_f)/x_4(t_f) \tag{12.38}$$

This addition makes the state equation third order:

$$\frac{d\mathbf{x}}{dt} = \frac{d}{dt}\begin{pmatrix} x_1 \\ x_3 \\ x_4 \end{pmatrix} = \frac{d}{dt}\begin{pmatrix} XV \\ V \\ t \end{pmatrix} = \begin{pmatrix} \mu x_1 \\ 0 \\ 1 \end{pmatrix} + \begin{pmatrix} 0 \\ 1 \\ 0 \end{pmatrix}F \quad \begin{pmatrix} x_1(0) \\ x_3(0) \\ x_4(0) \end{pmatrix} = \begin{pmatrix} x_{10} \\ x_{30} \\ 0 \end{pmatrix} \tag{12.39}$$

This additional state variable requires an additional adjoint variable, λ_4, and an extra term in the Hamiltonian:

$$H = \lambda_1\mu x_1 + \lambda_3 F + \lambda_4 = \lambda_1\mu x_1 + \lambda_4 + \lambda_3 F \overset{\Delta}{=} H_2 + \phi F = 0 \tag{12.40}$$

The adjoint variables must satisfy the differential equations

$$\frac{d\boldsymbol{\lambda}}{dt} = \frac{d}{dt}\begin{bmatrix} \lambda_1 \\ \lambda_3 \\ \lambda_4 \end{bmatrix} = -\begin{bmatrix} \partial H/\partial x_1 \\ \partial H/\partial x_3 \\ \partial H/\partial x_4 \end{bmatrix} = -\begin{bmatrix} \lambda_1\mu - \lambda_1 x_1\mu'/x_3 Y_{X/S} \\ \lambda_1 x_1\mu'(S_F - g/x_3)/x_3 \\ 0 \end{bmatrix} \tag{12.41}$$

and the boundary conditions are obtained from Eq. (9.28) by matching coefficients:

$$\delta P = \delta \left[\frac{x_1(t_f)}{x_4(t_f)} \right] = \frac{1}{x_4(t_f)} \delta x_1(t_f) - \frac{x_1(t_f)}{x_4(t_f)} \delta x_4(t_f) = \sum_{i=1, i \neq 3}^{4} \lambda_i(t_f) \delta x_i(t_f)$$

$$- \sum_{i=1, i \neq 3}^{4} \lambda_i(0) \delta x_i(0)^0 = \lambda_1(t_f) \delta x_1(t_f) + \lambda_2(t_f) \delta x_2(t_f)^0 + \lambda_4(t_f) \delta x_4(t_f)$$

$$(12.42)$$

or

$$\lambda(t_f) = \begin{bmatrix} \lambda_1(t_f) \\ \lambda_2(t_f) \\ \lambda_4(t_f) \end{bmatrix} = \frac{\partial P}{\partial \mathbf{x}(t_f)} = \begin{bmatrix} 1/x_4(t_f) \\ \text{free} \\ -x_1(t_f)/x_4^2(t_f) \end{bmatrix} \qquad (12.43)$$

Therefore, the Hamiltonian at the final time is

$$H(t_f) = \lambda_1(t_f) \mu(t_f) x_1(t_f) + \lambda_4(t_f) = \mu(t_f) x_1(t_f)/x_4(t_f) - x_1(t_f)/x_4^2(t_f) = 0$$

$$(12.44)$$

The final time is thus not infinite, but finite, as determined by the condition imposed by Eq. (12.45):

$$\mu(t_f) = \frac{1}{x_4(t_f)} = \frac{1}{t_f} \quad \Rightarrow t_f = \frac{1}{\mu(t_f)} \qquad (12.45)$$

Thus, the appearance of final time in the performance index eliminates the possibility of infinite final time.

12.2.3.6 Fixed Final Time, t_f

We say that the final time is fixed when it is given a priori or specified. A free final time refers to a situation in which the final time is not specified and is free to be chosen. The Hamiltonian on the optimal trajectory is a nonzero constant because the final time is fixed. Therefore, Eq. (12.25) becomes

$$H^* = \lambda_1 \mu x_1 + \lambda_2 F \triangleq H_1 + \phi F \triangleq H^+ \qquad (12.46)$$

where H^+ is a nonzero unknown constant. At the final time when the reactor volume is full, and therefore $F = 0$ and $\lambda_1(t_f) = 1$ (Eq. (12.28)), the Hamiltonian is positive because $\mu(t_f) x_1(t_f)/x_4(t_f) > 0$:

$$H^*(t_f) = \lambda_1(t_f) \mu(t_f) x_1(t_f) = \mu(t_f) x_1(t_f) = H^+ > 0 \qquad (12.47)$$

According to Eq. (12.47), the Hamiltonian is positive at the final time. Because the Hamiltonian is continuous, the Hamiltonian is a positive constant for the entire time period:

$$H^* = H^+ > 0, \quad 0 \leq t \leq t_f \qquad (12.48)$$

Because $F(t)$ appears linearly in Eq. (12.46), the optimal feed rate consists of the following:

$$F = \begin{cases} F_{\max} & \text{when } \phi = \lambda_2(t) > 0 \\ F_{\min} & \text{when } \phi = \lambda_2(t) < 0 \\ F_{\sin} & \text{when } \phi = \lambda_2(t) = 0 \text{ over a finite time interval, } t_q < t < t_{q+1} \\ F_b = 0 & \text{when } x_2(t) = V_{\max} \end{cases} \quad (12.49)$$

The remaining tasks are to determine the singular feed rate and the sequence and time interval of each flow rate: the maximum, minimum, singular, and zero.

12.2.3.7 Optimal Singular Feed Rate on the Singular Arc

The singular feed flow rate is determined by recognizing that $\phi = \lambda_2 = 0$, and because it is identically zero over a finite time interval, its derivatives of all order until the feed rate appears explicitly must also vanish:

$$\phi = \lambda_2 = 0 \quad (12.50)$$

$$\frac{d\phi}{dt} = \frac{d\lambda_2}{dt} = -\lambda_1 x_1 \frac{\partial \mu}{\partial x_2} = -\lambda_1 x_1 \frac{d\mu}{dS} \frac{\partial S}{\partial x_2} = -\lambda_1 x_1 \frac{d\mu}{dS} \frac{\partial (g/x_2)}{\partial x_2}$$
$$= -\lambda_1 x_1 \frac{d\mu}{dS} \frac{x_2 \partial g/\partial x_2 - g}{x_2^2} = -\lambda_1 x_1 \frac{d\mu}{dS} \left(\frac{S_F - S}{V} \right) = 0 \quad (12.51)$$

where Eq. (12.22) was used to obtain Eq. (12.51), because $\lambda_2 = 0$, λ_1 cannot be zero. Otherwise, this problem becomes trivial. Two terms do not vanish: the cell mass $x_1 = XV \neq 0$ and $(S_F - S)/V \neq 0$. Therefore, the only term that can remain zero over a finite time interval is

$$\frac{d\mu}{dS} = 0 \quad (12.52)$$

A physical interpretation of this equation is that the substrate concentration must be chosen properly to maximize the specific growth rate. One interpretation of Eq. (12.52) is that to maximize the amount of terminal cell mass, the specific growth rate must be maximized. To maximize the specific growth rate, the substrate concentration must be held constant at the value at which the specific growth rate is maximum. This conclusion is consistent with the intuitive argument used to maximize the cell mass in Chapter 3.

The second time derivative is obtained by differentiating Eq. (12.51):

$$\frac{d^2\phi}{dt^2} = -\frac{d(\lambda_1 x_1)}{dt} \frac{d\mu}{dS} \left(\frac{x_3 S_F - g}{x_3^2} \right) - \lambda_1 x_1 \frac{d(d\mu/dS)}{dt} \left(\frac{x_3 S_F - g}{x_3^2} \right)$$
$$- \lambda_1 x_1 \frac{d\mu}{dS} \frac{d}{dt} \left[\frac{(S_F - S)}{V} \right] = 0 \quad (12.53)$$

The first and third terms in Eq. (12.53) vanish owing to Eq. (12.52). Therefore, the remaining term, the second term, must also vanish. Applying the chain rule, we obtain

$$0 = \frac{d}{dt} \left(\frac{d\mu}{dS} \right) = \frac{d^2\mu}{dS^2} \frac{dS}{dt} \quad (12.54)$$

Equation (12.54) implies that

$$\frac{dS}{dt} = 0 \qquad (12.55)$$

or the second derivative must vanish:

$$\frac{d^2\mu}{dS^2} = 0 \qquad (12.56)$$

However, owing to Eq. (12.52), $d\mu/dS = 0$, this represents an inflection point. Thus, it is rejected. Equation (12.55) implies that the substrate concentration is constant during the entire singular period. Thus, Eqs. (12.52) and (12.55) imply that during the entire singular period, the substrate concentration is maintained constant at S_m, corresponding to the maximum specific growth rate, $\mu(S_m) = \mu_{\max}$. Thus, the singular arc is a straight line, $S = S_m$.

To obtain the singular feed rate to maintain the substrate concentration constant at S_m, we rearrange the substrate balance equation (12.2):

$$S_F F - \frac{\mu_m X V}{Y_{X/S}} = \frac{d(S_m V)}{dt} = S_m \frac{dV}{dt} = S_m F \qquad (12.57)$$

Solving for the feed rate,

$$F_{\sin} = \frac{\mu_m X V}{Y_{X/S}(S_F - S_m)} = \frac{\sigma_m X V}{(S_F - S_m)} \qquad (12.58)$$

Equation (12.58) is the feed rate expression during the singular period that would keep the substrate concentration constant at S_m to maximize the specific growth rate.

The preceding analysis is valid for specific growth rates that are nonmonotonic, that is, exhibit a maximum. If the specific growth rate is a monotonically increasing function of the substrate concentration, as in the case of Monod form, so that the maximum μ occurs at the highest substrate concentration, the substrate should be fed at the maximum rate, until the reactor is full, and then it should be operated in a batch mode until a desired conversion is achieved. In other words, a batch operation is the best mode of operation when the specific growth rate is a monotonically increasing function of substrate concentration, as in Monod form. Conversely, if the specific growth rate is a monotonically decreasing function of substrate concentration, as in the case of substrate inhibition, with the maximum rate occurring near zero substrate concentration, the substrate should be fed to maintain the concentration near zero using the singular feed rate of Eq. (12.58).

Let us investigate the adjoint variables. Owing to Eqs. (12.52) and (12.55), the adjoint equation (Eq. (12.41)) reduces on the singular arc to the following:

$$\frac{d}{dt}\begin{bmatrix} \lambda_1 \\ \lambda_2 \end{bmatrix} = -\begin{pmatrix} \lambda_1 \mu \\ 0 \end{pmatrix} = -\begin{pmatrix} \lambda_1 \mu_{\max} \\ 0 \end{pmatrix} \qquad (12.59)$$

Hence, $\lambda_1(t)$ is an exponentially decaying function,

$$\lambda_1(t) = \lambda_1(t_s) \exp[-\mu_{\max}(t - t_s)] \qquad (12.60)$$

where t_s is the time at which the singular interval begins and $\lambda_1(t_s)$ is the value of λ_1 at t_s. Therefore, the sign of λ_1 does not change on the singular interval.

12.2.3.8 The Optimal Singular Feed Rate Is in Feedback Control Form and Is Exponential

The singular feed rate expression obtained earlier (Eq. (12.58)) is in the form of feedback control, requiring measurement of the cell mass in the reactor, XV, and knowledge of the kinetic information, μ_m, S_m, and $Y_{X/S}$ and feed concentration, S_F.

It is simple to show that the singular feed rate is an exponential function. Because μ is maintained at its maximum value μ_m, the cell mass balance equation (12.1) can be integrated from t_s to t:

$$\frac{d(XV)}{dt} = \mu_{max}XV, \quad \Rightarrow XV(t) = (XV)_s \exp[\mu_{max}(t - t_s)] \qquad (12.61)$$

Substitution of Eq. (12.61) into Eq. (12.58) yields

$$F_{sin} = \frac{\mu_{max}XV}{Y_{X/S}(S_F - S_m)} = \frac{\mu_{max}(XV)_s \exp(-\mu_{max}t_s)}{Y_{X/S}(S_F - S_m)} \exp(\mu_{max}t) = \alpha \exp(\mu_{max}t) \qquad (12.62)$$

where the subscript s is used to denote the start of a singular period. According to Eq. (12.61), the cell mass accumulates exponentially. Because the feed rate is an exponential function given by Eq. (12.62), the bioreactor volume also increases exponentially during the singular period, until it becomes full, and is obtained by integrating Eq. (12.3) with Eq. (12.62):

$$V(t) = V(t_s) + \int_{t_s}^{t} F_{sin} d\tau = \left[V(t_s) - \frac{\alpha}{\mu_m} \exp(\mu_{max}t_s) \right] + \frac{\alpha}{\mu_m} \exp(\mu_{max}t)$$

$$= a + b \exp(\mu_{max}t) \qquad (12.63)$$

12.2.3.9 Optimal Feed Rate Profiles

The remaining task is to investigate the optimal feed rate sequence. For the simplest problem under consideration, it is possible to deduce the sequence knowing from PMP and singular control theory that the specific growth rate must be maximized and that the feed rate can be maximum, minimum, or singular. All we know is that the feed rate must be of maximum, minimum, or intermediate (singular flow rate) value. Let us suppose that the feed rate is singular for the entire period from the start until the reactor is full. This implies that the initial substrate concentration must be chosen to equal the value at which the specific growth rate is maximum, that is, $S_0 = S_m$. If the initial substrate concentration is less than S_m, $S_0 < S_m$, one would initially apply the maximum feed rate to increase the substrate concentration to S_m, $S = S_m$. Conversely, if the initial substrate concentration were too high for some reason, $S_0 > S_m$, one would use the minimum feed rate (a batch period $F_{min} = 0$, initially operating in a batch mode to reduce the substrate concentration to S_m) to reach the singular arc. Thus, three possible initial conditions relative to S_m should be treated separately: (1) $S_0 = S_m$, (2) $S_0 < S_m$, and (3) $S_0 > S_m$. It is intuitively obvious that case 1 is the best choice. Nevertheless, we shall consider all three cases.

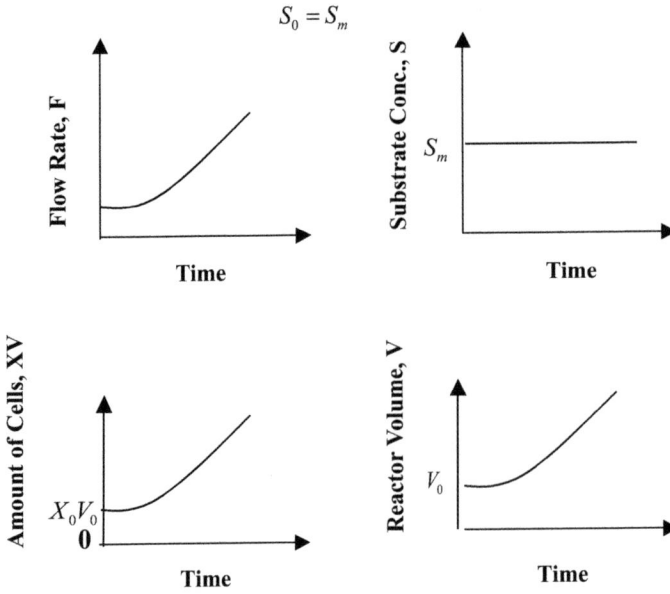

Figure 12.2. Time profiles for initial concentration $S_0 = S_m$.

Case 1: The initial substrate concentration is on the singular arc, that is, it equals the substrate concentration that maximizes the specific growth rate, $S_0 = S_m$ and $\mu(S_m) = \mu_{max}$.

This is the optimal choice of the initial substrate concentration. Because the initial substrate concentration is already on the singular arc, that is, the optimal substrate concentration S_m that maximizes the specific grow rate μ_{max}, the singular flow rate F_{sin} is applied from time zero to maintain the substrate concentration at S_m, until the bioreactor volume is full, t_{full}, when no feed is added and the reactor is operated in a batch mode until the final time, t_f. Therefore, the optimal feed policy can be summarized as follows:

$$
\begin{aligned}
S_0 &= S_m \quad \text{at } t = 0 \\
F &= \left(\begin{array}{l} F_{sin} = \dfrac{\mu_{max}(XV)_0}{Y_{X/S}(S_F - S_m)} \exp(\mu_{max} t), \; S \text{ held constant, } S = S_m, 0 < t \leq t_{full} \\ F_b = 0 \quad \text{batch operation,} \quad S \text{ decreases from } S_m \text{ to } S_f, t_{full} < t \leq t_f \end{array} \right)
\end{aligned}
$$

$$(12.64)$$

Typical time profiles of the optimum feed rate and substrate concentration are shown in Figure 12.2.

With this optimal feed policy, the total cell mass increases exponentially from time zero because the singular flow rate is applied throughout the filling stage and is obtained by integrating Eq. (12.1),

$$ XV = (XV)_0 \exp(\mu_{max} t) \tag{12.65} $$

while the bioreactor volume is obtained by integrating Eq. (12.3) with the singular feed rate (Eq. (12.62)):

$$V(t) = V_0 + \int_0^t \mu_{max} X_0 V_0 \exp(\mu_{max}\tau) d\tau / Y_{X/S}(S_F - S_m)$$

$$= V_0[Y_{X/S}(S_F - S_m) - X_0 + X_0 \exp(\mu_{max}t)]/Y_{X/S}(S_F - S_m) = \alpha + \beta \exp(\mu_{max}t) \tag{12.66}$$

According to Eq. (12.66), the bioreactor volume increases exponentially. The cell concentration is obtained by dividing Eq. (12.65) by Eq. (12.66):

$$X = \frac{XV}{V} = X_0 Y_{X/S}(S_F - S_m) \exp(\mu_{max}t)/[Y_{X/S}(S_F - S_m) - X_0 + X_0 \exp(\mu_{max}t)] \tag{12.67}$$

According to Eq. (12.67), the cell concentration can increase, decrease, or remain constant, depending on the sign of $Y_{X/S}(S_F - S_m) - X_0$,

$$\frac{dX}{dt} = \frac{X_0 Y_{X/S}(S_F - S_m)[Y_{X/S}(S_F - S_m) - X_0]\mu_{max} \exp(\mu_{max}t)}{[Y_{X/S}(S_F - S_m) - X_0 + X_0 \exp(\mu_{max}t)]^2} \tag{12.68}$$

so that

$$\frac{dX}{dt} \begin{cases} < 0 & \text{if } X_0 > Y_{X/S}(S_F - S_m), \text{ cell concentration decreases with time} \\ = 0 & \text{if } X_0 = Y_{X/S}(S_F - S_m), \text{ cell concentration remains constant} \\ > 0 & \text{if } X_0 < Y_{X/S}(S_F - S_m), \text{ cell concentration increases with time} \end{cases} \tag{12.69}$$

The cell concentration will decrease if the initial cell concentration is larger than the maximum cell concentration that can be obtained, $Y_{X/S}(S_F - S_m)$, while the cell concentration increases if the initial cell concentration is smaller than the maximum obtainable cell concentration. It is interesting to note that the cell concentration can remain constant at $X = X_0 = Y_{X/S}(S_F - S_m)$ during the entire period of singular feed rate, if the initial cell concentration and the feed substrate concentration are chosen properly to match the condition, $Y_{X/S}(S_F - S_m) = X_0$. The substrate concentration is maintained constant during the entire period of feeding. These phenomena were noticed in the exponentially fed fed-batch operation in Chapter 4.

During the batch period, $F = 0$ and $V = V_{max}$, $t_{full} \leq t \leq t_f$, and the cell and substrate concentrations are related by Eqs. (12.1) and (12.3),

$$\frac{d(SV)}{dt} = V_{max}\frac{dS}{dt} = -\sigma X V_{max} = -\frac{\mu X V_{max}}{Y_{X/S}} = -\frac{V_{max}}{Y_{X/S}}\frac{dX}{dt} S(t_{full}) = S_m \tag{12.70}$$

which may be integrated from t_{full} to t to yield

$$(S_m - S) = (1/Y_{X/S})(X - X_{full}) \quad t_{full} \leq t \leq t_f \tag{12.71}$$

Because the cell concentration is related to the substrate concentration during the batch period by Eq. (12.71) and the volume remains constant, the substrate balance equation (12.2) may be written as

$$\frac{dS}{dt} = -\frac{\mu(S)X}{Y_{X/S}} = -\frac{\mu(S)[X_m + Y_{X/S}(S_m - S)]}{Y_{X/S}} \quad t_{full} \leq t \leq t_f \tag{12.72}$$

which may be integrated numerically to obtain the time profile of substrate concentration.

12.2.3.10 A Special Case of Constant Substrate Concentration

As noted earlier, it is possible to pick the feed substrate concentration, the initial substrate concentration, and the inoculum concentration to satisfy $X_0 = Y_{X/S}(S_F - S_m)$ and $S(0) = S_m$ and apply the singular (exponential) feed rate, Eq. (12.62), to maintain constant the substrate and cell concentrations throughout the course of the singular feed rate period. Then, the cell and substrate concentrations remain constant, and the reactor volume increases exponentially during the singular flow rate period (entire filling period):

$$X(t) = X_0 = Y_{X/S}(S_F - S_m), \quad S(t) = S_m$$
$$F(t) = \frac{\mu_{max}X_0V_0\exp(\mu_{max}t)}{Y_{X/S}(S_F - S_m)}, \quad V(t) = \frac{F(t)}{\mu_{max}} \quad \text{for} \quad 0 < t \le t_{full} \quad (12.73)$$

Case 2: The initial substrate concentration is less than the substrate concentration at which the specific growth rate is maximum, $S_0 < S_m$.

Because the initial substrate concentration is less than the optimal substrate concentration S_m at which the specific growth rate is maximum, μ_{max}, the maximum feed rate, F_{max}, is applied from time zero to t_1 to bring the substrate concentration up to S_m, at which point the singular flow rate, F_{sin}, is applied until t_{full} to maintain the substrate concentration at S_m until the bioreactor is full, followed by a batch period:

$$S_0 < S_m \quad \text{at } t = 0$$
$$F = \begin{cases} F_{max} & 0 < t < t_1 & S \text{ increases from } S_0 \text{ to } S_m \\ F_{sin} & t_1 < t \le t_{full} & S \text{ held constant, } S = S_m \\ F_b = 0 & t_{full} < t \le t_f & S \text{ decreases from } S_m \text{ to } S_f \end{cases} \quad (12.74)$$

This sequence is generally known as the bang-singular-bang control as a feed rate profile consists of a period of maximum rate followed by a period of singular flow rate and a minimum feed rate of zero (a batch period). Typical time profiles of the optimal feed rate and the substrate and cell concentrations are shown in Figure 12.3.

Case 3: The initial substrate concentration is greater than the substrate concentration at which the specific growth rate is maximum, $S_0 > S_m$.

Because the initial substrate concentration is greater than the optimal substrate concentration, S_m, that maximizes the specific growth rate, μ_{max}, it must be brought down to S_m to be on the singular arc. Thus, no feed is applied initially, and a batch period is maintained until the concentration is reduced to S_m, at which point, the singular feed rate, F_{sin}, is applied to maintain the substrate concentration at S_m, until

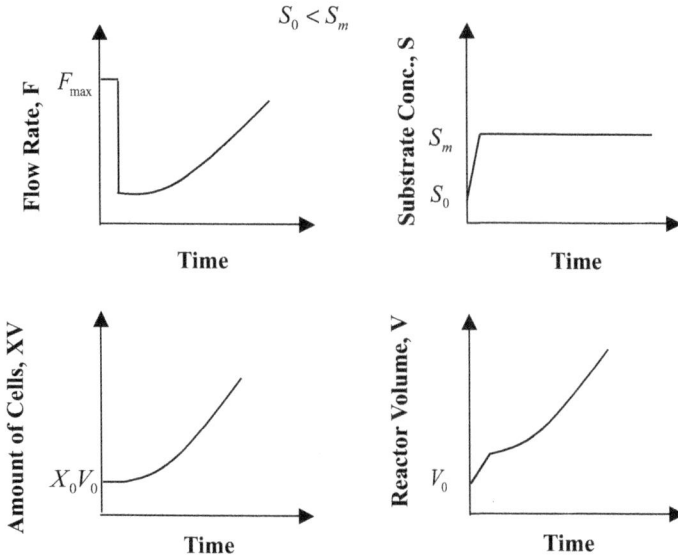

Figure 12.3. Time profiles for initial concentration $S_0 < S_m$.

the bioreactor volume is full. A batch period follows. This sequence is generally known as the bang-singular-bang control:

$$S_0 > S_m \quad \text{at } t = 0$$

$$F = \begin{cases} F_{\min} = 0 & 0 < t < t_2 & \text{Batch operation, } S \text{ decreases from } S_0 \text{ to } S_m \\ F_{\sin} & t_2 < t < t_{\text{full}} & \text{Singular operation, } S \text{ remains constant, } S = S_m \\ F_b = 0 & t_{\text{full}} < t \le t_f & \text{Batch operation, } S \text{ decreases from } S_m \text{ to } S_f \end{cases}$$

(12.75)

Typical time profiles of the optimal feed rate and the substrate concentration are shown in Figure 12.4. These feed rate profiles suggest that the initial substrate concentration should be properly chosen so that it is on the singular arc, that is, to coincide with S_m to obtain the optimal result. It should also be noted that all of the preceding results are obtained assuming that there is no constraint in the residual substrate concentration. In practice, however, the final substrate concentration may have to be much less than S_m for better utilization of substrate or to minimize the residual substrate concentration so that it would not lead to difficult separation steps. Thus, there is a batch period during which the substrate concentration is reduced to a desired value. This will also be the case if one is treating toxic waste and wishes to reduce its concentration to a desired value.

Let us investigate the Hamiltonian now. At the final time, the volume is full so that $F = 0$. Therefore, the last term in Eq. (12.35) vanishes, and in the absence of cell lysis or death, the Hamiltonian reduces to

$$H^*(t_f) = \lambda_1(t_f)\mu(t_f)X(t_f)V(t_f) = \mu(t_f)X(t_f) > 0$$

(12.76)

Thus, the Hamiltonian on the optimal path is a positive constant.

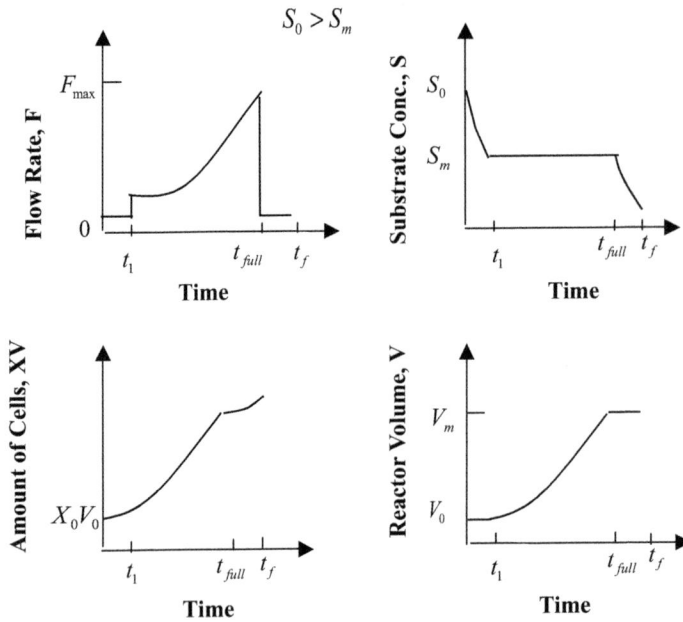

Figure 12.4. Time profiles for initial concentration $S_0 > S_m$.

At this point, it is proper to consider the effects of various operational parameters: the maximum feed rate, the final time, the initial substrate concentration, and the inoculum size.

12.2.4 Effects of Operating Parameters

12.2.4.1 Effect of Constraint on Maximum Feed Rate
When the upper bound on the feed rate is sufficiently large, as assumed earlier, the general sequence of feed rate is bang (upper or lower) → singular → batch or singular → batch. However, if the upper bound is not sufficiently large, the singular feed rate, being exponential or semiexponential, can reach the upper bound before the reactor is full, and therefore, the singular feed rate is short-lived and the upper bound is applied until the reactor is full. Therefore, the sequence may consist of bang (upper or lower) → singular → upper bang → batch. When the initial substrate concentration is on the singular arc S_m, the optimal starting condition, the feed rate sequence is singular → upper bang → batch. These cases are illustrated in Figure 12.5.

12.2.4.2 Effect of Final Time
When the final time is too short, it is possible to have a sequence of bang → singular → upper bang → batch. Because one has to use the maximum volume available, V_{max}, the minimum final time is the time to fill the reactor with the maximum feed rate, $t_{f,min} = (V_{max} - V_0)/F_{max}$. In fact, when the specified final time is less than $t_{f,min}$, then the feed rate sequence is an upper bang → batch (a rapid fill followed by a batch). It is also possible that the specified final time is not enough to continue applying the singular feed rate and must be followed by an upper bang.

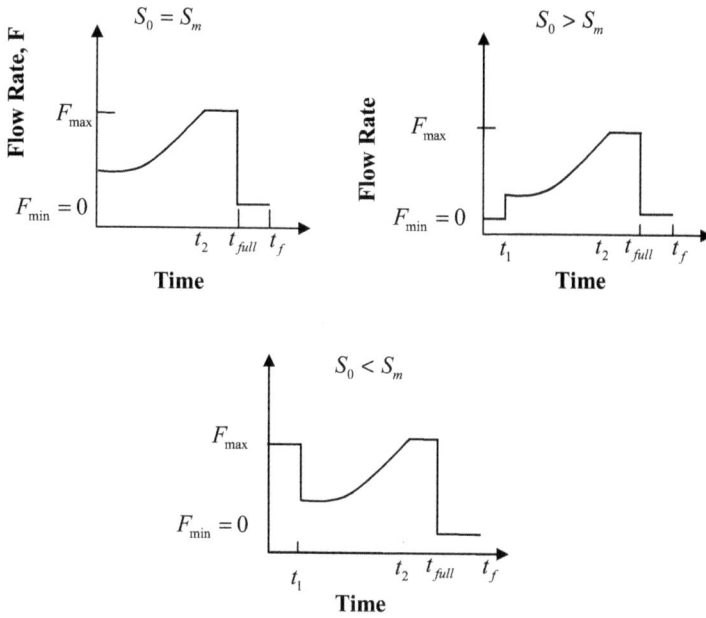

Figure 12.5. Time profiles when F_{max} is too small.

Thus, the sequence may be bang (upper or lower) → singular → upper bang → batch. Figure 12.6 illustrates various cases.

It is important to design a fed-batch system with a proper pumping capacity so that the singular feed rate does not run into the maximum pumping capacity, thus reducing the performance from the optimal. It is equally important to select the best

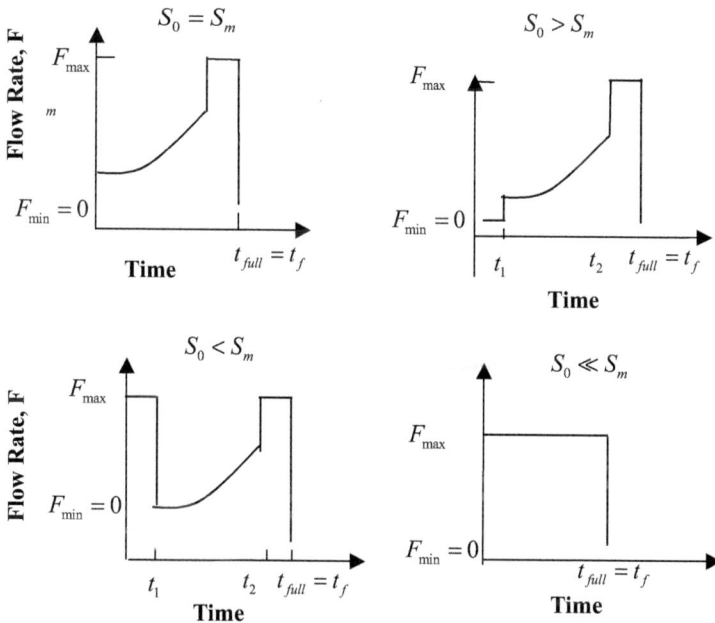

Figure 12.6. Time profiles when t_f is too small.

final time so that the optimal operation can be fully realized. In addition, we have all along assumed that the feed rate can be increased instantaneously to the maximum or minimum value, therefore allowing a bang-bang type of feed rate change. However, it may not be possible to change the magnitude instantaneously if the maximum rate of change is limited. In this situation, one may have to place a constraint on the rate of change, such as $dF/dt \leq (dF/dt)_{max}$, and replace the bang (maximum value) part of the feed rate by a pang (maximum rate) $\rightarrow$ bang (maximum value) feed rate.

12.2.4.3 Effect of Initial Substrate Concentration

When the initial substrate concentration is chosen properly so that the process is right on the singular arc, there is no need to introduce a large amount of substrate into the bioreactor using F_{max}, nor is there a need to have a batch period $F_{min} = 0$ to reduce the substrate concentration because the initial state is right on the singular arc, $S(0) = S_m$. Therefore, the optimum initial conditions must be on the singular arc. The initial conditions constitute the initial inoculum concentration, the initial bioreactor volume, and the initial substrate concentration. It is intuitive that the higher the inoculum concentration, the better will be the result of fed-batch operation. Therefore, as far as the initial inoculum concentration is concerned, it is limited by practical physical capability. Conversely, there are optimum initial volume and substrate concentrations.

12.2.4.2 Effects of Initial Conditions and Final Time on the Substrate Feed Rate

The optimal feed rate profiles for large and small final times, large and small F_{max}, and low, appropriate, and large initial substrate concentrations are depicted in Figure 12.7.

Figure 12.7 clearly illustrates the need to pick a sufficiently large final time, a sufficient pump capacity, F_{max}, and an appropriate initial substrate concentration so that the optimal substrate feed rate profile can be implemented fully without running into the constraints.

EXAMPLE 12.E.1: CELL MASS MAXIMIZATION FOR CONSTANT CELL MASS YIELD A constant-yield model[18] is given:

$$\mu(S) = \frac{S}{0.03 + S + 0.5S^2}, \quad Y_{X/S} = 0.5 \qquad (12.E.1.1)$$

Figure 12.E.1.1 shows the plot of specific growth rate as functions of substrate concentration. The specific growth rate increases with the substrate concentration, reaching the peak value of $\mu_{max} = 0.803$ g/L/hr at the substrate concentration of $S_m = 0.245$ g/L and then decreases with further increases in the substrate concentration.

For this model, we consider maximizing the cell mass at both free and fixed final times so that $P[\mathbf{x}(t_f)] = x_1(t_f)$. The Hamiltonian to be maximized is

$$\underset{F*(t)}{\text{Max}} H = \boldsymbol{\lambda}^T[\mathbf{a}(\mathbf{x}) + \mathbf{b}F] = \lambda_1 \mu x_1 + \lambda_2 \quad F = H_1 + \phi F \qquad (12.E.1.2)$$

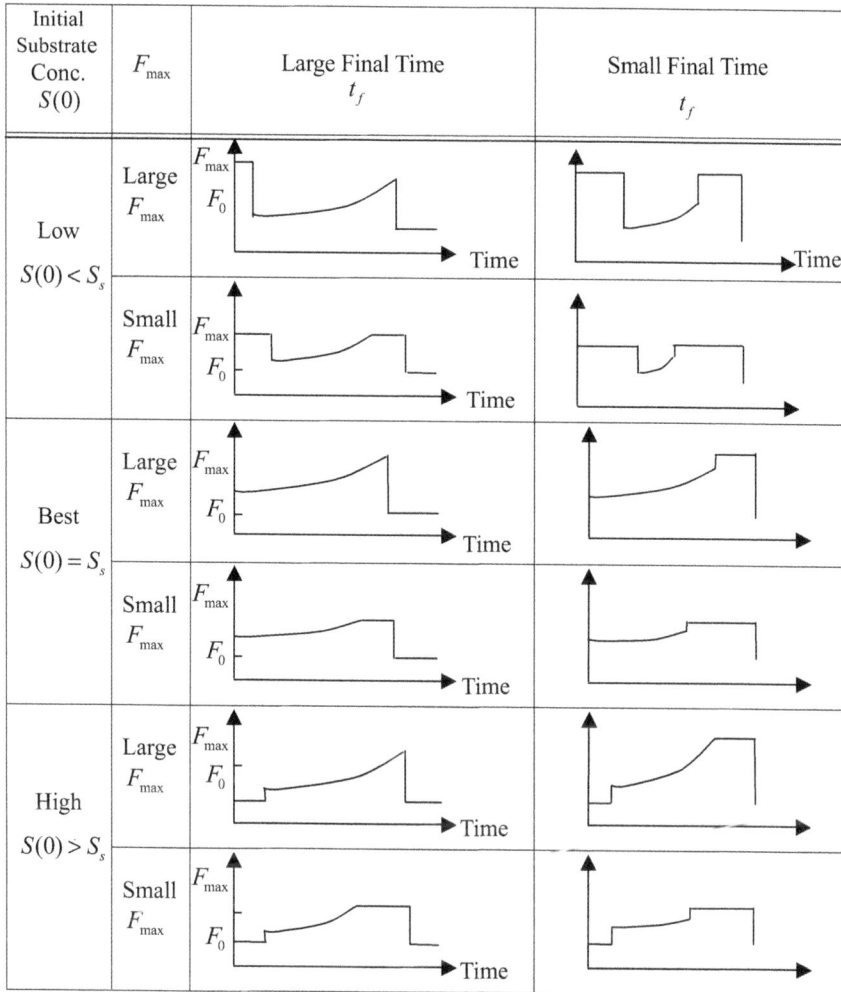

Figure 12.7. Effect on feed rate profile of initial condition $S(0)$, $F_{\max}$, and final time.

The state and adjoint equations and boundary conditions are

$$\frac{d\mathbf{x}}{dt} = \frac{d}{dt}\begin{pmatrix} x_1 \\ x_2 \end{pmatrix} = \frac{d}{dt}\begin{pmatrix} XV \\ V \end{pmatrix} = \begin{pmatrix} \mu x_1 \\ 0 \end{pmatrix} + \begin{pmatrix} 0 \\ 1 \end{pmatrix} F \quad \begin{pmatrix} x_1(0) = x_{10} \\ x_2(0) = x_{20} \end{pmatrix} \quad (12.\text{E}.1.3)$$

$$\frac{d\boldsymbol{\lambda}}{dt} = \frac{d}{dt}\begin{bmatrix} \lambda_1 \\ \lambda_2 \end{bmatrix} = -\begin{bmatrix} \partial H/\partial x_1 \\ \partial H/\partial x_2 \end{bmatrix} = -\begin{pmatrix} \lambda_1 \mu - \lambda_1 x_1 \mu'/x_2 Y_{X/S} \\ \lambda_1 x_1 \mu'(S_F - g/x_2) \end{pmatrix} \quad (12.\text{E}.1.4)$$

and

$$\boldsymbol{\lambda}(t_f) = \begin{bmatrix} \lambda_1(t_f) \\ \lambda_2(t_f) \end{bmatrix} = \frac{\partial \mathbf{P}}{\partial \mathbf{x}(t_f)} = \begin{bmatrix} 1 \\ \text{unknown constant} \end{bmatrix} \quad (12.\text{E}.1.5)$$

For free final time, the solution given by Eq. (12.31) states that the final time approaches infinity, and therefore, any mode of feeding operation, including a batch operation, gives the same amount of cells.

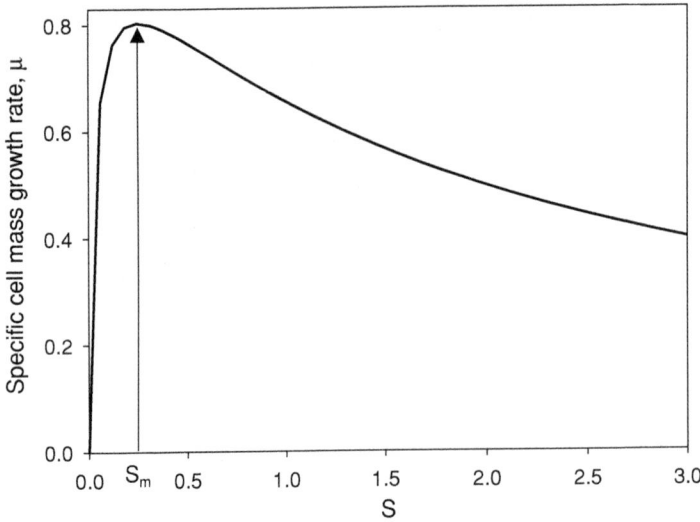

Figure 12.E.1.1. Specific growth rate versus substrate concentration.

For fixed final time, we see that this is case A in Figure 12.11 (later in this chapter) because the specific growth rate is a nonmonotonic function of substrate concentration with the peak value of $\mu = 0.803$ at $S_m = 0.245$. On the singular arc, the switching function and its derivative vanish:

$$\phi = \lambda_2 = 0$$
$$\dot{\phi} = \dot{\lambda}_2 = -\lambda_1 x_1 \mu'(S_F - g/x_2) = 0 \qquad (12.E.1.6)$$

Because $\lambda_1 x_1 (S_F - g/x_2) \neq 0$, it is necessary that

$$\mu' = d\mu/dS = 0 \qquad (12.E.1.7)$$

Thus, the singular feed rate must maximize the specific growth rate by keeping the substrate concentration constant.

The best initial condition is $S_0 = S_m = 0.245$, and the singular feed rate (an exponential feed rate) is applied from the beginning to the end until the bioreactor is full. Then a batch operation takes place until the specified final time is met. Thus, the feed rate sequence consists of

$S_0 = 0.245$ g/L   at $t = 0$

$$F = \begin{pmatrix} F_{\text{sin}} = \dfrac{0.803(XV)_0}{0.5(S_F - 0.245)}\exp(0.803t), \ S \text{ constant at } 0.245 \text{ g/L}, \ 0 < t \leq t_{\text{full}} \\[2mm] 0 \text{ Batch operation} \qquad\qquad\qquad S \text{ decreases from } 0.245 \text{ to } S_f, t_{\text{full}} < t \leq t_f \end{pmatrix}$$

$$(12.E.1.8)$$

This optimal feed rate profile and the substrate concentration profile are shown in Figure 12.E.1.2.

Conversely, if the initial substrate concentration is greater than or less than the value of 0.245, we apply a batch period ($F_{\text{min}} = 0$) or the maximum feed rate F_{max}, respectively, to bring the substrate concentration to the singular arc, $S_m = 0.245$, as soon as possible. Then the singular feed rate, an exponential feed rate, is applied until the bioreactor volume is full. Once the bioreactor volume is full, a batch mode

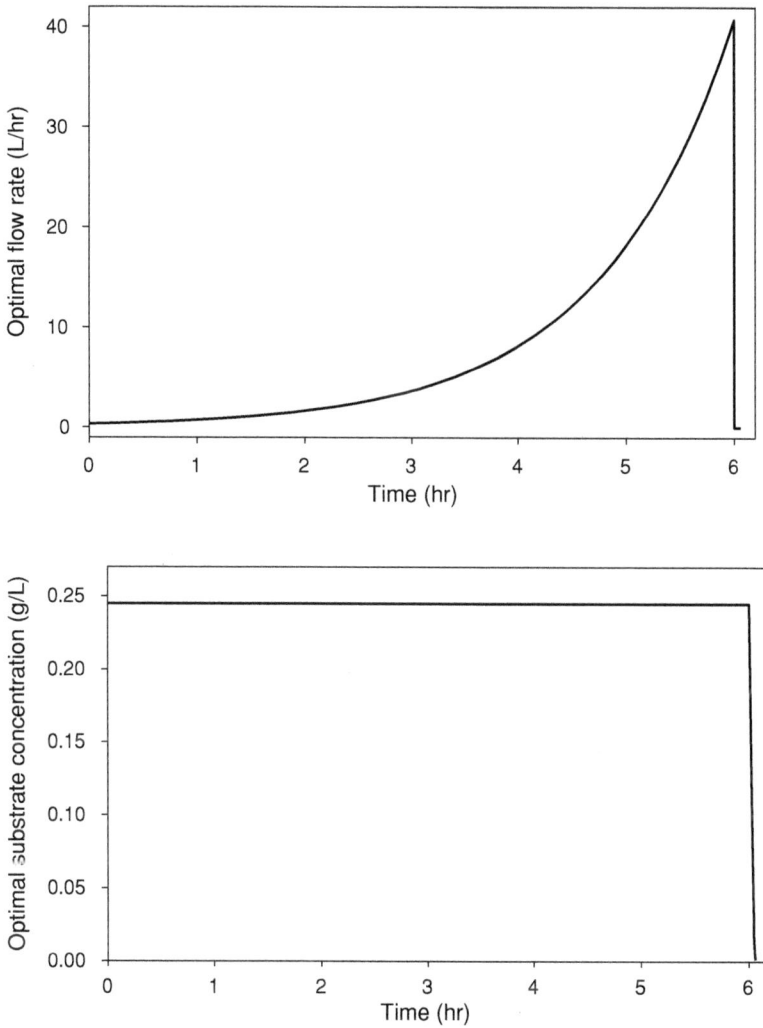

Figure 12.E.1.2. Optimal time profiles[18] for initial concentration $S_0 = S_m$.

of operation without any feed until the specified final time is reached. These feed rate profiles are given by

$$S_0 < S_M = 0.245 \quad \text{at } t = 0$$

$$F = \begin{cases} F_{\max} & 0 < t < t_1 & \text{maximum fee rate,} & S \text{ increases from } S_0 \text{ to } 0.245 \\ F_{\sin} & t_1 < t \le t_{\text{full}} & \text{exponential feed rate,} & S \text{ held constant at } S = 0.245 \\ F_b = 0 & t_{\text{full}} < t \le t_f & \text{batch operation,} & S \text{ decreases from } 0.245 \text{ to } S_f \end{cases}$$

$$(12.E.1.9)$$

$$S_0 > S_M = 0.245 \quad \text{at } t = 0$$

$$\text{or} \quad F = \begin{cases} F_{\min} = 0 & 0 < t < t_2 & \text{a batch operation,} & S \text{ decreases from } S_0 \text{ to } 0.245 \\ F_{\sin} & t_2 < t < t_{\text{full}} & \text{a singular operation,} & S \text{ constant at } S = 0.245 \\ F_b = 00 & t_{\text{full}} < t \le t_f & \text{a batch operation,} & S \text{ decreases from } 0.245 \text{ to } S_f \end{cases}$$

$$(12.E.1.10)$$

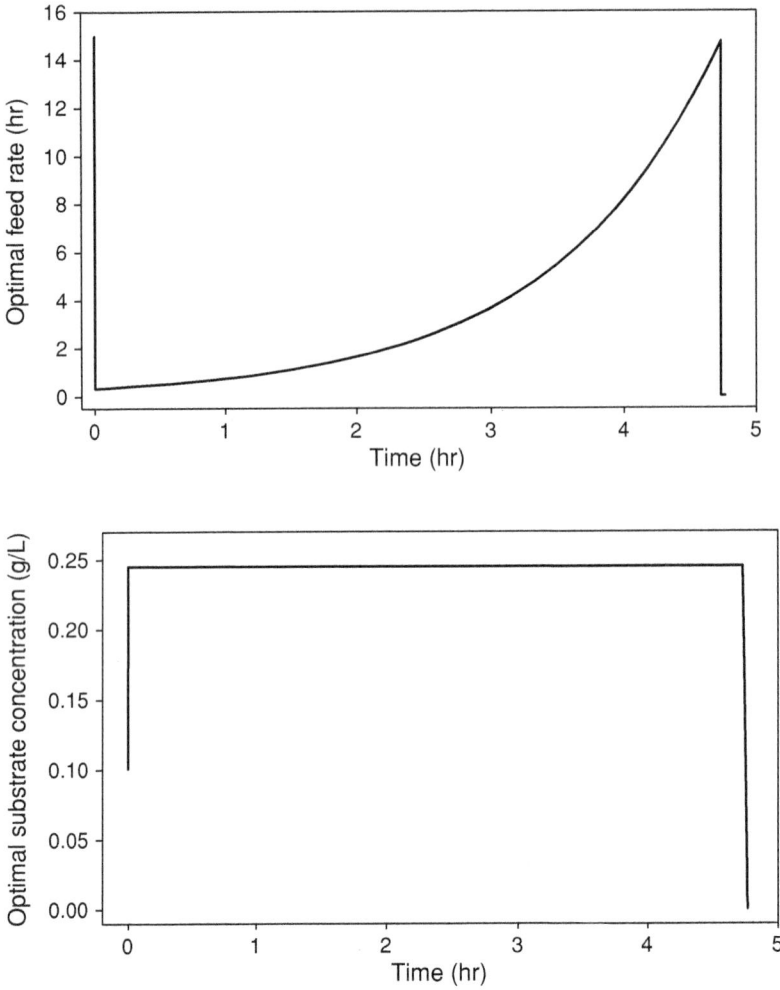

Figure 12.E.1.3. Optimal time profiles for initial concentration $S_0 < S_m$.

Various profiles corresponding to Figures 12.2–12.4 are given in Figures 12.E.1.3 and 12.E.1.4.

12.2.5 Variable-Yield Coefficients

We consider the general case of cell mass yield coefficient that is an arbitrary function of substrate concentration, $Y_{X/S}(S)$. The state equation, the adjoint equation, and boundary conditions are given by Eqs. (12.7), (12.16), and (12.18), respectively:

$$\frac{d}{dt}\begin{pmatrix} x_1 \\ x_2 \\ x_3 \end{pmatrix} = \frac{d}{dt}\begin{pmatrix} XV \\ SV \\ V \end{pmatrix} = \begin{pmatrix} \mu x_1 \\ -\sigma x_1 \\ 0 \end{pmatrix} + \begin{pmatrix} 0 \\ S_F \\ 1 \end{pmatrix} F, \quad \begin{pmatrix} x_1(0) = X_0 V_0 \\ x_2(0) = S_0 V_0 \\ x_3(0) = V_0 \end{pmatrix} \quad (12.77)$$

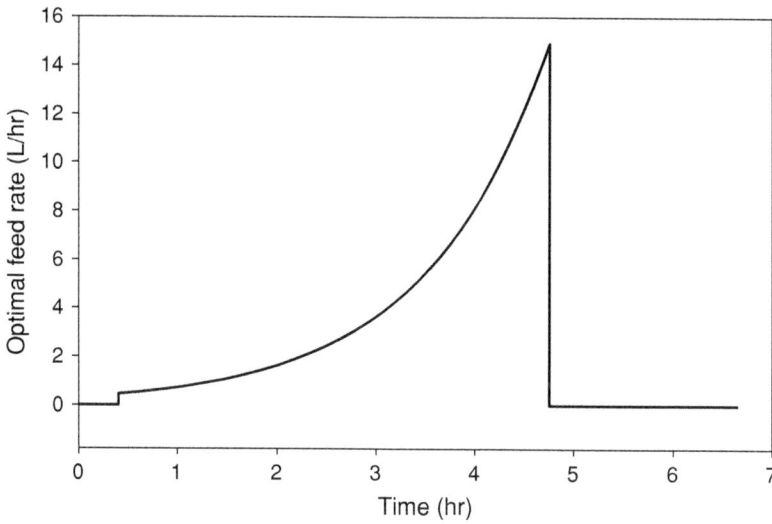

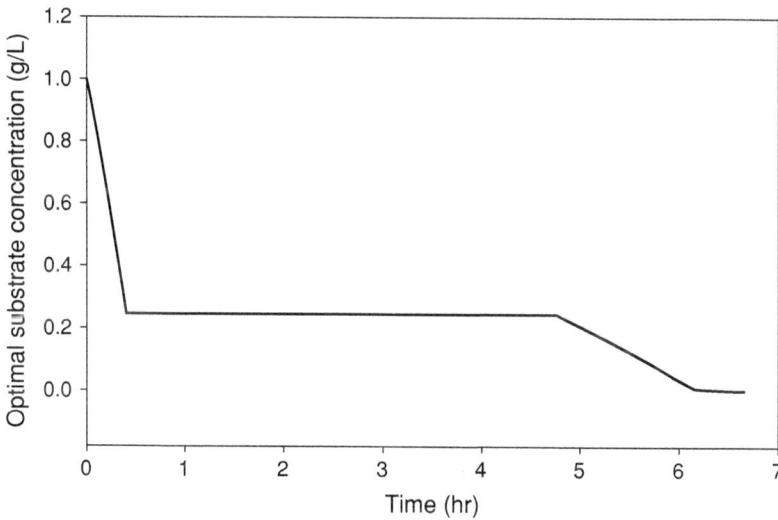

Figure 12.E.1.4. Optimal time profiles for initial concentration $S_0 > S_m$.

and

$$\frac{d\lambda}{dt} = \begin{bmatrix} d\lambda_1/dt \\ d\lambda_2/dt \\ d\lambda_3/dt \end{bmatrix} = -\begin{bmatrix} \partial H/\partial x_1 \\ \partial H/\partial x_2 \\ \partial H/\partial x_3 \end{bmatrix} = -\begin{bmatrix} \lambda_1\mu - \lambda_2\sigma \\ (\lambda_1\mu' - \lambda_2\sigma')x_1/x_3 \\ -(\lambda_1\mu' - \lambda_2\sigma')x_1x_2/x_3^2 \end{bmatrix} \quad (12.78)$$

The boundary conditions are

$$\delta P = \underline{(1)\delta x_1(t_f)} + \underline{\underline{(0)\delta x_2(t_f)}} + (0)\delta x_3(t_f) = \sum_{i=1}^{3}\lambda_i(t_f)\delta x_i(t_f) - \sum_{i=1}^{3}\lambda_i(0)\underline{\delta x_i(0)}^0$$

$$= \underline{\lambda_1(t_f)\delta x_1(t_f)} + \underline{\underline{\lambda_2(t_f)\delta x_2(t_f)}} + \lambda_3(t_f)\underline{\delta x_3(t_f)}^0$$

$$\boldsymbol{\lambda}(t_f) = \begin{bmatrix} \lambda_1(t_f) \\ \lambda_2(t_f) \\ \lambda_3(t_f) \end{bmatrix} = \begin{bmatrix} \partial P/\partial x_1(t_f) \\ \partial P/\partial x_3(t_f) \\ \text{free} \end{bmatrix} = \begin{bmatrix} 1 \\ 0 \\ \text{unknown constant} \end{bmatrix} \tag{12.79}$$

The Hamiltonian to be maximized is

$$\underset{F*(t)}{\text{Max}} \, H = (\lambda_1\mu - \lambda_2\sigma)x_1 + (S_F\lambda_2 + \lambda_3)F \overset{\Delta}{=} H_1 + \phi F \tag{12.80}$$

$$H_1 \overset{\Delta}{=} (\lambda_1\mu - \lambda_2\sigma)x_1, \quad \phi \overset{\Delta}{=} S_F\lambda_2 + \lambda_3$$

12.2.5.1 Singular Feed Rate
The singular feed rate is determined by recognizing that because $\phi = 0$, its higher-order derivatives must also vanish. In other words, we have

$$\phi = S_F\lambda_2 + \lambda_3 = 0 \tag{12.81}$$

$$\frac{d\phi}{dt} = S_F\frac{d\lambda_2}{dt} + \frac{d\lambda_3}{dt} = (\lambda_1\mu' - \lambda_2\sigma')\frac{x_1}{x_3}\left(\frac{x_2}{x_3} - S_F\right) = 0 \tag{12.82}$$

Because $x_1/x_3 = XV/V = X \neq 0$ and $x_2/x_3 - S_F = S - S_F \neq 0$, Eq. (12.82) implies that

$$(\lambda_1\mu' - \lambda_2\sigma') = 0 \tag{12.83}$$

The second derivative is

$$\frac{d^2\phi}{dt^2} = \left[\frac{x_1}{x_3}\left(\frac{x_2}{x_3} - S_F\right)\right]\left[\frac{d\lambda_1}{dt}\mu' + \lambda_1\frac{d\mu'}{dt} - \frac{d\lambda_2}{dt}\sigma' - \lambda_2\frac{d\sigma'}{dt}\right] = \left[\frac{x_1}{x_3}\left(\frac{x_2}{x_3} - S_F\right)\right]$$

$$\left[-(\lambda_1\mu - \lambda_2\sigma)\mu' + \lambda_1\mu''\frac{dS}{dt} + \overset{0}{\cancel{(\lambda_1\mu' - \lambda_2\sigma')}}\frac{x_1}{x_3}\sigma' - \lambda_2\sigma''\frac{dS}{dt}\right] = 0 \tag{12.84}$$

Equation (12.84) reduces to

$$\frac{dS}{dt} = \frac{(\lambda_1\mu - \lambda_2\sigma)\mu'}{(\lambda_1\mu'' - \lambda_2\sigma'')} \tag{12.85}$$

We can expand the left-hand side of Eq. (12.2), substitute Eq. (12.3) into it, and rearrange to obtain

$$F = \frac{\sigma XV + VdS/dt}{S_F - S} \tag{12.86}$$

Substituting Eq. (12.85) into Eq. (12.86) yields an equation that must be satisfied by the singular feed rate:

$$F_{\text{sin}} = \frac{\sigma XV}{(S_F - S)} + \frac{(\lambda_1\mu - \lambda_2\sigma)\mu'}{(\lambda_1\mu'' - \lambda_2\sigma'')(S_F - x_2/x_3)}V$$

$$= \frac{1}{(S_F - S)}\left\{\sigma X + \frac{[\mu - (\lambda_2/\lambda_1)\sigma]\mu'}{[\mu'' - (\lambda_2/\lambda_1)\sigma'']}\right\}V \tag{12.87}$$

Thus, the singular feed rate is obtained from the second derivative of the switching function. As discussed later, for fed-batch operation with a single feed stream, the second derivative yields the singular feed rate. This type of problem is known as the singular problem of order one.[16] It should be noted that if $\mu' = 0$, as in the case of a constant-yield coefficient, Eq. (12.87) reduces, as it should, to Eq. (12.58). To eliminate the adjoint variables that appear in Eq. (12.87), we substitute Eq. (12.83) into Eq. (12.87) and rearrange the result to obtain the singular feed rate in feedback mode:

$$F_{\text{sin}} = \frac{1}{(S_F - S)} \left\{ \sigma X - \frac{(\mu/\sigma)'(\sigma/\sigma')^2 \mu'}{(\mu'/\sigma')'} \right\} V \tag{12.88}$$

We should note that the first term on the right-hand side of Eq. (12.88) is responsible for maintaining the substrate concentration $S(t)$ constant, and the second term provides a correction that would vary the substrate concentration throughout the singular feed rate period.

The final time (process time) may be specified a priori, as one may want to maximize the cell mass at a specified final time, or left free, unspecified, and to be optimized. These two cases of free and fixed final times are treated subsequently.

12.2.5.2 Free Final Time

The Hamiltonian on the optimal path is constant. When the final time is free, the Hamiltonian is identically zero, providing an additional equation:

$$H^* = (\lambda_1 \mu - \lambda_2 \sigma)x_1 + (\lambda_2 + S_F \lambda_3)F = H_1 + \phi F = 0 \tag{12.89}$$

According to Eq. (12.18), at the final time, $\lambda_1(t_f) = 1$ and $\lambda_3(t_f) = 0$ and $F = 0$ so that the Hamiltonian reduces to

$$H(t_f)^* = [\lambda_1(t_f)^{=1} \mu(t_f) - \lambda_2(t_f)^{=0} \sigma(t_f)]x_1(t_f) = \mu(t_f)x_1(t_f) = 0 \tag{12.90}$$

Because $x_1(t_f) = X(t_f)V(t_f) \neq 0$, this implies that the specific growth rate at the final time must vanish:

$$\mu(t_f) = 0 \tag{12.91}$$

Because specific growth rates are zero at zero substrate concentration, Eq. (12.91) implies that at the final time, the substrate concentration must be zero. This suggests that the final time is infinite because theoretically, it would take infinite time for the substrate to be completely consumed.

The switching function is

$$\phi = S_F \lambda_2 + \lambda_3 \tag{12.92}$$

The feed rate can take on the maximum, minimum, singular, and zero values, depending on the sign of the switching curve, as indicated by Eq. (12.19). The remaining task is to determine the singular feed rate and the sequence and time intervals of the maximum, minimum, singular, and zero feed rates.

12.2.5.3 Feed Rate on the Singular Arc

During the singular feed rate period, the switching function is identically zero and the volume constraint is satisfied by assumption so that Eq. (12.80) reduces to

$$H^* = (\lambda_1 \mu - \lambda_2 \sigma) x_1 = 0 \tag{12.93}$$

Because $x_1 = XV$ cannot be zero,

$$(\lambda_1 \mu - \lambda_2 \sigma) = 0 \tag{12.94}$$

The singular feed rate is determined by recognizing that $\phi = 0$, and because it is identically zero over a finite time interval, its derivatives of all orders (usually up to the second order) must also vanish. We obtained earlier the results of setting the first- and second-order derivatives to zero:

$$\phi = S_F \lambda_2 + \lambda_3 = 0 \tag{12.95}$$

$$(\lambda_1 \mu' - \lambda_2 \sigma') = 0 \tag{12.96}$$

$$- (\lambda_1 \mu - \lambda_2 \sigma) \mu' + \left(\lambda_1 \mu'' - \lambda_2 \sigma'' \right) \frac{dS}{dt} = 0 \tag{12.97}$$

Owing to Eq. (12.94), Eq. (12.97) reduces to

$$\left(\lambda_1 \mu'' - \lambda_3 \sigma'' \right) \frac{dS}{dt} = 0 \tag{12.98}$$

Equation (12.98) implies that either $dS/dt = 0$ or $(\lambda_1 \mu'' - \lambda_2 \sigma'') = 0$. Assuming the latter to hold, and with Eqs. (12.94), (12.96), and (12.98), we obtain

$$\begin{bmatrix} \mu & -\sigma & 0 \\ \mu' & -\sigma' & 0 \\ \mu'' & -\sigma'' & 0 \end{bmatrix} \begin{bmatrix} \lambda_1 \\ \lambda_2 \\ \lambda_3 \end{bmatrix} = 0 \tag{12.99}$$

This represents a trivial solution for the adjoint variables, and therefore, $(\lambda_1 \mu'' - \lambda_2 \sigma'') \neq 0$. Therefore, the solution to Eq. (12.97) is

$$\frac{dS}{dt} = 0 \tag{12.100}$$

Thus, on the singular arc, if it exists, the substrate concentration would be kept constant at a yet unknown value. To determine this unknown constant, Eqs. (12.94) and (12.96) are examined. These two equations represent two homogenous equations for two adjoint variables:

$$\begin{bmatrix} \mu - \sigma \\ \mu' - \sigma' \end{bmatrix} \begin{pmatrix} \lambda_1 \\ \lambda_2 \end{pmatrix} = \begin{pmatrix} 0 \\ 0 \end{pmatrix} \tag{12.101}$$

For λs to have a nontrivial solution, the determinant must vanish:

$$0 = -\mu \sigma' + \mu' \sigma = \frac{\mu' \sigma - \mu \sigma'}{\sigma^2} \sigma^2 = \left(\frac{\mu}{\sigma} \right)' \sigma^2 = Y'_{X/S} \sigma^2 \Rightarrow Y'_{X/S} = 0 = \frac{\partial (\mu/\sigma)}{\partial S} \tag{12.102}$$

Equation (12.102) states that *when the final time is free and the cell mass yield coefficient is a function of substrate concentration only, the optimal singular feed rate maximizes the cell mass yield*. This result is totally consistent with the idea that when the final time is infinite, the growth rate is immaterial as there is ample time, and only the yield matters in maximizing the cell mass. *Thus, the substrate concentration should be maintained constant on the singular arc at $S = S_M$, at which the yield is maximum, $Y_{X/S}(S_M) = Y_{max}$.*

The singular feed rate to maintain $S = S_M$ to maximize the yield is determined from the substrate and mass balances (Eqs. (12.2) and (12.3)):

$$\frac{d(SV)}{dt} = S_M \frac{dV}{dt} = S_M F = S_F F - \frac{\mu(S_M) XV}{Y_{X/S}} \quad S(0) = S_0 \quad (12.103)$$

Solving for the singular flow rate, we obtain

$$F_{sin} = \frac{\mu(S_M)/Y_{X/S}(S_M)}{S_F - S_M} XV = \alpha XV \quad (12.104)$$

The singular feed rate is in feedback form. Knowledge of the terms on the right-hand side of Eq. (12.104) allows calculation of the feed rate. In other words, the optimal singular feed rate is calculated by monitoring the cell concentration X and the bioreactor volume V, knowing the kinetic parameters, $Y_{X/S}$, $\mu(S_M)$, and S_M, and the feed concentration S_F. The singular feed rate of Eq. (12.104) is an exponential feed rate because the cell mass grows exponentially as the substrate concentration is kept constant at S_M. This can be made clear by recognizing that the substrate concentration is maintained at S_M and by integrating the cell balance equation (12.1):

$$\frac{d(XV)}{dt} = \mu(S_M)XV \quad \Rightarrow \quad XV = X_0 V_0 \exp[(\mu(S_M)t] \quad (12.105)$$

Substitution of Eq. (12.105) into Eq. (12.104) yields

$$F_{sin} = \alpha XV = \alpha X_0 V_0 \exp[(\mu(S_M)t] = \beta \exp[(\mu(S_M)t] \quad (12.106)$$

12.2.5.4 Optimal Feed Rate Profiles

Because the yield is maximized when the final time is free, we can construct the optimal feed rate profile. The initial substrate concentration S_0 relative to the substrate concentration at which the yield is maximum, S_M, dictates the initial feed rate. If the initial substrate concentration is less than that at which the yield is maximum, $S_0 < S_M$, the maximum feed rate must be used to bring the substrate concentration to S_M so that the yield is maximum and then the singular control must be applied to maximize the yield by maintaining the substrate concentration at S_M until the reactor is full. Once the reactor is full, a batch operation continues until the final time which is infinite. Thus, a candidate feed rate structure is Bang (F_{max} or $F_{min} = 0$) – Singular (F_{sin}) – Batch ($F = 0$). We shall use this candidate structure to handle three classes of initial conditions, (1) $S_0 = S_M$, (2) $S_0 > S_M$, and (3) $S_0 < S_M$.

> *Case 1: The initial substrate concentration is equal to that at which the yield is maximum, $S_0 = S_M$.*

This is the optimal case (optimal initial substrate concentration). Because the initial substrate concentration corresponds to the maximum yield, the optimal feed rate

$$S_0 = S_M$$

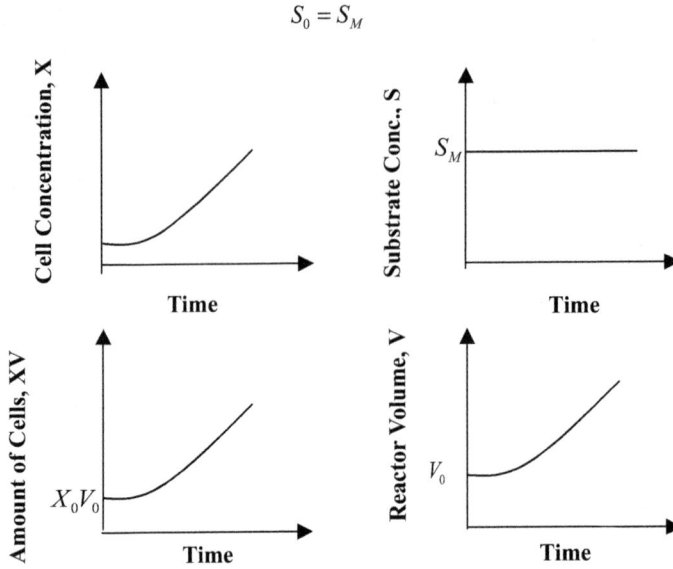

Figure 12.8. Time profiles for initial concentration $S_0 = S_M$.

profile should begin with the singular feed rate that keeps the substrate concentration constant at S_M. The optimal feed rate profile consists of a singular feed rate from $t = 0$ until the bioreactor is full and a batch period afterward until the substrate is practically all consumed.

$$S_0 = S_M \quad \text{at } t = 0 \quad Y_{X/S}(S_M) = Y_{X/S\max}$$

$$F = \begin{pmatrix} F_{\sin} = \dfrac{\mu(S_M)XV}{Y_{X/S}(S_F - S_M)} = \alpha X_0 V_0 \exp[(\mu(S_M)t], & \text{Exponential feed at } S = S_M, \\ & 0 < t \leq t_{\text{full}} \\ 0 \quad \text{Batch operation, } S \text{ decreases from } S_M \text{ to } S_f, t_{\text{full}} < t \leq t_f \end{pmatrix}$$

$$(12.107)$$

This sequence of control, which maximizes the yield coefficient, is similar to the case of constant yield in which the specific growth rate was maximized. Profiles of various concentrations, volume, and optimal feed rate are very similar to the case of constant yield and are shown Figure 12.8.

> *Case 2: The initial substrate concentration is less than that at which the yield is maximum, $S_0 < S_M$.*

Because the initial substrate concentration is less than S_M, the substrate concentration at which the yield is maximum, one must bring it up to S_M as fast as possible by applying the maximum feed rate and maintaining the concentration at S_M until the reactor is full by feeding the substrate using the singular feed rate shown above. Then, the bioreactor should be operated in a batch mode until the substrate is consumed completely. Typical profiles are summarized below and shown in Figure 12.9.

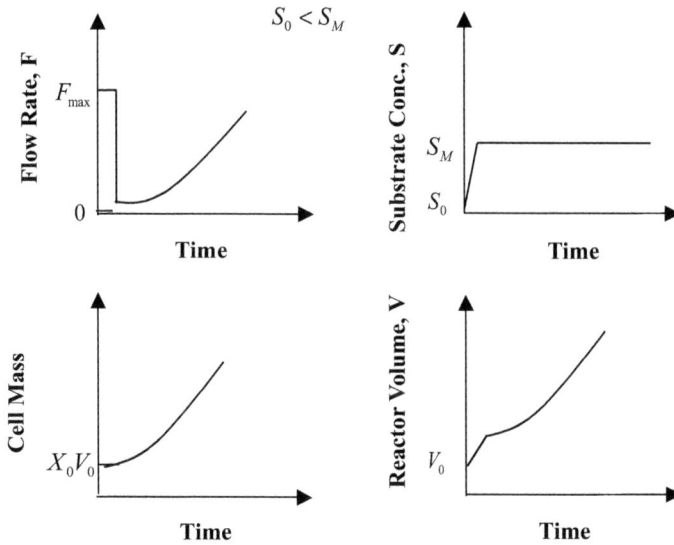

Figure 12.9. Time profiles for initial concentration $S_0 < S_M$.

$S_0 < S_M$ at $t = 0$ $Y_{X/S}(S_M) = Y_{X/S\text{max}}$

$$
F
\begin{pmatrix}
F_{\text{max}} & \text{Maximum feed rate, } S_0 \text{ increases to } S_M, 0 < t < t_1 \\[2mm]
F_{\text{sin}} = \dfrac{\mu(S_M)XV}{Y_{X/S}(S_F - S_M)}, & \text{Exponential, maximum yield at } S = S_M, t_1 < t \le t_{\text{full}} \\[2mm]
F_b = 0 & \text{Batch, } S_M \text{ decreases to } S_f, t_{\text{full}} < t \le t_f
\end{pmatrix}
$$

(12.108)

Case 3: The initial substrate concentration is greater than that at which the yield is maximum, $S_0 > S_M$.

Because the initial substrate concentration is greater than S_M, the substrate concentration at which the yield is maximum, one must reduce it as fast as possible to S_M by operating in a batch mode and maintaining the concentration at S_M until the reactor is full by feeding the substrate using the singular feed rate shown earlier. Thus, the bioreactor should be operated in a batch mode until the substrate is consumed to a desired level, S_f:

$S_0 > S_M$ at $t = 0$ $Y_{X/S}(S_M) = Y_{X/S\text{max}}$

$$
F =
\begin{pmatrix}
F_{\text{min}} = 0, & \text{Minimum feed rate (batch), decreases from } S_0 \text{ to } S_M, 0 < t < t_2 \\[2mm]
F_{\text{sin}} = \dfrac{\mu(S_M)XV}{Y_{X/S}(S_F - S_M)}, & \text{Exponential, maximizes yield at } S = S_M, t_2 < t \le t_{\text{full}} \\[2mm]
F_b = 0 & \text{Batch operation, } S \text{ decreases from } S_M \text{ to } S_f, t_{\text{full}} < t \le t_f
\end{pmatrix}
$$

(12.109)

Typical profiles are shown in Figure 12.10.

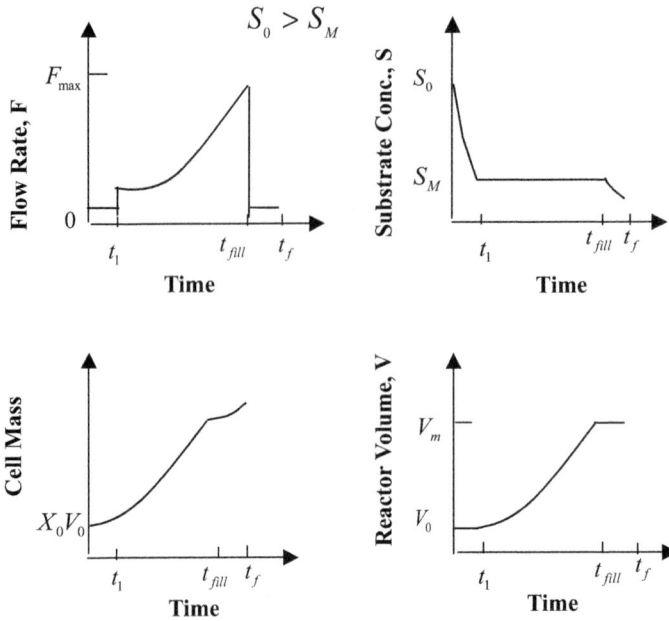

Figure 12.10. Various time profiles for initial concentration $S_0 > S_M$.

12.2.5.5 Summary for Maximum Cell Production at Free Final Time and for Variable-Yield Coefficient

When the final time does not appear in the performance index and therefore is free, the final time approaches infinity and the singular feed rate maximizes the yield by maintaining the substrate concentration constant at the value at which the yield is maximum. Therefore, the optimal initial condition for the substrate is the value at which the yield is maximum, S_M. The initial substrate concentration that is either higher or lower than S_M is suboptimal:

1. The final time is infinity, $t_f \to \infty$.
2. The singular feed rate maximizes the yield coefficient, $dY_{X/S}/dS = 0$, $Y_{X/S}(S_M) = Y_{max}$.
3. The optimal initial substrate concentration is that at which the yield coefficient is maximum, $S = S_M$. The lower initial concentration is the second best as it can be brought to S_M quickly by F_{max}.
4. The optimal feed rate profile is a singular feed rate from the start to the end when the bioreactor volume is full. The singular feed rate is an exponential function and is in feedback mode (Eq. (12.104)).
5. The optimal feed rate given is in feedback mode, requiring the measurement of cell concentration and bioreactor volume and the knowledge of feed substrate concentration, peak substrate concentration, and kinetic information. The bioreactor volume increases exponentially.

12.2.5.6 Fixed Final Time

Sometimes the final time is specified by production requirements or through experience. Because of the fixed final time, the Hamiltonian is an unknown nonzero constant,

$$H^* = (\lambda_1 \mu - \lambda_2 \sigma)x_1 + (S_F \lambda_2 + \lambda_3)F = H_1 + \phi F = H^+$$
$$H_1 = (\lambda_1 \mu - \lambda_2 \sigma)x_1, \quad \phi = S_F \lambda_2 + \lambda_3 \tag{12.110}$$

where H^+ is a unknown constant. Therefore, unlike the case of free final time, we have one less equation to work with when the final time is fixed. The switching function is the same as before. The adjoint differential equations and the final conditions are given by Eqs. (12.16) and (12.18). The feed rate can take on the maximum, minimum (zero), singular, and zero values ($F_b = 0$), according to Eq. (12.19). The remaining tasks are to determine the singular control and the sequence and time intervals of the maximum, minimum, singular, and zero feed rates. The Hamiltonian at the final time at which there is no feed, $F(t_f) = 0$, is obtained from Eq. (12.14) by recognizing that $\mu(t_f)x_1(t_f) > 0$, $\lambda_1(t_f) = 1$, and $\lambda_2(t_f) = 0$:

$$H(t_f) = \left[(\lambda_1(t_f)^{=1} \mu(t_f) - \lambda_2(t_f)^{=0} \sigma(t_f) \right] x_1(t_f) = \mu(t_f)x_1(t_f) > 0 \tag{12.111}$$

Equation (12.111) states that the Hamiltonian at the final time is positive. Because the Hamiltonian is continuous, this implies that the Hamiltonian is positive, $H^* > 0$, for all times. During the singular period, the switching function is identically zero so that

$$H^* = H_1 + \phi F = H_1 = (\lambda_1 \mu - \lambda_2 \sigma)x_1 > 0 \tag{12.112}$$

Because $x_1 > 0$, this implies that

$$(\lambda_1 \mu - \lambda_2 \sigma) > 0 \tag{12.113}$$

12.2.5.7 Feed Rate on Singular Arc

The singular feed rate is determined by recognizing that $\phi = 0$, and because it is identically zero over a finite time interval, its derivatives of all orders (until F appears explicitly) must also vanish. The results obtained earlier are

$$\phi = S_F \lambda_2 + \lambda_3 = 0 \tag{12.81}$$

$$(\lambda_1 \mu' - \lambda_2 \sigma') = 0 \tag{12.83}$$

$$-(\lambda_1 \mu - \lambda_2 \sigma)\mu' + (\lambda_1 \mu'' - \lambda_2 \sigma'')\frac{dS}{dt} = 0 \tag{12.85}$$

Equation (12.83) provides a relationship between two adjoint variables:

$$\lambda_2/\lambda_1 = \mu'/\sigma' \tag{12.83a}$$

Owing to Eq. (12.83), the adjoint equation (12.16) reduces on the singular arc to

$$\frac{d\lambda}{dt} = \begin{bmatrix} d\lambda_1/dt \\ d\lambda_2/dt \\ d\lambda_3/dt \end{bmatrix} = - \begin{bmatrix} \lambda_1 \mu - \lambda_2 \sigma \\ 0 \\ 0 \end{bmatrix} \tag{12.114}$$

The solution to Eq. (12.85) can be obtained by forcing each term in it to vanish or by forcing dS/dt to satisfy the equation, that is,

1. $(\lambda_1\mu - \lambda_2\sigma)\mu' = 0$ and $(\lambda_1\mu'' - \lambda_2\sigma'')dS/dt = 0$ or
2. $dS/dt = (\lambda_1\mu - \lambda_2\sigma)\mu'/(\lambda_1\mu'' - \lambda_2\sigma'')$.

For condition 1, we note from Eq. (12.113) that $(\lambda_1\mu - \lambda_2\sigma) > 0$, and therefore we have two possibilities: (1) $\mu' = 0$ and $\lambda_1\mu'' - \lambda_2\sigma'' = 0$ or (2) $\mu' = 0$ and $dS/dt = 0$. Owing to Eq. (12.83), $\mu' = 0$ implies that $\sigma' = 0$. Therefore, the peaks in μ and σ must occur at the same substrate concentration, that is, $\mu(S_m) = \mu_{max}$ and $\sigma(S_m) = \sigma_{max}$. As we shall see later, the sufficient condition given by the generalized Legendre–Clebsch (GLC) condition eliminates the possibility of $\lambda_1\mu'' - \lambda_2\sigma'' = 0$, except for a degenerate case in which μ is a linear function of σ, $\mu = a\sigma + b$, where a and b are constants. Therefore, $\mu' = 0$ and $\lambda_1\mu'' - \lambda_2\sigma'' = 0$ may be acceptable for a special case only. The other possibility, $\mu' = 0$ and $dS/dt = 0$, implies that $\sigma' = 0$ owing to Eq. (12.83) so that μ and σ must peak at the same substrate concentration at $S = S_m$, and therefore, the substrate concentration is kept at $S = S_m$ to maximize the specific growth rate. Thus, condition 1 implies that when $\mu = a\sigma + b$ and $\mu(S_m) = \mu_{max}$ and $\sigma(S_m) = \sigma_{max}$, the optimal policy is to maximize the specific growth rate.

Thus, the two possibilities for condition 1 are similar and represent a degenerate case of the variable-yield coefficient. As we saw in the case of fixed final time for constant cell mass yield, the optimal singular feed rate calls for maximizing μ. However, when the cell mass yield is variable, this can only be true if the fixed time is extremely small.

The other alternative, condition 2, is more general; dS/dt is picked to satisfy Eq. (12.85):

$$\frac{dS}{dt} = \frac{(\lambda_1\mu - \lambda_2\sigma)\mu'}{\lambda_1\mu'' - \lambda_2\sigma''} = \frac{[\mu - (\lambda_2/\lambda_1)\sigma]\mu'}{\mu'' - (\lambda_2/\lambda_1)\sigma''} = \frac{[\mu - (\mu'/\sigma')\sigma]\mu'}{\mu'' - (\mu'/\sigma')\sigma''}$$
$$= -\frac{(\mu/\sigma)'(\sigma/\sigma')^2\mu'}{(\mu'/\sigma')'} = -\frac{Y'_{X/S}(\sigma/\sigma')^2\mu'}{(\mu'/\sigma')'} \quad (12.115)$$

As seen previously, two possibilities exist for the singular arc: (1) one that maximizes the specific growth rate of cells, $\mu'(S_m) = 0$, by maintaining the substrate concentration constant at $S = S_m$, and (2) the other that maximizes a combination of the specific growth rate of cells and the cell mass yield by varying the substrate concentration, $dS/dt = -[Y'_{X/S}/(\mu'/\sigma')'](\sigma/\sigma')^2\mu'$. Here we find that either the specific growth rate or a weighted combination of the specific growth rate of cells and cell yield is maximized. When the specific growth rate is maximized at S_m, the feed rate that maintains the substrate concentration constant at S_m is obtained from the substrate balance equation (Eq. (12.2)):

$$\frac{d(SV)}{dt} = S_m\frac{dV}{dt} = S_mF = S_FF - \mu(S_m)XV/Y_{X/S}(S_m)$$
$$\Rightarrow F_{sin} = \frac{\mu(S_m)XV}{Y_{X/S}(S_m)(S_F - S_m)} = \frac{\mu_{max}XV}{Y_{X/S}(S_m)(S_F - S_m)} \quad (12.116)$$

The singular feed rate of Eq. (12.116) represents an isolated case in which the final time is relatively short and the initial conditions are proper ($S_0 = S_m$) so that the singular feed rate is applied from the beginning until the reactor is full, that is, the final time is equal to the fill time, $t_f = t_{full}$. Final times less than this value represent

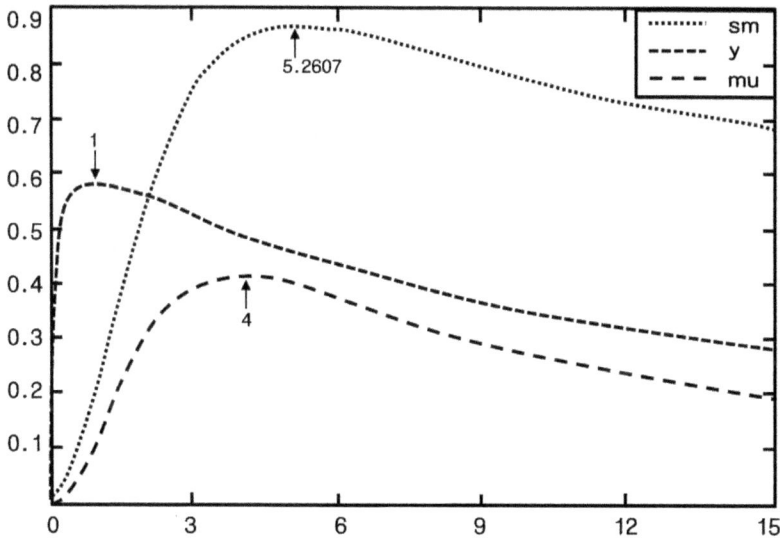

Figure 12.E.2.1. A plot of specific rate vs. substrate concentration.

insufficient time and lead to situations depicted in Figure 12.8. In other words, there is not enough time to continue the singular feed rate to fill up the reactor $(V(t_f) = V_{max})$, and therefore, the maximum feed rate is applied to meet the volume constraint. Otherwise, a combination of specific growth rate and cell mass yield is maximized by the following feed rate, which varies the substrate concentration according to Eq. (12.88):

$$F_{sin} = \frac{\sigma XV + V\,dS/dt}{(S_F - S)} = \frac{V}{(S_F - S)}\left\{\sigma X - \frac{-\mu'(\sigma/\sigma')^2(Y_{X/S})'}{(\mu'/\sigma')'}\right\} \qquad (12.117)$$

For all other initial conditions $(S_0 \neq S_m)$ and sufficient final times, the singular feed rate of Eq. (12.117) is applied so that *a combination of μ and $Y_{X/S}$ is maximized* throughout the course of operation.

EXAMPLE 12.E.2: CELL MASS MAXIMIZATION FOR VARIABLE YIELD AND FIXED FINAL TIME[6]
The kinetic information for this variable yield process is:

$$\text{Variable yield} \quad \mu = \frac{3S}{(S^3 + 10S + 128)}, \quad Y_{x/s} = \frac{7S}{(S^2 + 10S + 1)},$$

$$\sigma = \frac{3S(S^2 + 10S + 1)}{7(S^3 + 10S + 128)}$$

Let us consider a fed-batch culture of cell mass production with the preceding specific rates and variable-yield coefficient, where the specific rates of cell growth, substrate consumption, and cell mass yield are all nonmonotonic functions of substrate concentration, each showing a maximum. The final time is fixed at 10 hrs. Figure 12.E.2.1 shows that μ peaks at $S = 4\,\text{g/L}$, σ at $S = 5.261\,\text{g/L}$, and $Y_{X/S}$ at $S = 1\,\text{g/L}$. It is clear from Figure 12.9 that this example corresponds to case D of Figure 12.11 (later in this chapter) and that the singular region lies between $S = 1$ and $S = 4$.

The state equations are

$$\frac{d}{dt}\begin{pmatrix} x_1 \\ x_2 \\ x_3 \end{pmatrix} = \frac{d}{dt}\begin{pmatrix} XV \\ SV \\ V \end{pmatrix} = \begin{pmatrix} \mu x_1 \\ -\sigma x_1 \\ 0 \end{pmatrix} + \begin{pmatrix} 0 \\ S_F \\ 1 \end{pmatrix}F, \quad \begin{pmatrix} x_1(0) = X_0 V_0 \\ x_2(0) = S_0 V_0 \\ x_3(0) = V_0 \end{pmatrix} \quad (12.E.2.1)$$

and the Hamiltonian to be maximized is

$$\underset{F*(t)}{\text{Max }} H = (\lambda_1 \mu - \lambda_2 \sigma)x_1 + (S_F \lambda_2 + \lambda_3)F \overset{\Delta}{=} H_1 + \phi F \qquad (12.E.2.2)$$

The adjoint equations and boundary conditions are

$$\frac{d\boldsymbol{\lambda}}{dt} = \begin{bmatrix} d\lambda_1/dt \\ d\lambda_2/dt \\ d\lambda_3/dt \end{bmatrix} = -\begin{bmatrix} \partial H/\partial x_1 \\ \partial H/\partial x_2 \\ \partial H/\partial x_3 \end{bmatrix} = -\begin{bmatrix} \lambda_1 \mu - \lambda_3 \sigma \\ (\lambda_1 \mu' - \lambda_3 \sigma')x_1/x_2 \\ -(\lambda_1 \mu' - \lambda_3 \sigma')x_1 x_3/x_2^2 \end{bmatrix} \qquad (12.E.2.3)$$

$$\boldsymbol{\lambda}(t_f) = \begin{bmatrix} \lambda_1(t_f) \\ \lambda_2(t_f) \\ \lambda_3(t_f) \end{bmatrix} = \begin{bmatrix} \partial P/\partial x_1(t_f) \\ \partial P/\partial x_2(t_f) \\ \text{free} \end{bmatrix} = \begin{bmatrix} 1 \\ 0 \\ \text{unknown constant} \end{bmatrix} \qquad (12.E.2.4)$$

The singular feed rate is determined by recognizing that because $\phi = S_F \lambda_2 + \lambda_3 = 0$, its higher-order derivatives must also vanish. The singular feed rate is (Eq. (12.117))

$$F_{\text{sin}} = \frac{\left[\mu - \dfrac{\mu'}{\sigma'}\sigma\right]\mu' + \left[\mu'' - \dfrac{\mu'}{\sigma'}\sigma''\right]\sigma x_1/x_2}{\left[\mu'' - \dfrac{\mu'}{\sigma'}\sigma''\right](S_F - x_3/x_2)/x_2}V = \frac{-V\left(\dfrac{\mu}{\sigma}\right)'\left(\dfrac{\sigma}{\sigma'}\right)^2\mu' + \left(\dfrac{\mu'}{\sigma'}\right)'\sigma X}{\left(\dfrac{\mu'}{\sigma'}\right)'(S_F - S)}V$$

$$(12.E.2.5)$$

Thus, the singular feed rate is a feedback control requiring the measurement of the state variables, cell mass X, the culture volume V, and the substrate concentration S, in addition to knowledge of specific rates μ, σ, and π and their derivatives μ', σ', σ'', and μ''.

The switching function takes on the sequence of positive $\rightarrow$ zero $\rightarrow$ negative values, and therefore, the general optimal feed rate sequence is $F_{\text{max}} \rightarrow F_{\text{singular}} \rightarrow F_{\text{min}}$. The substrate concentration decreases (from near 4 to near 1) over the singular feed period.

A modified multiple shooting method is used to solve this problem, and the results are shown in Figures 12.E.2.2 and 12.E.2.3. Figure 12.E.2.2 shows the optimal profiles and the switching function. Furthermore, the adjoint variables at the final time match the boundary conditions of PMP. The singular feed rate as shown in Figure 12.E.2.2 is semiexponential.

In addition, Figure 12.E.2.3 shows that the specific growth rate of cell μ decreases and cell mass yield coefficient $Y_{x/s}$ increases over the singular feed rate period, as predicted theoretically for the singular period. These numerical results are in complete agreement with the preceding elucidated theoretical results.

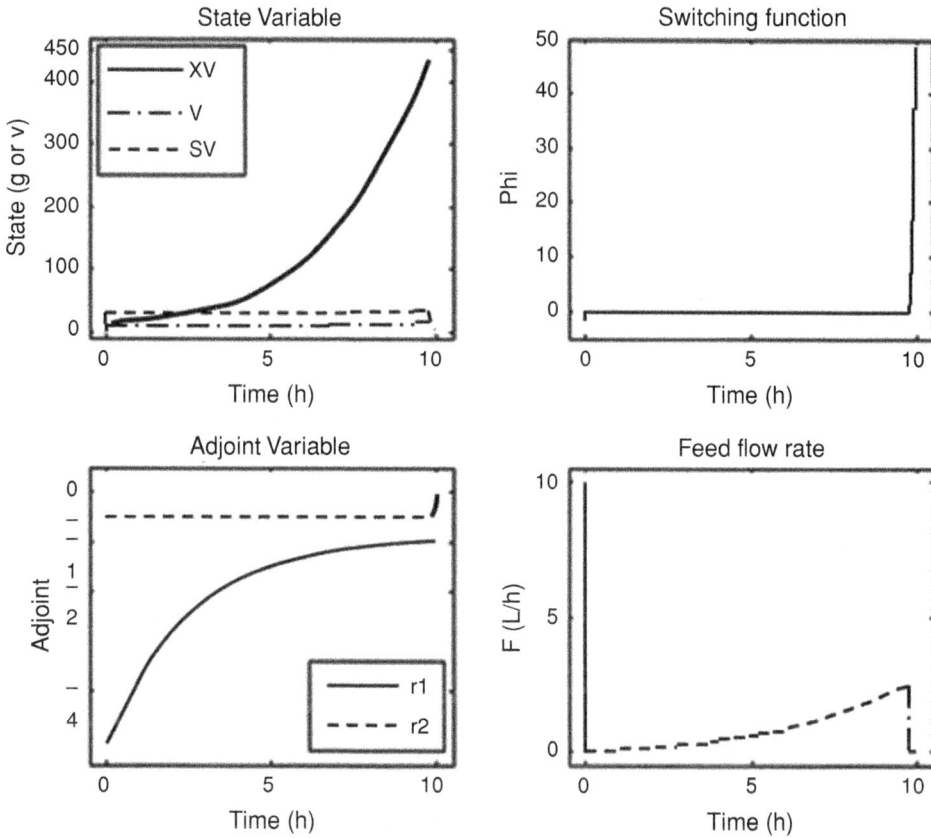

Figure 12.E.2.2. Optimal profiles, state and adjoint variables, switching function, and feed rate.

12.2.6 Assessment of Singular Regions

It would be nice to have singular regions identified in terms of two key parameters of cell mass production: the specific growth rate and the yield coefficient, $\mu(S)$ and $Y_{X/S}(S)$. If we could construct a map of singular regions, then we would be able to assess singular control without having to resort to numerical solutions. We will do this by incorporating not only the necessary but also the sufficient conditions.

Sufficient condition for a singular control is provided by the generalized Legendre–Clebsch condition:[14]

$$\frac{\partial}{\partial F}\frac{d^p}{dt^p}\frac{\partial H}{\partial F} = \frac{\partial}{\partial F}\frac{d^p\phi}{dt^p} = 0 \quad p\text{ odd}$$

$$(-1)^q\frac{\partial}{\partial F}\frac{d^{2q}}{dt^{2q}}\frac{\partial H}{\partial F} = (-1)^q\frac{\partial}{\partial F}\frac{d^{2q}\phi}{dt^{2q}} \geq 0 \tag{12.118}$$

where q is an order of singularity, the order of nonvanishing derivatives of the Hamiltonian in which a control variable appears explicitly. It can be shown[14] that fed-batch processes with the substrate feed rate as the manipulated variables represent $q = 1$ so that

$$\frac{\partial}{\partial F}\frac{d^2}{dt^2}H_F = \frac{\partial}{\partial F}\frac{d^2\phi}{dt^2} \leq 0 \tag{12.119}$$

Bang-Singular-Bang

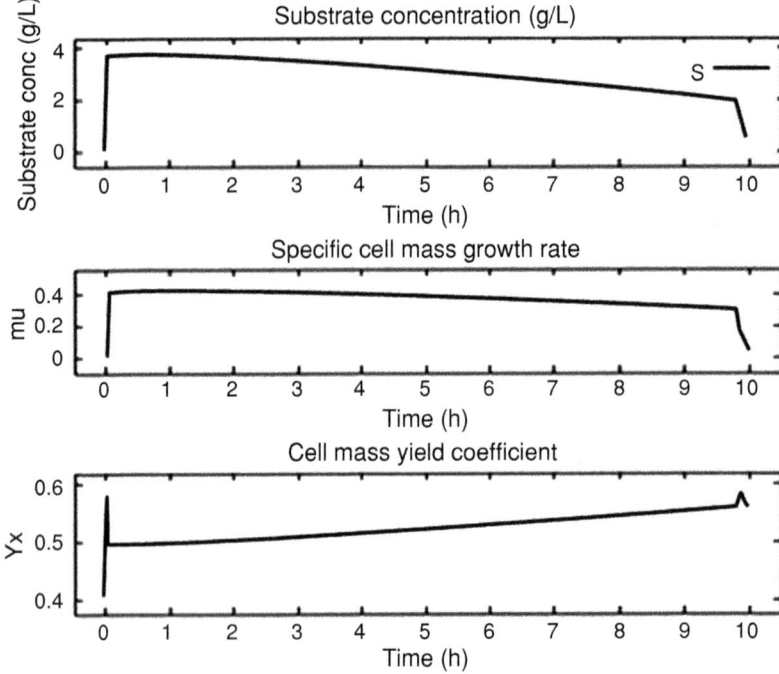

Figure 12.E.2.3. Time behavior of S, μ, and $Y_{X/S}$.

Substitution of Eq. (12.86) into Eq. (12.84) and rearrangement yield

$$\frac{d^2\phi}{dt^2} = -(\lambda_1\mu - \lambda_2\sigma)\mu' + (\lambda_1\mu'' - \lambda_2\sigma'')[F(S_F - x_2/x_3) - \sigma x_1]/x_3 \quad (12.120)$$

Application of Eq. (12.118) to Eq. (12.120) yields

$$\frac{\partial}{\partial F}\frac{d^2\phi}{dt^2} = (S_F - x_2/x_3)(\lambda_1\mu'' - \lambda_2\sigma'')/x_3 \leq 0 \quad (12.121)$$

Because $(S_F - x_2/x_3) \neq 0$ and $x_3 \neq 0$, Eq. (12.121) implies that

$$\lambda_1\mu'' - \lambda_3\sigma'' \leq 0 \quad (12.122)$$

Now, we investigate the possibility of the equality in Eq. (12.122) by assuming that

$$(\lambda_1\mu'' - \lambda_3\sigma'') = 0 \quad (12.123)$$

The preceding conditions for the singular arc, Eqs. (12.81), (12.83), and (12.123), represent three homogeneous equations in three adjoint variables:

$$\phi = S_F\lambda_2 + \lambda_3 = 0 \quad (12.81)$$

$$(\lambda_1\mu' - \lambda_2\sigma') = 0 \quad (12.83)$$

$$(\lambda_1\mu'' - \lambda_2\sigma'') = 0 \quad (12.123)$$

For a nonzero solution, the following determinant of coefficients must vanish:

$$\begin{vmatrix} 0 & S_F & 1 \\ \mu' & -\sigma' & 0 \\ \mu'' & -\sigma'' & 0 \end{vmatrix} = \sigma'\mu'' - \sigma''\mu' = \left[\frac{\sigma'\mu'' - \sigma''\mu'}{\sigma'^2}\right]\sigma'^2 = \left[\frac{\mu'}{\sigma'}\right]'\sigma'^2 = 0 \quad (12.124)$$

which implies that

$$\left[\frac{\mu'}{\sigma'}\right]' = 0 \quad \Rightarrow \frac{\mu'}{\sigma'} = \alpha \quad \Rightarrow \mu = \alpha\sigma + \beta \quad (12.125)$$

where α and β are constants. Thus, if the specific growth rate is a linear function of the specific substrate consumption rate, the GLC may be met. Therefore, this represents a special case, a case in which specific rates of both cell growth and substrate consumption peak at the same substrate concentration, $S_m = S_M$. For the general case, therefore, the following must hold:

$$(\lambda_1\mu'' - \lambda_2\sigma'') < 0 \quad (12.126)$$

Examination of Eq. (9.117) shows that the singular feed rate for the variable-yield case does not maximize the specific growth rate, is not a pure exponential function, and will not keep the substrate concentration constant during the operation. The substrate concentration will vary during the singular period. However, in a special case in which the specific rates peak at the same substrate concentration, $\mu'(S_m) = \sigma'(S_m) = 0$, the singular feed rate keeps the substrate concentration constant at $S = S_m$.

12.2.6.1 Substrate Concentration Profile during the Singular Feed Rate Period for the Variable-Yield Case

During the singular feed rate period, the substrate concentration can increase or decrease with time. Let us analyze this situation more formally. During the singular period, the following equations are valid, as seen previously:

$$(\lambda_1\mu - \lambda_2\sigma) > 0 \quad (12.113)$$

$$(\lambda_1\mu'' - \lambda_2\sigma'') < 0 \quad (12.126)$$

$$\frac{dS}{dt} = \frac{(\lambda_1\mu - \lambda_2\sigma)\mu'}{\lambda_1\mu'' - \lambda_2\sigma''} \quad (12.115)$$

To gain knowledge of how the substrate concentration changes with time during the singular time interval, we examine the sign of the right-hand side of Eq. (12.115). Taking the sign function of Eq. (12.115), we obtain, in view of Eqs. (12.113) and (12.126),

$$\text{sign}\left(\frac{dS}{dt}\right) = \text{sign}[(\lambda_1\mu - \lambda_2\sigma)]\text{sign}[\lambda_1\mu'' - \lambda_2\sigma'']\text{sign}(\mu') = -\text{sign}\left(\frac{d\mu}{dS}\right) \quad (12.127)$$

According to Eq. (12.127), the signs of dS/dt and $d\mu/dS$ are opposite to each other. In other words, in the region where the specific growth rate decreases with the substrate concentration, the substrate concentration would increase with time, whereas in the region where the specific growth rate increases with the substrate concentration, the substrate concentration would decrease with time. Thus, during the singular feed rate period, the substrate concentration increases or decreases with time at the expense of specific growth rate. The natural question that arises here is whether, then, the singular feed rate must improve something at the expense of reducing the growth rate.

To investigate this, we rearrange Eq. (12.126) with the aid of Eq. (12.83a) to obtain

$$\lambda_1[\mu'' - (\mu'/\sigma')\sigma''] = \lambda_1 \frac{(\sigma'\mu'' - \sigma''\mu')}{(\sigma')} \frac{\sigma'}{\sigma'} = \lambda_1 \left(\frac{\mu'}{\sigma'}\right)' \sigma' < 0 \qquad (12.128)$$

To eliminate λ_1 from the preceding, we must investigate the sign of $\lambda_1(t)$. Inspection of Eqs. (12.113) and (12.114) shows that $d\lambda_1/dt < 0$, that is, λ_1 can only decrease with time. Because the singular feed rate continues until the reactor volume is full and the final condition given by Eq. (12.18) is 1, we conclude that $\lambda_1(t)$ is positive for the entire singular time period, $t_s \le t \le t_f$. Therefore, Eq. (12.128) implies that

$$(\mu'/\sigma')'\sigma' < 0 \qquad (12.129)$$

We apply the same procedure to Eq. (12.113):

$$(\lambda_1\mu - \lambda_2\sigma) = \lambda_1[\mu - (\mu'/\sigma')\sigma] = \lambda_1[(\mu\sigma' - \mu'\sigma)/\sigma'](\sigma/\sigma')^2$$
$$= -\lambda_1(\mu/\sigma)'(\sigma^2/\sigma') > 0 \qquad (12.130)$$
$$\Rightarrow (\mu/\sigma)'\sigma^2/\sigma' < 0$$

Equations (12.129) and (12.130) yield another relationship:

$$\text{sign}\left[\left(\frac{\mu}{\sigma}\right)'\right] = \text{sign}\left(Y'_{X/S}\right) = \text{sign}\left[\left(\frac{\mu'}{\sigma'}\right)'\right] \qquad (12.131)$$

We now turn to assessing the sign of λ_3. Using Eqs. (12.81) and (12.83), we can express the first two adjoint variables, λ_1 and λ_2, in terms of the third, λ_3:

$$[\lambda_1, \lambda_2] = \left[-\frac{\sigma'}{\mu'S_F}\lambda_3, -\frac{1}{S_F}\lambda_3\right] \qquad (12.132)$$

The possibility of λ_3 changing its sign during the singular period is eliminated because $d\lambda_3/dt = 0$ in Eq. (12.114), and therefore λ_3 is either a positive or a negative constant on the singular arc. Let us assume that it is positive, $\lambda_3(t) > 0$. Substitution of Eq. (12.132) into Eq. (12.126) yields

$$\lambda_1\mu'' - \lambda_2\sigma'' = \left(-\frac{\sigma'\mu''}{\mu'} + \frac{\mu'}{\mu'}\sigma''\right)\frac{\lambda_3}{S_F} = -\left(\frac{\mu''\sigma' - \mu'\sigma''}{\mu'}\right)\frac{\lambda_3}{S_F} > 0 \quad \Rightarrow \quad \frac{\mu''\sigma' - \mu'\sigma''}{\mu'} < 0$$
$$(12.133)$$

Equation (13.115) can be rearranged to read

$$\frac{dS}{dt} = -\frac{\sigma\mu' - \sigma'\mu}{\sigma^2}\sigma^2\mu'/(\mu''\sigma' - \mu'\sigma'') = -Y'_{X/S}\frac{\sigma^2\mu'}{\mu''\sigma' - \mu'\sigma''} \qquad (12.134)$$

Taking sign functions of both sides of Eq. (12.134) and recognizing Eq. (12.133), we obtain

$$\text{sign}\left[\frac{dS}{dt}\right] = -\text{sign}[Y'_{X/S}]\text{sign}\left[\frac{\mu'}{\mu''\sigma' - \mu'\sigma''}\right] = \text{sign}[Y'_{X/S}] \qquad (12.135)$$

Equation (12.135) states that the substrate concentration increases (decreases) with time during the singular feed period in the direction of increasing (decreasing) cell mass yield coefficient. Combining Eq. (12.127) with Eq. (12.135), we obtain

$$\text{sign}\left(\frac{dS}{dt}\right) = \text{sign}\left(Y'_{X/S}\right) = -\text{sign}(\mu') \qquad (12.136)$$

Equation (12.136) states that during the singular period, the substrate concentration increases (decreases) with respect to time in the direction of increasing (decreasing) yield of cell mass with respect to substrate concentration and decreasing (increasing) specific growth rate of cells with respect to substrate concentration. Thus, Eq. (12.136) *defines the admissible regions of singular control*.

The preceding conclusion is based on the assumption that $\lambda_3 > 0$. Thus, we must show that λ_3 cannot be negative. Assuming $\lambda_3 < 0$ and following the same procedure used earlier, it has been shown[6] that it is not possible to have $\lambda_3 < 0$.

12.2.6.2 Sufficient Conditions for Admissible Singular Feed Rates for Cell Mass Maximization at Fixed Final Time

On the basis of Eq. (12.136), we can state the following rule of thumb for fixed final time problems. The regions in which a singular feed rate is admissible are those in which the slope of the yield coefficient $Y'_{X/S}$ is either zero or opposite in sign to that of the slope of the specific growth rate μ. When no such region can be identified from the kinetic information, the singular feed rate is inadmissible. On the singular arc, the substrate concentration increases or decreases with time to increase the cell yield at the expense of the cell growth rate. When the yield is constant, the substrate concentration is held constant. This is clearly illustrated in Figure 12.9 for nonmonotonic specific growth rate of cells.

12.2.7 Singular Regions Characterized by Kinetic Parameters, $\mu(S)$ and $Y_{X/S}(S)$

The specific growth rates and the yield coefficients are plotted against the substrate concentration in Figure 12.11. The regions in which the substrate concentration can vary during the singular period are identified, as are the directions of its change (increasing or decreasing with respect to time) by arrows. The regions are those in which the slope of the yield curve is opposite in sign to the slope of the specific growth rate curve. Identification of singular regions makes the numerical computation much simpler. The results can be summarized as follows:[6]

1. It is clear that when the yield coefficient is constant (Figure 12.11A), the substrate concentration is held at a constant value during the singular feed period, at the value at which the specific growth rate is maximum.
2. If the yield coefficient is a monotonically decreasing function of substrate concentration (Figure 12.11B), the substrate concentration is decreased during the

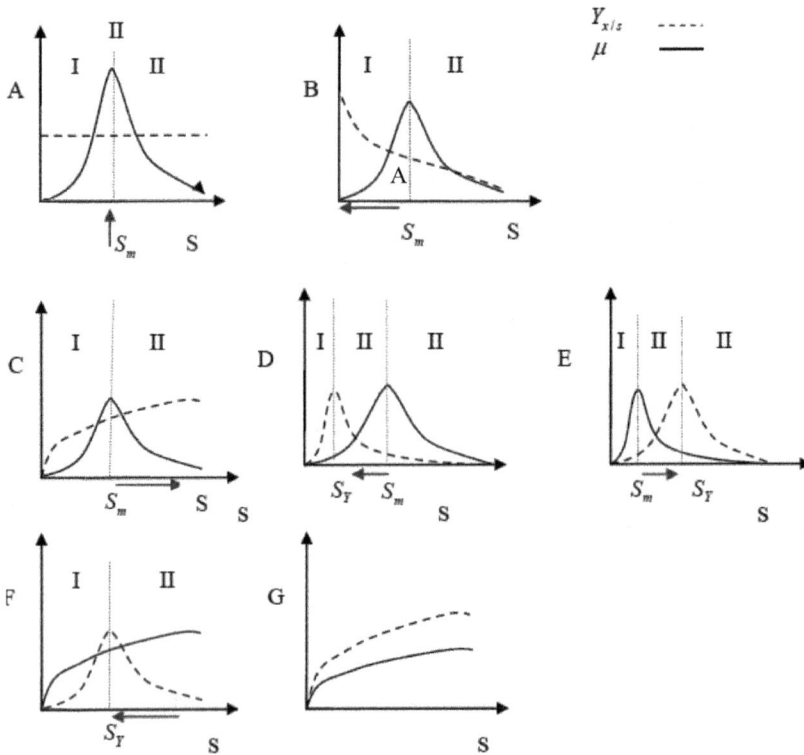

Figure 12.11. Singular regions (II) and variations in substrate concentration during singular interval. Solid arrows indicate the direction of the change in S with respect to time. A = nonmonotonic μ and constant $Y_{x/s}$; B = nonmonotonic μ and monotonically decreasing $Y_{x/s}$; C = nonmonotonic μ and monotonically increasing $Y_{x/s}$; D = nonmonotonic μ and $Y_{x/s}$ peaking before μ_{max}; E = nonmonotonic μ and $Y_{x/s}$ peaking after μ_{max}; F = monotonic μ and nonmonotonic $Y_{x/s}$; G = monotonic μ and $Y_{x/s}$.

singular feed period to increase the cell mass yield at the expense of decreasing specific growth rate.

3. Conversely, when the yield coefficient is a monotonically increasing function of substrate concentration (Figure 12.11C), the substrate concentration is increased to improve the cell mass yield at the expense of specific growth rate.

4. The substrate concentration is decreased if the peak in the yield occurs at a lower substrate concentration than that of the specific growth rate (Figure 12.11D).

5. The substrate concentration is increased if the peak in the yield occurs at higher substrate concentration than that of the specific growth rate (Figure 12.11E).

6. When the specific growth rate is a monotonically increasing function of substrate concentration, while the yield coefficient is nonmonotonic (Figure 12.11F), then the substrate concentration is increased to improve the yield.

7. When both the specific growth rate and the cell mass yield are monotonically increasing functions of substrate concentration (Figure 12.11G), no singular region exists, and a batch operation would be best.

For monotonic specific growth rates, only when the yield coefficient shows a maximum, a fed-batch operation to maintain the substrate concentration constant at the value at which the yield is maximum is the optimal mode of operation. When both the specific growth rate and the yield coefficient are monotonic (no peaks), the optimal mode of operation is batch operations.

All of the preceding are the results for fixed times. The singular feed rate manipulates the substrate concentration in the reactor to optimize the rate, the yield, or a weighted combination of the two. Therefore, it is anticipated that if the final time is large, the specific growth rate is not critical owing to ample available time allowed, and instead, the yield is maximized, while the rate would be maximized if the final time were relatively short.

If the final time is free and does not appear in the performance index, then the final time is immaterial, and one may choose a large value approaching infinity. Then, the rate is not important because one has all the time needed, and therefore, the yield must be optimized. Therefore, one would maximize the yield by maintaining the substrate concentration at $S = S_M$, corresponding to the maximum yield. At this substrate concentration, the specific growth rate is small, and therefore, longer time is needed. In other words, if the time is not critical, the yield should be maximized, whereas if the final time is short, the specific growth rate should be maximized to meet the short final time. The optimal singular feed rate policy weighs the specific growth rate and the cell mass yield, weighting more the specific growth rate when the final time is small and weighting more the yield if the final time is large. The optimal feed rate maximizes the balance between the growth rate and the yield.

12.2.7.1 Summary for Maximum Cell Mass Production at Fixed Final Time for the Case of a Variable-Yield Coefficient

When the final time is specified a priori, the singular fed rate maximizes a combination of the specific growth rate and cell mass yield by varying the substrate concentration in the direction of increasing cell mass yield and decreasing direction of specific cell growth rate. In fact, the singular feed rate exits in the region in which the sign of $d\mu/dS$ is opposite to that of $dY_{X/S}/dS$:

1. The singular feed rate maximizes a combination of the cell growth rate and cell mass yield coefficient.
2. The optimal initial substrate concentration is that which lies on the singular subarc. All other initial substrate concentrations require additional maximum or minimum feed rates prior to the singular feed rate.
3. The optimal feed rate sequence, in general, consists of F_{max} (or F_{min}) $\rightarrow F_{sin} \rightarrow F_{min}(= 0)$. The singular feed rate continues to the end when the bioreactor volume is full. The singular feed rate is not an exponential function and is in feedback mode (Eq. (12.117)).
4. The optimal feed rate given is in feedback mode, requiring the measurement of cell concentration and bioreactor volume and the knowledge of feed substrate concentration, peak substrate concentration, and kinetic information. The bioreactor volume increases exponentially.

12.2.7.2 Comparison of Free and Fixed Time Problems for Cell Mass Maximization

Table 12.1 compares the case of a constant-yield coefficient with that of a variable-yield coefficient. In particular, it provides the singular feed rates and the behavior of

Table 12.1. *Comparison of free and fixed final time problems for cell mass maximization*

Criterion	Cell mass maximization	
	Constant yield	Variable yield
Objective function	$\text{Max } X(t_f)V(t_f)$, $\text{Max } x_1(t_f)$	$\text{Max } X(t_f)V(t_f)$, $\text{Max } x_1(t_f)$
State equations	$\dfrac{d}{dt}\begin{pmatrix} x_1 \\ x_3 \\ x_4 \end{pmatrix} = \dfrac{d}{dt}\begin{pmatrix} XV \\ V \\ t \end{pmatrix} = \begin{pmatrix} \mu x_1 \\ 0 \\ 1 \end{pmatrix} + \begin{pmatrix} 0 \\ 1 \\ 0 \end{pmatrix} F$ $x_2 = x_{20} + S_F(x_3 - x_{30}) - (x_1 - x_{10})/Y_{X/S}$	$\dfrac{d}{dt}\begin{pmatrix} x_1 \\ x_2 \\ x_3 \\ x_4 \end{pmatrix} = \dfrac{d}{dt}\begin{pmatrix} XV \\ SV \\ V \\ t \end{pmatrix} = \begin{pmatrix} \mu x_1 \\ -\sigma x_1 \\ 0 \\ 1 \end{pmatrix} + \begin{pmatrix} 0 \\ S_F \\ 1 \\ 0 \end{pmatrix} F$
Hamiltonian	$H = \lambda_1 \mu x_1 + \lambda_3 F + 1$	$H = (\lambda_1 \mu - \lambda_2 \sigma)x_1 + (S_F \lambda_2 + \lambda_3)F + 1$
Adjoint equations	$\dfrac{d}{dt}\begin{bmatrix} \lambda_1 \\ \lambda_3 \end{bmatrix} = -\begin{bmatrix} \lambda_1 \mu + \lambda_1 x_1 \partial\mu/\partial x_1 \\ \lambda_1 x_1 \partial\mu/\partial x_3 \end{bmatrix}$	$\begin{bmatrix} d\lambda_1/dt \\ d\lambda_2/dt \\ d\lambda_3/dt \end{bmatrix} = -\begin{bmatrix} \lambda_1 \mu - \lambda_2 \sigma \\ (\lambda_1 \mu' - \lambda_2 \sigma')x_1/x_3 \\ -(\lambda_1 \mu' - \lambda_2 \sigma')x_1 x_2/x_3^2 \end{bmatrix}$
B. C. on adjoint variables	$\lambda_1(t_f) = -1$	$\lambda_1(t_f) = -1, \lambda_2(t_f) = 0$
Item maximized	Free t_f: none Fixed t_f: specific growth rate μ	Free t_f: cell mass yield $Y_{X/S}$ Fixed t_f: weighted sum of μ and $Y_{X/S}$
Feed rate	Minimum, maximum, or singular	Minimum, maximum, or singular
Form of singular feed rate	Free t_f: no singular feed rate Fixed t_f: exponential $F_{\sin} = \dfrac{\mu(S_m)XV}{Y_{X/S}(S_F - S_m)}$	Free t_f: exponential $F_{\sin} = \dfrac{\mu(S_M)XV}{Y_{X/S}(S_F - S_M)}$ Fixed t_f: semiexponential $F_{\sin} = \dfrac{\sigma XV}{(S_F - S)} - \dfrac{V\mu'Y'(\sigma/\sigma')^2(\mu'/\sigma')'}{(S_F - S)}$
Substrate concentration	Free t_f: any value Fixed t_f: remains constant at $S = S_m$, $\mu(S_m) = \mu_{\max}$	Free t_f: constant, $S_M Y'_{X/S}(S_M) = 0$ Fixed t_f: variable

substrate concentration during the singular period for free and fixed final times and for constant-yield and variable-yield coefficients.

12.3 Maximization of Cellular Productivity, $X(t_f)V(t_f)/t_f$

It is apparent that the solution to the maximum cellular productivity problem $P = X(t_f)V(t_f)/t_f$ can be obtained by solving the maximum cell mass problem $X(t_f)V(t_f)$ repeatedly for various values of final time t_f, calculating the corresponding productivity $X(t_f)V(t_f)/t_f$ and picking the one that gives the maximum value. Therefore, it is instructive to consider the problem of maximizing the biomass productivity, $X(t_f)V(t_f)/t_f$.

Because the performance index has to be a function of the state variables at the final time, we introduce an auxiliary state variable so that the performance index becomes a function of final state variables only:

$$x_4(t_f) = t_f \tag{12.137}$$

Equivalently, a differential equation,

$$\frac{dx_4}{dt} = 1 \quad x_4(0) = 0 \tag{12.138}$$

is added to Eq. (12.7) to obtain

$$\frac{d\mathbf{x}}{dt} = \mathbf{a}(\mathbf{x}) + \mathbf{b}F, \quad \frac{d}{dt}\begin{pmatrix} x_1 \\ x_2 \\ x_3 \\ x_4 \end{pmatrix} = \frac{d}{dt}\begin{pmatrix} XV \\ SV \\ V \\ t \end{pmatrix} = \begin{pmatrix} \mu x_1 \\ -\sigma x_1 \\ 0 \\ 1 \end{pmatrix} + \begin{pmatrix} 0 \\ S_F \\ 1 \\ 0 \end{pmatrix}F \tag{12.139}$$

The performance index is now a function of the final state only:

$$P = X(t_f)V(t_f)/t_f = x_1(t_f)/x_4(t_f) \tag{12.140}$$

Because the final time appears in the performance index, this is a free final time problem. Fixing the final time would reduce the problem to cell mass maximization. The feed flow rate and volume constraints remain intact:

$$0 \le F(t) \le F_{\max} \tag{12.5}$$

$$x_3(t_f) = V_{\max} \tag{12.10}$$

The mass balance equations are given by Eq. (12.139), and the Hamiltonian is zero because the final time is free,

$$H = (\lambda_1\mu - \lambda_2\sigma)x_1 + \lambda_4 + (S_F\lambda_2 + \lambda_3)F \overset{\Delta}{=} H_1 + \phi F = 0 \tag{12.141}$$

and the switching function is

$$\phi = S_F\lambda_2 + \lambda_3 \tag{12.142}$$

Because the Hamiltonian is linear in F, the feed flow rate can take on the maximum, minimum, singular, and zero values, depending on the sign of the switching function, ϕ:

$$F = \begin{cases} F_{\max} & \text{when } \phi = (S_F\lambda_2 + \lambda_3) > 0 \\ F_{\min} = 0 & \text{when } \phi = (\lambda S_{F2} + \lambda_3) < 0 \\ F_{\sin} & \text{when } \phi = (S_F\lambda_2 + \lambda_3) = 0 \quad t_q < t < t_{q+1} \\ F_b = 0 & \text{when } x_3(t) = V_{\max} \end{cases} \tag{12.143}$$

The adjoint vector must satisfy the following equation:

$$\frac{d\lambda}{dt} = \begin{bmatrix} d\lambda_1/dt \\ d\lambda_2/dt \\ d\lambda_3/dt \\ d\lambda_4/dt \end{bmatrix} = -\frac{\partial H}{\partial \mathbf{x}} = -\begin{bmatrix} \lambda_1\mu - \lambda_2\sigma \\ (\lambda_1\mu' - \lambda_2\sigma')x_1/x_3 \\ -(\lambda_1\mu' - \lambda_2\sigma')x_1 x_2/x_3^2 \\ 0 \end{bmatrix} \tag{12.144}$$

The boundary conditions are obtained from the transversality conditions by proper matching:

$$\delta P = \delta\begin{bmatrix} x_1(t_f) \\ \overline{x_4(t_f)} \end{bmatrix} = \underline{\frac{1}{x_4(t_f)}\delta x_1(t_f)} - \underline{\frac{x_1(t_f)}{x_4^2(t_f)}\delta x_4(t_f)} = \sum_{i=1}^{4} \lambda_i(t_f)\delta x_i(t_f)$$

$$-\sum_{i=1}^{4} \lambda_i(0)\underbrace{\delta x_i(0)}_{=0} = \lambda_1(t_f)\delta x_1(t_f) + \lambda_2(t_f)\delta x_2(t_f) + \lambda_3(t_f)\underbrace{\delta x_3(t_f)}_{=0}$$

$$+ \underline{\underline{\lambda_4(t_f)\delta x_4(t_f)}}$$

$$\Rightarrow \quad \lambda(t_f) = \frac{\partial P}{\partial \mathbf{x}}(t_f) = \begin{bmatrix} \lambda_1(t_f) \\ \lambda_2(t_f) \\ \lambda_3(t_f) \\ \lambda_4(t_f) \end{bmatrix} = \begin{bmatrix} 1/x_4(t_f) \\ 0 \\ \text{unknown constant} \\ -x_1(t_f)/x_4^2(t_f) \end{bmatrix} \tag{12.145}$$

From Eqs. (12.144) and (12.145), we note that $\lambda_4(t)$ is a constant,

$$\lambda_4(t) = \lambda_4(t_f) = -x_1(t_f)/x_4^2(t_f) = -X(t_f)V(t_f)/t_f^2 \tag{12.146}$$

so that the Hamiltonian is

$$H = (\lambda_1\mu - \lambda_2\sigma)x_1 + (S_F\lambda_2 + \lambda_3)F - x_1(t_f)/x_4^2(t_f) \tag{12.141}$$

It may be instructive to consider the case of constant and variable cell mass yield coefficients separately so that one can see the difference between the maximization of cell mass and maximization of cellular productivity.

12.3.1 Constant Cell Mass Yield Coefficient

For constant cell mass yields, the process can be described by two differential mass balance equations and one algebraic equation, as we have done earlier for cell mass maximization:

$$SV = x_2 = x_{20} + S_F(x_3 - x_{30}) - \frac{1}{Y_{X/S}}(x_1 - x_{10}) = g(x_1, x_3) \tag{12.22}$$

Owing to Eq. (12.22), the state equation (Eq. (12.139)) reduces to

$$\frac{d\mathbf{x}}{dt} = \frac{d}{dt}\begin{pmatrix} x_1 \\ x_3 \\ x_4 \end{pmatrix} = \frac{d}{dt}\begin{pmatrix} XV \\ V \\ t \end{pmatrix} = \begin{pmatrix} \mu x_1 \\ 0 \\ 1 \end{pmatrix} + \begin{pmatrix} 0 \\ 1 \\ 0 \end{pmatrix} F$$

$$d\mathbf{x}/dt = \mathbf{a}(\mathbf{x}) + \mathbf{b}F \tag{12.147}$$

The Hamiltonian (Eq. (12.141)) reduces to

$$\min_{F*(t)} H = \lambda^{\mathrm{T}}[\mathbf{a}(\mathbf{x}) + \mathbf{b}F] = \lambda_1 \mu x_1 + \lambda_3 F + \lambda_4 \overset{\Delta}{=} H_1 + \phi F = 0 \tag{12.148}$$

where

$$\phi(t) = \lambda_3(t) \tag{12.149}$$

The adjoint variables must satisfy the following differential equations (Eq. (12.144)):

$$\frac{d}{dt}\begin{pmatrix} \lambda_1 \\ \lambda_3 \\ \lambda_4 \end{pmatrix} = -\frac{\partial H}{\partial \mathbf{x}} = -\begin{pmatrix} \lambda_1 \mu + \lambda_1 x_1 \partial\mu/\partial x_1 \\ \lambda_1 x_1 \partial\mu/\partial x_3 \\ 0 \end{pmatrix} \tag{12.150}$$

where the required partial derivatives are

$$\frac{\partial\mu(S)}{\partial x_1} = \frac{\partial\mu(S)}{\partial S}\frac{\partial(S)}{\partial x_1} = \frac{d\mu}{dS}\frac{\partial(g/x_3)}{\partial x_1} = \mu'\frac{(\partial g/\partial x_1)}{x_3} = -\frac{\mu'}{Y_{X/S}x_3}$$

$$\frac{\partial\mu(S)}{\partial x_3} = \frac{\partial\mu(S)}{\partial S}\frac{\partial(S)}{\partial x_3} = \frac{d\mu}{dS}\frac{\partial(g/x_3)}{\partial x_3} = \mu'\frac{x_3(\partial g/\partial x_3) - g}{x_3^2} = \mu'\frac{(S_F - g/x_3)}{x_3} \tag{12.151}$$

Substitution of Eq. (12.151) into Eq. (12.150) yields

$$\frac{d}{dt}\begin{pmatrix} \lambda_1 \\ \lambda_3 \\ \lambda_4 \end{pmatrix} = -\frac{\partial H}{\partial \mathbf{x}} = -\begin{pmatrix} \lambda_1 \mu - \lambda_1 x_1 \mu'/x_3 Y_{X/S} \\ \lambda_1 x_1 \mu' \left(S_F/x_3 - g/x_3^2\right) \\ 0 \end{pmatrix} \tag{12.152}$$

The boundary conditions are obtained from the transversality condition (Eq. (12.145)):

$$\lambda(t_f) = \begin{pmatrix} \lambda_1(t_f) \\ \lambda_3(t_f) \\ \lambda_4(t_f) \end{pmatrix} = \frac{\partial P}{\partial \mathbf{x}}(t_f) = \begin{pmatrix} 1/x_4(t_f) \\ \text{unknown constant} \\ -x_1(t_f)/x_4^2(t_f) \end{pmatrix} \tag{12.153}$$

Equations (13.152) and (13.153) imply that λ_4 is a constant:

$$\lambda_4(t) = \lambda_4(t_f) = -x_1(t_f)/x_4^2(t_f) = -X(t_f)V(t_f)/t_f^2 \tag{12.154}$$

Therefore, we can simplify Eqs. (13.152) and (13.153) and work with two adjoint variables:

$$\frac{d}{dt}\begin{pmatrix} \lambda_1 \\ \lambda_3 \end{pmatrix} = -\begin{pmatrix} \lambda_1 \mu - \lambda_1 x_1 \mu'/x_3 Y_{X/S} \\ \lambda_1 x_1 \mu'(S_F - g/x_3)/x_3 \end{pmatrix} \quad \lambda(t_f) = \begin{pmatrix} \lambda_1(t_f) \\ \lambda_3(t_f) \end{pmatrix} = \begin{pmatrix} 1/x_4(t_f) \\ \text{unknown constant} \end{pmatrix} \tag{12.155}$$

Because $F(t)$ appears linearly in Eq. (12.148), the optimal feed rate that maximizes the Hamiltonian depends on the sign of the switching function, $\phi(t) = \lambda_3(t)$:

$$F = \begin{cases} F_{\max} & \text{when } \phi = \lambda_3(t) > 0 \\ F_{\min} & \text{when } \phi = \lambda_3(t) < 0 \\ F_{\sin} & \text{when } \phi = \lambda_3(t) = 0 \text{ over a finite time interval, } t_q < t < t_{q+1} \\ F_b = 0 & \text{when } x_3(t) = V_{\max} \end{cases} \tag{12.156}$$

The remaining tasks are to determine the singular feed rate and the sequence and time intervals of the maximum, minimum, singular, and zero feed rates.

12.3.1.1 Optimal Feed Rate on Interior Singular Arc

In the interior arc where the singular control is in effect, the switching function is identically zero, $\phi = \lambda_3 = 0$, so that Eq. (12.148) reduces to

$$H^* = \lambda_1 \mu x_1 + \lambda_4 = \lambda_1 \mu x_1 - x_1(t_f)/t_f^2 = 0 \tag{12.157}$$

On the singular arc, $\phi = 0$, $d\phi/dt = 0$, and $d^2\phi/dt^2 = 0$ so that

$$\frac{d\phi}{dt} = \frac{d\lambda_3}{dt} = -\frac{\mu' \lambda_1 x_1 (S_F - g/x_3)}{x_3} = 0 \tag{12.158}$$

In Eq. (12.158), $x_1 = XV \neq 0$ and $S_F - g/x_3 = S_F - S \neq 0$, and λ_1 cannot be zero. If it were, then all adjoint variables would be identically zero, making the problem trivial. Therefore, Eq. (12.158) is satisfied by setting

$$\mu' = \frac{d\mu}{dS} = 0 \tag{12.159}$$

Equation (12.159) states that the specific growth rate is maximized on the singular arc, $\mu(S_m) = \mu_{\max}$. In other words, the substrate concentration must be maintained constant, $S = S_m$, corresponding to the maximum value of the specific growth rate. The feed rate required to maintain the constant value of S_m is obtained from the substrate balance equation, as we have done previously:

$$F_{\sin} = \frac{\mu_m XV}{Y_{X/S}(S_F - S_m)} = \frac{\sigma_m XV}{(S_F - S_m)} \tag{12.160}$$

The second derivative of the switching function is

$$\frac{d^2\phi}{dt^2} = -\frac{d}{dt}\left[\frac{\lambda_1 x_1 \mu'(S_F - g/x_3)}{x_3}\right] = -\frac{d}{dt}\left[\frac{\lambda_1 x_1 (S_F - g/x_3)}{x_3}\right]\mu'$$

$$-\frac{d\mu'}{dt}\left[\frac{\lambda_1 x_1 (S_F - S)}{V}\right] = 0 \tag{12.161}$$

Because the first term is zero owing to Eq. (12.159) and $\lambda_1 x_1 (S_F - S)/V \neq 0$,

$$\frac{d\mu'}{dt} = \mu'' \frac{dS}{dt} = 0 \tag{12.162}$$

Equation (12.162) is satisfied if

$$\frac{dS}{dt} = 0 \tag{12.163}$$

Equation (12.163) states that the singular flow rate must maintain the substrate concentration constant so that the specific growth is maximized. Denoting as S_m

the substrate concentration corresponding to the maximum specific growth rate, the singular arc is

$$S = S_m \tag{12.164}$$

The singular feed rate is then readily obtained from the substrate balance equation (12.2) by recognizing that the substrate concentration must be maintained constant at S_m (Eq. (12.160)).

The adjoint variables on the singular arc are

$$d\lambda_1/dt = -\lambda_1 \mu = -\lambda_1 \mu_{max} \quad \Rightarrow \lambda_1(t) = \lambda_1(t_s) \exp[-\mu_{max}(t - t_s)]$$
$$\lambda_4 = x_1(t_f)/t_f^2 \tag{12.165}$$

Thus, $\lambda_1(t)$ decays exponentially, while λ_4 is a constant on the singular arc. At the final time when the bioreactor is full, the volume constraint is met, and $F = 0$ so that

$$H^* = \lambda_1(t_f)\mu(t_f)x_1(t_f) + \lambda_4(t_f) = 0 \tag{12.166}$$

Substitution of the final adjoint variables in Eq. (13.155) into Eq. (12.166) yields

$$H^* = [1/x_4(t_f)]\mu(t_f)x_1(t_f) - x_1(t_f)/x_4^2(t_f)$$
$$= [x_1(t_f)/x_4(t_f)][\mu(t_f) - 1/x_4(t_f)] = 0 \tag{12.167}$$

Solving for the optimum final time in Eq. (12.167), we obtain

$$t_f = 1/\mu(t_f) \tag{12.168}$$

This equation tells us that the optimal final time must equal the reciprocal of the specific growth rate.

The singular feed rate given by Eq. (12.160) is in feedback mode; that is, by monitoring the total amount of cell mass, XV, and knowing the kinetic information, μ_m, S_m, $Y_{X/S}$, and the feed substrate concentration, S_F, we can calculate the singular feed rate, F_{sin}. That this feed rate is an exponential function in time can be shown readily by recognizing that because the substrate concentration is maintained at S_m, the total amount of cell mass increases exponentially, as seen by integrating the cell mass balance equation (Eq. (12.1)):

$$\frac{d(XV)}{dt} = \mu XV = \mu_m XV \Rightarrow XV(t) = XV(t_s)\exp[\mu_m(t - t_s)]$$
$$= XV(t_s)\exp(-\mu_m t_s)\exp(\mu_m t) \tag{12.169}$$

where t_s is used to denote that the evaluation is done at the start time of the singular period. Substitution of Eq. (12.169) into Eq. (12.160) yields an exponential function for the singular feed rate:

$$F_{sin} = \frac{\mu_m XV}{(S_F - S)Y_{X/S}} = \left[\frac{\mu_m XV(t_s)\exp(-\mu_m t_s)}{(S_F - S_m)Y_{X/S}}\right]\exp(\mu_m t) = \alpha \exp(\mu_m t)$$
$$\alpha = \mu_m XV(t_s)\exp(-\mu_m t_s)/(S_F - S_m)Y_{X/S} \tag{12.170}$$

12.3.1.2 The Optimal Feed Rate Profile for Cellular Productivity
Maximization, Constant Yield

Knowing that the optimal policy maximizes the specific growth rate, it is now possible to construct the optimal feed rate profile. First, the initial substrate concentration should be chosen to give the maximum specific growth rate, $S(0) = S_m$, so that one can start the singular feed from the start until the bioreactor is full. A batch period follows for some time until the final time meets the constraint given by Eq. (12.168). In this way, the entire period of operation is singular and is spent maximizing the specific growth rate. In fact, the situation is identical to the case of maximizing cell mass for constant yield with fixed final time (Section 12.2.3.6). Therefore, the details with various initial substrate concentrations are not repeated here. Suffice it to say that if the initial substrate concentration is less than S_m, then the operation starts with the maximum flow rate to bring the concentration to S_m. If the initial substrate concentration is higher than S_m, then it should be brought to S_m as soon as possible by a batch operation ($F_{min} = 0$). Then the singular feed rate is applied until the reactor is full, at which point, a batch operation takes over until the final time at which it is equal to the reciprocal of the specific growth rate.

Although the functional form of the feed rate profile is identical to the case of cell mass production, the switching time and final time may be different. This is obvious as the boundary conditions on the adjoint vectors are different.

In the preceding treatment, we introduced for convenience an auxiliary state variable x_4 so that we can handle the objective function that contains t_f and the corresponding adjoint variable λ_4, which turned out to be a constant given by Eq. (12.154):

$$\lambda_4(t) = \lambda_4(t_f) = -x_1(t_f)/x_4^2(t_f) \tag{12.154}$$

This is identical to the case of cell mass maximization. However, the remaining boundary conditions, Eq. (12.153), differ from those of the cell mass production:

$$\lambda_1(t_f) = 1/x_4(t_f) = 1/t_f > 0$$
$$\lambda_3(t_f) = \text{free} \tag{12.153}$$

Namely, the boundary condition $\lambda_1(t_f)$ was 1 for the case of cell mass maximization, whereas here it is the inverse of the final time for the case of cell mass productivity maximization. It is also instructive to see that the Hamiltonian for the case of cell mass productivity maximization differs from that of cell mass maximization by a constant, $1/t_f$:

$$\text{Max}_{F*(t)} H = \lambda_1 \mu x_1 + \lambda_3 F + \lambda_4 = \lambda_1 \mu x_1 + \lambda_3 F - (1/t_f) \tag{12.171}$$

Therefore, the maximization of H for the cell mass productivity is equivalent to the maximization of H for cell mass maximization. In other words, the solution to the optimum cell mass productivity can be obtained by solving the cell mass maximization at various fixed times, calculating the corresponding cell mass productivities and then picking the best productivity.

EXAMPLE 12.E.3: CELL MASS PRODUCTIVITY MAXIMIZATION FOR CONSTANT YIELD A specific cell growth rate and constant-yield model[18] is given here, same as the example of cell mass maximization with constant yield:

$$\mu(S) = \frac{S}{0.03 + S + 0.5S^2}, \; Y_{X/S} = 0.5 \qquad (12.E.1.1)$$

Specific growth rate was plotted in Figure 12.E.1.1 as a function of substrate concentration. The initial state variables and parameter conditions are as follows:

$$[X, \; S, \; V](t_0) = [0.5 \, \text{g/L}, \; 0.245 \, \text{g/L}, \quad 2 \, \text{L}]$$

$$[S_F, \; V_{\max}, \; F_{\max}] = [10 \, \text{g/L}, \; 20 \, \text{L}, \quad 17 \, \text{L/hr}] \qquad (12.E.3.1)$$

The initial substrate concentration S_0 is S_m, corresponding to maximum cell mass growth rate, and is on a singular arc. As illustrated, the optimal feed rate strategy is to start the singular feed from the start until the bioreactor is full, followed by a batch period for some time until the final time when the constraint given by Eq. (12.168) is met. As seen in Figure 12.E.3.1, the singular feed, the exponential curve given by Eq. (12.170), is extended to the switching time 5.717 hr, at which the reactor is full, maintaining substrate concentration S_m, and then a batch $F_{\min}$ (no flow) is followed until the final time 5.75 hr, when the final condition given by Eq. (12.168) is met.

12.3.2 Variable Cell Mass Yield Coefficient

Let us now look at the productivity maximization for the variable-yield case. The mass balance equations are given by Eq. (12.139) and the Hamiltonian is zero (Eq. (12.141)) because the final time is free. The switching function is given by Eq. (12.142), and the flow rate that can take on the maximum, minimum, singular, and zero values is given by Eq. (12.143). The adjoint vector must satisfy Eq. (12.144) and the boundary conditions given by Eq. (12.145). We note that $\lambda_4(t)$ is a constant:

$$\lambda_4(t) = \lambda_4(t_f) = -x_1(t_f)/x_4^2(t_f) = -X(t_f)V(t_f)/t_f^2 \qquad (12.153)$$

Therefore, the adjoint equations, Eq. (12.144), reduce to

$$\begin{bmatrix} d\lambda_1/dt \\ d\lambda_2/dt \\ d\lambda_3/dt \end{bmatrix} = - \begin{bmatrix} \lambda_1\mu - \lambda_2\sigma \\ (\lambda_1\mu' - \lambda_2\sigma')x_1/x_3 \\ -(\lambda_1\mu' - \lambda_2\sigma')x_1x_2/x_3^2 \end{bmatrix} \qquad (12.172)$$

This is identical to the case of cell mass maximization (Eq. (12.16)). The boundary conditions, Eq. (12.153), reduce to

$$\begin{bmatrix} \lambda_1(t_f) \\ \lambda_2(t_f) \\ \lambda_3(t_f) \end{bmatrix} = \begin{bmatrix} 1/x_4(t_f) = 1/t_f \\ 0 \\ \text{unknown constant} \end{bmatrix} \qquad (12.173)$$

This is different from that of cell mass maximization (Eq. (12.18)). The Hamiltonian is

$$H = (\lambda_1\mu - \lambda_3\sigma)x_1 + (S_F\lambda_2 + \lambda_3)F - x_1(t_f)/t_f^2 = 0 \qquad (12.174)$$

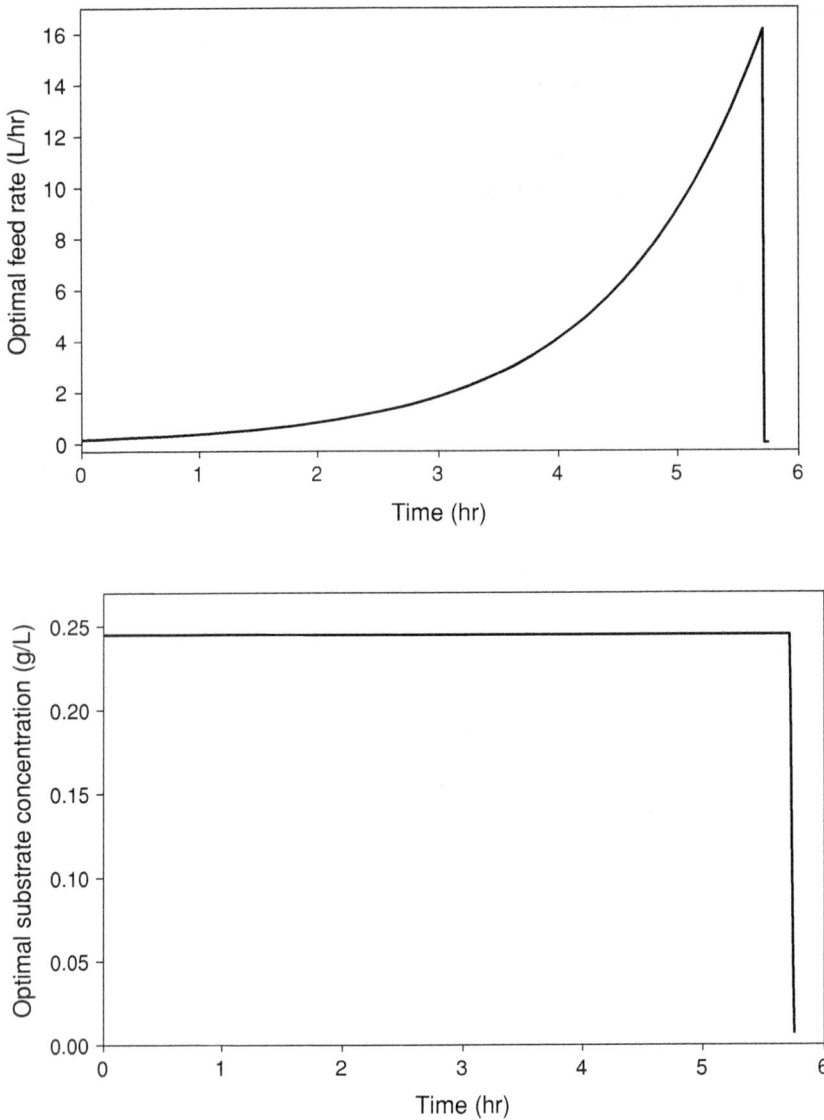

Figure 12.E.3.1. Optimal profiles, state and adjoint variables, switching function, and feed rate.

At the final time at which the volume constraint is met so that $F = 0$, the Hamiltonian is

$$H(t_f) = [\mu(t_f)/t_f - \lambda_3(t_f)\sigma(t_f)]x_1(t_f) - x_1(t_f)/t_f^2 = 0$$

$$\Rightarrow t_f = \frac{Y_{X/S}(t_f)}{2\lambda_3(t_f)} \pm \sqrt{\frac{Y_{X/S}(t_f)}{4\lambda_3(t_f)} - \frac{1}{\sigma(t_f)}} \qquad (12.175)$$

Thus, the final time is finite.

12.3.2.1 Feed Rate Profile on Singular Arc

Because the switching function and the adjoint equations for maximization of cell productivity (variable cell mass yield) (Eqs. (12.142) and (12.144)) are functionally the same as those for maximization of the cell mass (variable cell mass yield) (Eqs. (12.15) and (12.16)), we can conclude the same structural results for the maximization of cell productivity as for the maximization of cell mass. The singular control is obtained by differentiating the switching function, Eq. (12.142), as before:

$$\frac{d\phi}{dt} = S_F \frac{d\lambda_2}{dt} + \frac{d\lambda_3}{dt} = 0 \tag{12.176}$$

Substitution of Eq. (12.144) into Eq. (12.176) yields

$$\frac{d\phi}{dt} = \left(S_F - \frac{x_2}{x_3}\right)\frac{x_1}{x_3}(\lambda_1\mu' - \lambda_2\sigma') = 0 \tag{12.177}$$

The only way Eq. (12.177) can be satisfied is if

$$\lambda_1\mu' - \lambda_2\sigma' = 0 \tag{12.178}$$

The singular flow rate is obtained by setting to zero the second derivative of the switching function, or equivalently, the derivative of Eq. (12.178):

$$0 = \frac{d^2\phi}{dt^2} = \frac{d}{dt}(\lambda_1\mu' - \lambda_2\sigma') = -(\lambda_1\mu - \lambda_2\sigma)\mu' + (\lambda_1\mu'' - \lambda_2\sigma'')\frac{dS}{dt} \tag{12.179}$$

Because this is the same expression as that for cell mass maximization (fixed t_f and variable yield), the singular feed rate is the same as Eq. (12.88):

$$F_{\sin} = \frac{\sigma XV - \left[\dfrac{\mu'(\mu\sigma' - \mu'\sigma)}{\mu''\sigma' - \mu'\sigma''}\right]V}{(S_F - S)} = \frac{V}{(S_F - S)}\left\{\sigma X - \frac{\mu'Y'(\sigma/\sigma')^2}{(\mu'/\sigma')'}\right\} \tag{12.180}$$

The adjoint equations reduce to

$$\begin{bmatrix} d\lambda_1/dt \\ d\lambda_2/dt \\ d\lambda_3/dt \end{bmatrix} = -\begin{bmatrix} \lambda_1\mu - \lambda_2\sigma \\ 0 \\ 0 \end{bmatrix} \tag{12.181}$$

Therefore, the adjoint variables λ_2 and λ_3 are constants on the singular arc.

We note here that the adjoint equations and the switching functions for maximum cell productivity for the case of a variable-yield coefficient are identical but the boundary conditions on the adjoint variables are different than those of the cell mass production. Therefore, the numerical values of switching times for the maximum productivity would be different than those of the maximum cell production. Consequently, as stated earlier, the solution for the maximization of cell productivity can be obtained by repeated maximization of cell mass at various fixed final times, calculating the corresponding cell mass productivities by dividing the total cell mass obtained by the final time and picking the one that gives the maximum productivity.

EXAMPLE 12.E.4: EXAMPLE OF CELLULAR PRODUCTIVITY MAXIMIZATION FOR PROCESSES WITH VARIABLE-YIELD COEFFICIENTS Maximize the cellular productivity of the following fed-batch process with a variable-yield coefficient:

$$\mu = Se^{-S}, \quad \sigma = Se^{-S}(S^2 - S + 1), \quad Y_{X/S} = 1/(S^2 - S + 1) \tag{12.E.4.1}$$

The specific growth rate is nonmonotonic so that the maximum rate occurs at the substrate concentration of $S = 1$ and the yield coefficient decreases monotonically with the substrate concentration, peaking at near zero. The initial conditions are

$$X(0) = 1.5, \quad S(0) = 0, \quad V(0) = 100 \text{ L}$$

The operational conditions are

$$S_F = 3 \text{ g/L}, \quad F_{\max} = 70 \text{ L/hr}, \quad V_{\max} = 300 \text{ L}$$

As a solution, the mass balance equation is given by Eq. (12.139):

$$\frac{d}{dt}\begin{pmatrix} x_1 \\ x_2 \\ x_3 \\ x_4 \end{pmatrix} = \begin{pmatrix} \mu x_1 \\ -\sigma x_1 \\ 0 \\ 1 \end{pmatrix} + \begin{pmatrix} 0 \\ S_F \\ 1 \\ 0 \end{pmatrix} F = \frac{d}{dt}\begin{pmatrix} XV \\ SV \\ V \\ t \end{pmatrix} = \begin{pmatrix} Se^{-S}x_1 \\ -Se^{-S}(S^2 - S + 1)x_1 \\ 0 \\ 1 \end{pmatrix} + \begin{pmatrix} 0 \\ 3 \\ 1 \\ 0 \end{pmatrix} F$$

(12.E.4.2)

where $x_1 = XV$, $x_2 = SV$, $x_3 = V$, and $x_4 = t$. The objective is to maximize the performance index, the cell mass productivity,

$$\text{Max}[P = X(t_f)V(t_f)/t_f = x_1(t_f)/x_4(t_f)]$$

(12.E.4.3)

by manipulating the feed rate. The Hamiltonian is

$$H = (\lambda_1 \mu - \lambda_2 \sigma)x_1 + \lambda_4 + (S_F\lambda_2 + \lambda_3)F = \left\{ \lambda_1 \frac{x_2}{x_3}e^{-x_2/x_3} - \lambda_2 \frac{x_2}{x_3}e^{-x_2/x_3}\left[\left(\frac{x_2}{x_3}\right)^2 \right.\right.$$

$$\left.\left. - \frac{x_2}{x_3} + 1 \right]\right\} x_1 + \lambda_4 + (S_F\lambda_2 + \lambda_3)F \triangleq H_1 + \phi F = 0$$

(12.E.4.4)

and the switching function is

$$\phi = 3\lambda_2 + \lambda_3$$

(12.E.4.5)

Because the Hamiltonian is linear in F, the feed flow rate can take on the maximum, minimum, singular, and zero values, depending on the sign of the switching function, ϕ:

$$F = \begin{cases} F_{\max} & \text{when } \phi = (3\lambda_2 + \lambda_3) > 0 \\ F_{\min} = 0 & \text{when } \phi = (3\lambda_2 + \lambda_3) < 0 \\ F_{\text{sin}} & \text{when } \phi = (3\lambda_2 + \lambda_3) = 0 \ t_q < t < t_{q+1} \\ F_b = 0 & \text{when } x_3(t) = V_{\max} = 300 \end{cases}$$

(12.E.4.6)

The adjoint vector must satisfy the following equation:

$$\frac{d\lambda}{dt} = \begin{bmatrix} d\lambda_1/dt \\ d\lambda_2/dt \\ d\lambda_3/dt \\ d\lambda_4/dt \end{bmatrix} = -\frac{\partial H}{\partial \mathbf{x}} = -\begin{bmatrix} \lambda_1 \mu - \lambda_2 \sigma \\ (\lambda_1 \mu' - \lambda_2 \sigma')x_1/x_3 \\ -(\lambda_1 \mu' - \lambda_2 \sigma')x_1 x_2/x_3^2 \\ 0 \end{bmatrix}$$

(12.E.4.7)

where $\mu' = e^{-S}(1 - S)$ and $\sigma' = e^{-S}(1 - 3S + 4S^2)$. The boundary conditions are obtained from the transversality conditions:

$$\boldsymbol{\lambda}(t_f) = \frac{\partial P}{\partial \mathbf{x}}(t_f) = \begin{bmatrix} \lambda_1(t_f) \\ \lambda_2(t_f) \\ \lambda_3(t_f) \\ \lambda_4(t_f) \end{bmatrix} = \begin{bmatrix} 1/x_4(t_f) = 1/t_f \\ 0 \\ \text{unknown constant} \\ -x_1(t_f)/x_4^2(t_f) \end{bmatrix} \quad (12.E.4.8)$$

From Eqs. (12.E.4.7) and (12.E.4.8), we note that $\lambda_4(t)$ is a constant:

$$\lambda_4(t) = \lambda_4(t_f) = -x_1(t_f)/x_4^2(t_f) = -X(t_f)V(t_f)/t_f^2 \quad (12.E.4.9)$$

Therefore, the Hamiltonian is

$$H = (\lambda_1 \mu - \lambda_2 \sigma)x_1 + (3\lambda_2 + \lambda_3)F - x_1(t_f)/x_4^2(t_f) \quad (12.E.4.10)$$

Therefore, the adjoint equations (Eq. (12.E.4.7)) reduce to

$$\begin{bmatrix} d\lambda_1/dt \\ d\lambda_2/dt \\ d\lambda_3/dt \end{bmatrix} = - \begin{bmatrix} \lambda_1 \mu - \lambda_2 \sigma \\ (\lambda_1 \mu' - \lambda_2 \sigma')x_1/x_3 \\ -(\lambda_1 \mu' - \lambda_2 \sigma')x_1 x_2/x_3^2 \end{bmatrix} \quad (12.E.4.11)$$

This is identical to the case of cell mass maximization (Eq. (12.16)). The boundary conditions (Eq. (12.E.4.8)) reduce to

$$\begin{bmatrix} \lambda_1(t_f) \\ \lambda_2(t_f) \\ \lambda_3(t_f) \end{bmatrix} = \begin{bmatrix} 1/x_4(t_f) = 1/t_f \\ 0 \\ \text{unknown constant} \end{bmatrix} \quad (12.E.4.12)$$

The singular control is obtained by differentiating the switching function, Eq. (12.E.4.5), as before:

$$\frac{d\phi}{dt} = 3\frac{d\lambda_2}{dt} + \frac{d\lambda_3}{dt} = 0 \quad (12.E.4.13)$$

Substitution of Eq. (12.E.4.11) into Eq. (12.E.4.13) yields

$$\frac{d\phi}{dt} = \left(3 - \frac{x_2}{x_3}\right)\frac{x_1}{x_3}(\lambda_1 \mu' - \lambda_2 \sigma') = 0 \quad (12.E.4.14)$$

The only way Eq. (12.E.4.14) can be satisfied is if

$$\lambda_1 \mu' - \lambda_2 \sigma' = 0 \quad (12.E.4.15)$$

The singular flow rate is obtained by setting to zero the second derivative of the switching function, or equivalently, the derivative of Eq. (12.E.4.14):

$$0 = \frac{d^2\phi}{dt^2} = \frac{d}{dt}(\lambda_1 \mu' - \lambda_2 \sigma') = -(\lambda_1 \mu - \lambda_2 \sigma)\mu' + (\lambda_1 \mu'' - \lambda_2 \sigma'')\frac{dS}{dt} \quad (12.E.4.16)$$

Because this is the same expression as that for cell mass maximization (fixed t_f and variable yield), the singular feed rate is the same as Eq. (12.88):

$$F_{\text{sin}} = \frac{\sigma X V - \left[\dfrac{\mu'(\mu\sigma' - \mu'\sigma)}{\mu''\sigma' - \mu'\sigma''}\right]V}{(S_F - S)} = \frac{V}{(S_F - S)}\left\{\sigma X - \frac{\mu'Y'(\sigma/\sigma')^2}{(\mu'/\sigma')'}\right\} \quad (12.E.4.17)$$

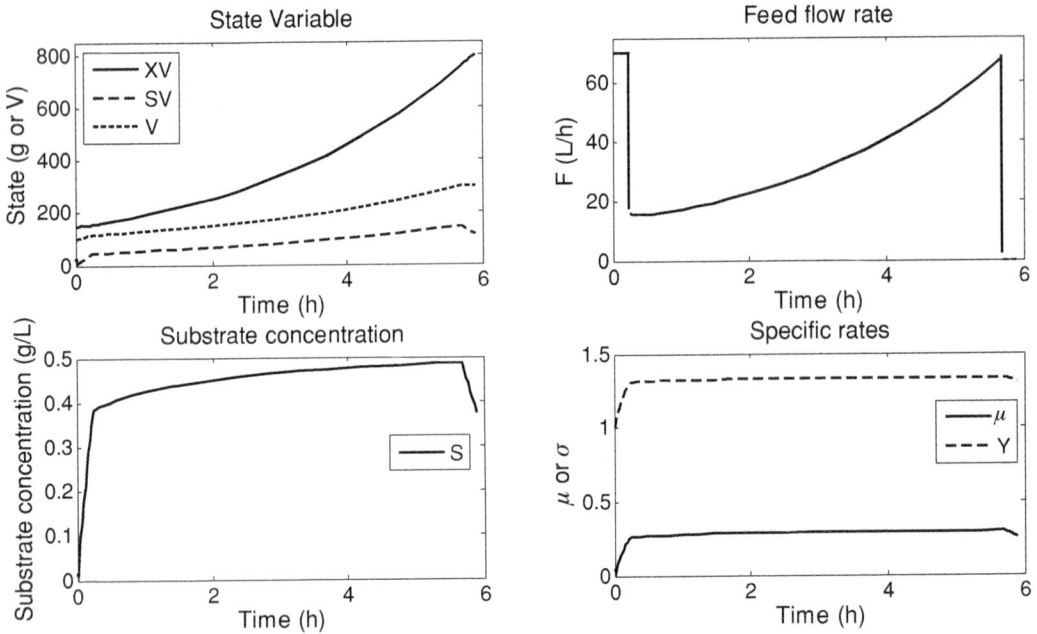

Figure 12.E.4.1. Optimal time profiles for maximum cellular mass productivity. Amounts of cell mass and substrate and culture volume, feed rate profile, substrate concentration, and specific rates.

The adjoint equations reduce to

$$\begin{bmatrix} d\lambda_1/dt \\ d\lambda_2/dt \\ d\lambda_3/dt \end{bmatrix} = - \begin{bmatrix} \lambda_1\mu - \lambda_2\sigma \\ 0 \\ 0 \end{bmatrix} \qquad (12.\text{E}.4.18)$$

Therefore, the adjoint variables λ_2 and λ_3 are constants on the singular arc.

Using the numerical procedure detailed in Chapter 11, this example problem is optimized. The results are shown in Figure 12.E.4.1, where various species time profiles are shown. The feed rate profile is $F_{\max} \rightarrow F_{\sin} \rightarrow F_b = 0$. The maximum cellular productivity obtained was 124.2 g/hr, the final cell mass was 794.3 g, and the final time was 6.43 hrs. During the singular feed rate period, the substrate concentration is varied to maximize a weighted sum of the specific growth rate and the yield coefficient.

12.4 Cell Mass Productivity Maximization through Time Optimal Formulation

It is instructive to consider an alternate route to maximize the cellular productivity by determining the minimum time to obtain a fixed amount of cell mass, that is, *a minimum time problem*. The objective is to produce a specified amount of cell mass $V(t_f)X(t_f)$ in the shortest time possible,

$$\underset{F^*(t)}{\text{Min}}\, \text{P} = t_f \quad \Rightarrow \underset{F^*(t)}{\text{Max}}(\text{P} = -t_f) \qquad (12.182)$$

by manipulating the feed flow rate. To express the performance index in terms of final state variables, we introduce, as before, the fourth state variable: $x_4(t) = t$; $dx/dt = 1, x_4(0) = 0$.

When the cell mass yield is constant, specifying the final amount of cell mass fixes also the terminal amount of substrate. In addition, the reactor volume is known, and therefore, specifying the final amount of cell mass fixes also the final substrate and therefore the state (see Eq. (12.13)). Thus, the final target is a point in three-dimensional space:

$$\begin{pmatrix} x_1(0) = X(0)V(0) \\ x_2(0) = S(0)V(0) \\ x_3(0) = V(0) \\ x_4(0) = 0 \end{pmatrix} \rightarrow \rightarrow \begin{pmatrix} x_1(t_f) = X(t_f)V(t_f) \\ x_2(t_f) = S(t_f)V(t_f) \\ x_3(t_f) = V(t_f) \\ x_4(t_f) = \text{free} \end{pmatrix} \quad (12.183)$$

The final conditions on the corresponding adjoint variables are obtained from the transversality condition (Eq. (9.27)):

$$\delta P = -1\delta x_4(t_f) = \lambda_1(t_f)\underbrace{\delta x_1(t_f)}_{=0} + \lambda_2(t_f)\underbrace{\delta x_2(t_f)}_{=0} + \lambda_3(t_f)\underbrace{\delta x_3(t_f)}_{=0} + \underline{\lambda_4(t_f)\delta x_4(t_f)}$$

$$\begin{pmatrix} \lambda_1(t_f) = \text{unknown constant} \\ \lambda_2(t_f) = \text{unknown constant} \\ \lambda_3(t_f) = \text{unknown constant} \\ \lambda_4(t_f) = -1 \end{pmatrix} \quad (12.184)$$

However, when the yield is variable, the final amount of substrate is not fixed by the final amount of cell mass and must be chosen properly to yield a solution, that is, the terminal state must be reachable from the initial state and the amount of substrate that goes into the bioreactor. Thus, when the yield coefficient is a function of substrate concentration and the final amount of substrate is left free (thus, the final state is partially fixed), that is, the target is a set,

$$\begin{pmatrix} x_1(0) = X(0)V(0) \\ x_2(0) = S(0)V(0) \\ x_3(0) = V(0) \\ x_4(0) = 0 \end{pmatrix} \rightarrow \rightarrow \begin{pmatrix} x_1(t_f) = X(t_f)V(t_f) \\ x_2(t_f) \leq \delta(\text{undefined}) \\ x_3(t_f) = V(t_f) \\ x_4(t_f) = \text{unknown constant} \end{pmatrix} \quad (12.185)$$

In this case, the final values of the adjoint variables are

$$\begin{pmatrix} \lambda_1(t_f) = \text{unknown constant} \\ \lambda_2(t_f) = \text{unknown constant} \\ \lambda_3(t_f) = \text{unknown constant} \\ \lambda_4(t_f) = -1 \end{pmatrix} \quad (12.186)$$

As before, we consider first the case of a constant-yield coefficient, starting with fixed final conditions.

12.4.1 Constant Cell Mass Yield Coefficient and Fixed Final Conditions

For a constant cell yield, we recall that only two state variables are independent. To handle the final time that appears in the performance index, we introduced an auxiliary state variable, $x_4 (dx_4/dt = 1, x_4(0) = 0)$, to work with a third-order\break system:

$$\frac{d\mathbf{x}}{dt} = \frac{d}{dt} \begin{pmatrix} x_1 \\ x_3 \\ x_4 \end{pmatrix} = \frac{d}{dt} \begin{pmatrix} XV \\ V \\ t \end{pmatrix} = \begin{pmatrix} \mu x_1 \\ 0 \\ 1 \end{pmatrix} + \begin{pmatrix} 0 \\ 1 \\ 0 \end{pmatrix} F$$

$$\begin{aligned} x_1(0) &= X(0)V(0) & x_1(t_f) &= x_{1f} \\ x_3(0) &= V(0) & x_3(t_f) &= V_{\max} \\ x_4(0) &= 0 & x_4(t_f) &= t_f \end{aligned} \qquad (12.187)$$

Note that the final state is specified completely; that is, this is a fixed point problem. As usual, the Hamiltonian is

$$H = \lambda_1 \mu x_1 + \lambda_3 F + \lambda_4 \qquad (12.188)$$

and the adjoint variables must satisfy the following:

$$\frac{d\boldsymbol{\lambda}}{dt} = \frac{d}{dt} \begin{bmatrix} \lambda_1 \\ \lambda_3 \\ \lambda_4 \end{bmatrix} = -\frac{\partial H}{\partial \mathbf{x}} = - \begin{bmatrix} \lambda_1 \mu + \lambda_1 x_1 \partial \mu / \partial x_1 \\ \lambda_1 x_1 \partial \mu / \partial x_3 \\ 0 \end{bmatrix} = - \begin{bmatrix} \lambda_1 \mu - \lambda_1 x_1 \mu' / x_3 Y_{X/S} \\ \lambda_1 x_1 \mu' (S_F - g/x_3)/x_3 \\ 0 \end{bmatrix}$$

$$(12.189)$$

The performance index is

$$\text{Max}[P = -x_4(t_f)] \qquad (12.190)$$

The final conditions are obtained from the transversality condition:

$$\delta \mathrm{P} = -\delta x_4(t_f) = \sum_{i=1,3,4} \lambda_i(t_f)\delta x_i(t_f) - \sum_{i=1,3,4} \lambda_i(0)\,\delta x_i(0)^{=0}$$

or

$$-1\delta x_4(t_f) = \lambda_1(t_f)\,\partial x_1(t_f)^{=0} + \lambda_3(t_f)\,\delta x_3(t_f)^{=0} + \lambda_4(t_f)\delta x_4(t_f) \qquad (12.191)$$

Therefore,

$$\begin{bmatrix} \lambda_1(t_f) \\ \lambda_3(t_f) \\ \lambda_4(t_f) \end{bmatrix} = \begin{bmatrix} \text{unknown constant} \\ \text{unknown constant} \\ -1 \end{bmatrix} \qquad (12.192)$$

We deduce from Eqs. (12.189) and (12.192) that λ_4 is -1:

$$\lambda_4(t) = \lambda_4(t_f) = -1 \qquad (12.193)$$

Thus, the Hamiltonian is

$$H = \lambda_1 \mu XV + \lambda_3 F - 1 \qquad (12.194)$$

Note that the Hamiltonian for this minimum time problem differs from that of cell mass maximization by only a constant, -1. Therefore, the structure of the optimal control for the minimum time problem remains the same as that for the maximum cell mass problem.

12.4.1.1 Feed Rate Profile on Singular Arc

The switching function is $\phi = \lambda_3$, and therefore, the feed rate depends on λ_3:

$$F = \begin{cases} F_{max} & \text{when } \phi = \lambda_3(t) > 0 \\ F_{min} & \text{when } \phi = \lambda_3(t) < 0 \\ F_{sin} & \text{when } \phi = \lambda_3(t) = 0 \text{ over a finite time interval, } t_q < t < t_{q+1} \\ F_b = 0 & \text{when } x_3(t) = V_{max} \end{cases}$$

(12.195)

As usual, the singular flow rate is obtained from $\phi = 0$, $\dot{\phi} = 0$, and $\ddot{\phi} = 0$:

$$\phi = \lambda_3 = 0$$
$$\frac{d\phi}{dt} = \frac{d\lambda_3}{dt} = -\lambda_1 x_1 \frac{\partial \mu}{\partial x_3} = 0$$

(12.196)

Because $x_1 \neq 0$ and $\lambda_1 \neq 0$ (because $\lambda_3 = 0$, both adjoint variables cannot be zero, otherwise, the problem becomes trivial), therefore,

$$0 = \frac{\partial \mu}{\partial x_3} = \frac{\partial \mu}{\partial S} \frac{\partial S}{\partial x_3} = \frac{\partial \mu}{\partial S} \frac{\partial (x_2/x_3)}{\partial x_3} = \mu' \frac{(S_F x_3 - x_2)}{x_3^2} = \mu' \frac{(S_F - S)}{V}$$

(12.197)

Because $(S_F - S)/V \neq 0$, the specific growth rate must be maximized:

$$\mu' = \frac{d\mu}{dS} = 0$$

(12.198)

Thus, during the singular period, the specific growth rate must be maximized, the same conclusion we obtained for the case of free final time cell mass maximization. The feed rate to maintain the substrate concentration that maximizes the specific growth rate, S_m, is obtained from the substrate balance equation, as in the case of maximization of cell mass with constant cell yield:

$$F_{sin} = \frac{\mu(S_m)XV}{(S_F - S_m)Y_{X/S}} = \alpha XV$$

(12.199)

On the singular arc, the adjoint equations are reduced to

$$\frac{d}{dt} \begin{bmatrix} \lambda_1 \\ \lambda_3 \\ \lambda_4 \end{bmatrix} = - \begin{bmatrix} \lambda_1 \mu + \lambda_1 x_1 \partial \mu / \partial x_1 \\ \lambda_1 x_1 \partial \mu / \partial x_3 \\ 0 \end{bmatrix} = - \begin{bmatrix} \lambda_1 \mu - \lambda_1 x_1 \mu' / (Y_{X/S} x_3) \\ 0 \\ 0 \end{bmatrix} = - \begin{bmatrix} \lambda_1 \mu_m \\ 0 \\ 0 \end{bmatrix}$$

(12.200)

Therefore, on the singular arc, λ_3 and λ_4 are constants, whereas λ_1 is an exponential function:

$$\lambda_1(t) = \lambda_1(t_s) \exp[-\mu_m(t - t_s)]$$

(12.201)

The construction of the feed flow rate profile is the same as before. One must choose the initial conditions, the substrate concentration, to coincide with that at which the specific growth rate is at maximum and maintain the concentration during the singular period until the bioreactor is full. If the initial substrate cannot be chosen at will, then the substrate concentration should be brought as fast as possible to match the concentration at which the specific growth rate is maximum.

12.4.1.2 Optimum Feed Rate Profile

When the initial and final states are fixed, the two state equations, dx_1/dt and dx_2/dt (dx_4/dt is independent of the feed rate), can be integrated using a sequence of feed rates; a period of maximum (or minimum $= 0$) flow from 0 to t_s, followed by the singular feed rate of Eq. (12.199) until the reactor is full, t_{full}, and finally, a batch period until the final condition(s) is met, t_f. This results in two unknown switching times, t_s and t_{full}, in two equations $(x_1(t_f) = x_{1f}, x_2(t_f) = V_{max})$. Thus, we can determine the two switching times. A degenerate case arises when the initial conditions lie on the singular arc so that the singular feed rate is applied from the initial time and the two equations lead to only one switching from the singular feed to no feed (batch).

12.4.1.3 A Target Point versus a Target Set

In the preceding development for the target point, the final conditions on the adjoint variables were used explicitly. Thus, the solution to a fixed point is also applicable to the problem of a fixed set. This is expected because for constant yield, fixing the final amount of cell mass and final volume also fixes the final amount of substrate through the mass balance equation (12.13). Therefore, fixing the final values of any two state variables is equivalent to fixing all three state variables. Thus, for constant yields, the solution to the fixed final point is also the solution for the fixed final set. When the yield coefficient is variable, Eq. (12.22) cannot be used, and therefore, fixing any two state variables at the final time is not equivalent to setting all three state variables.

Therefore, we see that the problems of cell mass maximization, cell mass production rate, and minimum time are all related. Repeated cell mass maximization at various fixed final times can be used to obtain the solutions for the maximum cell production rate and also for the minimum time problem. By repeatedly solving the minimum time problem at various amounts of final cell mass, one can obtain the solution for the maximum cell mass production rate. One could have predicted this outcome by inspection of the objective of each problem:

For cell mass maximization, Max $[X(t_f)V(t_f)]$ at various final times t_f

For minimum time, Max $[-t_f$, final time] at various final cell mass, $X(t_f)V(t_f)$

For maximum cell mass productivity, Max $[X(t_f)V(t_f)/t_f]$

EXAMPLE 12.E.5: EXAMPLE OF MINIMUM TIME FOR PROCESSES WITH CONSTANT-YIELD COEFFICIENTS Consider a process with the following kinetic information:

$$\mu(S) = S/(0.03 + S + 0.5S^2), \quad Y_{X/S} = 0.5 = \mu/\sigma$$

The initial conditions are $[XV, SV, V](t = 0) = [2\,\text{g}, 2\,\text{g}, 2\,\text{L}]$, and the operational conditions are $S_F = 10\,\text{g/L}$, $F_{max} = 15\,\text{L/hr}$, and $V_{max} = 20\,\text{L}$. The objective is to achieve the process final conditions:

$$[XV, SV, V](t = t_f) = [91\,\text{g}, 4.14\,\text{g}, 20\,\text{L}]$$

in the minimum time possible.

As presented previously, for the case of a constant cell yield, only two state variables are independent, and we introduced an auxiliary state variable to handle the final time that appears in the performance index, x_4: $(dx_4/dt = 1, x_4(0) = 0)$,

$$\frac{d\mathbf{x}}{dt} = \frac{d}{dt}\begin{pmatrix} x_1 \\ x_3 \\ x_4 \end{pmatrix} = \frac{d}{dt}\begin{pmatrix} XV \\ V \\ t \end{pmatrix} = \begin{pmatrix} \mu x_1 \\ 0 \\ 1 \end{pmatrix} + \begin{pmatrix} 0 \\ 1 \\ 0 \end{pmatrix} F \qquad \begin{array}{ll} x_1(0) = 2\,\text{g} & x_1(t_f) = 91\,\text{g} \\ x_3(0) = 2\,\text{L} & x_3(t_f) = 20\,\text{L} \\ x_4(0) = 0 & x_4(t_f) = \text{free} \end{array}$$

$$(12.E.5.1)$$

The Hamiltonian is

$$H = \lambda_1 \mu x_1 + \lambda_3 F + \lambda_4 \qquad (12.E.5.2)$$

where the adjoint variables must satisfy the following:

$$\frac{d\boldsymbol{\lambda}}{dt} = \frac{d}{dt}\begin{bmatrix} \lambda_1 \\ \lambda_3 \\ \lambda_4 \end{bmatrix} = -\frac{\partial H}{\partial \mathbf{x}} = -\begin{bmatrix} \lambda_1 \mu - \lambda_1 x_1 \mu'/(x_3 Y_{X/S}) \\ \lambda_1 x_1 \mu'(S_F - g/x_3)/x_3 \\ 0 \end{bmatrix} \qquad (12.E.5.3)$$

where

$$g(x_1, x_3) = SV = x_2 - x_{20} + S_F(x_3 - x_{30}) - \frac{1}{Y_{X/S}}(x_1 - x_{10})$$

$$= 2 + 15(x_3 - 2) - 2(x_1 - 2) \qquad (12.E.5.4)$$

The performance index is

$$\text{Max}[P = -x_4(t_f)] \qquad (12.E.5.5)$$

The final conditions are obtained from the transversality condition:

$$\begin{bmatrix} \lambda_1(t_f) \\ \lambda_3(t_f) \\ \lambda_4(t_f) \end{bmatrix} = \begin{bmatrix} \text{unknown constant} \\ \text{unknown constant} \\ -1 \end{bmatrix} \qquad (12.E.5.6)$$

We deduce from Eqs. (12.E.5.3) and (12.E5.6) that λ_4 is –1:

$$\lambda_4(t) = \lambda_4(t_f) = -1 \qquad (12.E.5.7)$$

Thus, the Hamiltonian is

$$H = \lambda_1 \mu XV + \lambda_3 F - 1 \qquad (12.E.5.8)$$

As noted earlier, the structure of the optimal control for the minimum time problem remains the same as that for the maximum cell mass problem:

$$F = \begin{cases} F_{\max} & \text{when } \phi = \lambda_3(t) > 0 \\ F_{\min} & \text{when } \phi = \lambda_3(t) < 0 \\ F_{\sin} & \text{when } \phi = \lambda_3(t) = 0 \text{ over a finite time interval, } t_q < t < t_{q+1} \\ F_b = 0 & \text{when } x_3(t) = V_{\max} \end{cases}$$

$$(12.E.5.9)$$

From Eq. (12.E5.3), the adjoint equations are

$$\frac{d\lambda_1}{dt} + [\mu - 2\mu' x_1/x_3]\lambda_1 = 0, \quad \lambda_1(t_f) = a_1(\text{unknown}) \qquad (12.E.5.10)$$

$$\frac{d\lambda_3}{dt} + \lambda_1 x_1 \mu'(10 - g/x_3)/x_3 = 0, \quad \lambda_3(t_f) = a_3(\text{unknown}) \qquad (12.E.5.11)$$

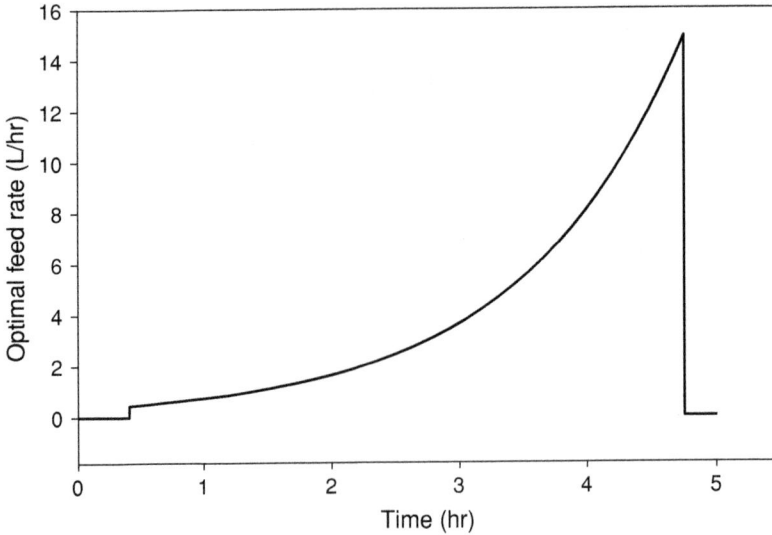

Figure 12.E.5.1. Optimal feed rate profile for minimum time problem.

where

$$\mu(S) = S/(0.03 + S + 0.5S^2) = x_2 x_3/(0.03x_3^2 + x_2 x_3 + 0.5x_2^2),$$
$$\mu' = \mu^2(0.03x_3 - 0.5x_2)/x_2$$

and the state equations are

$$\frac{dx_1}{dt} = \mu x_1, x_1(0) = 2, x_2(t_f) = 91 \qquad (12.E.5.12)$$

$$\frac{dx_3}{dt} = F, \quad x_3(0) = 2, x_3(t_f) = 20 \qquad (12.E.5.13)$$

Using the shooting method described in detail in Chapter 11, the adjoint equations, Eqs. (12.E.5.10) and (12.E.5.11), and the state equations, Eqs. (12.E.5.12) and (12.E.5.13), are solved simultaneously to obtain a numerical solution of the minimum time problem to obtain the results.

The results are given in Figures 12.E.5.1 and 12.E.5.2. The optimal feed rate profile is shown in Figure 12.E.5.1. The optimal feed rate consists of 0.405 hrs of batch period, $F_{min} = 0$, followed by a period of singular feed rate (almost exponential feed rate), which lasts until 4.75 hrs, when the bioreactor volume is full (20 L), and is finally followed by a batch period (F_b) until the final time of 5 hrs.

Figure 12.E.5.2 shows the substrate concentration profile. During the initial batch period until $t = 0.405$ hrs, the substrate concentration decreases as no feed is applied, and on the singular arc, the substrate concentration remains constant at $S = 0.06$ g/L, corresponding to the maximum specific growth rate.

It ought to be apparent here that the initial substrate concentration should have been chosen at $S = 0.06$ g/L so that the initial batch period would have been eliminated and the feed rate would be a singular rate for the entire time period until the reactor was full. In this way, the specific growth rate would have been maximized during the entire period of filling the bioreactor.

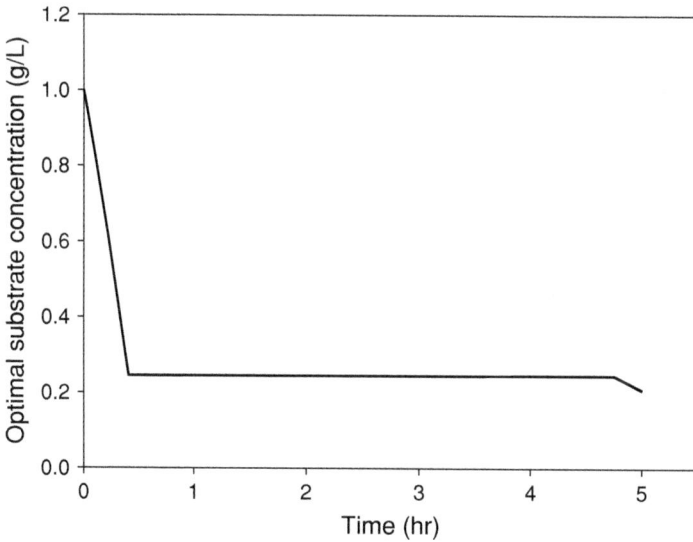

Figure 12.E.5.2. Optimal substrate concentration profile for minimum time problem.

12.4.2 Variable Cell Mass Yield Coefficient and Fixed Final Conditions

When the yield coefficient is not constant, all three mass balance equations are independent. Adding an auxiliary state variable x_4, we have a fourth-order state equation:

$$\frac{d\mathbf{x}}{dt} = \mathbf{a}(\mathbf{x}) + \mathbf{b}F, \quad \frac{d}{dt}\begin{pmatrix} x_1 \\ x_2 \\ x_3 \\ x_4 \end{pmatrix} = \frac{d}{dt}\begin{pmatrix} XV \\ SV \\ V \\ t \end{pmatrix} = 0\begin{pmatrix} \mu x_1 \\ -\sigma x_1 \\ 1 \\ 1 \end{pmatrix} + \begin{pmatrix} 0 \\ S_F \\ 1 \\ 0 \end{pmatrix}F \quad (12.139)$$

The objective is to produce a specified amount of cell mass in the shortest time possible:

$$\underset{F^*(t)}{\text{Max}}[\mathbf{P} = (-t_f) = -x_4(t_f)] \quad (12.190)$$

The Hamiltonian is

$$\begin{aligned} H &= \lambda_1\mu x_1 + \lambda_2(S_F F - \sigma x_1) + \lambda_3 F + \lambda_4 \\ &= (\lambda_1\mu - \lambda_2\sigma)x_1 + \lambda_4 + (S_F\lambda_2 + \lambda_3)F \triangleq H_1 + \phi F = 0 \end{aligned} \quad (12.202)$$

where

$$H_1 = (\lambda_1\mu - \lambda_2\sigma)x_1 + \lambda_4, \quad \phi = S_F\lambda_2 + \lambda_3 \quad (12.203)$$

and the adjoint equations are

$$\begin{bmatrix} d\lambda_1/dt \\ d\lambda_2/dt \\ d\lambda_3/dt \\ d\lambda_4/dt \end{bmatrix} = -\begin{bmatrix} \partial H/\partial x_1 \\ \partial H/\partial x_2 \\ \partial H/\partial x_3 \\ \partial H/\partial x_4 \end{bmatrix} = -\begin{bmatrix} \lambda_1\mu - \lambda_2\sigma \\ (\lambda_1\mu' - \lambda_2\sigma')x_1/x_3 \\ -(\lambda_1\mu' - \lambda_2\sigma')x_1 x_2/x_3^2 \\ 0 \end{bmatrix} \quad (12.204)$$

The final conditions on the adjoint variables are determined from

$$\delta \mathrm{P} = \underline{-1} \delta x_4(t_f) = \sum_{i=1}^{4} \lambda_i(t_f)\delta x_i(t_f) - \sum_{i=1}^{4} \lambda_i(0)\cancel{\delta x_i(0)}^{0}$$

$$= \lambda_1(t_f)\cancel{\delta x_1(t_f)}^{0} + \lambda_2(t_f)\delta x_2(t_f) + \lambda_3(t_f)\cancel{\delta x_3(t_f)}^{0} + \underline{\lambda_4(t_f)\delta x_4(t_f)} \qquad (12.205)$$

where $\delta x_i(0) = 0$ ($i = 1, 2, 3, 4$) because all the initial state variables are specified. Because $x_1(t_f)$ and $x_2(t_f)$ are specified, $\delta x_1(t_f) = 0$ and $\delta x_2(t_f) = 0$, and $\delta x_3(t_f)$ is zero as $x_3(t_f) = V_{\max}$ is specified. Therefore, the final conditions on the adjoint variables for specified final state variables are

$$\begin{bmatrix} \lambda_1(t_f) \\ \lambda_2(t_f) \\ \lambda_3(t_f) \\ \lambda_4(t_f) \end{bmatrix} = \begin{bmatrix} \text{unknown constant} \\ \text{unknown constant} \\ \text{unknown constant} \\ \partial P/\partial x_4(t_f) = -1 \end{bmatrix} \qquad (12.206)$$

From Eqs. (12.204) and (12.206), we conclude that λ_4 is a constant:

$$\lambda_4(t) = \lambda_4(t_f) = -1 \qquad (12.207)$$

Therefore, the Hamiltonian, Eq. (12.202), reduces to

$$H = (\lambda_1\mu - \lambda_2\sigma)x_1 - 1 + (S_F\lambda_2 + \lambda_3)F = 0 \qquad (12.208)$$

12.4.2.1 Feed Rate Profile on a Singular Arc

Because the switching function is identically zero on the singular arc, the Hamiltonian and adjoint equations, Eqs. (12.208) and (12.204), are reduced to

$$H = (\lambda_1\mu - \lambda_2\sigma)x_1 - 1 = 0 \qquad (12.209)$$

$$\begin{bmatrix} \lambda_1(t_f) \\ \lambda_2(t_f) \\ \lambda_3(t_f) \end{bmatrix} = \begin{bmatrix} \text{unknown constant} \\ \text{unknown constant} \\ \text{unknown constant} \end{bmatrix} \qquad (12.210)$$

The switching function and its first time derivative are

$$\phi = S_F\lambda_2 + \lambda_3 = 0$$

$$\frac{d\phi}{dt} = S_F\frac{d\lambda_2}{dt} + \frac{d\lambda_3}{dt} = -(\lambda_1\mu' - \lambda_2\sigma')\frac{x_1}{x_3} - S_F(\lambda_1\mu' - \lambda_2\sigma')\frac{x_1x_2}{x_3^2}$$

$$= -\left[(\lambda_1\mu' - \lambda_2\sigma')\frac{x_1}{x_3}(S_F - S)\right] = 0 \qquad (12.211)$$

Because $(x_1/x_3)(S_F - S)] \neq 0$, we conclude that

$$\lambda_1\mu' - \lambda_2\sigma' = 0 \qquad (12.212)$$

This is exactly the same as Eq. (12.178) for the case of maximization of cellular productivity. Therefore, one anticipates the same solution here. Indeed, that is the case. The adjoint equation, Eq. (12.204), reduces to

$$
\begin{bmatrix} d\lambda_1/dt \\ d\lambda_2/dt \\ d\lambda_3/dt \end{bmatrix} = -\begin{bmatrix} \lambda_1\mu - \lambda_2\sigma \\ 0 \\ 0 \end{bmatrix}
\tag{12.213}
$$

The singular flow rate is obtained by setting to zero the second derivative of the switching function, or equivalently, the derivative of Eq. (12.212),

$$
\begin{aligned}
0 &= \frac{d}{dt}(\lambda_1\mu' - \lambda_2\sigma') = \frac{d\lambda_1}{dt}\mu' + \lambda_1\mu''\frac{dS}{dt} - \frac{d\lambda_2}{dt}\sigma' - \lambda_2\sigma''\frac{dS}{dt} \\
&= -\lambda_1\left[\left(\mu - \frac{\lambda_2}{\lambda_1}\sigma\right)\mu' + \left(\mu'' - \frac{\lambda_2}{\lambda_1}\sigma''\right)\frac{dS}{dt}\right]
\end{aligned}
\tag{12.214}
$$

which is the same as Eq. (12.179) in the case of maximum cellular productivity. Thus, the solution is the same in form as the maximization of cellular productivity, only differing in switching times as the boundary conditions are different. Substitute Eq. (12.212) into Eq. (12.214) to obtain

$$
\begin{aligned}
0 &= \left[\left(\mu - \frac{\lambda_2}{\lambda_1}\sigma\right)\mu' + \left(\mu'' - \frac{\lambda_2}{\lambda_1}\sigma''\right)\frac{dS}{dt}\right] \\
&= \left[\left(\frac{\sigma\mu' - \sigma'\mu}{\sigma'}\right)\mu' + \left(\frac{\sigma'\mu'' - \mu'\sigma''}{\sigma'}\right)\frac{dS}{dt}\right]
\end{aligned}
\tag{12.215}
$$

Because dS/dt is related to the flow rate through the substrate balance equation, we substitute Eq. (12.86) to obtain the singular feed rate

$$
F_{\sin} = \frac{\sigma XV + \left[\dfrac{\mu'(\mu\sigma' - \mu'\sigma)}{\mu''\sigma' - \mu'\sigma''}\right]V}{(S_F - S)} = \frac{V}{(S_F - S)}\left\{\sigma X - \frac{\mu'Y'(\sigma/\sigma')^2}{(\mu'/\sigma')'}\right\}
\tag{12.216}
$$

This singular feed rate we obtain for the time-optimal solution case (Eq. (12.216)) is exactly the same as we obtained for the case of cellular productivity maximization for variable yield (Eq. (12.180)). The question remaining is, is the feed rate profile that minimizes the time to obtain a fixed amount of cell mass identical to that which maximizes the productivity? The structure is identical, but numerically, these two solutions are not exactly the same. The difference is in the final conditions on the adjoint variables. Here they are not known, whereas the final conditions are partially known in the case of maximum cell mass productivity:

$$
\begin{pmatrix} \lambda_1(t_f)\text{free (unknown)} \\ \lambda_3(t_f)\text{free (unknown)} \\ \lambda_4(t_f)\text{free (unknown)} \end{pmatrix} \text{versus} \begin{bmatrix} \lambda_1(t_f) \\ \lambda_2(t_f) \\ \lambda_3(t_f) \end{bmatrix} = \begin{bmatrix} 1/t_f \\ 0 \\ \text{unknown constant} \end{bmatrix}
$$

The Hamiltonian for this problem of minimizing the time is

$$
H = (\lambda_1\mu - \lambda_2\sigma)x_1 + (S_F\lambda_2 + \lambda_3)F - 1
\tag{12.217}
$$

versus

$$
H = (\lambda_1\mu - \lambda_2\sigma)x_1 + (S_F\lambda_2 + \lambda_3)F + x_1(t_f)/t_f^2
\tag{12.218}
$$

for maximization of cellular productivity. These two show that they differ by constants only. The adjoint equations are the same (Eq. (12.172) vs. Eq. (12.144)).

Therefore, the solution to the minimum time with fixed final conditions can differ from that of the maximum cellular productivity; that is, the switching time from F_{max} (or F_{min}) to F_{sin}, or vice versa, may be different, as may be the final time.

When the yield coefficient is constant so that the process is described by two differential equations, Table 12.2 shows that the singular control is an exponential function and that the substrate concentration remains constant during the singular feed rate period. However, when the yield coefficient is a function of substrate concentration so that the process is described by three differential equations, the singular flow rate is not exponential, and the substrate concentration varies with time during the period.

12.5 Specific Rates as Functions of Substrate and Cell Concentrations

Specific rates are sometimes found to depend not only on substrate concentrations but also on product concentrations. Some metabolites are known to inhibit specific rates; for example, ethanol is known to inhibit μ, σ, and π. It has been reported that high cell concentrations of fungi lead to slow growth rates, and the Contois[17] model is used to describe the specific growth rate at high cell concentrations:

$$\mu(S, X) = \frac{\mu_m S}{KX + S}, \quad \sigma(S, X) = \frac{\mu(S, X)}{Y_{X/S}} \tag{12.219}$$

The state equations, the constraint equations, the performance index, and the Hamiltonian remain intact, as in Section 12.2.2:

$$\underset{F*(t)}{\text{Max}} H = (\lambda_1 \mu - \lambda_2 \sigma) x_1 + (S_F \lambda_2 + \lambda_3) F \overset{\Delta}{=} H_1 + \phi F \tag{12.14}$$

where

$$H_1 = (\lambda_1 \mu - \lambda_2 \sigma) x_1, \quad \phi = (S_F \lambda_2 + \lambda_3) \tag{12.15}$$

The adjoint equations do change to

$$\begin{bmatrix} d\lambda_1/dt \\ d\lambda_2/dt \\ d\lambda_3/dt \end{bmatrix} = - \begin{bmatrix} \partial H/\partial x_1 \\ \partial H/\partial x_2 \\ \partial H/\partial x_3 \end{bmatrix} = - \begin{bmatrix} (\lambda_1 \partial \mu/\partial x_1 - \lambda_2 \partial \sigma/\partial x_1) x_1 + (\lambda_1 \mu - \lambda_2 \sigma) \\ (\lambda_1 \partial \mu/\partial x_2 - \lambda_2 \partial \sigma/\partial x_2) x_1 \\ (\lambda_1 \partial \mu/\partial x_3 - \lambda_2 \partial \sigma/\partial x_3) x_1 \end{bmatrix}$$

$$\tag{12.220}$$

The final conditions on the adjoint variables are the same:

$$\boldsymbol{\lambda}(t_f) = \begin{bmatrix} \lambda_1(t_f) \\ \lambda_2(t_f) \\ \lambda_3(t_f) \end{bmatrix} = \begin{bmatrix} \partial P/\partial x_1(t_f) \\ \partial P/\partial x_3(t_f) \\ \text{unknown constant} \end{bmatrix} = \begin{bmatrix} 1 \\ 0 \\ \text{unknown constant} \end{bmatrix} \tag{12.18}$$

12.5.1 Optimal Feed Rate

To maximize the Hamiltonian, the feed rate must take on the maximum value when $\phi > 0$, the minimum when $\phi < 0$, and zero at the final time when the reactor volume is full. But when $\phi \equiv 0$ over a finite interval or intervals, PMP fails to provide a

Table 12.2. *Comparison between constant- and variable-yield problems for cellular productivity maximization*

Criterion	Cell mass productivity	
	Constant yield	Variable yield
Objective function	Max $X(t_f)V(t_f)/t_f$	Max $X(t_f)V(t_f)/t_f$
State equations	$\dfrac{d}{dt}\begin{pmatrix} x_1 \\ x_3 \\ x_4 \end{pmatrix} = \begin{pmatrix} \mu x_1 \\ G \\ 1 \end{pmatrix} + \begin{pmatrix} 0 \\ 1 \\ 0 \end{pmatrix} F$	$\dfrac{d}{dt}\begin{pmatrix} x_1 \\ x_2 \\ x_3 \\ x_4 \end{pmatrix} = \begin{pmatrix} \mu x_1 \\ -\sigma x_1 \\ 0 \\ 1 \end{pmatrix} + \begin{pmatrix} 0 \\ S_F \\ 1 \\ 0 \end{pmatrix} F$
Hamiltonian	$H = \lambda_1 \mu X V + \lambda_3 F - 1$	$H = (\lambda_1 \mu - \lambda_2 \sigma)x_1 + (S_F \lambda_2 + \lambda_3)F + \lambda_4$
Adjoint equations	$\begin{bmatrix} d\lambda_1/dt \\ d\lambda_3/dt \\ d\lambda_4/dt \end{bmatrix} = -\begin{bmatrix} \lambda_1\mu - \lambda_2\sigma \\ -(\lambda_1\mu' - \lambda_2\sigma')x_1 x_2/x_3^2 \\ 0 \end{bmatrix}$	$\begin{bmatrix} d\lambda_1/dt \\ d\lambda_2/dt \\ d\lambda_3/dt \\ d\lambda_4/dt \end{bmatrix} = -\begin{bmatrix} \lambda_1\mu - \lambda_2\sigma \\ -(\lambda_1\mu' - \lambda_2\sigma')x_1 x_2/x_3^2 \\ (\lambda_1\mu' - \lambda_2\sigma')x_1/x_3 \\ 0 \end{bmatrix}$
Boundary condition on adjoint variables	$\lambda_1(t_f) = -1/t_f, \lambda_3(t_f) = 0$	$\lambda_1(t_f) = -1/t_f, \lambda_2(t_f) = 0, \lambda_3(t_f) = 0$
Item maximized	Specific growth rate μ	Weighted sum of μ and $Y_{X/S}$
General form of feed rates	Concatenation of minimum, maximum, or singular	Concatenation of minimum, maximum, or singular
Form singular feed rate	Semiexponential	Semiexponential
Variation of S during singular feed rate interval	Maintained constant	Varies with time

solution. The feed rate profile is

$$
F = \begin{cases}
F_{\max} & \text{when } \phi = (S_F\lambda_2 + \lambda_3) > 0 \\
F_{\min} = 0 & \text{when } \phi = (S_F\lambda_2 + \lambda_3) < 0 \\
F_{\sin} & \text{when } \phi = (S_F\lambda_2 + \lambda_3) \equiv 0 \text{ over } t_q < t < t_{q+1} \\
F_b = 0 & \text{when } x_3(t) = V_{\max}
\end{cases} \quad (12.19)
$$

The remaining task is to determine the singular feed rate and the sequence and intervals for the maximum, minimum, and singular feed rates.

12.5.2 Constant Cell Mass Yield Coefficient, $Y_{X/S}$

Repeating what is in Section 12.2.3, the mass balance equation is

$$
\frac{d\mathbf{x}}{dt} = \frac{d}{dt}\begin{pmatrix} x_1 \\ x_3 \end{pmatrix} = \frac{d}{dt}\begin{pmatrix} XV \\ V \end{pmatrix} = \begin{pmatrix} \mu x_1 \\ 0 \end{pmatrix} + \begin{pmatrix} 0 \\ 1 \end{pmatrix}F \quad (12.23)
$$

where μ is a function of $S = SV/V$ and X:

$$
\mu(S, X) = \mu(SV/V, X) = \mu(x_2/x_3, x_1) = \mu[g(x_1, x_3)/x_3, x_1] \quad (12.24)
$$

where $g(x_1, x_3)$ is given by Eq. (12.22). The Hamiltonian (Eq. (12.14)) reduces to

$$
\underset{F^*(t)}{\text{Max}} H = \boldsymbol{\lambda}^{\mathbf{T}}[\mathbf{a}(\mathbf{x}) + \mathbf{b}F] = \lambda_1\mu x_1 + \lambda_3 F \stackrel{\Delta}{=} H_1 + \phi F \quad (12.25)
$$

where

$$
H_1 = \lambda_1\mu x, \quad \phi = \lambda_3 \quad (12.26)
$$

The differential equation for the adjoint variables, Eq. (12.16), reduces to

$$
\frac{d\boldsymbol{\lambda}}{dt} = \frac{d}{dt}\begin{bmatrix} \lambda_1 \\ \lambda_3 \end{bmatrix} = -\begin{bmatrix} \dfrac{\partial H}{\partial x_1} \\ \dfrac{\partial H}{\partial x_3} \end{bmatrix} = -\begin{pmatrix} \lambda_1\mu + \lambda_1 x_1\dfrac{\partial\mu}{\partial x_1} \\ \lambda_1 x_1\dfrac{\partial\mu}{\partial x_3} \end{pmatrix} \quad (12.27)
$$

and the boundary conditions are obtained from Eq. (12.18):

$$
\boldsymbol{\lambda}(t_f) = \begin{bmatrix} \lambda_1(t_f) \\ \lambda_3(t_f) \end{bmatrix} = \frac{\partial P}{\partial \mathbf{x}}(t_f) = \begin{bmatrix} 1 \\ \text{unknown constant} \end{bmatrix} \quad (12.28)
$$

12.5.3 Free Final Time, t_f

If the final time is free (also to be optimized), then the Hamiltonian must vanish everywhere, according to PMP:

$$
H^* = \lambda_1\mu x_1 + \lambda_3 F \stackrel{\Delta}{=} H_1 + \phi F = 0 \quad (12.25)
$$

At the final time, the feed rate is zero because the volume is full. Thus, the last term in Eq. (12.25) vanishes, leaving only the first term. Because $\lambda_1(t_f) = 1$, according to Eq. (12.28),

$$
H^*(t_f) = H_1^*(t_f) = \lambda_1(t_f)\mu(t_f)x_1(t_f) = \mu(t_f)x_1(t_f) = 0 \quad (12.30)
$$

Because $x_1(t_f) = XV \neq 0$, the specific growth rate at the final time must be zero:

$$\mu(t_f) = 0 \qquad (12.31)$$

For the specific growth rate to vanish, the substrate concentration must approach zero so that

$$\mu(t_f) = 0 = \mu(S \to 0) = \mu(t_f \to \infty) \qquad (12.31a)$$

Equation (12.31a) states that substrate concentration at the final time must approach zero, which in turn implies that *the final time must also approach infinity*, $t_f \to \infty$. The result is exactly the same as in Section 12.2.3.1. Because the yield coefficient is constant, *any mode of operation will lead to the same number of cells at infinite final time*:

$$XV(t_f) = X_0 V_0 + S_0 V_0 Y_{X/S} + [V(t_f) - V_0] S_F Y_{X/S} \qquad (12.33)$$

12.5.3.1 Optimal Feed Rate Profile

There is no singular arc, and any mode of feed rate leads to the same amount of cell mass at the final time, which approaches infinity. As in Section 12.2.3.6 the Hamiltonian on the optimal trajectory is a nonzero constant because the final time is fixed, not free:

$$H^* = \lambda_1 \mu x_1 + \lambda_3 F \overset{\Delta}{=} H_1 + \phi F = H^+ \qquad (12.46)$$

where H^+ is a nonzero unknown constant. At the final time when the reactor volume is full and therefore $F = 0$, the Hamiltonian is positive because $\mu(t_f) x_1(t_f) > 0$:

$$H^*(t_f) = \lambda_1(t_f)\mu(t_f)x_1(t_f) = \mu(t_f)x_1(t_f) = H^+ > 0 \qquad (12.47)$$

Because the Hamiltonian is continuous, the Hamiltonian is a positive constant for the entire time period:

$$H^* = H^+ > 0, \quad 0 \le t \le t_f \qquad (12.48)$$

Because $F(t)$ appears linearly in Eq. (12.46), the optimal feed rate consists of the following:

$$F = \begin{cases} F_{max} & \text{when } \phi = \lambda_3(t) > 0 \\ F_{min} = 0 & \text{when } \phi = \lambda_3(t) < 0 \\ F_{sin} & \text{when } \phi = \lambda_3(t) = 0 \text{ over a finite time interval, } t_q < t < t_{q+1} \\ F_b = 0 & \text{when } x_3(t) = V_{max} \end{cases}$$

$$(12.49)$$

12.5.3.2 Feed Rate on the Singular Arc

The singular flow rate is determined by recognizing that $\phi = \lambda_3 = 0$, and because it is identically zero over a finite time interval, its derivatives of all orders until the feed rate appears explicitly must also vanish:

$$\phi = \lambda_3 = 0 \qquad (12.50)$$

$$\frac{d\phi}{dt} = \frac{d\lambda_3}{dt} = -\lambda_1 x_1 \frac{\partial \mu}{\partial x_3} = 0 \qquad (12.51)$$

Because $\lambda_3 = 0$, λ_1 cannot be zero. Otherwise, this problem is trivial. The cell mass is nonzero, $x_1 = XV \neq 0$. Therefore, the only term that can remain zero over a finite time interval is

$$\frac{\partial \mu}{\partial x_3} = \frac{\partial \mu}{\partial S}\frac{\partial S}{\partial x_3} + \frac{\partial \mu}{\partial X}\frac{\partial X}{\partial x_3} = \frac{\partial \mu}{\partial S}\frac{\partial [g/x_3]}{\partial x_3} + \frac{\partial \mu}{\partial X}\frac{\partial (x_1/x_3)}{\partial x_3} = \frac{\partial \mu}{\partial S}\frac{x_3 S_F - g}{x_3^2}$$

$$-\frac{\partial \mu}{\partial X}\frac{x_1}{x_3^2} = \frac{\partial \mu}{\partial S}\frac{S_F - g/x_3}{x_3} - \frac{\partial \mu}{\partial X}\frac{x_1/x_3}{x_3} = \left[\frac{\partial \mu}{\partial S}(S_F - S) - \frac{\partial \mu}{\partial X}X\right]\frac{1}{V} = 0 \quad (12.221)$$

Therefore, the solution to Eq. (12.221) is

$$(\partial \mu/\partial S)/(\partial \mu/\partial X) = X/(S_F - S) \quad (12.222)$$

A physical interpretation of this equation is that *the substrate concentration must be varied to satisfy Eq. (12.222) on the singular arc*. The second time derivative is obtained by differentiating Eq. (12.51):

$$\frac{d^2\phi}{dt^2} = \frac{d}{dt}\left(\frac{d\phi}{dt}\right) = \frac{d}{dt}\left(-\lambda_1 x_1 \frac{\partial \mu}{\partial x_3}\right) = -\frac{\partial \mu}{\partial x_3}^{=0}\frac{d(\lambda_1 x_1)}{dt} - \lambda_1 x_1 \frac{d}{dt}\left(\frac{\partial \mu}{\partial x_3}\right) = 0$$

$$\Rightarrow \frac{d}{dt}\left(\frac{\partial \mu}{\partial x_3}\right) = \frac{d}{dt}\left[\frac{\partial \mu}{\partial S}(S_F - S) - \frac{\partial \mu}{\partial X}X\right]\frac{1}{V} = \frac{1}{V}\frac{\partial^2\mu}{\partial S^2}\frac{dS}{dt}(S_F - S) - \frac{1}{V}\frac{\partial \mu}{\partial S}\frac{dS}{dt}$$

$$-\frac{\partial \mu}{\partial S}(S_F - S)\frac{F}{V^2} - \frac{\partial^2\mu}{\partial^2 X}\frac{dX}{dt}\frac{1}{V} - \frac{\partial \mu}{\partial X}\frac{dX}{dt}\frac{1}{V} + \frac{\partial \mu}{\partial X}X\frac{F}{V^2} = 0$$

$$(12.223)$$

which is solved for the singular feed rate

$$F_{\text{sin}} = \frac{\left[\dfrac{\partial^2\mu}{\partial S^2}(S_F - S) - \dfrac{\partial \mu}{\partial S}\right]\dfrac{dS}{dt} - \left(\dfrac{\partial^2\mu}{\partial^2 X} - \dfrac{\partial \mu}{\partial X}\right)\dfrac{dX}{dt}}{\left[\dfrac{\partial \mu}{\partial S}(S_F - S) - \dfrac{\partial \mu}{\partial X}X\right]}V \quad (12.224)$$

Equation (12.223) defines the singular arc; that is, the cell and substrate concentrations must obey Eq. (12.223). In other words, the singular feed rate must keep the cell and substrate concentrations to follow Eq. (12.223). Equation (12.224) states that the singular feed rate is a nonlinear feedback involving the time derivative of substrate and cell concentrations, the cell and substrate concentrations, the bioreactor volume, and the specific growth rate and substrate consumption rate.

For Contois kinetics, $\mu = \mu_m S/(KX + S)$, and therefore,

$$(\partial \mu/\partial S)/(\partial \mu/\partial X) = -X/S \quad (12.225)$$

Equation (12.225) does not agree with Eq. (12.222), and therefore, there is no singular feed. A simple batch operation is called for this case of Contois kinetics.

REFERENCES

1. Pontryagin, L. S., Boltyanski, V. G., Gamkrelidge, R. V., and Mischenko, E. F. 1962. *The Mathematical Theory of Optimal Processes*. Wiley-Interscience.
2. Berber, R., Pertev, C., and Tucker, M. 1999. Optimization of feeding profiles for baker's yeast production by dynamic programming. *Bioprocess Engineering* 20: 263–269.

3. Modak, J. M., and Lim, H. C. 1989. Simple nonsingular control approach to fed-batch fermentation optimization. *Biotechnology Bioengineering* 33: 11–15.
4. Weigand, W. A. 1981. Maximum cell productivity by repeated fed-batch culture, constant yield case. *Biotechnology Bioengineering* 23: 249–266.
5. San, K.-Y., and Stephanopoulos, G. 1989. Optimization of fed-batch penicillin fermentation: A case of singular optimal control with state constraints. *Biotechnology Bioengineering* 34: 72–78.
6. Shin, S. H., and Lim, H. C. 2007. Cell-mass maximization in fed-batch culture: Sufficient conditions for singular arc and optimal feed rate profiles. *Bioprocess and Biosystems Engineering* 29: 335–347.
7. Shin, S. H., and Lim, H. C. 2007. Optimization of metabolite production in fed-batch culture: Use of sufficiency and characteristics of singular arc and properties of adjoint vector in numerical computation. *Industrial Engineering Chemistry Research* 46: 2526–2534.
8. Shin, H., and Lim, H. C. 2007. Maximization of metabolite in fed-batch cultures: Sufficient conditions for singular arc and optimal feed rate profiles. *Biochemical Engineering Journal* 37: 62–74.
9. Lee, J. H., Lim, H. C., and Hong, J. 1997. Application of non-singular transformation to on-line optimal control of poly-β-hydroxybutyrate fermentation. *Journal of Biotechnology* 55: 135–150.
10. Miele, A. 1962. *Optimization Technique*, ed. Leitman, G. Academic Press.
11. Ohno, H., Nakanishi, E., and Takamatsu, T. 1976. Optimal control of a semi-batch fermentation. *Biotechnology Bioengineering* 18: 847–864.
12. Lee, J. H., Lim, H. C., Yoo, Y. H., and Park, Y. H. 1999. Optimization of feed rate profile for the monoclonal antibody production. *Bioprocess Engineering* 20: 137–146.
13. Lee, J. H., Lim, H. C., and Kim, S. I. 2001. A nonsingular optimization approach to the feed rare profile optimization of fed-batch cultures. *Bioprocess and Biosystems Engineering* 24: 115–125.
14. Jayant, A., and Pushpavanam, S. 1998. Optimization of a biochemical fed-batch reactor transition from a non-singular to a singular problem. *Industrial Engineering Chemistry Research* 37: 4314–4321.
15. Pushpavanam, S., Rao, S., and Kahn, I. 1999. Optimization of a biochemical fed-batch reactor using sequential quadratic programming. *Industrial Engineering Chemistry Research* 38: 1998–2004.
16. Kelley, H. J. 1965. A transformation approach to singular subarcs in optimal trajectory and control problems. *Journal of the Society for Industrial Applied Mathematics, Control* 2: 234–241.
17. Contois, D. 1959. Relationship between population density and specific growth rate of continuous cultures. *Journal of General Microbiology* 21: 40–50.
18. Weigand, W. A., Lim, H. C., Creagan, C., and Mohler, R. 1979. Optimization of a repeated fed-batch reactor for maximum cell productivity. *Biotechnology and Bioengineering Symposium* 9: 335–348.
19. Modak, J. M., and Lim, H. C. 1989. Optimal operation of fed-batch bioreactors with two control variables. *Chemical Engineering Journal* 42: B15–B24.

13 Optimization for Metabolite Production

In the previous chapter, we treated the simplest case of producing cell mass as the product. We now consider metabolite production by fed-batch culture. We begin with a metabolite production process described by four mass balance equations (four differential equations) of cell mass, substrate, product, and total mass. More complex processes described by more than four balance equations are then treated. Many industrially important processes are very complex so that more than four dynamic balances are necessary to describe them.

In this chapter, we will consider optimization of metabolite processes that are described by specific rates that are (1) functions of substrate concentration only and (2) functions of both substrate and product concentrations. Both constant- and variable-yield coefficients are treated. The objective functions that are both independent and dependent on the final time are also treated. We consider the simplest case first to gain some insight that can aid us in analyzing and solving more complex processes.

13.1 Product Formation Models

There are a number of models for product formation, ranging from the simplest to more complex, incorporating inhibition effects, mass transfer, and cell morphology.

Simple models consist of four differential equations resulting from mass balances of cell, substrate, and product and one overall balance. These fourth-order models ignore effects of intermediates, by-products, inducers, inhibitors, and mass transfer and morphology of cells. We begin with the simplest minimal model for metabolite production:

$$\frac{d(XV)}{dt} = \mu XV \qquad XV(0) = V_0 X_0 \tag{13.1}$$

$$\frac{d(PV)}{dt} = \pi XV \qquad PV(0) = V_0 P_0 \tag{13.2}$$

$$\frac{dV}{dt} = F \qquad V(0) = V_0 \tag{13.3}$$

$$\frac{d(SV)}{dt} = S_F F - \sigma XV = S_F F - \left(\frac{\mu}{Y_{X/S}} + \frac{\pi}{Y_{P/S}} + m\right) XV \qquad S(0)V(0) = V_0 S_0 \tag{13.4}$$

where X, S, and P are the concentrations of cells, substrate, and product, respectively; μ, σ, and π are the specific rates of cell growth, substrate consumption, and product formation, respectively; $Y_{X/S}$ and $Y_{P/S}$ are the yield coefficients for cell mass and product, respectively; and m is the maintenance coefficient. As before, $F(t)$ is the volumetric feed rate of the substrate, S_F is the feed substrate concentration, which is assumed to be constant, and $V(t)$ is the culture volume.

Complex models incorporate effects of intermediates, by-products, inducers and inhibitors, mass transfer effects, and morphology of cells. Therefore, these models usually involve more than four mass balances. Most industrially important processes, such as amino acids and antibiotics and processes with recombinant cells, require more than four mass balance equations. These models are too complex to obtain analytical expressions for the singular feed rates, and one must rely on numerical work. The details will be given later, when we deal with these models for optimization studies.

Specific rates of cell growth, substrate consumption, and product formation μ, σ, and π are normally functions of the limiting substrate concentration, $\mu(S)$, $\sigma(S)$, and $\pi(S)$. Sometimes they are also functions of both limiting substrate and product concentrations, $\mu(S, P)$, $\sigma(S, P)$, and $\pi(S, P)$. For amino acid production, μ may be a function of S, while σ and π may be functions of μ: $\mu(S)$, $\sigma(\mu)$, and $\pi(\mu)$. For alcohol and antibiotic productions, μ, σ, and π may be functions of both S and P; $\mu(S, P)$, $\sigma(S, P)$, and $\pi(S, P)$, or specifically, S and P affect independently; or $\mu(S, P) = \mu_1(S)\mu_2(P)$, $\sigma(S, P) = \sigma_1(S)\sigma_2(P)$, and $\pi(S, P) = \pi_1(S)\pi_2(P)$. In rare situations, the specific rates may depend on the cell concentration as well, for example, high cell density processes in which the cell concentration is so high that the activity of water is empirically modeled by cell concentration in the specific growth rate. The yield coefficients for cell and product, $Y_{X/S}$ and $Y_{P/S}$, are most often considered constant, and in other situations, they may be functions of the limiting substrate concentration, $Y_{X/S}(S)$ and $Y_{P/S}(S)$. We shall consider first the simple case of specific rates that depend on the substrate concentration only and yield coefficients that are constants.

13.2 General Optimization Problem for Metabolites

The objective is to maximize a performance index such as the amount of product $V(t_f)P(t_f)$, the product productivity $V(t_f)P(t_f)/t_f$, the yield of a desired product $Y_{P/S}$, or a profit $P = \$_P V(t_f)P(t_f) - \$_S S_F[V(t_f) - V(0)] - \$_M t_f$, where $\$_P$, $\$_S$, and $\$_M$ are the unit prices of product and substrate and the maintenance cost per unit time, respectively, and t_f is the final operational time.

13.2.1 Choice of Manipulated Variables

As discussed in Chapter 12, inspection of the preceding mass balance equations (13.1)–(13.4) shows that a number of potential variables can be manipulated to optimize the performance of fed-batch cultures. The first obvious choice is the substrate feed flow rate $F(t)$. Another possibility is to vary the feed substrate concentration $S_F(t)$.[1,2] Although it is physically difficult to realize variable feed concentrations, in theory, one can generate the feed substrate concentration to vary with time by mixing two streams from two tanks with variable speed pumps. Another possibility is the mass flow rate of substrate[3] $FS_F(t)$. Yet another possibility is the fermentor volume $V(t)$.[4] The specific growth rate was also considered as a manipulated variable.[5] To avoid the singular control problem, the substrate concentration[1,6] $S(t)$ was chosen as the manipulated variable or the performance index was augmented with a nonlinear function of the feed rate.[7] The choice of S_F, S, and FS_F as manipulated variables to optimize the performance index has not taken into account in most cases the dependence of specific rates on concentrations of cell mass and product and the substrate consumption dynamics, and the results are suboptimal when compared with the results obtained with the substrate feed rate $F(t)$ as the manipulated variable.[8] Thus, as in Chapter 12, we choose the feed rate of substrate as the manipulated variable for fourth-order models, whereas for higher-order models, we consider a special transformation technique to convert singular problems into nonsingular problems.

13.2.2 Substrate Feed Rate as Manipulated Variable

With the exception of substrate concentration as the manipulated variable, all cases lead to the singular control problem because these manipulated variables appear linearly in the substrate balance equation. Because of difficulty in obtaining analytical and numerical solutions to singular control problems, attempts have been made to modify the performance index by introducing an arbitrary nonlinear function of the feed rate with a variable parameter so that the problem becomes nonsingular and then repeatedly solving the same nonsingular problem by decreasing asymptotically the parameter value to a zero value, thus obtaining the singular solution.[7] However, such approaches have been met with a convergence problem and failed to lead to a solution to the singular control problem. Therefore, we shall consider first the substrate feed rate as the manipulated variable without an arbitrary modification of the performance index.

As shown in Chapter 9, there is a method that transforms the singular problem into a nonsingular problem and utilizes the gradient method to obtain the optimal substrate concentration profile as the manipulated variable. This transformation technique becomes very practical when the process to be optimized is complex owing to a large number of mass balance equations and more than one manipulated variable. However, this approach leads to integrodifferential equations, which are often difficult to solve.

13.2.3 Optimization Problem Formulation

The objective is to determine the optimal substrate feed rate profile $F(t)$ that maximizes (or minimizes) a performance index that depends on the final outcome of fed-batch operation, such as the amount of cells produced $(XV)(t_f)$, the product formed $(PV)(t_f)$ and the unused substrate remaining $(SV)(t_f)$, and the final time t_f (operational costs are assumed to be proportional to the final time). For maximization, we have

$$\underset{F(t)}{\text{Max}}\, P[(XV)(t_f), (SV)(t_f), (PV)(t_f), t_f] \qquad (13.5)$$

This performance index reflects the values associated with the final cell mass and metabolites and operational costs and perhaps penalizes the residual amount of substrate, which may add to separation and purification costs. The performance index that depends on the final time t_f is readily converted into a standard problem that depends only on the final values of the state variables by augmenting the state variables with an additional state variable x_5 ($dx_5/dt = 1, x_5(0) = 0$) so that $t_f = x_5(t_f)$. Therefore, we consider two cases: performance indices with and without explicit dependence on the final time, t_f.

There are volume (state variable) and flow rate (manipulated variable) constraints:

$$V(t_f) = V_{max} \tag{13.6}$$

and

$$0 = F_{min} \le F(t) \le F_{max} \tag{13.7}$$

Equation (13.6) states that the available maximum effective culture volume should be utilized, and Eq. (13.7) puts the upper and lower limits on the substrate feed flow rate. In terms of state variables, the preceding problem posed by Eqs. (13.1)–(13.4), (13.6), and (13.7) is represented by the following augmented state equations:

$$\frac{d}{dt}\begin{bmatrix} x_1 \\ x_2 \\ x_3 \\ x_4 \\ x_5 \end{bmatrix} = \frac{d}{dt}\begin{bmatrix} XV \\ PV \\ V \\ SV \\ t \end{bmatrix} = \begin{bmatrix} \mu x_1 \\ \pi x_1 \\ 0 \\ -\sigma x_1 \\ 1 \end{bmatrix} + \begin{bmatrix} 0 \\ 0 \\ 1 \\ S_F \\ 0 \end{bmatrix} F \quad \begin{bmatrix} x_1(0) \\ x_2(0) \\ x_3(0) \\ x_4(0) \\ x_5(0) \end{bmatrix} = \begin{bmatrix} XV(0) \\ PV(0) \\ V(0) \\ SV(0) \\ t(0) \end{bmatrix} = \begin{bmatrix} x_{10} \\ x_{20} \\ x_{30} \\ x_{40} \\ 0 \end{bmatrix}$$

$$\tag{13.8}$$

$$\frac{d\mathbf{x}}{dt} = \mathbf{a}(\mathbf{x}) + \mathbf{b}F \qquad \mathbf{x}(0) = \mathbf{x_0}$$

where $x_1 = XV$, $x_2 = PV$, $x_3 = V$, $x_4 = SV$ and $x_5 = t$. The specific substrate consumption rate is made up of three terms: cell growth, product formation, and maintenance,

$$\sigma(S) = \frac{\mu(S)}{Y_{X/S}} + \frac{\pi(S)}{Y_{P/S}} + m \tag{13.9}$$

where $S = SV/V = x_4/x_3$. The performance index to be maximized is in terms of augmented state variables at the final time,

$$\underset{F(t)}{\text{Max}}\, P[\mathbf{x}(t_f)] \tag{13.10}$$

subject to the following fermentor volume and feed flow rate constraints:

$$x_3(t_f) = V_{max} \tag{13.11}$$

and

$$0 = F_{min} \le F(t) \le F_{max} \tag{13.7}$$

13.3 Necessary Conditions for Optimality for Metabolite Production

Following Pontryagin's maximum principle (PMP),[43] we begin by formulating the Hamiltonian, $H = \boldsymbol{\lambda}^{\mathrm{T}}(\mathbf{a} + \mathbf{b}F)$, which is a scalar product of the adjoint variables $\boldsymbol{\lambda}$ and the right-hand side of the state equation, $\mathbf{a}(\mathbf{x}) + \mathbf{b}F$.

13.3.1 Hamiltonian and Adjoint Vector

Forming the scalar product of Eq. (13.8) with the adjoint vector $\boldsymbol{\lambda}$, we obtain

$$
\begin{aligned}
H(\mathbf{x}, \boldsymbol{\lambda}, F) &= \boldsymbol{\lambda}^{\mathrm{T}}(\mathbf{a} + \mathbf{b}F) = \lambda_1 \mu x_1 + \lambda_2 \pi x_1 + \lambda_3 F + \lambda_4(-\sigma x_1 + S_F F) + \lambda_5 \\
&= [(\lambda_1 \mu + \lambda_2 \pi - \lambda_4 \sigma)x_1 + \lambda_5] + (\lambda_3 + S_F \lambda_4)F \triangleq H_1 + \phi F, \\
H_1 &= [(\lambda_1 \mu + \lambda_2 \pi - \lambda_4 \sigma)x_1 + \lambda_5], \quad \phi = (\lambda_3 + S_F \lambda_4)
\end{aligned}
\tag{13.12}
$$

According to PMP, instead of the performance index, the Hamiltonian is maximized by picking the optimal feed flow rate profile, $F^*(t)$:

$$
\underset{F^*(t)}{\mathrm{Max}}[H(\mathbf{x}, \boldsymbol{\lambda}, F) = H_1 + \phi F]
\tag{13.13}
$$

Over the optimal path, the Hamiltonian is an unknown constant, H^*:

$$
H(\mathbf{x}^*, \boldsymbol{\lambda}^*, F^*) = H^*
\tag{13.14}
$$

When the final time is free, the Hamiltonian is zero over the optimal path:

$$
H(\mathbf{x}^*, \boldsymbol{\lambda}^*, F^*) = 0
\tag{13.15}
$$

The boundary conditions on the adjoint variables are known to be constants, zeros, or free (unknown constants), depending on whether the corresponding final state variable does or does not appear in the performance index or is constrained. As in Chapter 9, the transversality conditions are used to obtain the boundary conditions (appropriately underlined items are matched):

$$
\delta P[\mathbf{x}(t_f)] = \sum_{i=1}^{n}\left[\frac{\partial P}{\partial x_i(t_f)}\right]\delta x_i(t_f) = \sum_{i=1}^{n}\lambda_i(t_f)\delta x_i(t_f) - \sum_{i=1}^{n}\lambda_i(0)\underbrace{\delta x_i(0)}_{=0}
\tag{9.27}
$$

or

$$
\begin{aligned}
\frac{\partial P}{\partial x_1(t_f)}\delta x_1(t_f) &+ \frac{\partial P}{\partial x_2(t_f)}\delta x_2(t_f) + \frac{\partial P}{\partial x_3(t_f)}\underbrace{\delta x_3(t_f)}_{=0} + \frac{\partial P}{\partial x_4(t_f)}\delta x_4(t_f) + \frac{\partial P}{\partial x_5(t_f)}\delta x_5(t_f) \\
&= \lambda_1(t_f)\delta x_1(t_f) + \lambda_2(t_f)\delta x_2(t_f) + +\lambda_3(t_f)\underbrace{\delta x_3(t_f)}_{=0} + \lambda_4(t_f)\delta x_4(t_f) + \lambda_5(t_f)\delta x_5(t_f)
\end{aligned}
$$

Owing to the final state constraint (Eq. (13.11)), $\delta x_3(t_f) = 0$, the final condition on λ_3 is free (undetermined constant), and therefore, the terminal boundary conditions for the adjoint variables are obtained by matching terms with equal number of bars:

$$\lambda(t_f) = \begin{bmatrix} \lambda_1(t_f) \\ \lambda_2(t_f) \\ \lambda_3(t_f) \\ \lambda_4(t_f) \\ \lambda_5(t_f) \end{bmatrix} = \begin{bmatrix} \partial P/\partial x_1(t_f) \\ \partial P/\partial x_2(t_f) \\ \text{free(unknown)} \\ \partial P/\partial x_4(t_f) \\ \partial P/\partial x_5(t_f) \end{bmatrix} \tag{13.16}$$

The adjoint equation must satisfy the following differential equation:

$$\frac{d\lambda}{dt} = \begin{bmatrix} \dfrac{d\lambda_1}{dt} \\ \dfrac{d\lambda_2}{dt} \\ \dfrac{d\lambda_3}{dt} \\ \dfrac{d\lambda_4}{dt} \\ \dfrac{d\lambda_5}{dt} \end{bmatrix} = -\frac{\partial H}{\partial \mathbf{x}} = -\begin{bmatrix} \dfrac{\partial H}{\partial x_1} \\ \dfrac{\partial H}{\partial x_2} \\ \dfrac{\partial H}{\partial x_3} \\ \dfrac{\partial H}{\partial x_4} \\ \dfrac{\partial H}{\partial x_5} \end{bmatrix} = -\begin{bmatrix} \mu & \pi & 0 & -\sigma & 0 \\ 0 & 0 & 0 & 0 & 0 \\ -\dfrac{\mu' x_1 x_4}{x_3^2} & -\dfrac{\pi' x_1 x_4}{x_3^2} & 0 & \dfrac{\sigma' x_1 x_4}{x_3^2} & 0 \\ \mu'\dfrac{x_1}{x_3} & \pi'\dfrac{x_1}{x_3} & 0 & \dfrac{-\sigma' x_1}{x_4} & 0 \\ 0 & 0 & 0 & 0 & 0 \end{bmatrix} \begin{bmatrix} \lambda_1 \\ \lambda_2 \\ \lambda_3 \\ \lambda_4 \\ \lambda_5 \end{bmatrix}$$

$$\tag{13.17}$$

where the primes denote the derivative with respect to S, that is, $\mu' = d\mu/dS$. Inspection of Eq. (13.17) shows that λ_2 and λ_5 are constants in the entire operational time period, including the singular arc:

$$\begin{bmatrix} \lambda_2 \\ \lambda_5 \end{bmatrix} = \begin{bmatrix} \partial P/\partial x_2(t_f) \\ \partial P/\partial x_5(t_f) \end{bmatrix} \tag{13.18}$$

Because the Hamiltonian is linear in F, maximization of the Hamiltonian depends on the sign of the coefficient of F, the *switching function* ϕ. The maximum feed rate should be used when the sign of ϕ is positive, whereas the minimum flow rate ($= 0$) should be used when the sign is negative. However, if the switching function is identically zero over a finite time interval or intervals, PMP does not provide a solution. This finite time interval is termed the singular interval, and the feed rate is termed the singular feed rate. Thus, the feed rate can take the following forms:

$$F = \begin{cases} F_{\max} & \text{when } \phi = (\lambda_3 + S_F \lambda_4) > 0 \\ F_{\min} = 0 & \text{when } \phi = (\lambda_3 + S_F \lambda_4) < 0 \\ F_{\sin} & \text{when } \phi = (\lambda_3 + S_F \lambda_4) \equiv 0 \text{ over } t_q < t < t_{q+1} \\ F_b = 0 & \text{when } x_3(t) = V_{\max} \end{cases} \tag{13.19}$$

Besides the intervals in which the flow rates are maximum or minimum, we consider an interior singular region in which the switching function is identically zero and the volume constraint is inactive, $x_3(t) < V_{\max}$, and the boundary region in which the volume is full, $x_3(t) = V_{\max}$.

13.3.2 Optimal Feed Rate for Boundary Arc, $x_3(t) = V_{max}$

On the boundary arc (t_{full}, t_f), the volume is full, and therefore, $x_3 - V_{max} = 0$, the feed flow stops, $F_b = 0$, so that the Hamiltonian (13.12) reduces to

$$H = [\lambda_1\mu + \pi\,\partial P/\partial x_2(t_f) - \lambda_4\sigma]x_1 + \lambda_5 \quad (t_{full}, t_f) \tag{13.20}$$

The boundary control is in feedback mode because the switching from the interior singular arc to the boundary arc is triggered when the bioreactor volume is full.

13.3.3 Optimal Feed Rate for Interior Singular Arc, $x_3(t) < V_{max}$

On the interior singular arc, where the volume constraint is inactive, the switching function is identically zero:

$$\phi = \boldsymbol{\lambda}^T\mathbf{b} = \lambda_3 + S_F\lambda_4 = 0 \tag{13.12}$$

The constant Hamiltonian is

$$H = [\lambda_1\mu + \pi\,\partial P/\partial x_2(t_f) - \lambda_4\sigma]x_1 + \lambda_5 \quad (t_{sin}, t_{full}) \tag{13.21}$$

If the final time is free to be chosen, the Hamiltonian is identically zero. Conversely, if the final time is fixed, the Hamiltonian is a constant (but unknown). Because the Hamiltonian is continuous, it must also be either zero or constant, H^*, on the singular arc. Because the switching function is identically zero over finite time intervals, its derivatives must also vanish (generally, the first and second derivatives vanish because the order of singularity is usually one, and therefore, higher-order derivatives vanish redundantly):

$$\frac{d\phi}{dt} = \frac{d\lambda_3}{dt} + S_F\frac{d\lambda_4}{dt} = 0 \tag{13.22}$$

Substitution of Eq. (13.17) into Eq. (13.22) yields

$$\frac{d\phi}{dt} = (\lambda_1\mu' + \lambda_2\pi' - \lambda_4\sigma')\left(\frac{x_4}{x_3} - S_F\right)\frac{x_1}{x_3} = 0 \tag{13.23}$$

Because $(x_1/x_3)(x_4/x_3 - S_F) = X(S - S_F) \neq 0$, the remaining terms must vanish:

$$\lambda_1\mu' + \lambda_2\pi' - \lambda_4\sigma' = 0 \tag{13.24}$$

Differentiating once again the switching function, we obtain

$$\frac{d^2\phi}{dt^2} = \frac{d}{dt}\left(\frac{d\phi}{dt}\right) = \left(\frac{x_1}{x_3}\right)\left(\frac{x_4}{x_3} - S_F\right)\frac{d}{dt}(\lambda_1\mu' + \lambda_2\pi' - \lambda_4\sigma')$$

$$= \left(\frac{x_1}{x_3}\right)\left(\frac{x_4}{x_3} - S_F\right)\left[-(\lambda_1\mu + \lambda_2\pi - \lambda_4\sigma)\mu' + (\lambda_1\mu'' + \lambda_2\pi'' - \lambda_4\sigma'')\frac{dS}{dt}\right] = 0$$

$$\Rightarrow -(\lambda_1\mu + \lambda_2\pi - \lambda_4\sigma)\mu' + (\lambda_1\mu'' + \lambda_2\pi'' - \lambda_4\sigma'')\frac{dS}{dt} = 0 \tag{13.25}$$

The general solution to Eq. (13.25) is

$$dS/dt = (\lambda_1\mu + \lambda_2\pi - \lambda_4\sigma)\mu'/(\lambda_1\mu'' + \lambda_2\pi'' - \lambda_4\sigma'') \tag{13.26}$$

Equation (13.26) implies that the substrate concentration is varied with time on the singular arc. The singular feed rate is obtained by substituting Eq. (13.26) into the substrate balance equation (13.8), which is rearranged as follows:

$$\frac{d(SV)}{dt} = V\frac{dS}{dt} + S\frac{dV}{dt} = V\frac{dS}{dt} + SF = FS_F - \sigma XV$$

$$\Rightarrow F = \frac{V\,dS/dt + \sigma XV}{(S_F - S)} \tag{13.27}$$

Substitution of Eq. (13.26) into Eq. (13.27) yields

$$F_{\text{sin}} = \frac{V\,dS/dt + \sigma XV}{S_F - S} = \frac{(\lambda_1\mu + \lambda_2\pi - \lambda_4\sigma)\mu'V}{(\lambda_1\mu'' + \lambda_2\pi'' - \lambda_4\sigma'')(S_F - S)} + \frac{\sigma XV}{(S_F - S)} \tag{13.28}$$

It is convenient at this point to consider two types of problems separately: specific rates that are (1) functions of the substrate concentration only, $\mu(S)$, $\pi(S)$, $\sigma(S)$, and (2) functions of both substrate and product concentrations, $\mu(S, P)$, $\pi(S, P)$, and $\sigma(S, P)$. We begin with the simpler situation.

13.4 Substrate Concentration–Dependent Specific Rates

The specific rates are functions of the limiting substrate concentration, $\mu(S)$, $\pi(S)$, and $\sigma(S)$, and there is no maintenance requirement, $m = 0$. For this case, the adjoint equation (Eq. (13.17)) reduces to

$$\frac{d\lambda}{dt} = \begin{bmatrix} d\lambda_1/dt \\ d\lambda_2/dt \\ d\lambda_3/dt \\ d\lambda_4/dt \\ d\lambda_5/dt \end{bmatrix} = - \begin{bmatrix} \lambda_1\mu + \lambda_2\pi - \lambda_4\sigma \\ 0 \\ -(\lambda_1\mu' + \lambda_2\pi' - \lambda_4\sigma')\left(x_1x_4/x_3^2\right) \\ (\lambda_1\mu' + \lambda_2\pi' - \lambda_4\sigma')(x_1/x_3) \\ 0 \end{bmatrix} \tag{13.29}$$

The final conditions on the adjoint variables remain intact (Eq. (13.16)). Thus, it is apparent that λ_2 and λ_5 are constants:

$$\lambda_2 = \lambda_2(t_f) = \partial P/\partial x_2(t_f)$$
$$\lambda_5 = \lambda_5(t_f) = \partial P/\partial x_5(t_f) \tag{13.30}$$

On the Boundary Arc. The volume is full and the feed rate stops; $x_3 = V_{\text{max}}$ and $F_b = 0$. The Hamiltonian is same as before (Eq. (13.20)):

$$H = [\lambda_1\mu + \lambda_2\pi\,\partial P/\partial x_2(t_f) - \lambda_4\sigma]x_1 + \lambda_5 \quad (t_{\text{full}}, t_f) \tag{13.31}$$

In view of Eq. (13.24), the adjoint equation, Eq. (13.17), reduces to

$$\frac{d\lambda}{dt} = \begin{bmatrix} d\lambda_1/dt \\ d\lambda_2/dt \\ d\lambda_3/dt \\ d\lambda_4/dt \\ d\lambda_5/dt \end{bmatrix} = - \begin{bmatrix} \lambda_1\mu + \lambda_2\pi - \lambda_4\sigma \\ 0 \\ 0 \\ 0 \\ 0 \end{bmatrix} \quad t_{\text{sin}} \le t \le t_{\text{full}} \tag{13.32}$$

Inspection of Eq. (13.28) shows that the singular feed rate is not in the form of feedback control as it contains the adjoint variables.

One additional equation is available if the final time t_f is free, which makes the Hamiltonian zero, $H = 0$. Hence, we consider two different cases, free and fixed final times, and two different types of performance indices, which lead to different final conditions on the adjoint variables. The yield coefficients can be constants, and sometimes they are also the functions of substrate concentration. When the yield coefficients are constant, it is possible to eliminate one of the differential equations with an algebraic equation, a stoichiometric equation, so that the system order is effectively reduced by one and, therefore, the singular feed rate can be obtained in feedback mode even for the fixed final time for which $H \neq 0$.

13.4.1 Constant-Yield Coefficients and No Maintenance Requirement

When the yield coefficients, $Y_{X/S}$ and $Y_{P/S}$, are constant and there is no maintenance term ($m = 0$) in Eq. (13.9),

$$\sigma(S) = \mu(S)/Y_{X/S} + \pi(S)/Y_{P/S} \tag{13.33}$$

Because of the constant-yield coefficients, the fourth-order system described by four mass balance equations can be reduced to a third-order system described by three differential equations and one algebraic equation. Equations (13.1)–(13.3) are substituted into Eq. (13.4), and Eq. (13.33) is used to obtain

$$\frac{d(SV)}{dt} = S_F F - \sigma XV = S_F \frac{dV}{dt} - \frac{\mu XV}{Y_{X/S}} - \frac{\pi XV}{Y_{P/S}} = S_F \frac{dV}{dt} - \frac{d(XV)}{Y_{X/S}dt} - \frac{d(PV)}{Y_{P/S}dt} \tag{13.34}$$

Because the yield coefficients are constant, each term in Eq. (13.34) is integrated to obtain an algebraic equation among four state variables (equivalent to a stoichiometric equation):

$$SV - (SV)_0 = S_F(V - V_0) - [XV - (XV)_0]/Y_{X/S} - [PV - (PV)_0]/Y_{P/S} \tag{13.35}$$

In terms of the state variables, Eq. (13.35) is

$$SV = x_4 = (x_{40} + x_{10}/Y_{X/S} + x_{20}/Y_{P/S} - x_{30}S_F) - x_1/Y_{X/S} - x_2/Y_{P/S} + x_3 S_F$$
$$\stackrel{\Delta}{=} g(x_1, x_2, x_3) \tag{13.36}$$

According to Eq. (13.36), the total amount of residual substrate is a function of the total amounts of cell mass and product and the bioreactor volume. Therefore, we can describe the process with three differential balance equations and the preceding algebraic equation. With this choice of state variables, the specific growth and product formation rates that are functions of substrate concentration are now regarded as functions of $x_1, x_2,$ and x_3. The substrate concentration on which the specific rates depend is

$$S = \frac{SV}{V} = \frac{x_4}{x_3} = \frac{g(x_1, x_2, x_3)}{x_3} \stackrel{\Delta}{=} h(x_1, x_2, x_3) \tag{13.37}$$

Thus, $S = x_4/x_3 = h(x_1, x_2, x_3)$, and it can be replaced by $h(x_1, x_2, x_3)$ to obtain a reduced state equation from Eq. (13.8):

$$\frac{d}{dt}\begin{bmatrix} XV \\ PV \\ V \\ t \end{bmatrix} = \frac{d}{dt}\begin{bmatrix} x_1 \\ x_2 \\ x_3 \\ x_5 \end{bmatrix} = \begin{bmatrix} \mu(S)x_1 = \mu(h)x_1 \\ \pi(S)x_1 = \pi(h)x_1 \\ 0 \\ 1 \end{bmatrix} + \begin{bmatrix} 0 \\ 0 \\ 1 \\ 0 \end{bmatrix} F \quad \begin{bmatrix} x_1(0) \\ x_2(0) \\ x_3(0) \\ x_5(0) \end{bmatrix} = \begin{bmatrix} x_{10} \\ x_{20} \\ x_{30} \\ 0 \end{bmatrix}$$
$$\tag{13.38}$$

The Hamiltonian is linear in F:

$$H = (\lambda_1 \mu + \lambda_2 \pi)x_1 + \lambda_5 + \lambda_3 F = H_1 + \phi F \qquad (13.39)$$

The switching function is

$$\phi = \lambda_3 \qquad (13.40)$$

The adjoint equations are

$$
\begin{bmatrix} d\lambda_1/dt \\ d\lambda_2/dt \\ d\lambda_3/dt \\ d\lambda_4/dt \end{bmatrix} = -
\begin{bmatrix}
(\lambda_1 \mu + \lambda_2 \pi) + x_1(\lambda_1 \partial \mu/\partial x_1 + \lambda_2 \partial \pi/\partial x_1) \\
x_1(\lambda_1 \partial \mu/\partial x_2 + \lambda_2 \partial \pi/\partial x_2) \\
x_1(\lambda_1 \partial \mu/\partial x_3 + \lambda_2 \partial \pi/\partial x_3) \\
0
\end{bmatrix}
$$

$$
= -
\begin{bmatrix}
(\lambda_1 \mu + \lambda_2 \pi) - (\lambda_1 \mu' + \lambda_2 \pi')x_1/Y_{X/S}x_3 \\
-(\lambda_1 \mu' + \lambda_2 \pi')x_1/Y_{P/S}x_3 \\
(\lambda_1 \mu' + \lambda_2 \pi')(S_F - g/x_3)x_1/x_3 \\
0
\end{bmatrix} \qquad (13.41)
$$

and the boundary conditions are

$$
\lambda(t_f) = \left(\frac{\partial P}{\partial \mathbf{x}(t_f)} \right) = \begin{bmatrix} \dfrac{\partial P}{\partial x_1(t_f)} & \dfrac{\partial P}{\partial x_2(t_f)} & \text{free} & \dfrac{\partial P}{\partial x_4(t_f)} \end{bmatrix}^T \qquad (13.42)
$$

where the third adjoint variable at the final time is free (an unknown constant) owing to the volume constraint (Eq. (13.11)). The optimal feed rate profile depends on the switching function and consists of the following:

$$
F = \begin{bmatrix}
F_{\max} & \text{if } \phi = \lambda_3 > 0 \\
F_{\min} = 0 & \text{if } \phi = \lambda_3 < 0 \\
F_{\sin} & \text{if } \phi = \lambda_3 = 0 \text{ over finite interval(s)} \\
F_b = 0 & \text{if } x_3 = V_{\max}
\end{bmatrix} \qquad (13.43)
$$

We must now consider the performance index. We consider first the case of free final time that does not appear in the performance index.

13.4.1.1 Performance Index Independent of Final Time, $\partial P/\partial t_f = 0$

Consider a performance index that does not depend explicitly on the final time, a maximization of the final amount of product:

$$P[\mathbf{x}(t_f)] = x_2(t_f) \qquad (13.44)$$

Then, there is no need for the augmented state variable x_5, and therefore, the state equation (Eq. (13.38)) reduces to

$$
\frac{d}{dt} \begin{bmatrix} XV \\ PV \\ V \end{bmatrix} = \frac{d}{dt} \begin{bmatrix} x_1 \\ x_2 \\ x_3 \end{bmatrix} = \begin{bmatrix} \mu(S)x_1 = \mu(h)x_1 \\ \pi(S)x_1 = \pi(h)x_1 \\ 0 \end{bmatrix} + \begin{bmatrix} 0 \\ 0 \\ 1 \end{bmatrix} F \qquad \begin{bmatrix} x_1(0) \\ x_2(0) \\ x_3(0) \end{bmatrix} = \begin{bmatrix} x_{10} \\ x_{20} \\ x_{30} \end{bmatrix}
$$
$$(13.45)$$

The Hamiltonian, Eq. (13.39), reduces to

$$H = \lambda_1 \mu x_1 + \lambda_2 \pi x_1 + \lambda_3 F \qquad (13.46)$$

The adjoint equations are obtained from Eq. (13.41), and the boundary conditions are obtained from Eq. (13.42):

$$\frac{d\lambda}{dt} = -\begin{bmatrix} (\lambda_1\mu + \lambda_2\pi) - (\lambda_1\mu' + \lambda_2\pi')\dfrac{x_1}{Y_{X/S}x_3} \\[2mm] (\lambda_1\mu' + \lambda_2\pi')\dfrac{x_1}{Y_{P/S}x_3} \\[2mm] (\lambda_1\mu' + \lambda_2\pi')(S_F - g/x_3)\dfrac{x_1}{x_3} \end{bmatrix} \quad \begin{bmatrix} \lambda_1(t_f) \\ \lambda_2(t_f) \\ \lambda_3(t_f) \end{bmatrix} = \begin{bmatrix} 0 \\ 1 \\ \text{free} \end{bmatrix} \quad (13.47)$$

We will consider first the free (unspecified) final time and then the fixed (specified) final time.

13.4.1.1.1 FREE (UNSPECIFIED) FINAL TIME, t_f NOT IN THE PERFORMANCE INDEX. Let us first investigate the implication of free final time and the performance index that is free of the final time. At the final time, the feed rate is zero and the volume constraint is met so that the Hamiltonian, Eq. (13.46), reduces, and because it is continuous everywhere, we have

$$H(t) = H(t_f) = [\lambda_1(t_f)\mu(t_f) + \lambda_2(t_f)\pi(t_f)]x_1(t_f) = 0 \qquad (13.48)$$

Because $x_1(t_f) = X(t_f)V(t_f)$ represents the total cell mass in the bioreactor, it cannot be identically zero, unless the cells lyse. Therefore

$$\lambda_1(t_f)\mu(t_f) + \lambda_2(t_f)\pi(t_f) = 0 \qquad (13.49)$$

The performance index is independent of $x_1(t_f)$, and therefore, $\lambda_1(t_f)$ is zero and $\lambda_2(t_f) = 1$. Thus, the following must hold to satisfy Eq. (13.49):

$$H(t_f) = \pi(t_f) = 0 \qquad (13.50)$$

According to Eq. (13.50), the specific product formation rate must vanish. If the product does not decay owing to, say, hydrolysis, then the specific rate is nonnegative for all substrate concentrations and zero only at zero substrate concentration. Therefore, Eq. (13.50) implies that *the final time is infinity* so that the substrate is completely consumed and, therefore, the specific rate is zero. Conversely, the (net) specific rates can be zero at a nonzero value of substrate concentration if cells and product decay, as we have seen in Chapter 6:

$$\mu^{\text{net}}(S)XV = \mu(S)XV - k_xXV \qquad (6.41)$$

$$\pi^{\text{net}}(S)XV = \pi(S)XV - k_pPV \qquad (6.43)$$

Then, the net specific rates can be zero at a finite final time. In these situations, it is obvious, then, that the final time cannot be infinite because the product concentration would decrease to zero, and therefore, the performance index would not be optimized. It must therefore be finite, and the final time may be the time at which the net production rate, the difference between the rate of formation and the rate of decay, is zero. In the absence of product decay, the final time becomes infinite.

When the performance index does not depend on the final time and if the cells and product concentration cannot decrease (no cell lysis and no product decay), then the final time approaches infinity to consume every bit of substrate in the reactor to produce cells and product. In reality, the final time should be large enough to allow substrate concentration to reach an arbitrarily low value for reasons of economy and less separation cost. As we have seen in previous chapters, in general, optimizations of any reaction revolve around maximization of the rate, yield, or a combination of rate and yield. When the final time (operation time) is large, the rate is not as important as there is enough time and therefore the yield is more pertinent, whereas if the final time is short, the rate is more important than the yield. When the operation time is in between, a combination of rate and yield may be important.

Adjoint Variables. Because the final time is free, the Hamiltonian is identically zero:

$$H = (\lambda_1 \mu + \lambda_2 \pi)x_1 + \lambda_3 F = 0 \tag{13.51}$$

The adjoint variables and boundary condition are given by Eq. (13.47).

Optimal Feed Rate on Interior Singular Arc. On the singular arc on which the switching function is identically zero, we have

$$\phi = \lambda_3 = 0 \tag{13.52}$$

The Hamiltonian for free final time is zero, and on the singular arc, the Hamiltonian, Eq. (13.51), reduces to

$$H = (\lambda_1 \mu + \lambda_2 \pi)x_1 = 0 \tag{13.53}$$

Because $x_1 = XV \neq 0$,

$$\lambda_1 \mu + \lambda_2 \pi = 0 \tag{13.54}$$

Because the switching function is identically zero over a finite time interval, its higher-order time derivative must also vanish:

$$\frac{d\phi}{dt} = \frac{d\lambda_3}{dt} = -(\lambda_1 \mu' + \lambda_2 \pi')(S_F - g/x_3)\frac{x_1}{x_3} = 0 \tag{13.55}$$

Because $(S_F - g/x_3)x_1/x_3 = (S_F - S)X \neq 0$, the remaining term in Eq. (13.55) must vanish:

$$(\lambda_1 \mu' + \lambda_2 \pi') = 0 \tag{13.56}$$

Combining Eqs. (13.54) and (13.56), we have that both λ_1 and λ_2 are identically zero, or the determinant must vanish:

$$\begin{bmatrix} \mu & \pi \\ \mu' & \pi' \end{bmatrix}\begin{bmatrix} \lambda_1 \\ \lambda_2 \end{bmatrix} = \begin{bmatrix} 0 \\ 0 \end{bmatrix} \tag{13.57}$$

First, checking the determinant,

$$(\mu\pi' - \mu'\pi) = 0 \Rightarrow \frac{(\mu\pi' - \mu'\pi)}{\mu^2} = 0 \Rightarrow \frac{d}{dS}\left(\frac{\pi}{\mu}\right) = 0 \tag{13.58}$$

Equation (13.58) implies that on the singular arc, the ratio of the specific rates, product formation to cell growth, is maximized. Alternatively, owing to Eqs. (13.54) and (13.56), the adjoint variables are constants:

$$
\begin{bmatrix} d\lambda_1/dt \\ d\lambda_2/dt \\ d\lambda_3/dt \end{bmatrix} = - \begin{bmatrix} (\lambda_1\mu + \lambda_2\pi) - (\lambda_1\mu' + \lambda_2\pi')x_1/Y_{X/S}x_3 \\ (\lambda_1\mu' + \lambda_2\pi')x_1/Y_{P/S}x_3 \\ (\lambda_1\mu' + \lambda_2\pi')(S_F - g/x_3)x_1/x_3 \end{bmatrix}
$$

$$
= \begin{bmatrix} 0 \\ 0 \\ 0 \end{bmatrix} \qquad \begin{bmatrix} \lambda_1(t_f) \\ \lambda_2(t_f) \\ \lambda_3(t_f) \end{bmatrix} = \begin{bmatrix} 0 \\ 1 \\ \text{free} \end{bmatrix} \tag{13.47}
$$

Thus, on the singular arc, all adjoint variables λ_1, λ_2, and λ_3 are constants. Thus, this is a trivial case.

Because the final time is infinite and the yield coefficients are constant, it should not matter how we feed the substrate into the reactor. The total amount of substrate fed into the reactor is obtained by integrating the overall balance (Eq. (13.3)):

$$
[V(t_f) - V_0]S_F = \left[\int_0^{t_f} F(t)dt \right] S_F \tag{13.59}
$$

Because the left-hand side is specified by the problem statement, a same amount of substrate is added to the fermentor regardless of feed rate profile to fill the fermentor. Because the yield coefficients are constant and there is no time limit, the amounts of cells and metabolite produced by the added substrate are simply the amounts owing to the substrate added and the initial amount of substrate:

$$
XV(t_f) - XV(0) = (V_f - V_0)S_FY_{X/S} + S_0V_0Y_{X/S}
$$
$$
PV(t_f) - PV(0) = (V_f - V_0)S_FY_{P/S} + S_0V_0Y_{P/S} \tag{13.60}
$$

Because the right-hand sides are specified by the optimization problem formulation, the outcome of fermentor operation is identical regardless of how the fermentor was fed. It follows, then, that the performance is identical. In other words, because the operational time is infinite and the yield coefficients are constant, any mode of feed, be it maximum, exponential, batch, or any arbitrary function of time, leads to the same value of performance index.

Optimal Feed Rate Profile for Free t_f, Constant Yield, and $\partial P/\partial t_f = 0$. As shown, there is no single optimal feed rate profile. The solution is not unique. Any method of filling the fermentor should yield the same performance. Simply fill the reactor and let the culture take its course over a substantially long time period. Obviously, this is an ill-posed problem.

13.4.1.1.2 FIXED FINAL TIME, t_f. When the final time is specified, the Hamiltonian is not zero but a nonzero unknown constant:

$$
H = (\lambda_1\mu + \lambda_2\pi)x_1 + \lambda_3F = H^* \neq 0 \tag{13.61}
$$

Thus, we have one less equation for the adjoint variables as compared to the free final time problem. The dynamic state and adjoint equations and the boundary conditions are the same as those for the free final time (Eqs. (13.45–13.47)).

Optimal Feed Rate on the Interior Singular Arc. The switching function and its time derivative are the same as those for the free final time:

$$\phi = \lambda_3 = 0 \tag{13.52}$$

$$(\lambda_1 \mu' + \lambda_2 \pi') = 0 \tag{13.56}$$

The Hamiltonian on the interior arc on which the volume constraint is inactive and $\phi = 0$ is obtained from Eq. (13.61):

$$H = (\lambda_1 \mu + \lambda_2 \pi) x_1 = H^* \neq 0 \tag{13.62}$$

where H^* is an unknown constant. Therefore, we conclude that

$$H = (\lambda_1 \mu + \lambda_2 \pi) x_1 \neq 0 \Rightarrow \lambda_1 \mu + \lambda_2 \pi \neq 0 \tag{13.63}$$

Substitution of Eq. (13.56) into Eq. (13.47) yields

$$\frac{d\lambda}{dt} = \begin{bmatrix} d\lambda_1/dt \\ d\lambda_2/dt \\ d\lambda_3/dt \end{bmatrix} = \begin{bmatrix} -(\lambda_1 \mu + \lambda_2 \pi) \\ 0 \\ 0 \end{bmatrix} \tag{13.64}$$

Equation (13.64) implies that λ_2 and λ_3 are constants on the singular arc. The singular feed rate is obtained from the second time derivative of the switching function with the aid of Eqs. (13.64) and (13.56):

$$\begin{aligned}
\frac{d^2\phi}{dt^2} &= \frac{d}{dt}\left[-(\lambda_1 \mu' + \lambda_2 \pi') \frac{(S_F - h)x_1}{x_3} \right] \\
&= -(S_F - h)X \left[\frac{d\lambda_1}{dt}\mu' + \frac{d\lambda_2}{dt}\pi' + (\lambda_1 \mu'' + \lambda_2 \pi'') \frac{dS}{dt} \right] \\
&= -(S_F - h)X\lambda_1 \left[-(\mu' - \mu'\pi/\pi') + (\mu'' - \mu'\pi''/\pi')\frac{dS}{dt} \right] = 0 \quad (13.65)
\end{aligned}$$

or

$$-(\mu - \mu'\pi/\pi')\mu' + (\mu'' - \mu'\pi''/\pi')\frac{dS}{dt} = 0 \tag{13.66}$$

Equation (13.66) can be satisfied by the following cases:

$$\begin{aligned}
&1.\ \mu - \mu'\pi/\pi' = 0 \text{ and } \mu'' - \mu'\pi''/\pi' = 0 \\
&2.\ \mu - \mu'\pi/\pi' = 0 \text{ and } dS/dt = 0 \\
&3.\ \mu' = 0 \text{ and } \mu'' - \mu'\pi''/\pi' = 0 \\
&4.\ \mu' = 0 \text{ and } dS/dt = 0 \\
&5.\ dS/dt = (\mu - \mu'\pi/\pi')\mu'/(\mu'' - \mu'\pi''/\pi')
\end{aligned} \tag{13.67}$$

In view of Eq. (13.64), case 1 implies that all adjoint variables are constant, and therefore, the switching function is a constant and there is no singular period. Case 2 shows that the substrate concentration is time invariant, and the first condition implies that the specific rates must be proportional to each other, $\pi/\mu = a$. Owing to Eq. (13.56), case 3 implies that

$$\mu' = 0 \Rightarrow \mu' = 0, \pi' = 0 \tag{13.68}$$

Equation (13.68) implies that μ and π must peak at the same value of S, and therefore, this cannot be a general solution. Case 4 implies that

$$\mu' = 0, \; dS/dt = 0 \text{ and } \mu - \mu'\pi/\pi' = 0 \Rightarrow \quad \pi' = 0 (\Rightarrow \mu = c\pi) \text{ and constant } S$$

(13.69)

Equations (13.68) and (13.69) represent special cases in which specific rates are proportional and the substrate concentration is kept constant at the value at which the specific rates are maximum:

$$S = S_m, \; \mu'(S_m) = 0 \quad \text{and} \quad \pi'(S_m) = 0$$

(13.70)

The general case, case 5, implies that

$$\frac{dS}{dt} = \frac{(\mu - \mu'\pi/\pi')\mu'}{(\mu'' - \mu'\pi''/\pi')} = \frac{(\pi/\mu)'\mu^2\mu'/\pi'}{(\pi'/\mu')'\mu^2/\pi'} = \frac{(\pi/\mu)'\mu^2}{(\pi'/\mu')'\mu'}$$

(13.71)

Thus, the substrate concentration is time variant, except the special cases, cases 2 and 4, in which the substrate concentration is to be kept constant, $dS/dt = 0$. The singular feed rate is obtained by substituting Eq. (13.71) into the substrate balance equation (13.4):

$$F_{\sin} = \frac{V\dfrac{dS}{dt} + \sigma XV}{S_F - S} = \frac{\left[\dfrac{(\pi/\mu)'\mu^2}{(\pi'/\mu')'\mu'} + \left(\dfrac{\mu}{Y_{X/S}} + \dfrac{\pi}{Y_{P/S}}\right)X\right]V}{S_F - S}$$

(13.72)

This singular feed rate is in feedback mode, and the substrate concentration is time variant during the singular period. However, we do not know at this stage the criterion that is being maximized by the singular feed rate. Therefore, it is difficult to assess the sequence of feed rates.

13.4.1.1.3 OPTIMAL FEED RATE PROFILE FOR FIXED t_f, CONSTANT YIELD, $\partial P/\partial t_f = 0$. As

seen, we have three types of solutions: a general case of arbitrary functional forms of $\mu(S)$ and $\pi(S)$ and two special cases in which μ and π are linearly related; $\pi(S) = a\mu(S)$ or $\mu(S) = a\pi(S) + b$.

General Case. We consider the general situation first. With the singular feed rate obtained earlier (Eq. (13.72)), we proceed to construct the optimal feed rate profile for the entire period of operation, 0 to t_f. The idea behind this is that one should force the process to reach the singular arc as soon as possible $(0, t_{\sin})$ and then apply the singular feed rate until the bioreactor is full $(t_{\sin}, t_{full})$. Ideally, the final time must match the filling time, $t_f = t_{full}$, so that there should not be a batch period following the singular period. Of these three times, $t_{\sin}$ is the only unknown because t_f is given and t_{full} is obtained during integration step because we know the singular feed rate and the maximum volume. Thus, one has to iterate on $t_{\sin}$ to obtain the maximum value of the performance index. A numerical scheme involves integrating the state equations with the given initial values from $t = 0$ to $t = t_{\sin}$ using either F_{max} or $F_{min} = 0$ (depending on whether the initial substrate concentration is less than or greater than the singular substrate concentration) and then integrating using $F_{\sin}$ from $t_{\sin}$ until the reactor volume is full, t_{full}. Finally, the state equations are integrated from t_{full} to the given t_f using $F_b = 0$. This process is repeated for various values of t_{sin}; the one that yields the highest value of the performance index is the optimum t_{sin}, and the feed rate profile with the optimum t_{sin} is the optimum feed rate profile. We denote the unknown target point on the singular arc by $S(t_{sin})$ and proceed to consider three different cases of initial conditions: (1) the initial substrate concentration lying on the singular arc, $S(0) = S_{sin}(t_{sin} = 0)$, (2) the initial substrate concentrations higher than those on the singular arc, $S(0) > S(t_{sin})$, and (3) the initial substrate concentrations lower than those on the singular arc, $S(0) < S(t_{sin})$:

1. Because the initial substrate concentration is on the singular arc, the optimal feed rate profile consists of singular flow rate, Eq. (13.72), from $t = 0$ until the bioreactor is full, t_{full}, and a batch period, $F_b = 0$, from t_{full} to the fixed final time, t_f. We note that this initial condition is optimal because there is no need to apply the maximum or minimum flow rate to steer the process onto the singular arc. However, this initial condition is not known a priori. Therefore, an iteration scheme is needed to search the initial condition that would maximize the performance index with the sequence $F_{sin} \rightarrow F_b = 0$. One would presumably prepare the initial substrate concentration to match this and proceed with the singular feed rate until the bioreactor volume is full. This situation represents the best operational policy because there is no bang period as the initial conditions are already on the singular arc. This is depicted in Figure 13.1.

2. Because the initial substrate concentration is larger than those on the singular arc, no feed is supplied ($F_{min} = 0$) from $t = 0$ to $t = t_{sin}$ to decrease the substrate concentration to reach the singular arc S_{sin} as soon as possible. Once the substrate concentration reaches the singular arc, the singular flow rate F_{sin} takes over until the bioreactor is full (t_{sin} to t_{full}). This should be followed by a batch period, $F_b = 0$, until the given final time, $F_{min} = 0 \rightarrow F_{sin} \rightarrow F_b = 0$. Therefore, the only unknown in this case is t_{sin}. A search scheme is used to determine the optimal t_{sin} to maximize the performance index.

3. Because the initial substrate concentration is less than S_{sin} on the singular arc, the feed rate must begin with the maximum value, F_{max}, from $t = 0$ to $t = t_{sin}$ to increase the substrate concentration to reach the singular arc as soon as possible. Once the substrate concentration reaches the singular arc, the singular flow rate F_{sin} takes over until the bioreactor is full (t_{sin} to t_{full}). This should be followed by a batch period, $F_b = 0$, until the fixed final time t_f, $F_{max} \rightarrow F_{sin} \rightarrow F_b = 0$. Therefore, the only unknown in this case is t_{sin}. A search scheme to determine the optimal t_{sin} to maximize the performance index is needed.

Special Cases Covered Previously. As shown earlier when the specific growth rate and the specific product formation rate are linearly related, we know the singular arc $S = S_m$ and the singular feed rate (Eq. (13.72)). In this situation, the sequence of optimum feed rates is quite simple, and no iteration is required. If $S(0) < S_m$, one begins with F_{max} until $S = S_m$, followed by F_{sin} until the reactor is full ($x_3(t_{full}) = V_{max}$), and finally by $F_b = 0$, a batch process, until the given final time, t_f. Conversely, if $S(0) > S_m$, then the feed sequence begins with $F_{min} = 0$, a batch process, until $S(t) = S_m$, followed by F_{sin} until the reactor is full, $x_3(t_{full}) = V_{max}$, and finally, $F_b = 0$,

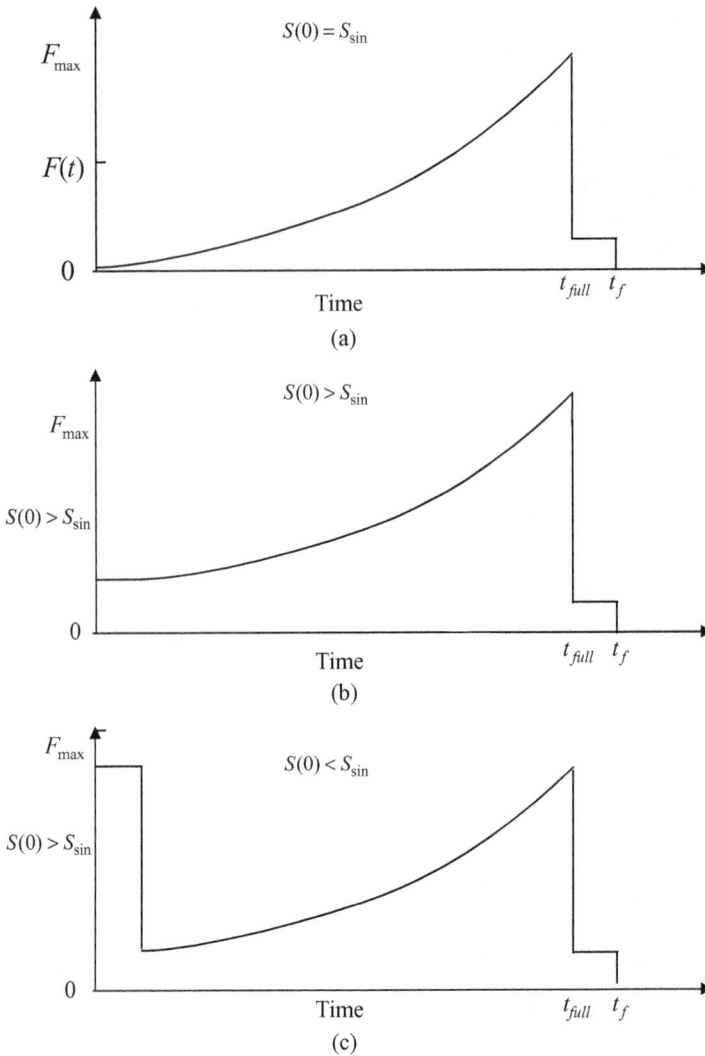

Figure 13.1. Optimal feed rate profiles for maximum final amount of product, fixed final time, constant yields.

a batch process, until the given final time, t_f. Finally, if $S(0) = S_m$, the feed rate begins with F_{sin} until the reactor is full ($x_3(t_{\text{full}}) = V_{\text{max}}$), followed by $F_b = 0$, a batch process, until the given final time, t_f. Hence, no search scheme is needed.

EXAMPLE 13.E.1: LYSINE FERMENTATION As an example of constant-yield coefficients, $Y_{X/S}$ and $Y_{P/S}$, and no maintenance, $m = 0$, we consider the model of Ohno et al.[11] The specific rates, the operational parameters, and the initial conditions are given as follows:

$$\mu = 0.124S, \pi = -384\mu^2 + 134\mu, \sigma = \mu/0.135,$$
$$[S_F, F_{\text{max}}, F_{\text{min}}, V_{\text{max}}] = [2.7\,\text{wt.\%}, 2\,\text{L/hr}, 0, 20\,\text{L}],$$
$$[X_0, S_0, P_0, V_0] = [0.06\,\text{g/L}, 2.7\,\text{wt.\%}, 0\,\text{g/L}, 4\,\text{L}]$$

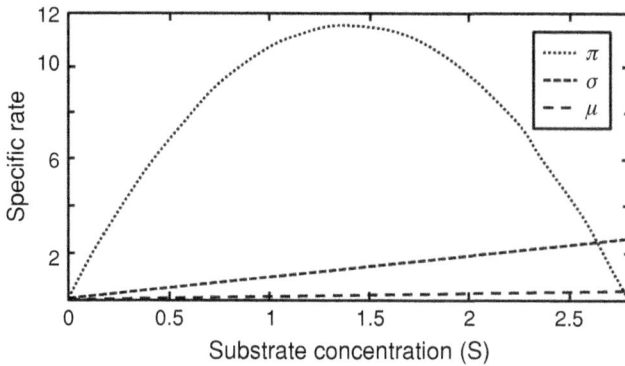

Figure 13.E.1.1. Specific rates from Ohno et al.[11]

In this model, the specific rates of cell growth and substrate consumption μ and σ are monotonic, whereas the specific product formation rate π is a nonmonotonic function of substrate concentration. The maximum value of π is 11.69 and occurs at $S = 1.407$ g/L. The yield coefficients are $Y_{X/S} = 0.135$, $Y_{P/S} = \pi/\sigma = 18.09 - 6.428S$, $m = 0$. The objective is to maximize the final amount of lysine. Therefore, the performance index is

$$P = P(t_f)V(t_f) = x_2(t_f)$$

As apparent from Figure 13.E.1.1, the derivatives of the specific rates are first positive, $\mu' > 0$, $\sigma' > 0$ and $\pi' > 0$, until $S = 1.407$, when they are zero, and then negative thereafter. Thus, $\mu' > 0$, $\sigma' > 0$ and $\pi' > 0$ in the substrate concentration range of $S = 0$ to $S = 1.407$. It is also clear that $\sigma'/\sigma = \mu'/0.135/\mu/0.135 = \mu'/\mu$.

Following Section 13.4.1.1, we have

$$\frac{d}{dt}\begin{bmatrix} XV \\ PV \\ V \end{bmatrix} = \frac{d}{dt}\begin{bmatrix} x_1 \\ x_2 \\ x_3 \end{bmatrix} = \begin{bmatrix} \mu(S)x_1 = 0.124hx_1 \\ \pi(S)x_1 = [-384(0.124h)^2 + 134(0.124h)]x_1 \\ 0 \end{bmatrix} + \begin{bmatrix} 0 \\ 0 \\ 1 \end{bmatrix} F$$

$$\begin{bmatrix} x_1(0) \\ x_2(0) \\ x_3(0) \end{bmatrix} = \begin{bmatrix} x_{10} \\ x_{20} \\ x_{30} \end{bmatrix} = \begin{bmatrix} 0.24g \\ 0 \\ 4l \end{bmatrix} \qquad (13.E.1.1)$$

where

$$h(x_1, x_2, x_3) = x_4/x_3 = (x_{40} + x_{10}/Y_{X/S} + x_{20}/Y_{P/S} - x_{30}S_F)/x_3$$
$$- x_1/x_3Y_{X/S} - x_2/x_3Y_{P/S} + S_F \qquad (13.E.1.2)$$

The Hamiltonian is constant:

$$H = \lambda_1\mu x_1 + \lambda_2\pi x_1 + \lambda_3 F = H^* \qquad (13.E.1.3)$$

The adjoint equations and the boundary conditions are as follows:

$$\frac{d\lambda}{dt} = - \begin{bmatrix} [\lambda_1\mu + \lambda_2\pi] - [\lambda_1\mu' + \lambda_2\pi']\dfrac{x_1}{x_3}/Y_{X/S} \\[2mm] [\lambda_1\mu' + \lambda_2\pi']x_1/x_3 Y_{P/S} \\[2mm] [\lambda_1\mu' + \lambda_2\pi'](S_F - g/x_3)\dfrac{x_1}{x_3} \end{bmatrix},$$

$$\begin{bmatrix} \lambda_1(t_f) \\ \lambda_2(t_f) \\ \lambda_3(t_f) \end{bmatrix} = \begin{bmatrix} \partial P/\partial x_1(t_f) \\ \partial P/\partial x_2(t_f) \\ \text{free} \end{bmatrix} = \begin{bmatrix} 0 \\ 1 \\ \text{free} \end{bmatrix} \qquad (13.\text{E}.1.4)$$

The switching function and its derivative are

$$\phi = \lambda_3 = 0 \qquad (13.\text{E}.1.5)$$

$$\frac{d\phi}{dt} = \frac{d\lambda_3}{dt} = -(\lambda_1\mu' + \lambda_2\pi')(S_F - g/x_3)\frac{x_1}{x_3} = 0 \qquad (13.\text{E}.1.6)$$

which are satisfied by

$$(\lambda_1\mu' + \lambda_2\pi') = \lambda_1\left(\mu' + \frac{\lambda_2}{\lambda_1}\pi'\right) = 0 \qquad (13.\text{E}.1.7)$$

The Hamiltonian on the interior singular arc on which the volume constraint is inactive and $\phi = 0$ (Eq. (13.52)) is

$$H = (\lambda_1\mu + \lambda_2\pi)x_1 = H^* \qquad (13.\text{E}.1.8)$$

where H^* is an unknown constant. Therefore, we conclude that

$$H = (\lambda_1\mu + \lambda_2\pi)x_1 \neq 0 \quad \Rightarrow \lambda_1\mu + \lambda_2\pi \neq 0 \qquad (13.\text{E}.1.9)$$

Substitution of Eq. (13.E.1.7) into Eq. (13.E.1.4) yields

$$\frac{d\lambda}{dt} = \begin{bmatrix} d\lambda_1/dt \\ d\lambda_2/dt \\ d\lambda_3/dt \end{bmatrix} = \begin{bmatrix} -(\lambda_1\mu + \lambda_2\pi) \\ 0 \\ 0 \end{bmatrix} \qquad (13.\text{E}.1.10)$$

Thus, λ_2 and λ_3 are constants on the singular arc.

The singular feed rate is obtained from the second time derivative of the switching function:

$$\frac{d^2\phi}{dt^2} = -(S_F - g/x_3)X\lambda_1\left[-\mu'\frac{(\mu\pi' - \mu'\pi)}{\pi'} + \frac{(\pi'\mu'' - \mu'\pi'')}{\pi'}\right]\frac{dS}{dt} = 0 \quad (13.\text{E}.1.11)$$

or

$$\frac{dS}{dt} = \frac{(\pi/\mu)'}{(\pi'/\mu')'}\left(\frac{\mu}{\mu'}\right)^2\mu' \qquad (13.\text{E}.1.12)$$

Substitution of Eq. (13.E.1.12) into the substrate balance equation yields

$$F_{\sin} = \frac{\left[\dfrac{(\pi/\mu)'}{(\pi'/\mu')'}\left(\dfrac{\mu}{\mu'}\right)^2\mu' + \left(\dfrac{\mu}{Y_{X/S}} + \dfrac{\pi}{Y_{P/S}}\right)X\right]V}{S_F - S} = \frac{(0.919X - 0.062S)SV}{(2.7 - S)}$$

$$(13.\text{E}.1.13)$$

According to Eq. (13.E.1.13), the singular feed rate is in feedback mode. The optimal feed rate profile is now constructed following the procedure described earlier in Section 13.4.1.1.

Because the initial substrate concentration, $S_F = 2.7$, is larger than those on the singular arc, less than 1.407, no feed is supplied ($F_{min} = 0$) from $t = 0$ to $t = t_{sin}$ to decrease the substrate concentration to reach the singular arc S_{sin} as soon as possible. Once the substrate concentration reaches the singular arc, the singular flow rate given in Eq. (13.E.1.13) takes over until the bioreactor is full (t_{sin} to t_{full}). This should be followed by a batch period, $F_b = 0$, until the given final time, t_f; $F_{max} \rightarrow F_{sin} \rightarrow F_b = 0$. Therefore, the only unknown in this case is t_{sin}. A search scheme to determine the optimal t_{sin} to maximize the performance index is used to obtain the optimal switching time, t_1. The results are given in Figure 13.E.1.2. The total process time is 39 hrs, and the switching times t_{sin} and t_{full} are 1.74 and 33.0 hrs, respectively. The final state variables are $[XV, SV, PV, V](t_f) = [7.6\,g,\ 0.95\,g,\ 774.6\,g,\ 20.0\,L]$.

13.4.1.2 Performance Index Dependent on Free Final Time, $P[x(t_f), t_f]$

Because the performance index depends on the final time t_f, $\partial P/\partial t_f \neq 0$, the state variables are augmented with an additional state variable, x_5, so that the performance index is in terms of the final states. The final time is free to be chosen. The state equations, the Hamiltonian, and the switching function are the same as Section 13.4.1:

$$\frac{d}{dt}\begin{bmatrix} XV \\ PV \\ V \\ t \end{bmatrix} = \frac{d}{dt}\begin{bmatrix} x_1 \\ x_2 \\ x_3 \\ x_5 \end{bmatrix} = \begin{bmatrix} \mu(S)x_1 = \mu[h(x_1, x_2, x_3)]x_1 \\ \pi(S)x_1 = \pi[h(x_1, x_2, x_3)]x_1 \\ 0 \\ 1 \end{bmatrix} + \begin{bmatrix} 0 \\ 0 \\ 1 \\ 0 \end{bmatrix} F \quad \begin{bmatrix} x_1(0) \\ x_2(0) \\ x_3(0) \\ x_5(0) \end{bmatrix} = \begin{bmatrix} x_{10} \\ x_{20} \\ x_{30} \\ 0 \end{bmatrix}$$

(13.38)

The Hamiltonian is linear in F:

$$H = \lambda_1 \mu x_1 + \lambda_2 \pi x_1 + \lambda_5 + \lambda_3 F \overset{\Delta}{=} H_1 + \phi F, \ H_1 = \lambda_1 \mu x_1 + \lambda_2 \pi x_1 + \lambda_5, \ \phi = \lambda_3$$

(13.39)

The switching function is

$$\phi = \lambda_3 \tag{13.40}$$

The adjoint equations are

$$\begin{bmatrix} d\lambda_1/dt \\ d\lambda_2/dt \\ d\lambda_3/dt \\ d\lambda_4/dt \end{bmatrix} = -\begin{bmatrix} (\lambda_1\mu + \lambda_2\pi) - (\lambda_1\mu' + \lambda_2\pi')x_1/Y_{X/S}x_3 \\ -(\lambda_1\mu' + \lambda_2\pi')x_1/Y_{P/S}x_3 \\ (\lambda_1\mu' + \lambda_2\pi')(S_F - g/x_3)x_1/x_3 \\ 0 \end{bmatrix}$$

(13.41)

where $\mu_i = \partial\mu/\partial x_i$ and $\pi_i = \partial\pi/\partial x_i$. The final conditions are given by Eq. (13.73).

$$[\lambda_1(t_f), \lambda_2(t_f), \lambda_3(t_f), \lambda_5(t_f)] = [\partial P/\partial x_1(t_f), \partial P/\partial x_2(t_f), \text{Free}, \partial P/\partial x_5(t_f)] \tag{13.73}$$

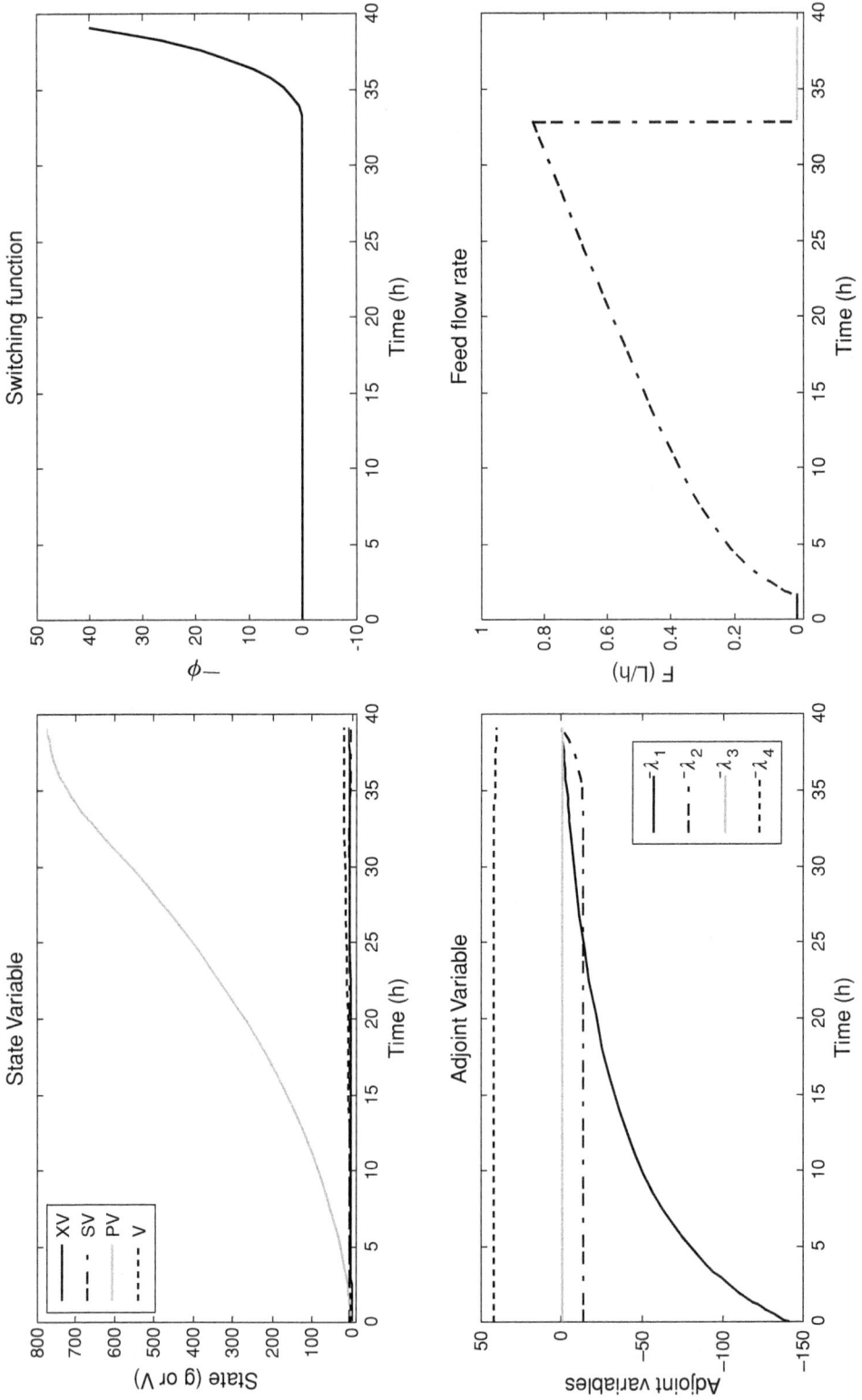

Figure 13.E.1.2. Various time profiles.

Because the Hamiltonian is linear in F and the switching function is λ_3, the Hamiltonian is maximized by the following feed rates:

$$
F = \begin{bmatrix}
F_{\max} & \text{if} & \phi = \lambda_3 > 0 \\
F_{\min} = 0 & \text{if} & \phi = \lambda_3 < 0 \\
F_{\sin} & \text{if} & \phi = \lambda_3 \equiv 0 \text{ over finite interval(s)} \\
F_b = 0 & \text{if} & x_3 = V_{\max}
\end{bmatrix}
\tag{13.43}
$$

The adjoint equations and boundary conditions are given by Eqs. (13.41) and (13.73). We note that λ_5 is a constant:

$$
\lambda_5(t) = \lambda_5(t_f) = \partial P/\partial x_5(t_f)
\tag{13.74}
$$

We consider specific performance indices, beginning with minimum time problems.

13.4.1.2.1 MINIMUM TIME PROBLEMS. For time-optimal problems, the initial conditions and final conditions are usually provided, and the objective is to go from the initial state to the final state in minimum time. Therefore, to minimize the final (operational) time, we equivalently maximize the negative value of the final time:

$$
P[\mathbf{x}(t_f)] = -t_f = -x_5(t_f)
\tag{13.75}
$$

Sometimes the final conditions may not be specified completely. Instead, the final target set may be specified, for example, the total amounts of product, the residual substrate, and the volume may be specified, but the amount of cells produced may not be specified but rather allowed to float:

$$
\begin{bmatrix}
x_1(0) \\
x_2(0) \\
x_3(0) \\
x_5(0)
\end{bmatrix}
=
\begin{bmatrix}
x_{10} \\
x_{20} \\
x_{30} \\
0
\end{bmatrix},
\qquad
\begin{bmatrix}
x_1(t_f) \\
x_2(t_f) \\
x_3(t_f) \\
x_5(t_f)
\end{bmatrix}
=
\begin{bmatrix}
x_{1f} \\
x_{2f} \\
x_{3f} \\
\text{free}
\end{bmatrix}
\tag{13.76}
$$

The boundary conditions on the adjoint variables with the initial and final states specified by Eq. (13.76) and in the presence of the terminal constraint are obtained by matching coefficients, as shown:

$$
\left(\frac{\partial P}{\partial \mathbf{x}(t_f)}\right)^T \delta\mathbf{x}(t_f) = \sum_{i=1,2,3,5} \frac{\partial P}{\partial x_i(t_f)}\delta x_i(t_f) = \sum_{i=1,2,3,5} \lambda_i(t_f)\delta x_i(t_f) - \sum_{i=1,2,3,5} \lambda_i(0)\underbrace{\delta x_i(0)}_{=0}
$$

$$
\Rightarrow \frac{\partial PI}{\partial x_5(t_f)}\delta x_5(t_f) = (-1)\delta x_5(t_f)
$$

$$
= \lambda_1(t_f)\underbrace{\delta x_1(0)}_{=0} + \lambda_2(t_f)\underbrace{\delta x_2(0)}_{=0} + \lambda_3(t_f)\underbrace{\delta x_3(0)}_{=0} + \lambda_5(t_f)\delta x_5(t_f)
$$

$$
\Rightarrow [\lambda_1(t_f), \lambda_2(t_f), \lambda_3(t_f), \lambda_5(t_f)] = [a_1, a_2, a_3, -1]
\tag{13.77}
$$

where $a_1, a_2,$ and a_3 are unknown constants. Equations (13.74) and (13.77) imply that the fifth adjoint variable is a constant:

$$
\lambda_5(t) = \lambda_5(t_f) = -1
\tag{13.78}
$$

Because λ_5 is a constant, we can use the original state variables, $x_1, x_2,$ and x_3, and the corresponding adjoint variables, $\lambda_1, \lambda_2,$ and λ_3.

Conversely, when the final amount of cell mass is not specified but left free, then the final conditions are

$$\begin{bmatrix} x_1(t_f) \\ x_2(t_f) \\ x_3(t_f) \end{bmatrix} = \begin{bmatrix} \text{free} \\ x_{2f} \\ x_{3f} \end{bmatrix} \tag{13.79}$$

and the final conditions on the adjoint variables are

$$\lambda(t_f) = \begin{bmatrix} \lambda_1(t_f) \\ \lambda_2(t_f) \\ \lambda_3(t_f) \end{bmatrix} = \left(\frac{\partial P}{\partial \mathbf{x}(t_f)} \right) = \begin{bmatrix} \partial P/\partial x_1(t_f) \\ a_2 \\ a_3 \end{bmatrix} \tag{13.80}$$

Optimal Feed Rate on the Interior Singular Arc. On the interior singular arc, the volume constraint is inactive and the switching function is identically zero:

$$\phi = \lambda_3 = 0 \tag{13.40}$$

The Hamiltonian, Eq. (13.39), reduces to

$$H = \lambda_1 \mu x_1 + \lambda_2 \pi x_1 - 1 = 0 \tag{13.81}$$

The first time derivatives of the switching function must also vanish:

$$\frac{d\phi}{dt} = \frac{d\lambda_3}{dt} = -(\lambda_1 \mu' + \lambda_2 \pi') \frac{(S_F - g)x_1}{x_3} = 0 \quad \Rightarrow (\lambda_1 \mu' + \lambda_2 \pi') = 0 \tag{13.82}$$

By substituting Eq. (13.82) into Eq. (13.41), we obtain the adjoint equation

$$\frac{d\lambda}{dt} = \begin{bmatrix} d\lambda_1/dt \\ d\lambda_2/dt \\ d\lambda_3/dt \end{bmatrix} = \begin{bmatrix} -(\lambda_1 \mu + \lambda_2 \pi) \\ 0 \\ 0 \end{bmatrix} \tag{13.83}$$

In view of Eq. (13.83), we conclude that λ_2 and λ_3 are constants.

Differentiating Eq. (13.82) once more, setting it to zero and substituting Eqs. (13.83) and (13.82) into the resultant, we obtain

$$\frac{d^2\phi}{dt^2} = -\frac{(S_F - g/x_3)x_1}{x_3}\left[\frac{d\lambda_1}{dt}\mu' + \frac{d\lambda_2}{dt}\pi' + (\lambda_1\mu'' + \lambda_2\pi'')\frac{dS}{dt} \right]$$

$$= -(\lambda_1\mu + \lambda_2\pi)\mu' + (\lambda_1\mu'' + \lambda_2\pi'')\frac{dS}{dt} = 0 \tag{13.84}$$

Equation (13.84) can be satisfied by the following cases:

$$\begin{array}{l} 1.\ (\lambda_1\mu + a_2\pi) = 0 \text{ and } (\lambda_1\mu'' + a_2\pi'') = 0 \\ 2.\ (\lambda_1\mu + a_2\pi) = 0 \text{ and } dS/dt = 0 \\ 3.\ \mu' = 0 \text{ and } (\lambda_1\mu'' + a_2\pi'') = 0 \\ 4.\ \mu' = 0 \text{ and } dS/dt = 0 \\ 5.\ dS/dt = (\lambda_1\mu + a_2\pi)/(\lambda_1\mu'' + a_2\pi'') \end{array} \tag{13.85}$$

In view of Eq. (13.81), cases 1 and 2 imply that all adjoint variables are constant, and therefore, the switching function is a constant, and there is no singular period. Case 3 implies that

$$\mu' = 0, \lambda_1\mu'' + a_2\pi'' = 0 \quad \text{and} \quad \lambda_1\mu' + a_2\pi' = 0, \quad \Rightarrow \mu' = 0, \pi' = 0 \tag{13.86}$$

Equation (13.86) implies that once again, μ and π must peak at the same value of S, and therefore, this cannot be a general solution. Case 4 implies that

$$\mu' = 0,\ dS/dt = 0 \text{ and } \lambda_1\mu'+\lambda_2\pi'= 0 \quad \Rightarrow \quad \pi' = 0(\Rightarrow \mu = c\pi) \text{ and constant } S$$

$$(13.87)$$

Once again, case 4 cannot be satisfied. Finally, case 5 implies that

$$
\begin{aligned}
\frac{dS}{dt} &= \frac{(\lambda_1\mu + \lambda_2\pi)\mu'}{\lambda_1\mu'' + \lambda_2\pi''} = \frac{[\mu - (\mu'/\pi')\pi]\mu'}{\mu'' - (\mu'/\pi')\pi''} = -\frac{\mu'\mu^2[(\mu\pi' - \mu'\pi)/\mu^2]}{(\mu')^2[(\mu'\pi'' - \mu''\pi')/(\mu')^2]} \\
&= -\frac{(\pi/\mu)'}{(\pi'/\mu')'}\left(\frac{\mu}{\mu'}\right)^2 \mu'
\end{aligned}
$$

$$(13.88)$$

where Eq. (13.82) is utilized. Thus, the general solution to Eq. (13.84) is Eq. (13.88), that is, the substrate concentration must be varied during the singular period in accordance with Eq. (13.88). The feed rate to satisfy Eq. (13.88) is obtained by substituting it into the substrate balance equation (13.27):

$$
F_{\sin} = \frac{\sigma XV + V\,dS/dt}{S_F - S} = \frac{\left(\dfrac{\mu}{Y_{X/S}} + \dfrac{\pi}{Y_{P/S}} + m\right)XV}{S_F - S} - \frac{V\left(\dfrac{\pi}{\mu}\right)' \mu'}{(S_F - S)\left(\dfrac{\pi'}{\mu'}\right)'}\left(\frac{\mu}{\mu'}\right)^2
$$

$$(13.89)$$

Thus, the singular feed rate given by Eq. (13.89) is in feedback mode; that is, knowledge of kinetic information (specific rates and kinetic parameters) and the measurements of state variables (substrate concentration, total amount of cells, and reactor volume) allow a feedback control of the singular feed rate.

13.4.1.2.2 OPTIMAL FEED RATE PROFILE FOR MINIMUM TIME PROBLEM. So far we know that the feed rate consists of a concatenation of various feed rates, $F_{\max}$, $F_{\min}$ and $F_{\sin}$. The sequence and the times at which the feed rate changes from one form to another are unknown at this point. What is known is the initial feed rate, which is $F_{\min}$ when the initial substrate concentration is high, $F_{\max}$ when the initial concentration is low, and $F_{\sin}$ when the initial concentration is chosen appropriately. However, we do not know the precise values of high, low, and appropriate substrate concentrations. Thus, we must postulate sequences of flow rates and numerically determine the best switching times, picking the one sequence that yields the best performance.

For this problem, we have three differential equations for three state variables, and therefore, we introduce three unknown parameters. They are the switching times t_1 (from $F_{\max}$ to $F_{\min}$) and $t_{\sin}$ (between $F_{\min}$ and $F_{\sin}$) and the final time, t_f. To determine these switching times, we integrate three differential equations with a candidate (assumed) sequence of feed rates with three switching times, for example, $F_{\max}(0, t_1) \rightarrow F_{\min}(t_1, t_{\sin}) \rightarrow F_{\sin}(t_{\sin}, t_{\text{full}}) \rightarrow F_b(t_{\sin}, t_f) = 0$ with the given initial conditions. This is possible because $F_{\sin}$ obtained from Eq. (13.89) is in terms of state variables only and free of adjoint variables. At the final time, the integrated result must match the given final condition (Eq. (13.79)). Thus, we iterate on the three switching times, t_1, t_{full}, and t_f, until the calculated final state matches the given final state. If the sequence guessed were wrong, there would be a convergence problem. Other feasible sequences of feed rate for the minimum time problems are as follows: $F_{max}(0, t_{sin}) \rightarrow F_{sin}(t_{sin}, t_{full}) \rightarrow F_b(t_{full}, t_f) = 0$, $F_{min}(0, t_{sin}) = 0 \rightarrow F_{sin}(t_{sin}, t_{full}) \rightarrow F_b(t_{full}, t_f) = 0$ and $F_{sin}(0, t_{full}) \rightarrow F_b(t_{full}, t_f) = 0$. Hence, we must try all sequences that are consistent and feasible and then pick the one that yields the best performance index.

EXAMPLE 13.E.2: MINIMUM TIME PROBLEM: A CONSTANT-YIELD CASE Specific rates and simulation conditions are the same as in Example 13.E.1, except for the performance index and the fixed final metabolite amount. In summary, the minimum time optimization problem is organized as follows: $P[\mathbf{x}(t_f)] = -t_f = -x_5(t_f)$, subject to

$$\mu = 0.124S, \pi = -384\mu^2 + 134\mu, \sigma = \mu/0.135, \ PV(t_f) = 760 \text{ g}$$

$$[S_F, F_{max}, F_{min}, V_{max}] = [2.7 \text{ wt.\%}, \ 2 \text{ L/hr}, 0.20 \text{ L}],$$

$$[X_0, S_0, P_0, V_0] = [0.06 \text{ g/L}, \ 2.7 \text{ wt.\%}, 0 \text{ g/L}, 4 \text{ L}]$$

Similar to the metabolite maximization problem of Example 13.E.1, the initial substrate concentration, $S_F = 2.7$, is larger than that on the singular arc, $S_m = 1.407$. Therefore, initially, the reactor is operated in batch mode ($F_{min}(0, t_{sin}) = 0$) to decrease the substrate concentration to reach the singular arc, S_{sin}, as soon as possible. The singular feed, $F_{sin}(t_{sin}, t_{full})$, follows until the bioreactor is full. The final batch period, $F_b = 0$, follows until the given final metabolite amount, $PV(t_f) = 760$ g, is met. The overall optimal feed rate strategy is $F_{min} = 0 \rightarrow F_{sin} \rightarrow F_b = 0$. The simulation results are given in Figure 13.E.2.1. The total process time is $t_f = 38.7$ hrs, and the switching times t_{sin} and t_{full} are 1.66 and 32.6 hrs, respectively. The final state variables, the total numbers of cells, substrates, and product, and the bioreactor volume are $[XV, \ SV, \ PV, \ V](t_f) = [7.47 \text{ g}, \ 0.953 \text{ g}, \ 760 \text{ g}, \ 20.0 \text{ L}]$.

It is apparent from the preceding discussions that the optimal feed rate sequence depends on the initial condition (substrate concentration). Therefore, it is very important to choose the proper initial substrate concentration so that the optimal sequence begins with the singular feed rate without either the maximum or minimum feed rate period. Thus, the selection of the best initial substrate concentration is essential. However, in general, the best initial substrate concentration is not known a priori and must be determined based on the model. Only in limited cases, such as when the specific product formation rate is proportional to the specific cell growth rate $\pi = a\mu$, and therefore, they both peak at the same substrate concentration, $S = S_m$, is the best initial condition $S = S_m$.

13.4.1.2.3 MAXIMUM PRODUCTIVITY PROBLEMS. We now consider maximizing the productivity. The performance index to be maximized is the total amount of product formed at the final time per unit operating time. Therefore,

$$P[\mathbf{x}(t_f)] = \frac{P(t_f)V(t_f)}{t_f} = \frac{x_2(t_f)}{x_5(t_f)} \tag{13.90}$$

The state equations and initial conditions remain the same (Eq. (13.38)), and the adjoint equations remain intact (Eq. (13.41)). However, the boundary conditions are

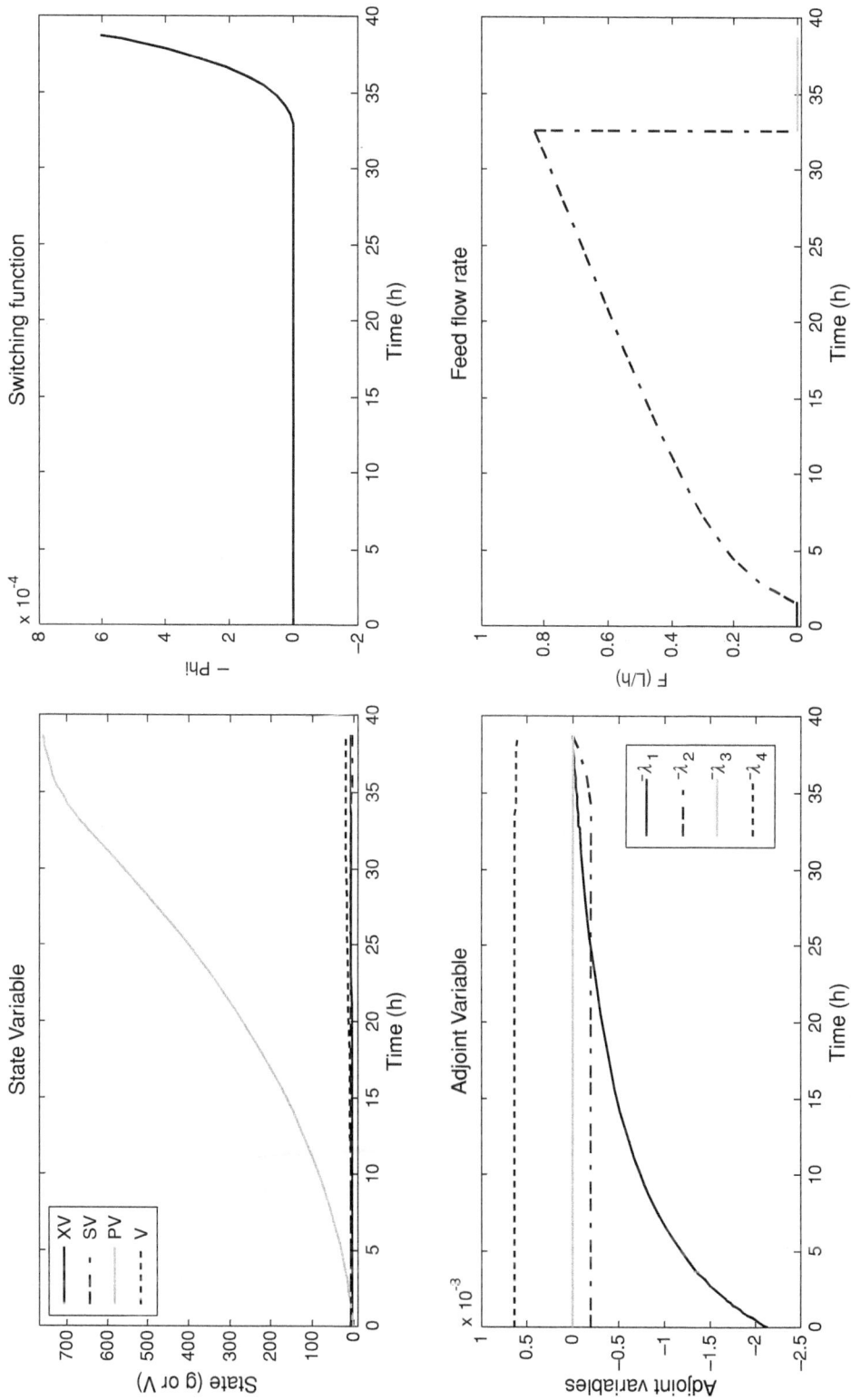

Figure 13.E.2.1. Various time profiles.

different:

$$\lambda(t_f) = \left[\frac{\partial P}{\partial \mathbf{x}(t_f)}\right] = \left[\frac{\partial P}{\partial x_1(t_f)} \ \frac{\partial P}{\partial x_2(t_f)} \ \frac{\partial P}{\partial x_3(t_f)} \ \frac{\partial P}{\partial x_5(t_f)}\right]^T = \left[0 \ \ \frac{1}{x_5(t_f)} \ \text{free} \ 0 \ \ \frac{-x_2(t_f)}{[x_5(t_f)]^2}\right]^T$$

(13.91)

The Hamiltonian is Eq. (13.39). The singular feed rate is given by Eq. (13.89).

Because the only differences between this section on productivity maximization and the previous section on the minimum time are the boundary conditions on the adjoint variables, the results presented for the minimum time hold here with minor modifications. In view of Eqs. (13.41) and (13.91), we have

$$\lambda_5(t) = \lambda_5(t_f) = x_2(t_f)/x_5^2(t_f) = x_2(t_f)/t_f^2$$

(13.92)

and because the final time is free,

$$H = \lambda_1 \mu x_1 + \lambda_2 \pi x_1 + \lambda_3 F + \lambda_5(t_f) = 0$$

(13.93)

At the final time at which the volume constraint is met and the feed rate is zero, the Hamiltonian is

$$H(t_f) = \lambda_1 \cancel{(t_f)}^{=0} \mu(t_f)x_1(t_f) + \lambda_2(t_f)\pi(t_f)x_1(t_f) + \lambda_5(t_f) = \frac{1}{t_f}\pi(t_f)x_1(t_f) - \frac{x_2(t_f)}{t_f^2}$$

$$= 0 \Rightarrow t_f = \frac{x_2(t_f)}{\pi(t_f)x_1(t_f)} = \frac{PV(t_f)}{\pi XV(t_f)} = \frac{PV(t_f)}{[d(PV)/dt]_{t=t_f}}$$

(13.94)

According to Eq. (13.94), the final time is equal to the total amount of metabolite at the final time divided by the rate of production of metabolite at the final time. Thus, the results presented for the minimum time apply here. Obviously, the numerical results would be different. Nevertheless, they are presented subsequently with little commentary.

Optimal Feed Rate on the Interior Singular Arc. On the interior singular arc, the switching function is identically zero,

$$\phi = \lambda_3 = 0$$

(13.40)

and the Hamiltonian, Eq. (13.93), reduces to

$$H = \lambda_1 \mu x_1 + \lambda_2 \pi x_1 + \lambda_5(t_f) = 0$$

(13.95)

The first and second time derivatives of the switching function must also vanish:

$$\frac{d\phi}{dt} = \frac{d\lambda_3}{dt} = -(\lambda_1\mu' + \lambda_2\pi')\frac{(S_F - g/x_3)x_1}{x_3} = 0 \quad \Rightarrow (\lambda_1\mu' + \lambda_2\pi') = 0$$

(13.96)

$$\frac{d^2\phi}{dt^2} = -(\lambda_1\mu + \lambda_2\pi)\mu' + (\lambda_1\mu'' + \lambda_2\pi'')\frac{dS}{dt} = 0$$

(13.97)

As shown for the minimum time problem, a general solution to this equation is obtained by solving Eq. (13.97) for dS/dt:

$$\frac{dS}{dt} = \frac{(\lambda_1\mu + \lambda_2\pi)\mu'}{\lambda_1\mu'' + \lambda_2\pi''} = \frac{[\mu - (\mu'/\pi')\pi]\mu'}{\mu'' - (\mu'/\pi')\pi''} = -\frac{(\pi/\mu)'}{(\pi'/\mu')'}\left(\frac{\mu}{\mu'}\right)^2 \mu'$$

(13.98)

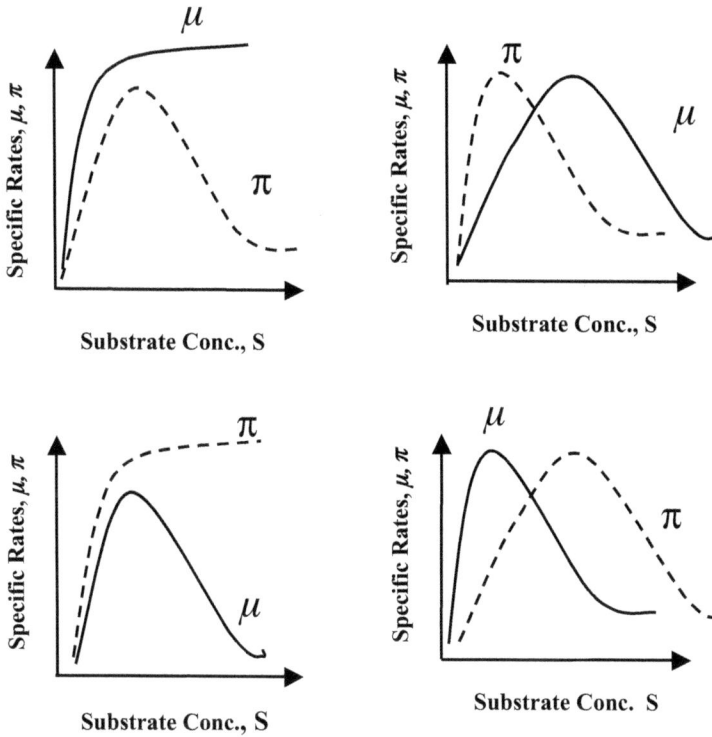

Figure 13.2. Various forms of specific rates as a function of substrate concentration.

Thus, according to Eq. (13.98), the substrate concentration must be varied during the singular period. The feed rate to satisfy Eq. (13.98) is obtained by substituting it into the substrate balance equation (13.4):

$$F_{\sin} = \frac{\sigma X V + V \, dS/dt}{S_F - S} = \frac{(\mu/Y_{X/S} + \pi/Y_{P/S} + m)XV}{S_F - S} - \frac{V}{(S_F - S)} \frac{(\pi/\mu)'\mu'}{(\pi'/\mu')'} \left(\frac{\mu}{\mu'}\right)^2$$

(13.99)

Thus, the singular feed rate given by Eq. (13.99) is in feedback form.

As in the case of the minimum time problem, we must iterate on the switching times, t_1, $t_{\sin}$, and t_f. The only difference is that the final time is determined from Eq. (13.94), whereas in the minimum time problem, the final time must satisfy the given final state, the final amount of product.

13.4.2 Variable-Yield Coefficients and Maintenance Requirement

When one or both yield coefficients are variable, say, functions of substrate concentration, or if there is a maintenance requirement, $m \neq 0$, one must work with all four dynamic equations as no reduction in the number of differential equations is possible. Variable yields imply that the specific growth rate and the specific metabolite production rate peak at different substrate concentrations or that specific rates may be monotonic. These situations are depicted in Figure 13.2. When both specific rates are monotonic, the optimal policy is batch operation, just as in the case of Monod.

13.4.2.1 Performance Index Independent of Final Time, $\partial P/\partial t_f = 0$

Because the performance index is independent of the final time, we do not need to augment with an additional state variable, x_5. Thus, the state equations are given by Eq. (13.8) without the last row of Eq. (13.8):

$$\frac{d\mathbf{x}}{dt} = \frac{d}{dt}\begin{bmatrix} x_1 \\ x_2 \\ x_3 \\ x_4 \end{bmatrix} = \frac{d}{dt}\begin{bmatrix} XV \\ PV \\ V \\ SV \end{bmatrix} = \begin{bmatrix} \mu x_1 \\ \pi x_1 \\ F \\ S_F F - \sigma x_1 \end{bmatrix} \begin{bmatrix} x_1(0) \\ x_2(0) \\ x_3(0) \\ x_4(0) \end{bmatrix} = \begin{bmatrix} x_{10} \\ x_{20} \\ x_{30} \\ x_{40} \end{bmatrix} \quad (13.8)$$

The objective is to maximize the performance index:

$$\underset{F(t)}{\text{Max}}\, P[\mathbf{x}(t_f)] \quad (13.10)$$

The following volume (state variable) and flow rate (manipulated variable) constraints remain intact:

$$x_3(t_f) = V_{\max} \quad (13.11)$$

and

$$0 = F_{\min} \leq F(t) \leq F\max \quad (13.7)$$

The substrate concentration in terms of state variables is $S = SV/V = x_4/x_3$.

The Hamiltonian is

$$H = (\lambda_1 \mu + \lambda_2 \pi - \lambda_4 \sigma)x_1 + (\lambda_3 + S_F\lambda_4)F = H_1 + \phi F \quad (13.100)$$

The switching function is

$$\phi = \lambda_3 + S_F\lambda_4 \quad (13.101)$$

The adjoint equations are

$$\frac{d}{dt}\begin{bmatrix} \lambda_1 \\ \lambda_2 \\ \lambda_3 \\ \lambda_4 \end{bmatrix} = -\begin{bmatrix} \partial H/\partial x_1 \\ \partial H/\partial x_2 \\ \partial H/\partial x_3 \\ \partial H/\partial x_4 \end{bmatrix} = -\begin{bmatrix} \lambda_1 \mu + \lambda_2 \pi - \lambda_4 \sigma \\ 0 \\ -(\lambda_1 \mu' + \lambda_2 \pi' - \lambda_4 \sigma')(x_1 x_4/x_3^2) \\ (\lambda_1 \mu' + \lambda_2 \pi' - \lambda_4 \sigma')(x_1/x_3) \end{bmatrix} \quad (13.102)$$

where

$$\sigma = \mu/Y_{X/S} + \pi/Y_{P/S} + m \quad (13.9)$$

The boundary conditions are

$$[\lambda_1(t_f)\; \lambda_2(t_f)\; \lambda_3(t_f)\; \lambda_4(t_f)] = [\partial P/\partial x_1(t_f)\; \partial P/\partial x_2(t_f)\; free\; \partial P/\partial x_4(t_f)]$$
$$(13.103)$$

Equations (13.102) and (13.103) imply that the second adjoint variable is a constant:

$$\lambda_2(t) = \lambda_2(t_f) = \partial P/\partial x_2(t_f) \quad (13.104)$$

The first-order derivative of the switching function is

$$\frac{d\phi}{dt} = \frac{d\lambda_3}{dt} + S_F \frac{d\lambda_4}{dt} = \frac{x_1}{x_3}[\lambda_1\mu' + \lambda_2\mu' - \lambda_4\sigma']\left(\frac{x_4}{x_3} - S_F\right) = 0$$
$$\Rightarrow \lambda_1\mu' + \lambda_2\pi' - \lambda_4\sigma' = 0 \qquad (13.105)$$

13.4.2.2 Free Final Time, t_f

Because the final time is free, the Hamiltonian must vanish:

$$H = (\lambda_1\mu + \lambda_2\pi - \lambda_4\sigma)x_1 + (\lambda_3 + S_F\lambda_4)F = 0 \qquad (13.106)$$

At the final time, there is no feed, and the volume constraint is met. Therefore, the last term in Eq. (13.106) vanishes. Because the Hamiltonian is continuous, its value at the final time must also be zero:

$$H(t_f) = [\lambda_1(t_f)\mu(t_f) + \lambda_2(t_f)\pi(t_f) - \lambda_4(t_f)\sigma(t_f)]x_1(t_f)$$
$$= \left[\frac{\partial P}{\partial x_1(t_f)}\mu(t_f) + \frac{\partial P}{\partial x_2(t_f)}\pi(t_f) - \frac{\partial P}{\partial x_4(t_f)}\sigma(t_f)\right]x_1(t_f) = 0 \qquad (13.107)$$

Because $x_1(t_f) = XV(t_f) \neq 0$,

$$\frac{\partial P}{\partial x_1(t_f)}\mu(t_f) + \frac{\partial P}{\partial x_2(t_f)}\pi(t_f) - \frac{\partial P}{\partial x_4(t_f)}\sigma(t_f) = 0 \qquad (13.108)$$

If the performance index contains explicitly all three state variables, $x_i(t_f), i = 1, 2, 4$, so that $\partial P/x_i(t_f) \neq 0, i = 1, 2, 4$, then the specific rates, $\mu(t_f), \pi(t_f)$, and $\sigma(t_f)$, must collectively vanish at the final time. If only one of the partial derivatives is zero, say, $\partial P/x_1(t_f) = 0$ (the performance index is free of $x_1(t_f)$), then the corresponding specific rate, $\mu(t_f)$, can be finite, but the remaining specific rates, $\pi(t_f)$ and $\sigma(t_f)$, must vanish at the final time. In either case, this can happen only when the substrate is completely consumed so that the specific rates are zero. This implies that the final time, which is assumed free, must approach infinity, $t_f \to \infty$. *Namely, when the final time does not appear explicitly in the performance index and is free, the final time approaches infinity*, a familiar event we have seen before.

On the Boundary Arc. On the boundary arc, (t_{full}, t_f), the feed rate is zero, $F_b = 0$, and the volume constraint, $V = V_m$, is met so that the Hamiltonian, Eq. (13.106), reduces to

$$H = (\lambda_1\mu + \lambda_2\pi - \lambda_4\sigma)x_1 = (\lambda_1\mu + [\partial P/\partial x_2(t_f)]\pi - \lambda_4\sigma)x_1 = 0$$
$$\Rightarrow \lambda_1\mu + \lambda_2\pi - \lambda_4\sigma = \lambda_1\mu + [\partial P/\partial x_2(t_f)]\pi - \lambda_4\sigma = 0 \quad t_{\text{full}} \leq t \leq t_f \quad (13.109)$$

From Eqs. (13.102) and (13.109), the time derivative of $\lambda_1(t)$ is equal to zero, and thus we see that the first adjoint variable is also a constant:

$$\lambda_1(t) = \partial P/\partial x_1(t_f) \quad t_{\text{full}} \leq t \leq t_f \qquad (13.110)$$

Optimal Feed Rate on Singular Arc. On the singular arc, the volume constraint is satisfied, and the switching function is identically zero. Therefore, the Hamiltonian, Eq. (13.106), becomes

$$H = (\lambda_1\mu + \lambda_2\pi - \lambda_4\sigma)x_1 = 0 \quad t_{\text{sin}} \leq t \leq t_{\text{full}} \qquad (13.111)$$

Because $x_1 \neq 0$, the quantity in the parentheses must be identically zero:

$$(\lambda_1 \mu + \lambda_2 \pi - \lambda_4 \sigma) = 0 \quad t_{\sin} \leq t \leq t_{\text{full}} \tag{13.112}$$

In view of Eq. (13.106), this equality holds over the time interval covering the singular period and the subsequent batch period, $t_{\sin} \leq t \leq t_f$, and because there is no jump[20] in the adjoint variables across the interior singular arc and the boundary arc, Eq. (13.110) must hold over the singular arc:

$$\lambda_1(t) = \partial P / \partial x_1(t_f) \quad t_{\sin} \leq t \leq t_f \tag{13.113}$$

The switching function and its derivatives are

$$\phi = \lambda_3 + S_F \lambda_4 = 0 \tag{13.114}$$

$$\frac{d\phi}{dt} = \frac{d\lambda_3}{dt} + S_F \frac{d\lambda_4}{dt} = (\lambda_1 \mu' + \lambda_2 \pi' - \lambda_4 \sigma')(S_F - x_4/x_3)(x_1/x_3) = 0 \tag{13.115}$$

Because $(S_F - x_4/x_3)(x_1/x_3) = (S_F - S)X \neq 0$, Eq. (13.115) implies that

$$\lambda_1 \mu' + \lambda_2 \pi' - \lambda_4 \sigma' = 0 \tag{13.116}$$

Substitution of Eqs. (13.112) and (13.116) into the dynamic adjoint equations, Eq. (13.102), shows that all adjoint variables are constants:

$$\frac{d}{dt} \begin{bmatrix} \lambda_1(t) \\ \lambda_2(t) \\ \lambda_3(t) \\ \lambda_4(t) \end{bmatrix} = \begin{bmatrix} 0 \\ 0 \\ 0 \\ 0 \end{bmatrix} \Rightarrow \begin{bmatrix} \lambda_1(t) \\ \lambda_2(t) \\ \lambda_3(t) \\ \lambda_4(t) \end{bmatrix} = \begin{bmatrix} \partial P/\partial x_1(t_f) \\ \partial P/\partial x_2(t_f) \\ a \\ b \end{bmatrix} \tag{13.117}$$

where a and b are unknown constants.

Determination of Singular Feed Rate from $\phi = 0$, $\dot{\phi} = 0$, *and* $\ddot{\phi} = 0$. The second derivative of the switching function is obtained by differentiating Eq. (13.115):

$$\frac{d^2\phi}{dt^2} = \left[\left(S_F - \frac{x_4}{x_3} \right) \left(\frac{x_1}{x_3} \right) \right] \frac{d}{dt} (\lambda_1 \mu' + \lambda_2 \pi' - \lambda_4 \sigma')$$

$$= \left[\frac{dS}{dt} (\lambda_1 \mu'' + \lambda_2 \pi'' - \lambda_4 \sigma'') + \left(\frac{d\lambda_1}{dt} \mu' + \frac{d\lambda_2}{dt} \pi' - \frac{d\lambda_4}{dt} \sigma' \right) \right] \left[\left(S_F - \frac{x_4}{x_3} \right) \left(\frac{x_1}{x_3} \right) \right]$$

$$= 0 \tag{13.118}$$

This implies that

$$\frac{dS}{dt} = -\frac{\dfrac{d\lambda_1}{dt} \mu' + \dfrac{d\lambda_2}{dt} \pi' - \dfrac{d\lambda_4}{dt} \sigma'}{\lambda_1 \mu'' + \lambda_2 \pi'' - \lambda_4 \sigma''} \tag{13.119}$$

In view of Eq. (13.117), Eq. (13.119) reduces to show that the substrate concentration is kept constant on the singular arc:

$$\frac{dS}{dt} = 0 \tag{13.120}$$

The value of the constant substrate concentration that needs to be kept constant is not known at this point. Let us denote the unknown constant substrate concentration as S_1. The singular feed rate to maintain the substrate concentration constant is obtained by utilizing Eq. (13.120) in the substrate balance equation (Eq. (13.4)) and is

$$\frac{d(VS)}{dt} = S_1 \frac{dV}{dt} = S_1 F = S_F F - \sigma XV \quad \Rightarrow F_{\sin} = \frac{\sigma(S_1)XV}{S_F - S_1} \quad (13.121)$$

To gain insight as to the unknown measure the singular feed rate is maximizing, we now analyze the singular arc.

Analysis of Singular Arc. Substituting Eq. (13.117) into Eqs. (13.112) and (13.116), we obtain

$$\begin{aligned} [\partial P/\partial x_1(t_f)]\mu + [\partial P/\partial x_2(t_f)]\pi - b\sigma = 0 \\ [\partial P/\partial x_1(t_f)]\mu' + [\partial P/\partial x_2(t_f)]\pi' - b\sigma' = 0 \end{aligned} \quad (13.122)$$

Solving Eq. (13.122) for b and equating to each other, we obtain

$$b = \lambda_4 = \frac{[\partial P/\partial x_1(t_f)]\mu' + [\partial P/\partial x_2(t_f)]\pi'}{\sigma'} = \frac{[\partial P/\partial x_1(t_f)]\mu + [\partial P/\partial x_2(t_f)]\pi}{\sigma} \quad (13.123)$$

This is rearranged to yield

$$\begin{aligned} [\partial P/\partial x_1(t_f)](\sigma\mu' - \sigma'\mu) &+ [\partial P/\partial x_2(t_f)](\sigma\pi' - \sigma'\pi) \\ &= \left\{ [\partial P/\partial x_1(t_f)]\frac{(\sigma\mu' - \sigma'\mu)}{\sigma^2} + [\partial P/\partial x_2(t_f)]\frac{(\sigma\pi' - \sigma'\pi)}{\sigma^2} \right\} \sigma^2 \\ &= \sigma^2 \frac{d}{dS}\left\{ [\partial P/\partial x_1(t_f)]\left(\frac{\mu}{\sigma}\right) + [\partial P/\partial x_2(t_f)]\left(\frac{\pi}{\sigma}\right) \right\} = 0 \\ &\Rightarrow \frac{d}{dS}\left[[\partial P/\partial x_1(t_f)]\left(\frac{\mu}{\sigma}\right) + [\partial P/\partial x_2(t_f)]\left(\frac{\pi}{\sigma}\right) \right] = 0 \end{aligned} \quad (13.124)$$

Equation (13.124) implies that the singular feed rate maximizes the weighted sum of cell and product yields:

$$\underset{S}{\mathrm{Max}}\left[\frac{\partial P}{\partial x_1(t_f)}\left(\frac{\mu}{\sigma}\right) + \frac{\partial P}{\partial x_2(t_f)}\left(\frac{\pi}{\sigma}\right) \right] \quad (13.125)$$

or

$$\underset{S}{\mathrm{Max}}\left[\frac{\partial P}{\partial x_1(t_f)}Y_{X/S} + \frac{\partial P}{\partial x_2(t_f)}Y_{P/S} \right] \quad (13.126)$$

The cell mass yield is weighted by the sensitivity of the performance index with respect to the total amount of cell mass, $\partial P/\partial x_1(t_f)$, while the product yield is weighted by the sensitivity of the performance index with respect to the total amount of product, $\partial P/\partial x_2(t_f)$. Therefore, at this point, it is instructive to consider specific performance indices.

13.4.2.2.1 CASE I PERFORMANCE INDEX DEPENDS ON ONLY $x_2(t_f)$.

This case corresponds to maximization of the product at the final time so that $P = x_2(t_f)$. Then $\partial P/\partial x_1(t_f) = 0$ and $\partial P/\partial x_2(t_f) = 1$, and Eq. (13.125) reduces to

$$\underset{S^*}{\mathrm{Max}}\left[\left(\frac{\pi}{\sigma}\right)\right] = \underset{S^*}{\mathrm{Max}}[Y_{P/S}] \quad (13.127)$$

Thus, the singular feed rate maximizes the specific product formation rate relative to the specific substrate consumption rate, in other words, the yield coefficient $Y_{P/S} = \pi/\sigma$ is maximized.

13.4.2.2.2 CASE II PERFORMANCE INDEX DEPENDS ON $x_1(t_f)$ AND $x_2(t_f)$. This is the case in which both the cell mass and metabolite are valuable so that the performance index weighs the total amounts of both the cell mass and metabolite, such as $P = [\alpha x_1(t_f) + x_2(t_f)]$, where $\alpha \ll 1$ is the value of cell mass relative to that of metabolite. In this case, the weighing factors are $\partial P/\partial x_1(t_f) = \alpha$ and $\partial P/\partial x_2(t_f) = 1$. Thus, the singular feed rate maximizes the weighted sum of the cell mass yield, $(\mu/\sigma) = Y_{X/S}$, and the product yield, $(\pi/\sigma) = Y_{P/S}$. Equation (13.125) becomes

$$\underset{S^*}{\text{Max}}\left[\alpha\left(\frac{\mu}{\sigma}\right) + \left(\frac{\pi}{\sigma}\right)\right] = \underset{S^*}{\text{Max}}[\alpha Y_{X/S} + Y_{P/S}] \qquad (13.128)$$

Because Eq. (13.128) is a function of substrate concentration only, let us denote the substrate concentration that maximizes it as $S = S_1$.

The singular feed rate expression can be obtained by analyzing Eq. (13.126), the fact that the singular feed rate maximizes the weighted sum of cell and product yields. We do know that the substrate concentration must be held at this value throughout the singular feed rate period. The singular feed rate that maintains $S = S_1$ is readily obtained from the substrate balance equation, Eq. (13.27), as

$$F_{\text{sin}} = \frac{\sigma(S_1)XV}{S_F - S_1} = \left[\frac{\mu(S_1)}{Y_{X/S}(S_1)} + \frac{\pi(S_1)}{Y_{P/S}(S_1)}\right]\frac{XV}{S_F - S_1} \qquad (13.129)$$

We have the singular feed rate in feedback mode, and it is also an exponential function of time, as we have seen a number of times. During the singular feed rate period, the substrate concentration is maintained at S_1, and consequently, the cell mass grows exponentially according to the cell mass balance equation (13.1):

$$XV(t) = XV(t_1)\exp\{\mu(S_1)(t - t_f)\} = XV(t_1)\exp[-\mu(S_1)t_f]\exp[\mu(S_1)t]$$
$$= \alpha\exp[\mu(S_1)t] \qquad (13.130)$$

where t_1 is the time at which the singular feed rate begins and also the time at which $S = S_1$. Therefore, the singular feed rate is an exponential feed rate obtained by substituting Eq. (13.130) into Eq. (13.129):

$$F_{\text{sin}} = \frac{\sigma(S_1)XV}{S_F - S_1} = \frac{\sigma(S_1)\alpha\exp[\mu(S_1)t]}{S_F - S_1} = \beta\exp[\mu(S_1)t] \qquad (13.131)$$

13.4.2.3 Optimal Feed Rate Profile

We must now construct the optimal profile for the entire operational time, $0 \le t \le t_f$. The procedure is summarized as follows:

1. First we determine the value of the substrate concentration S_1 that maximizes the weighted sum of cell mass and product yields or the product yield using the condition imposed by Eq. (13.127) or Eq. (13.126), $\partial[\ ]/\partial S = 0$.
2. Then, we check if the initial substrate concentration S_0 is equal to, greater than, or less than S_1 and classify the initial conditions into three cases: (1) $S_0 = S_1$, (2) $S_0 > S_1$, and (3) $S_0 < S_1$.

3. For case 1, apply the singular flow rate (F_{sin}) given by Eq. (13.129) from $t = 0$ until the bioreactor is full ($t = t_{full}$) and then run as a batch ($F_b = 0$) until the final time t_f.

   For case 2, run as a batch ($F_{min} = 0$) until the substrate concentration is reduced to S_1, $S = S_1$, and then apply the singular flow rate (F_{sin}) of Eq. (13.129) until the bioreactor is full, $t = t_{full}$; then, follow with a batch period ($F_b = 0$) until the final time t_f.

4. For numerical determination of the switching times t_1, t_{full}, and t_f, knowing the optimal feed rate structure, that is, the sequence and the form of feed rate for each interval (a period of maximum feed rate F_{max}, minimum feed rate $F_{min} = 0$, singular feed rate F_{sin}, and finally, a batch operation $F_b = 0$), we can integrate the four state differential equations, Eq. (13.8), with an initial guess of unknown parameters, t_1 and t_f (t_{full} becomes known during the integration because it is the time required to fill the reactor). As an example, consider case 3. Because the structure of the optimal feed rate profile is known, and the unknowns are the switching times, $F_{max}(0, t_1) \rightarrow F_{sin}(t_1, t_{full}) \rightarrow F_b = 0 (t_{full}, t_f)$, we integrate the four mass balance equations (13.8) with the given initial conditions using the optimal sequence given. Note that t_{full} is determined from V_{max}, V_0, and t_1, and therefore, only two switching times, t_1 and t_f, are to be optimally searched by carrying out a parameter optimization of t_1 and t_f to maximize the performance index (Eq. (13.128)). Other cases can be handled similarly by using the appropriate optimal feed rate sequences. The optimum initial substrate concentration is $S_0 = S_1$, and for this initial concentration, the singular flow rate is applied for the entire period of filling and then for a batch period until the final time.

13.4.2.4 Fixed Final Time

Because the final time is fixed, the Hamiltonian is an unknown constant:

$$H = (\lambda_1 \mu + \lambda_2 \pi - \lambda_4 \sigma)x_1 + (\lambda_3 + S_F\lambda_4)F = H^*(\text{constant}) \quad (13.132)$$

Thus, there is one less equation as compared to the free final time problem.

Optimal Feed Rate on Singular Arc. On the singular arc, the volume constraint is satisfied, and the switching function is identically zero. Therefore, the Hamiltonian becomes

$$H = (\lambda_1 \mu + \lambda_2 \pi - \lambda_4 \sigma)x_1 = H^* \quad (13.133)$$

Because $x_1 \neq 0$,

$$(\lambda_1 \mu + \lambda_2 \pi - \lambda_4 \sigma) = H^*/x_1 \quad (13.134)$$

Because H^* is an unknown constant, Eq. (13.131) cannot be used. The switching function and its derivative are

$$\phi = \lambda_3 + S_F\lambda_4 = 0 \quad (13.114)$$

$$\dot{\phi} = \lambda_1 \mu' + \lambda_2 \pi' - \lambda_4 \sigma' = 0 \quad (13.116)$$

Therefore, the adjoint equation, Eq. (13.102), reduces to

$$\frac{d}{dt}\begin{bmatrix} \lambda_1 \\ \lambda_2 \\ \lambda_3 \\ \lambda_4 \end{bmatrix} = -\begin{bmatrix} \partial H/\partial x_1 \\ \partial H/\partial x_2 \\ \partial H/\partial x_3 \\ \partial H/\partial x_4 \end{bmatrix} = -\begin{bmatrix} \lambda_1 \mu + \lambda_2 \pi - \lambda_4 \sigma \\ 0 \\ 0 \\ 0 \end{bmatrix} \qquad (13.135)$$

Equation (13.135) shows that the second, third, and fourth adjoint variables are constants on the singular arc.

The singular feed rate is obtained from the second derivative of the switching function, or the derivative of Eq. (13.116), with the aid of Eqs. (13.117) and (13.135):

$$\begin{aligned} \frac{d^2\phi}{dt^2} &= \frac{d}{dt}\{[\lambda_1\mu' + \lambda_2\pi' - \lambda_4\sigma'](S_F - S)x_1\} \\ &= \left[\frac{d\lambda_1}{dt}\mu' + (\lambda_1\mu'' + \lambda_2\pi'' - \lambda_4\sigma'')\frac{dS}{dt}\right](S_F - S)x_1 = 0 \end{aligned} \qquad (13.136)$$

Substitution of Eqs. (13.104) and (13.116) into Eq. (13.136) and solving for dS/dt, we obtain

$$\frac{dS}{dt} = \frac{[\lambda_1\mu + \lambda_2\pi - \lambda_4\sigma]\mu'}{\lambda_1\mu'' + \lambda_2\pi'' - \lambda_4\sigma''} = \frac{[\lambda_4(\mu/\sigma)'\sigma^2 + (\partial P/\partial x_2(t_f))(\pi/\mu)'\mu^2]\mu'}{\lambda_4(\mu'/\sigma')'(\sigma')^2 + (\partial P/\partial x_2(t_f))(\pi'/\mu')'(\mu')^2} \qquad (13.137)$$

where λ_4 is an unknown constant. Unfortunately, there is one unknown constant that appears in Eq. (13.137) so that we cannot assess how the substrate concentration varies on the singular arc and also makes the numerical computation a bit more complex. Substitution of the preceding into the substrate balance equation (13.27) yields the singular feed rate

$$F_{\text{sin}} = \frac{(dS/dt + \sigma X)V}{S_F - S} = \frac{[\lambda_4(\mu/\sigma)'\sigma^2 + (\partial P/\partial x_2(t_f))(\pi/\mu)'\mu^2]\mu'}{\left[\lambda_4\left(\dfrac{\mu'}{\sigma'}\right)'(\sigma')^2 + \dfrac{\partial P}{\partial x_2(t_f)}\left(\dfrac{\pi'}{\mu'}\right)'(\mu')^2\right](S_F - S)} + \frac{\sigma XV}{S_F - S}$$

$$(13.138)$$

Therefore, we must resort to a numerical scheme to determine the switching times. The procedure is exactly the same as before, except the final time is given.

13.4.2.4.1 NUMERICAL DETERMINATION OF THE SWITCHING TIME t_1. There are two unknown parameters, λ_4 and t_1. The sequence of feed rates is (1) $F_{\text{max}} \rightarrow F_{\text{sin}} \rightarrow F_b = 0$, (2) $F_{\text{min}} \rightarrow F_{\text{sin}} \rightarrow F_b = 0$, or (3) $F_{\text{sin}} \rightarrow F_b = 0$. Because which of these sequences is the optimal is not known a priori, the performance indices using each one of the sequences have to be evaluated, and the one that gives the maximum value is the optimal sequence with proper switching times. With the first sequence, the state equation, Eq. (13.8), is integrated from $t = 0$ to a guessed value of $t = t_{\text{sin}}$ using $F_{\text{max}}(0, t_{\text{sin}})$, followed by a guessed λ_4^0 value in the $F_{\text{sin}}(t_{\text{sin}}, t_{\text{full}})$ until the bioreactor is full, t_{full}, which is followed by $F_b = 0$ until the given t_f. The resulting performance index is evaluated and denoted by $P^0(t_{\text{sin}}^0, \lambda_4^0)$. Then, using slightly different values of switching time t_{sin} and λ_4, $t_{\text{sin}}^1 = t_{\text{sin}}^0 + \varepsilon_1^1$ and $\lambda_4^1 = \lambda_4^0 + \varepsilon_2^1$, the procedure is repeated to evaluate $P_1^1(t_{\text{sin}}^1, \lambda_4^0)$ and $P_2^1(t_{\text{sin}}^1, \lambda_4^1)$. The gradients $\partial P/\partial t_{\text{sin}} = [P_1^1(t_{\text{sin}}^1, \lambda_4^0) - P^0(t_{\text{sin}}^0, \lambda_4^0)]/\varepsilon_1^0$ and $\partial P/\partial \lambda_4 = [P_2^1(t_{\text{sin}}^1, \lambda_4^1) - P_2^0(t_{\text{sin}}^1, \lambda_4^0)]/\varepsilon_2^0$ are evaluated, and two-parameter search algorithms on t_{sin} and λ_4 are initiated to maximize the performance index. This procedure is repeated for the second and the third sequences, except the third sequence has only one unknown, t_{sin}, which should make the iteration one-dimensional. Of the three sequences, the one that gives the highest value of the performance index is the optimal solution.

13.4.2.5 Performance Index Dependent on Final Time

Because the performance index depends on the final time, we must work with all five state variables:

$$
\frac{d}{dt}\begin{bmatrix} x_1 \\ x_2 \\ x_3 \\ x_4 \\ x_5 \end{bmatrix} = \frac{d}{dt}\begin{bmatrix} XV \\ PV \\ V \\ SV \\ t \end{bmatrix} = \begin{bmatrix} \mu x_1 \\ \pi x_1 \\ 0 \\ -\sigma x_1 \\ 1 \end{bmatrix} + \begin{bmatrix} 0 \\ 0 \\ 1 \\ S_F \\ 0 \end{bmatrix} F \qquad \begin{bmatrix} x_1(0) \\ x_2(0) \\ x_3(0) \\ x_4(0) \\ x_5(0) \end{bmatrix} = \begin{bmatrix} XV(0) \\ PV(0) \\ V(0) \\ SV(0) \\ t(0) \end{bmatrix} = \begin{bmatrix} x_{10} \\ x_{20} \\ x_{30} \\ x_{40} \\ 0 \end{bmatrix}
$$

$$(13.8)$$

The Hamiltonian is zero because the final time is free:

$$H = (\lambda_1 \mu + \lambda_2 \pi - \lambda_4 \sigma)x_1 + (\lambda_3 + S_F \lambda_4)F + \lambda_5 = 0 \qquad (13.12)$$

The switching function is

$$\phi = \lambda_3 + S_F \lambda_4 \qquad (13.12)$$

Therefore, the optimal feed rate can take on the following values:

$$
F = \begin{cases} F_{\text{max}} & \text{when } \phi = (\lambda_3 + S_F \lambda_4) > 0 \\ F_{\text{min}} = 0 & \text{when } \phi = (\lambda_3 + S_F \lambda_4) < 0 \\ F_{\text{sin}} & \text{when } \phi = (\lambda_3 + S_F \lambda_4) \equiv 0 \text{ over } t_q < t < t_{q+1} \\ F_b = 0 & \text{when } x_3(t) = V_{\text{max}} \end{cases} \qquad (13.19)
$$

Let us look at the Hamiltonian at the final time, when the volume constraint is met and the feed rate is zero,

$$
H(t_f) = [\lambda_1(t_f)\mu(t_f) + \lambda_2(t_f)\pi(t_f) - \lambda_4(t_f)\sigma(t_f)]x_1(t_f) + \lambda_5(t_f)
$$
$$
= \left\{ \left[\frac{\partial P}{\partial x_1(t_f)}\right]\mu(t_f) + \left[\frac{\partial P}{\partial x_2(t_f)}\right]\pi(t_f) - \left[\frac{\partial P}{\partial x_4(t_f)}\right]\sigma(t_f) \right\} x_1(t_f) + \lambda_5(t_f) = 0
$$

$$(13.139)$$

which, unlike the problem of maximizing the performance index independent of a final time that is free, does not suggest an infinite time. *Namely, when the final time appears explicitly in the performance index and is free, the final time is finite.*

Optimal Feed Rate on Singular Arc. On the arc, the switching function is identically zero and the volume constraint is inactive so that the reduced Hamiltonian is also zero:

$$H = (\lambda_1 \mu + \lambda_2 \pi - \lambda_4 \sigma)x_1 + \lambda_5 = 0 \qquad (13.140)$$

The first time derivative of the switching function is zero:

$$\frac{d\phi}{dt} = \frac{d\lambda_3}{dt} + S_F \frac{d\lambda_4}{dt} = (\lambda_1 \mu' + \lambda_2 \pi' - \lambda_4 \sigma')\left(S_F - \frac{x_4}{x_3}\right)\left(\frac{x_1}{x_3}\right) = 0 \qquad (13.141)$$

Because $(S_F - x_4/x_3)x_1/x_3 \neq 0$,

$$(\lambda_1 \mu' + \lambda_2 \pi' - \lambda_4 \sigma') = 0 \tag{13.142}$$

The adjoint equations reduce to the following owing to Eq. (13.140):

$$\frac{d}{dt}\begin{bmatrix} \lambda_1 \\ \lambda_2 \\ \lambda_3 \\ \lambda_4 \\ \lambda_5 \end{bmatrix} = -\begin{bmatrix} \partial H/\partial x_1 \\ \partial H/\partial x_2 \\ \partial H/\partial x_3 \\ \partial H/\partial x_4 \\ \partial H/\partial x_5 \end{bmatrix} = -\begin{bmatrix} \lambda_1 \mu + \lambda_2 \pi - \lambda_4 \sigma \\ 0 \\ -(\lambda_1 \mu' + \lambda_2 \pi' - \lambda_4 \sigma')(x_1 x_4/x_3^2) \\ (\lambda_1 \mu' + \lambda_2 \pi' - \lambda_4 \sigma')x_1/x_3 \\ 0 \end{bmatrix}$$

$$= -\begin{bmatrix} \lambda_1 \mu + \lambda_2 \pi - \lambda_4 \sigma \\ 0 \\ 0 \\ 0 \\ 0 \end{bmatrix} \tag{13.143}$$

The final conditions on the adjoint variables are

$$[\lambda_1(t_f), \lambda_2(t_f), \lambda_3(t_f), \lambda_4(t_f), \lambda_5(t_f)]$$
$$= \left[\frac{\partial P}{\partial x_1(t_f)}, \frac{\partial P}{\partial x_2(t_f)}, \text{ a constant}, \frac{\partial P}{\partial x_4(t_f)}, \frac{\partial P}{\partial x_5(t_f)} \right] \tag{13.144}$$

Equations (13.143) and (13.144) imply that only the first adjoint variable varies with time and the remaining four are constants during the singular period. In view of Eqs. (13.143) and (13.144), the second and fifth adjoint variables are constant over the entire operational period:

$$\begin{bmatrix} \lambda_2 \\ \lambda_5 \end{bmatrix} = \begin{bmatrix} \partial P/\partial x_2(t_f) \\ \partial P/\partial x_5(t_f) \end{bmatrix} \tag{13.145}$$

The second time derivative of the switching function is

$$\frac{d^2\phi}{dt^2} = \left[\frac{d\lambda_1}{dt} \mu' + (\lambda_1 \mu'' + \lambda_2 \pi'' - \lambda_4 \sigma'') \frac{dS}{dt} \right] (S_F - S)x_1 = 0 \tag{13.146}$$

Solving Eq. (13.146) for dS/dt and substituting Eqs. (13.140), (13.142), and (13.145) into it yields

$$\frac{dS}{dt} = \frac{(\lambda_1 \mu + \lambda_2 \pi - \lambda_4 \sigma)\mu'}{\lambda_1 \mu'' + \lambda_2 \pi'' - \lambda_4 \sigma''}$$

$$= \frac{-[\partial P/\partial x_5(t_f)]\mu'/x_1}{\left\{ \dfrac{\partial P}{\partial x_2(t_f)} \left(\dfrac{\pi'}{\mu'}\right)' (\mu')^2 + \left[\left(\dfrac{\partial P}{\partial x_5(t_f)}\right) \dfrac{\mu'}{x_1} - \dfrac{\partial P}{\partial x_2(t_f)} \left(\dfrac{\pi}{\mu}\right)' \mu^2 \right] \left[\left(\dfrac{\mu'}{\sigma'}\right)' / \left(\dfrac{\mu}{\sigma}\right)' \left(\dfrac{\sigma'}{\sigma}\right)^2 \right] \right\}} \tag{13.147}$$

According to Eq. (13.147), the substrate concentration varies with time. It should be noted here that if the performance index does not depend on the final time,

$\partial P/\partial x_5(t_f) = 0$, then the substrate concentration is identically zero on the singular arc. Substitution of Eq. (13.147) into the substrate balance equation (13.27) to obtain the singular feed rate,

$$F_{\sin} = \frac{\sigma XV + V\,dS/dt}{S_F - S} = \frac{\sigma x_1}{S_F - x_4 x_3}$$

$$+ \frac{-[\partial P/\partial x_5(t_f)]\mu' x_3/x_1}{\left(S_F - \frac{x_4}{x_3}\right)\left\{\frac{\partial P}{\partial x_2(t_f)}\left(\frac{\pi}{\mu'}\right)'(\mu')^2 + \left[\left(\frac{\partial P}{\partial x_5(t_f)}\right)\frac{\mu'}{x_1} - \frac{\partial P}{\partial x_2(t_f)}\left(\frac{\pi}{\mu}\right)'\mu^2\right]\left[\left(\frac{\mu'}{\sigma'}\right)' / \left(\frac{\mu}{\sigma}\right)'\left(\frac{\sigma'}{\sigma}\right)^2\right]\right\}}$$

$$(13.148)$$

Once again, the expression for the singular feed rate appears very complicated, but the functional form is identical to the case of the performance index independent of the final time. Equation (13.148) is the singular feed rate in feedback form. However, we do not have any insight as to what the singular feed rate does. Therefore, we must resort to a numerical scheme to determine the switching times. The procedure is exactly the same as before.

13.4.2.5.1. MINIMUM TIME PROBLEMS. The performance index is the minimization of the final time, which is equivalent to maximizing the negative of the final time to obtain a fixed amount of product, $P(t_f)V(t_f) = x_{2f}$, using the given reactor volume:

$$P = -x_5(t_f) \qquad (13.149)$$

The state equations are given by Eq. (13.8), and the initial and final conditions are as follows:

$$\frac{d}{dt}\begin{bmatrix} x_1 \\ x_2 \\ x_3 \\ x_4 \\ x_5 \end{bmatrix} = \frac{d}{dt}\begin{bmatrix} XV \\ PV \\ V \\ SV \\ t \end{bmatrix} = \begin{bmatrix} \mu x_1 \\ \pi x_1 \\ 0 \\ -\sigma x_1 \\ 1 \end{bmatrix} + \begin{bmatrix} 0 \\ 0 \\ 1 \\ S_F \\ 0 \end{bmatrix} F$$

$$\begin{bmatrix} x_1(0) \\ x_2(0) \\ x_3(0) \\ x_4(0) \\ x_5(0) \end{bmatrix} = \begin{bmatrix} x_{10} \\ x_{20} \\ x_{30} \\ x_{140} \\ 0 \end{bmatrix} \qquad \begin{bmatrix} x_1(t_f) \\ x_2(t_f) \\ x_3(t_f) \\ x_4(t_f) \\ x_5(t_f) \end{bmatrix} = \begin{bmatrix} \text{free} \\ x_{2f} \\ x_{3f} \\ \text{free} \\ \text{free} \end{bmatrix} \qquad (13.150)$$

It is assumed that the final state can be reached from the initial state with the piecewise continuous feed rate; that is, the controllability is assumed.

The Hamiltonian is zero:

$$H = (\lambda_1 \mu + \lambda_2 \pi - \lambda_4 \sigma)x_1 + (\lambda_3 + S_F \lambda_4)F + \lambda_5 = 0 \qquad (13.12)$$

where the specific rates are functions of x_4/x_3. The switching function is

$$\phi = \lambda_3 + S_F \lambda_4 \qquad (13.12)$$

The optimal feed rate that maximizes the Hamiltonian is a concatenation of

$$
F = \begin{cases}
F_{max} & \text{when } \phi = (\lambda_3 + S_F \lambda_4) > 0 \\
F_{min} = 0 & \text{when } \phi = (\lambda_3 + S_F \lambda_4) < 0 \\
F_{sin} & \text{when } \phi = (\lambda_3 + S_F \lambda_4) \equiv 0 \text{ over } t_q < t < t_{q+1} \\
F_b = 0 & \text{when } x_3(t) = V_{max}
\end{cases}
\tag{13.19}
$$

The adjoint equations are

$$
\frac{d}{dt}\begin{bmatrix} \lambda_1 \\ \lambda_2 \\ \lambda_3 \\ \lambda_4 \\ \lambda_5 \end{bmatrix} = -\begin{bmatrix} \partial H/\partial x_1 \\ \partial H/\partial x_2 \\ \partial H/\partial x_3 \\ \partial H/\partial x_4 \\ \partial H/\partial x_5 \end{bmatrix} = -\begin{bmatrix} \lambda_1 \mu + \lambda_2 \pi - \lambda_4 \sigma \\ 0 \\ -(\lambda_1 \mu' + \lambda_2 \pi' - \lambda_4 \sigma')(x_1 x_4/x_3^2) \\ (\lambda_1 \mu' + \lambda_2 \pi' - \lambda_4 \sigma')x_1/x_3 \\ 0 \end{bmatrix}
\tag{13.151}
$$

The boundary conditions on the adjoint variables are obtained from the transversality,

$$
\sum_{i=1,2,3,5} \frac{\partial P}{\partial x_i(t_f)} \delta x_i(t_f) = \sum_{i=1,2,3,4,5} \lambda_i(t_f)\delta x_i(t_f) - \sum_{i=1,2,3,4,5} \lambda_i(0)\delta x_i(0)^{=0}
$$

$$
\Rightarrow -\delta x_5(t_f) = \lambda_1(t_f)\delta x_1(t_f) + \lambda_2(t_f)\delta x_2(t_f)^{=0} + \lambda_3(t_f)\delta x_3(t_f)^{=0}
$$

$$
+ \lambda_4(t_f)\delta x_4(t_f) + \lambda_5(t_f)\delta x_5(t_f)
$$

$$
\Rightarrow \left[\lambda_1(t_f), \lambda_2(t_f), \lambda_3(t_f), \lambda_4(t_f), \lambda_5(t_f)\right] = [0, a_1, a_2, 0, -1]
\tag{13.152}
$$

where a_1 and a_2 are unknown constants. It is apparent that the second and fifth adjoint variables are constant over the entire operational period $0 \le t \le t_f$:

$$
[\lambda_2(t)\lambda_5(t)] = [a_1 - 1]
\tag{13.153}
$$

Optimal Feed Rate on Singular Arc. Because this is a free final time problem, the Hamiltonian is zero. Therefore, Eq. (13.12) reduces to

$$
H = (\lambda_1 \mu + \lambda_2 \pi - \lambda_4 \sigma)x_1 - 1 = 0
\tag{13.154}
$$

Thus,

$$
\lambda_1 \mu + \lambda_2 \pi - \lambda_4 \sigma = 1/x_1
\tag{13.155}
$$

Owing to Eq. (13.142), the adjoint equation (13.151) reduces to

$$
\frac{d\lambda}{dt} = \begin{bmatrix} \lambda_1 \mu + \lambda_2 \pi - \lambda_4 \sigma \\ 0 \\ 0 \\ 0 \\ 0 \end{bmatrix}
\tag{13.156}
$$

Because of Eq. (13.151), $d^2\phi/dt^2 = 0$ implies that

$$
\frac{d\lambda_1}{dt}\mu' + (\lambda_1 \mu'' + \lambda_2 \pi'' - \lambda_4 \sigma'')\frac{dS}{dt} = 0
\tag{13.157}
$$

To obtain the expression for dS/dt, we substitute Eq. (13.156) into Eq. (13.157):

$$\frac{dS}{dt} = \frac{(\lambda_1\mu + \lambda_2\pi - \lambda_4\sigma)\mu'}{\lambda_1\mu'' + \lambda_2\pi'' - \lambda_4\sigma''} \qquad (13.158)$$

Substitution of Eqs. (13.142), (13.153), and (13.155) into Eq. (13.158) yields

$$\frac{dS}{dt} = \frac{(\lambda_1\mu + \lambda_2\pi - \lambda_4\sigma)\mu'}{\lambda_1\mu'' + \lambda_2\pi'' - \lambda_4\sigma''} = -\frac{(\sigma'-\sigma)/x_1}{\dfrac{\sigma'-\sigma''}{x_1} + a_1\left[\left(\dfrac{\pi}{\sigma}\right)'\sigma^2 + \pi''(\sigma'-\sigma) - \sigma''(\pi'-\pi)\right]} \qquad (13.159)$$

The presence of adjoint variables prevents us from assessing the time rate of change in substrate concentration during the singular period. However, Eq. (13.159) suggests that the substrate concentration is not necessarily constant during the period.

The singular feed rate is obtained by substituting Eq. (13.159) into the substrate balance equation (13.27):

$$F_{\text{sin}} = \frac{\left(\dfrac{dS}{dt} + \sigma X\right)V}{S_F - S}$$

$$= \frac{1}{S_F - S}\left[\frac{\dfrac{(\sigma'-\sigma)}{x_1}}{\dfrac{\sigma'-\sigma''}{x_1} + a_1\lfloor\left(\dfrac{\pi}{\sigma}\right)'\sigma^2 + \pi''(\sigma'-\sigma) - \sigma''(\pi'-\pi)\rfloor}V + \sigma XV\right] \qquad (13.160)$$

Numerical Solution. The optimal feed rate structure is given by Eq. (13.19). However, the exact sequence and the switching times are not known. In addition, the singular feed rate given by Eq. (13.160) contains one unknown constant a_1 and one switching time t_{sin}. Feasible sequences of feed rates are (1) $F_{\text{max}} \rightarrow F_{\text{sin}} \rightarrow F_b = 0$, (2) $F_{\text{min}} \rightarrow F_{\text{sin}} \rightarrow F_b = 0$, or (3) $F_{\text{sin}} \rightarrow F_b = 0$. The performance indices using each one of the sequences have to be evaluated, and the one that gives the maximum value is the optimal sequence with proper switching times. The procedure is much like that used in Section 13.4.1.2.2.

13.4.2.5.2. MAXIMUM PRODUCTIVITY PROBLEMS. The performance index to be maximized is the total amount of product formed per unit operating time:

$$P[\mathbf{x}(t_f)] = \frac{P(t_f)V(t_f)}{t_f} = \frac{x_2(t_f)}{x_5(t_f)} \qquad (13.90)$$

The state equations and the initial conditions are given by Eq. (13.150). The Hamiltonian is

$$H = (\lambda_1\mu + \lambda_2\pi - \lambda_4\sigma)x_1 + \lambda_5 + (\lambda_3 + S_F\lambda_4)F = H_1 + \phi F \qquad (13.12)$$

and the adjoint equations are

$$\frac{d}{dt}\begin{bmatrix} \lambda_1 \\ \lambda_2 \\ \lambda_3 \\ \lambda_4 \\ \lambda_5 \end{bmatrix} = -\begin{bmatrix} \partial H/\partial x_1 \\ \partial H/\partial x_2 \\ \partial H/\partial x_3 \\ \partial H/\partial x_4 \\ \partial H/\partial x_5 \end{bmatrix} = -\begin{bmatrix} \lambda_1\mu + \lambda_2\pi - \lambda_4\sigma \\ 0 \\ -(\lambda_1\mu' + \lambda_2\pi' - \lambda_4\sigma')(x_1 x_4/x_3^2) \\ (\lambda_1\mu' + \lambda_2\pi' - \lambda_4\sigma')x_1/x_3 \\ 0 \end{bmatrix} \qquad (13.161)$$

The boundary conditions are obtained from Eq. (13.91):

$$\lambda(t_f) = \begin{bmatrix} \dfrac{\partial P}{\partial x_1(t_f)} & \dfrac{\partial P}{\partial x_2(t_f)} & \dfrac{\partial P}{\partial x_3(t_f)} & \dfrac{\partial P}{\partial x_4(t_f)} & \dfrac{\partial P}{\partial x_5(t_f)} \end{bmatrix}^T = \begin{bmatrix} 0 & \dfrac{1}{x_5(t_f)} & \text{free} & 0 & \dfrac{-x_2(t_f)}{[x_5(t_f)]^2} \end{bmatrix}^T$$

$$(13.162)$$

From Eqs. (13.161) and (13.162), the second and fifth adjoint variables are constants through the entire operational period:

$$[\lambda_2 \quad \lambda_5] = \begin{bmatrix} \dfrac{1}{x_5(t_f)} & \dfrac{-x_2(t_f)}{[x_5(t_f)]^2} \end{bmatrix} \qquad (13.163)$$

The optimal feed rate that maximizes the Hamiltonian is a concatenation of

$$F = \begin{cases} F_{\max} & \text{when } \phi = \lambda_3 > 0 \\ F_{\min} = 0 & \text{when } \phi = \lambda_3 < 0 \\ F_{\sin} & \text{when } \phi = \lambda_3 \equiv 0 \text{ over } t_q < t < t_{q+1} \\ F_b = 0 & \text{when } x_3(t) = V_{\max} \end{cases} \qquad (13.19)$$

Because the only differences between this section on productivity maximization and the previous section on the minimum time are the boundary conditions on the adjoint variables, the results presented for the minimum time should hold here with minor modifications.

Because the final time is free,

$$H = (\lambda_1\mu + \lambda_2\pi - \lambda_4\sigma)x_1 + \lambda_5(t_f) + (\lambda_3 + S_F\lambda_4)F \overset{\Delta}{=} H_1 + \phi F = 0 \qquad (13.164)$$

At the final time at which the volume constraint is met and the feed rate is zero, the Hamiltonian is

$$H(t_f) = \lambda_1 \cancel{(t_f)}^{=0}\mu(t_f)x_1(t_f) + \lambda_2(t_f)\pi(t_f)x_1(t_f) - \lambda_4\cancel{(t_f)}^{=0}\sigma(t_f)x_1(t_f) + \lambda_5(t_f)$$
$$= [\pi(t_f)x_1(t_f) - x_2(t_f)/t_f]/t_f = 0 \quad \Rightarrow t_f = x_2(t_f)/x_1(t_f)\pi(t_f) \qquad (13.165)$$

Equation (13.165) sets the final time. The results presented for the minimum time apply here. Of course, numerical results would be different. Nevertheless, they are presented subsequently with little commentary.

Optimal Feed Rate on the Interior Singular Arc. On the interior singular arc, the switching function is identically zero,

$$\phi = \lambda_3 + S_F\lambda_4 \qquad (13.12)$$

and the Hamiltonian, Eq. (13.164), reduces to

$$H = (\lambda_1\mu + \lambda_2\pi - \lambda_4\sigma)x_1 + \lambda_5(t_f) = 0 \tag{13.166}$$

The first and second time derivatives of the switching function must also vanish:

$$\frac{d\phi}{dt} = \frac{d\lambda_3}{dt} = (\lambda_1\mu' + \lambda_2\pi' - \lambda_4\sigma')\frac{x_1 x_4}{x_3^2} = 0 \quad \Rightarrow \quad (\lambda_1\mu' + \lambda_2\pi' - \lambda_4\sigma') = 0$$

$$\tag{13.167}$$

$$\frac{d^2\phi}{dt^2} = \frac{d\lambda_1}{dt}\mu' + (\lambda_1\mu'' + \lambda_2\pi'' - \lambda_4\sigma'')\frac{dS}{dt} = 0 \tag{13.168}$$

In view of Eq. (13.167), we see that Eq. (13.161) yields

$$\frac{d}{dt}\begin{bmatrix} \lambda_1 \\ \lambda_2 \\ \lambda_3 \\ \lambda_4 \\ \lambda_5 \end{bmatrix} = -\begin{bmatrix} \partial H/\partial x_1 \\ \partial H/\partial x_2 \\ \partial H/\partial x_3 \\ \partial H/\partial x_4 \\ \partial H/\partial x_5 \end{bmatrix} = -\begin{bmatrix} \lambda_1\mu + \lambda_2\pi - \lambda_4\sigma \\ 0 \\ 0 \\ 0 \\ 0 \end{bmatrix} \quad t_{\text{sin}} \leq t_{\text{full}} \tag{13.169}$$

Owing to Eqs. (13.166) and (13.169), Eq. (13.168) reduces to

$$\frac{d^2\phi}{dt^2} = -\mu'\lambda_5(t_f)/x_1 + (\lambda_1\mu'' + \lambda_2\pi'' - \lambda_4\sigma'')\frac{dS}{dt} = 0 \tag{13.170}$$

As shown for the minimum time problem, a general solution to this equation is obtained by solving Eq. (13.170) for dS/dt and substituting Eqs. (13.163), (13.166) and (13.167) into the resultant.

$$\frac{dS}{dt} = \frac{\mu'\lambda_5(t_f)/x_1}{\lambda_2(t_f)\left[\pi' - \dfrac{[\mu''(\pi/\sigma)' + \sigma''(\pi/\mu)'(\mu/\sigma)^2]}{(\mu/\sigma)'}\right] - \lambda_5(t_f)\left[\dfrac{(\mu'\sigma')'}{x_1(\mu/\sigma)'\sigma^2}\right]}$$

$$\tag{13.171}$$

Thus, according to Eq. (13.171), the substrate concentration changes with time on the singular period. The feed rate to satisfy Eq. (13.171) is obtained by substituting it into the substrate balance equation (13.4):

$$F_{\text{sin}} = \frac{\sigma XV + V\,dS/dt}{S_F - S}$$

$$= \frac{(\mu/Y_{X/S} + \pi/Y_{P/S} + m)XV}{S_F - S}$$

$$+ \frac{\mu'\lambda_5(t_f)V/[x_1(S_F - S)]}{\lambda_2(t_f)\left[\pi' - \dfrac{[\mu''(\pi/\sigma)' + \sigma''(\pi/\mu)'(\mu/\sigma)^2]}{(\mu/\sigma)'}\right] - \lambda_5(t_f)\left[\dfrac{(\mu'\sigma')'}{x_1(\mu/\sigma)'\sigma^2}\right]}$$

$$\tag{13.172}$$

Thus, the singular feed rate given by Eq. (13.172) is in feedback mode because the adjoint variables $\lambda_2(t_f)$ and $\lambda_5(t_f)$ are constant parameters as shown in Eq. (13.163).

As in the case of minimum time, we must iterate on the switching times, t_1, t_{fill}, and t_f. The only difference is that the final time is determined from

Eq. (13.165), whereas in the minimum time problem, the final time must satisfy the given final value of the state.

Optimal Feed Rate Profile. The optimal feed rate profile is a concatenation of F_{max}, $F_{min} = 0$, F_{sin}, and $F_b = 0$. The exact sequence and the times at which the feed rate forms change are unknown. The general sequence is $F_{max}(0, t_1) \rightarrow F_{min} = 0 (t_1, t_{sin}) \rightarrow F_{sin}(t_{sin}, t_{full}) \rightarrow F_b = 0(t_{full}, t_f)$. We have two unknown times: t_1, the time at which F_{max} changes to $F_{min} = 0$, and t_{sin}, at which F_{min} changes to F_{sin}. It should be noted that t_{full} is obtained during the integration period when the bioreactor volume is full and the final time, t_f, is given by Eq. (13.165). Using the initial conditions, Eq. (13.38), integrate the state equations (13.38) forward in time using the preceding feed rate sequence in which t_1 and t_2 are the unknowns until the final time (Eq. (13.165)). Search the best values of t_1 and t_2 that maximize the performance index.

13.5 Substrate and Product Concentration–Dependent Specific Rates, $\mu(S, P)$, $\pi(S, P)$, and $\sigma(S, P)$

The specific rates of growth, product formation, and substrate consumption are now functions of the concentrations of substrate and product: $\mu(S, P)$, $\pi(S, P)$, and $\sigma(S, P)$. We consider only maximization of the total amount of metabolite at the final time, $\underset{F(t)}{Max}[P = P(t_f)V(t_f)]$. The state equations are

$$\frac{d}{dt}\begin{bmatrix} x_1 \\ x_2 \\ x_3 \\ x_4 \end{bmatrix} = \frac{d}{dt}\begin{bmatrix} XV \\ SV \\ PV \\ V \end{bmatrix} = \begin{bmatrix} \mu x_1 \\ -\sigma x_1 \\ \pi x_1 \\ 0 \end{bmatrix} + \begin{bmatrix} 0 \\ S_F \\ 0 \\ 1 \end{bmatrix} F \qquad \begin{bmatrix} x_1(0) \\ x_2(0) \\ x_3(0) \\ x_4(0) \end{bmatrix} = \begin{bmatrix} x_{10} \\ x_{20} \\ x_{30} \\ x_{40} \end{bmatrix} \qquad (13.173)$$

The objective, once again, is to maximize the performance index, the final amount of product:

$$\underset{F(t)}{Max}[P = P(t_f)V(t_f) = x_3(t_f)] \qquad (13.174)$$

The following volume and flow rate constraints are placed:

$$x_4(t_f) = V_{max} \qquad (13.11)$$

and

$$0 = F_{min} \leq F(t) \leq F_{max} \qquad (13.7)$$

The Hamiltonian is

$$H = (\lambda_1\mu - \lambda_2\sigma + \lambda_3\pi)x_1 + (S_F\lambda_2 + \lambda_4)F \triangleq H_1 + \phi F \qquad (13.175)$$

Therefore, the switching function is

$$\phi = S_F\lambda_2 + \lambda_4 \qquad (13.176)$$

The adjoint equations and boundary conditions are

$$\frac{d}{dt}\begin{bmatrix} \lambda_1 \\ \lambda_2 \\ \lambda_3 \\ \lambda_4 \end{bmatrix} = -\frac{\partial H}{\partial \mathbf{x}} = -\begin{bmatrix} \lambda_1 \mu - \lambda_2 \sigma + \lambda_3 \pi \\ (\lambda_1 \mu_s - \lambda_2 \sigma_s + \lambda_3 \pi_s)(x_1/x_4) \\ (\lambda_1 \mu_p - \lambda_2 \sigma_p + \lambda_3 \pi_p)(x_1/x_4) \\ [(\lambda_1 \mu_s - \lambda_2 \sigma_s + \lambda_3 \pi_s)(-x_2/x_4)+ \\ (\lambda_1 \mu_p - \lambda_2 \sigma_p + \lambda_3 \pi_p)(-x_3/x_4)](x_1/x_4) \end{bmatrix},$$

$$\begin{bmatrix} \lambda_1(t_f) \\ \lambda_2(t_f) \\ \lambda_3(t_f) \\ \lambda_4(t_f) \end{bmatrix} = \begin{bmatrix} 0 \\ 0 \\ 1 \\ \text{free} \end{bmatrix} \qquad (13.177)$$

where the subscripts s and p refer to the partial derivative with respect to substrate concentration and product concentration, respectively.

The Hamiltonian at the optimal condition is continuous and constant. At the final time at which $F = 0$, Eq. (13.175) reduces, owing to the boundary conditions of Eq. (13.177), to

$$H(t) = H(t_f) = [\lambda_1(t_f)\mu(t_f) - \lambda_2(t_f)\sigma(t_f) + \lambda_3(t_f)\pi(t_f)]x_1(t_f)$$
$$= \pi(t_f)x_1(t_f) > 0 \qquad (13.178)$$

Thus, the Hamiltonian is positive over the entire time $(0, t_f)$. Because the Hamiltonian is linear in the switching function ϕ, it is maximized according to the sign of ϕ:

$$F = \begin{bmatrix} F_{\max} & \text{when } \phi = S_F \lambda_2 + \lambda_4 > 0 \\ F_{\min} = 0 & \text{when } \phi = S_F \lambda_2 + \lambda_4 < 0 \\ F_{\sin} & \text{when } \phi = S_F \lambda_2 + \lambda_4 = 0 \quad \text{over finite interval(s)} \\ F_b = 0 & \text{when } x_4 = V_{\max} \end{bmatrix} \qquad (13.179)$$

The optimal feed rate profile $F(t)$ is a concatenation of $F_{\max}$, $F_{\min}$, $F_{\sin}$, and $F_b = 0$. What is not known at this point is the exact sequence and the times at which the switching takes place from one form to another. The initial conditions, the magnitude of $F_{\max}$, and the final time t_f play a key role in the structure of $F(t)$ and the switching times.

Singular Feed Rate. Because the switching function is identically zero over the finite time singular interval, its time derivatives must also vanish:

$$\phi = S_F \lambda_2 + \lambda_4 = 0 \qquad (13.180)$$

$$\frac{d\phi}{dt} = S_F \frac{d\lambda_2}{dt} + \frac{d\lambda_4}{dt} = -\{\lambda_1[\mu_s(S_F - S) - \mu_p P] - \lambda_2[\sigma_s(S_F - S) - \sigma_p P] + \lambda_3[\pi_s(S_F - S) - \pi_p P]\}x_1/x_4 = 0 \qquad (13.181)$$

Equation (13.181) can be further simplified by recognizing that

$$\chi_s(S_F - S) = -\partial\chi/\partial \ln(S_F - S) = -\partial\chi/\partial \ln \tilde{S}, \quad \chi = \mu, \pi, \sigma$$
$$\chi_p P = \partial\chi/\partial \ln P, \quad \chi = \mu, \pi, \sigma \qquad (13.182)$$

In terms of Eq. (13.182), Eq. (13.181) can be written as

$$\lambda_1 \left[\frac{\partial \mu}{\partial \ln \tilde{S}} + \frac{\partial \mu}{\partial \ln P} \right] - \lambda_2 \left[\frac{\partial \sigma}{\partial \ln \tilde{S}} + \frac{\partial \sigma}{\partial \ln P} \right] + \lambda_3 \left[\frac{\partial \pi}{\partial \ln \tilde{S}} + \frac{\partial \pi}{\partial \ln P} \right] \triangleq \lambda_1 \underset{\sim}{\mu} - \lambda_2 \underset{\sim}{\sigma} + \lambda_3 \underset{\sim}{\pi} = 0$$

$$(13.183)$$

The second derivative is

$$\frac{d^2\phi}{dt^2} = \dot{\lambda}_1 \underset{\sim}{\mu} - \dot{\lambda}_2 \underset{\sim}{\sigma} + \dot{\lambda}_3 \underset{\sim}{\pi} + (\lambda_1 \underset{\sim}{\mu}_s - \lambda_2 \underset{\sim}{\sigma}_s + \lambda_3 \underset{\sim}{\pi}_s)\frac{dS}{dt} + (\lambda_1 \underset{\sim}{\mu}_p - \lambda_2 \underset{\sim}{\sigma}_p + \lambda_3 \underset{\sim}{\pi}_p)\frac{dP}{dt} = 0$$

$$(13.184)$$

where the subscripts S and P denote partial derivatives with respect to S and P, respectively. The derivatives dS/dt and dP/dt are obtained from the state equation, Eq. (13.173):

$$\frac{d(SV)}{dt} = V\frac{dS}{dt} + SF = FS_F - \sigma XV \Rightarrow \frac{dS}{dt} = F(S_F - S)/V - \sigma X$$

$$\frac{d(PV)}{dt} = V\frac{dP}{dt} + PF = \pi XV \Rightarrow \frac{dP}{dt} = -PF/V + \pi X \qquad (13.185)$$

Substituting Eq. (13.185) and the adjoint equation, Eq. (13.177), into Eq. (13.184) and rearranging yields

$$F_{\sin} = \frac{-R\mu + [(R_p/\pi)\check{Y}Y_{P/S}^2 - (R_s/\sigma)\check{Y}]\sigma^2 X}{(R/x_4)^{\hat{\hat{}}}} \qquad (13.186)$$

where $R \triangleq \lambda_1 \underset{\sim}{\mu} - \lambda_2 \underset{\sim}{\sigma} + \lambda_3 \underset{\sim}{\pi}$, the single hat is $()^{\hat{}} \triangleq \partial()/\partial \ln \tilde{S} + \partial()/\partial \ln P$, and the double hat is used to denote the second derivative: $()^{\hat{\hat{}}} \triangleq \partial^2()/(\partial \ln \tilde{S})^2 + \partial^2()/(\partial \ln P)^2$.

Existence of Singular Arc. Assessment of time behavior of adjoint variables can lead to a very efficient way of predicting the optimal feed rate strategy. So far, all we know is that the optimal feed rate structure is a concatenation of $F_{\max}, F_{\min} = 0, F_{\sin}$, and $F_b = 0$, but the exact sequence is not known at this time. It would be extremely helpful if we could obtain sufficient conditions in terms of specific rates μ, σ, and π. In some cases, it has been possible to obtain the criteria free of the adjoint variables by carefully deducing and examining their time behavior.[12] The criteria in terms of specific rates are given in Table 13.1. In the references cited, there are many other criteria, but they contain the adjoint variables and therefore cannot be used a priori to accept or reject the singular arc; however, they can be checked a posteriori for numerical results. For example, when $\hat{\sigma} > 0, \hat{\mu} < 0$, and $\hat{\pi} < 0$, or $\hat{\sigma} < 0, \hat{\mu} > 0$ and $\hat{\pi} > 0$, then there is no singular arc.

To make use of the sufficiency given in Table 13.1, we need to calculate the special gradient of the specific rates: $\hat{\mu}, \hat{\sigma}$, and $\hat{\pi} (\hat{\theta} = \partial\theta/\partial \ln \tilde{S} + \partial\theta/\partial \ln P)$.

Table 13.1. *Sufficient conditions for singular arc*

$\hat{\sigma}$	$\hat{\mu}$	$\hat{\pi}$	Constraints	Singular arc	No.
+	+	+	$\hat{\sigma}/\sigma = \hat{\mu}/\mu > \hat{\pi}/\pi > 0$ or $\hat{\sigma}/\sigma > \hat{\mu}/\mu \geq \hat{\pi}/\pi > 0$	Yes	1
+	+	+	$\hat{\sigma}/\sigma \geq \hat{\mu}/\mu \geq \hat{\pi}/\pi > 0, \hat{\hat{\sigma}}/\hat{\sigma} \geq \hat{\hat{\mu}}/\hat{\mu} > \hat{\hat{\pi}}/\hat{\pi}$, except $\hat{\sigma}/\sigma = \hat{\mu}/\mu = \hat{\pi}/\pi$ and $\hat{\hat{\sigma}}/\hat{\sigma} = \hat{\hat{\mu}}/\hat{\mu} = \hat{\hat{\pi}}/\hat{\pi}$	Yes	2
+	+	+	$0 < \hat{\sigma}/\sigma \leq \hat{\mu}/\mu \leq \hat{\pi}/\pi$ or $\hat{\hat{\sigma}}/\hat{\sigma} \leq \hat{\hat{\mu}}/\hat{\mu} \leq \hat{\hat{\pi}}/\hat{\pi}$, except $\hat{\sigma}/\sigma = \hat{\mu}/\mu = \hat{\pi}/\pi$ and $\hat{\hat{\sigma}}/\hat{\sigma} = \hat{\hat{\mu}}/\hat{\mu} = \hat{\hat{\pi}}/\hat{\pi}$	No	3
+	+	+	$0 < \hat{\sigma}/\sigma \leq \hat{\pi}/\pi \leq \hat{\mu}/\mu$ or $\hat{\hat{\sigma}}/\hat{\sigma} \leq \hat{\hat{\pi}}/\hat{\pi} \leq \hat{\hat{\mu}}/\hat{\mu}$, except $\hat{\sigma}/\sigma = \hat{\mu}/\mu = \hat{\pi}/\pi$ and $\hat{\hat{\sigma}}/\hat{\sigma} = \hat{\hat{\mu}}/\hat{\mu} = \hat{\hat{\pi}}/\hat{\pi}$	No	4
−	−	−	$\hat{\pi}/\pi \leq \hat{\mu}/\mu \leq \hat{\sigma}/\sigma < 0$ or $\hat{\hat{\pi}}/\hat{\pi} \leq \hat{\hat{\mu}}/\hat{\mu} \leq \hat{\hat{\sigma}}/\hat{\sigma}$, except $\hat{\sigma}/\sigma = \hat{\mu}/\mu = \hat{\pi}/\pi$ and $\hat{\hat{\sigma}}/\hat{\sigma} = \hat{\hat{\mu}}/\hat{\mu} = \hat{\hat{\pi}}/\hat{\pi}$	No	5
−	−	−	$\hat{\mu}/\mu \leq \hat{\pi}/\pi \leq \hat{\sigma}/\sigma < 0$ or $\hat{\hat{\mu}}/\hat{\mu} \leq \hat{\hat{\pi}}/\hat{\pi} \leq \hat{\hat{\sigma}}/\hat{\sigma}$, except $\hat{\sigma}/\sigma = \hat{\mu}/\mu = \hat{\pi}/\pi$ and $\hat{\hat{\sigma}}/\hat{\sigma} = \hat{\hat{\mu}}/\hat{\mu} = \hat{\hat{\pi}}/\hat{\pi}$	No	6
−	−	−	$0 > \hat{\pi}/\pi \geq \hat{\mu}/\mu > \hat{\sigma}/\sigma$ or $0 > \hat{\pi}/\pi > \hat{\mu}/\mu = \hat{\sigma}/\sigma$	Yes	7
−	−	−	$0 > \hat{\pi}/\pi \geq \hat{\sigma}/\sigma \geq \hat{\mu}/\mu > 0, \hat{\hat{\sigma}}/\hat{\sigma} \geq \hat{\hat{\mu}}/\hat{\mu} > \hat{\hat{\pi}}/\hat{\pi}$, except $\hat{\sigma}/\sigma = \hat{\mu}/\mu = \hat{\pi}/\pi$ and $\hat{\hat{\sigma}}/\hat{\sigma} = \hat{\hat{\mu}}/\hat{\mu} = \hat{\hat{\pi}}/\hat{\pi}$	Yes	8
+	−	−		No	9
−	+	+		No	10

Source: Combined tables 1 and 2 from Shin and Lim.[12]

EXAMPLE 13.E.3: ETHANOL FROM GLUCOSE BY *S. CEREVISIAE* As an example of specific rates being functions of substrate concentration S and product concentration P, we can cite the fed-batch fermentation[13,14] for ethanol from glucose by *Saccharomyces cerevisiae*. The specific rates are as follows:

$$\mu = \frac{0.408S}{\left(1 + \dfrac{P}{16}\right)(0.22 + S)}, \quad \sigma = 10\mu = \frac{4.08S}{\left(1 + \dfrac{P}{16}\right)(0.22 + S)},$$

$$\pi = \frac{S}{\left(1 + \dfrac{P}{71.5}\right)(0.44 + S)} \tag{13.E.3.1}$$

The initial state values are $[XV, SV, PV, V]$ $(t = 0) = [0.1\,\text{g}, 0.001\,\text{g}, 0\,\text{g}, 1\,\text{L}]$, and the operational parameters (feed substrate concentration, final time, feed rate constraints, and maximum volume) are

$$\left[S_F, t_f, F_{\max}, F_{\min}, V_{\max}\right] = [150\,\text{g/L}, 39.5\,\text{hr}, 2\,\text{L/hr}, 0\,\text{L/hr}, 10\,\text{L}]$$

To make use of the sufficiency given in Table 13.1, we need to calculate the special gradient of the specific rates, $\hat{\mu}, \hat{\sigma},$ and $\hat{\pi}$ $(\hat{\theta} = \partial\theta/\partial \ln \tilde{S} + \partial\theta/\partial \ln P)$:

$$\hat{\mu} = \frac{\partial\mu}{\partial \ln \hat{S}} + \frac{\partial\mu}{\partial \ln P} = -(S_F - S)\frac{\partial\mu}{\partial S} + P\frac{\partial\mu}{\partial P}$$

$$= -\frac{1}{(1 + P/16)(0.22 + S)}\left[\frac{(0.22)(0.408)(S_F - S)}{(0.22 + S)} + \frac{(1/16)0.408S}{(1 + P/16)}\right] < 0$$

$$\hat{\sigma} = 10\hat{\mu}$$

$$\hat{\pi} = -\frac{\partial\pi}{\partial S}(S_F - S) + \frac{\partial\pi}{\partial P}P$$

$$= -\frac{1}{(1 + P/71.5)(0.44 + S)}\left[\frac{0.44(S_F - S)}{(0.44 + S)} + \frac{SP}{71.5(1 + P/71.5)}\right] < 0$$

$$\text{(13.E.3.2)}$$

Equation (13.E.3.2) shows that the logarithmic gradients $\hat{\mu}, \hat{\sigma},$ and $\hat{\pi}$ are all negative, and therefore, cases 7 and 8 in Table 13.1 are possibilities. We now evaluate the inequality conditions. Because $\sigma = 10\mu$, it is apparent that $\hat{\sigma}/\sigma = \hat{\mu}/\mu$. Therefore, number 7 in Table 13.1 is applicable, and the condition is

$$0 > \hat{\pi}/\pi > \hat{\mu}/\mu = \hat{\sigma}/\sigma \quad \Rightarrow \mu\hat{\pi} > \hat{\mu}\pi \quad \text{(13.E.3.3)}$$

Conversely,

$$(\pi/\hat{\mu}) = (\mu\hat{\pi} - \hat{\mu}\pi)/\mu^2 > 0 \quad \text{(13.E.3.4)}$$

which is due to Eq. (13.E.3.3). Likewise, it can be shown that

$$(\hat{\pi}/\hat{\mu}) > 0 \quad \text{(13.E.3.5)}$$

The contour plot of $(\pi/\hat{\mu})$, not shown here, indicates that the substrate concentration should be greater than 10, $S > 10$, to satisfy inequality (13.E.3.5) so that a singular feed rate is part of the optimal feed rate profile. The singular feed rate is given by Eq. (13.186):

$$F_{\sin} = \frac{-R\mu + [(R_p/\pi)\tilde{Y}Y_{P/S}^2 - (R_s/\sigma)\tilde{Y}]\sigma^2 X}{(R/x_4)\hat{Y}} \quad \text{(13.186)}$$

where $R = \lambda_1\mu - \lambda_2\sigma + \lambda_3\pi$ and the subscripts p and s stand for partial derivatives with respect to p and s, respectively.

Optimal Feed Rate Profile. Interpreting the logarithmic gradients to play the roles of the usual gradients of the specific rates, the feed rate should be manipulated to favor $\hat{\mu}$ first, then $(\pi/\hat{\mu})$. Thus, initially, the substrate concentration should be increased from its low initial value of 0.001 g/L using F_{max} to favor the cell growth $\hat{\mu}$ and then should be changed to favor the alcohol production $(\pi/\hat{\mu})$. Numerical results are shown in Figure 13.E.3.1. The optimal feed rate profile consists of F_{max}(shortperiod) $\rightarrow F_{min}(=0) \rightarrow F_{sin} \rightarrow F_b = 0$. The substrate concentration quickly rises and starts to decrease, remains almost constant over the period of singular feed, and rapidly decreases when the culture volume is full. The ethanol concentration rises rapidly during the singular feed period.

13.6 Recombinant Cell Products

To improve the rate of product formation and/or yield of desired product, recombinant cells are widely used. The most common vector is plasmids that contain the specific product genes,[15-18] although chromosomal integration and phage are also used. A specific plasmid forces the cells to synthesize the product originating from the specific gene. In general, the higher the copy number of plasmids, the higher the rate, yield, and productivity will be. The relation is not linear but a saturation type. However, there can be major drawbacks. The growth rates of plasmid-bearing cells (PBC) are usually depressed by the added metabolic burden placed on them by the cloned gene-originated product formation. Thus, the faster-growing plasmid-free cells (PFC) can inevitably replace PBC, making the reactor operation less productive and lower in product yield as time progresses. This problem needs to be avoided.[19] In addition to the depression of cell growth,[22] there is a problem associated with the segregational instability of plasmids.[20,21]

The copy number of a plasmid in a cell is regulated by the replication origin. However, the partition of the plasmids from mother cells to daughter cells during cell division under the par function is not precise and leads to uneven distribution of plasmids between the mother and daughter cells. Transposable DNA elements in cells can be inserted into chromosomes and plasmids. If they are inserted into the cloned specific gene from which the desired product is produced, the gene is modified and cannot produce the desired product. This phenomenon of structural instability is not treated here. PFC compete for limited nutrients in culture and eventually outgrow PBC.

13.6.1 Recombinant Cells with Plasmid Instability

To minimize the depression of the growth rate of PBC, a high ratio of the specific growth rate of PBC relative to that of PFC, μ^+/μ^-, should be maintained through manipulation of the substrate concentration. Some experimental results of temperature effect on the specific growth rate have also been reported.[23-25] However, the growth conditions usually affect not only the ratio μ^+/μ^- but also the plasmid stability. For example, the ratio μ^+/μ^- for *Escherichia coli* increases[27] in media limited in

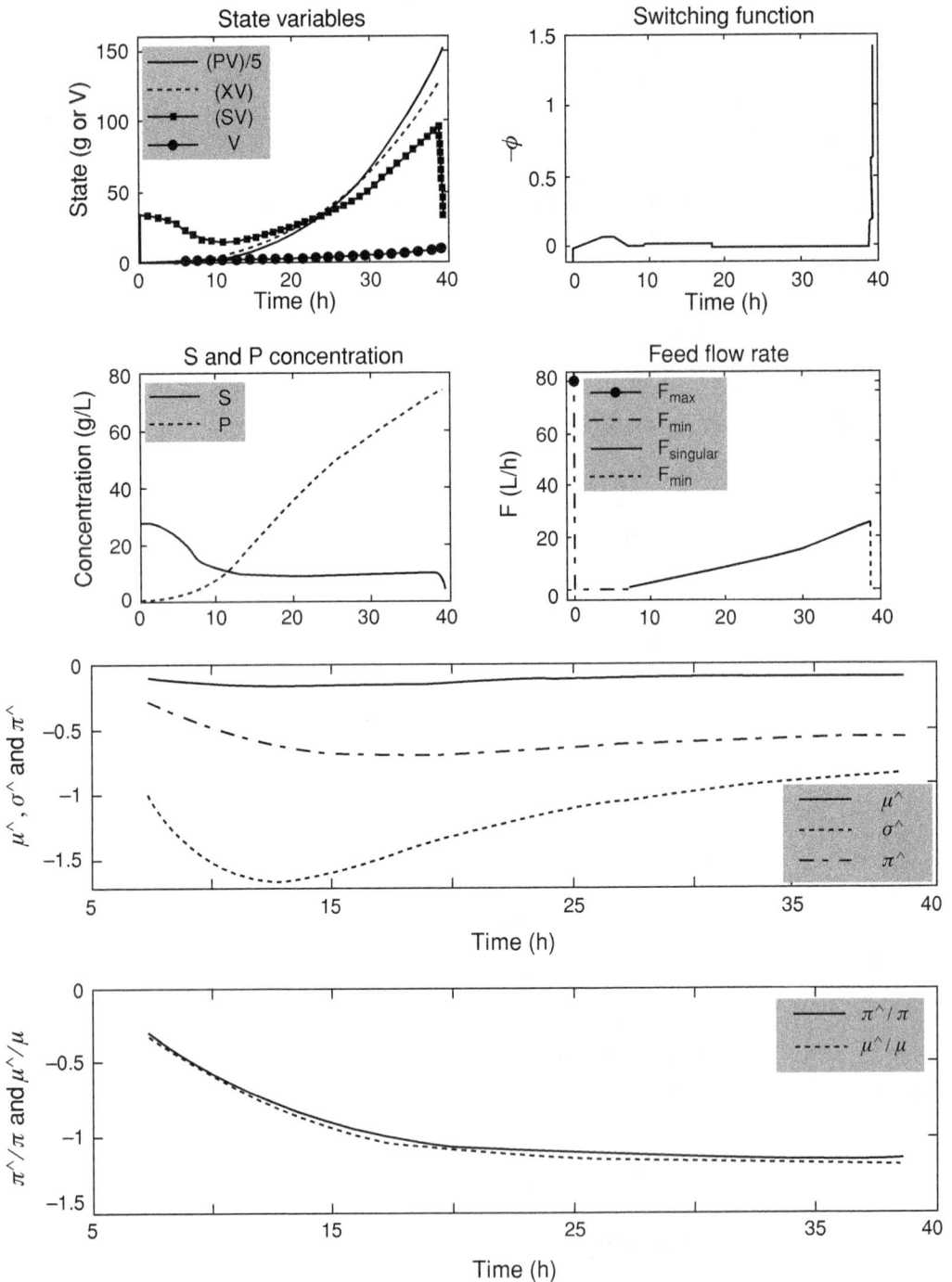

Figure 13.E.3.1. Time profiles of state variables, switching function, and feed rate (ethanol production by *S. cerevisiae*): $\mu = 0.408S/[(1 + P/16)(0.22 + S)]$, $\sigma = 10\mu$, $\pi = S/[(1 + P/71.5)(0.44 + S)]$, $[x_1, x_2, x_3, x_4](t_0) = [0.1\,\text{g}, 4.49\,\text{g}, 0\,\text{g}, 0.244\,\text{L}]$, $[S_F, t_f, F_{max}, F_{min}, V_{max}] = [150\,\text{g/L}, 39.5\,\text{hr}, 2\,\text{L/hr}, 0\,\text{L/hr}, 10\,\text{L}]$.

phosphate and magnesium, but the condition is adverse to plasmid stability.[28] Starvation of essential amino acids is reported to increase the plasmid copy number,[29] and the plasmid stability decreases with decreasing dilution rate in a continuous culture.[30]

A number of methods have been proposed to improve plasmid stability by conferring a competitive advantage to PBC over PFC. One method is to use a plasmid encoded to synthesize a bacteriocin, a protein that is toxic to PFC but not to PBC. This method has been used to improve the plasma stability[31–34] for *E. coli* and *Bacillus subtilis*. Another method is to use a plasmid that is encoded to produce antibiotics so that PBC with antibiotic resistance can survive in a culture medium while the PFC[35,36] without an antibiotic resistance are killed. Although powerful, the high cost of antibiotics and their removal costs are potential drawbacks. A new method was proposed[39] in which a plasmid is used to remove the substrate inhibition effect on PBC, and therefore only PFC are susceptible to the effect, leading to a higher ratio, μ^+/μ^-. An analysis of continuous cultures of recombinant methylotrophs has been reported.[38]

Despite being a powerful method to increase the yield and/or productivity of cell mass, metabolites, and recombinant product, application of the above methods to recombinant cells, in which the plasmid stability is controlled,[40–43] has been very limited. Although there are some experimental reports of optimal fed-batch operation for recombinant microorganisms, it is anticipated that more applications in this area will soon be developed. This chapter deals with the optimal fed-batch operation for maximum productivity or yield by PBC. First, we consider the case of recombinant cells with plasmid instability without cell death and, finally, with different cell death rates for PBC and PFC.

In this chapter, we provide a complete analysis of optimal solutions of fed-batch fermentation for recombinant product maximization for the Leudeking–Piret form of the specific product formation rate. The precise conditions under which a singular feed rate is part of the optimal feed rate sequences are characterized in terms of the relative ratio of the specific growth rates of PBC to PFC.

13.6.1.1 Problem Formulation

A few mathematical models have been proposed to describe the microbial fermentation process for plasmid-encoded proteins.[39] PBC and PFC compete for limited nutrients, and owing to a metabolic burden placed on PBC, their specific growth rate μ^+ is usually lower than that of PFC, μ^-. It is usually assumed that the rate of reversion of PBC to PFC is proportional to the specific growth rate of PBC, $r_{X^-} = \mu^- x_2 = \alpha r_{X^+} = \alpha \mu^+ x_1$, where α is a proportionality constant.[39]

As a first step in analyzing recombinant cells with plasmid instability, we adapt the simple model originally proposed[31] that lumps the cells into two groups, one bearing plasmids (PBC) and the other free of plasmids (PFC). In this model, the potential distribution of plasmids from zero copy to a maximum copy is ignored, and instead, the cells are grouped into only two types. The Leudeking–Piret equation is used for product formation. Considering the case of only one feed of limiting substrate (control variable) F, the dynamic behavior is represented by the following unsteady state mass balance equations for PBC, PFC, substrate, total mass, and product:

$$\frac{d}{dt}\begin{bmatrix} X^+V \\ X^-V \\ SV \\ V \\ PV \end{bmatrix} = \frac{d}{dt}\begin{bmatrix} x_1 \\ x_2 \\ x_3 \\ x_4 \\ x_5 \end{bmatrix} = \begin{bmatrix} (1-\alpha)\mu^+x_1 \\ \alpha\mu^+x_1 + \mu^-x_2 \\ -(\mu^+/Y_{x/s}^+)x_1 - (\mu^-/Y_{x/s}^-)x_2 \\ 0 \\ mx_1 + n\dot{x}_1 \end{bmatrix} + \begin{bmatrix} 0 \\ 0 \\ S_F \\ 1 \\ 0 \end{bmatrix} F,$$

$$\underline{x}(t_0) = \underline{x}_0 \quad F_{\min} \leq F \leq F_{\max} \quad x_4(t_f) = V_{\max} \qquad (13.187)$$

where $Y_{X/S}^+$ and $Y_{X/S}^-$ are cell mass yield coefficients for PBC and PFC, respectively, which may be either constants or functions of substrate concentration, and m and n are constants. The specific growth rates of PBC and PFC are assumed to be functions of substrate concentration only: $\mu^+(S) = \mu^+(x_3/x_4)$ and $\mu^-(S) = \mu^-(x_3/x_4)$. The amount of substrate needed to produce the product is assumed negligible ($Y_{P/S} = \infty$) as compared to that required for the cell formation. The bioreactor volume is to be fully utilized, $x_4(t_f) = V_{\max}$. The objective is to maximize the amount of product formed at the final time by manipulating the feed rate:

$$\underset{F(t)}{\text{Max}}[P = x_5(t_f)] \Rightarrow \underset{F(t)}{\text{Max}}\left[\int_0^{t_f}\dot{x}_5 dt\right] = \underset{F(t)}{\text{Max}}\left[\int_0^{t_f}(mx_1 + n\dot{x}_1)dt\right]$$

$$= \underset{F(t)}{\text{Max}}\left\{n[x_1(t_f) - x_{10}] + m\int_0^{t_f}x_1 dt\right\} \qquad (13.188)$$

Equation (13.188) states that the cell mass must be maximized for all times to maximize the amount of product at the final time. The performance index, Eq. (13.188), may be put into a standard form by introducing two auxiliary state variables x_6 and x_7:

$$\dot{x}_6 = x_1, \quad x_{60} = 0, \quad \dot{x}_7 = 1, \quad x_{70} = 0 \qquad (13.189)$$

Then, the complemented state equations are

$$\begin{bmatrix} \dot{x}_1 \\ \dot{x}_2 \\ \dot{x}_3 \\ \dot{x}_4 \\ \dot{x}_5 \\ \dot{x}_6 \\ \dot{x}_7 \end{bmatrix} = \begin{bmatrix} (1-\alpha)\mu^+x_1 \\ \alpha\mu^+x_1 + \mu^-x_2 \\ -(\mu^+x_1/Y_{x/s}^+) - (\mu^-x_2/Y_{x/s}^-) \\ 0 \\ mx_1 + n\dot{x}_1 = mx_1 + n(1-\alpha)\mu^+x_1 \\ x_1 \\ 1 \end{bmatrix} + \begin{bmatrix} 0 \\ 0 \\ S_F \\ 1 \\ 0 \\ 0 \\ 0 \end{bmatrix} F, \quad \begin{array}{l} \underline{x}(t_0) = \underline{x}_0 \\ 0 = F_{\min} \leq F \leq F_{\max} \\ x_4(t_f) = V_{\max} \end{array}$$

$$(13.190)$$

and the performance index becomes the standard form:

$$\underset{F(t)}{\text{Max}}\{P = n[x_1(t_f) - x_{01}] + mx_6(t_f)\} \qquad (13.191)$$

PMP[44] yields the Hamiltonian as

$$H = \underline{\lambda}^T\underline{f} = [\lambda_1(1-\alpha) + \lambda_2\alpha - \lambda_3/Y_{x/s}^+ + \lambda_5 n(1-\alpha)]\mu^+x_1 + (\lambda_2 - \lambda_3/Y_{x/s}^-)\mu^-x_2$$
$$+ \lambda_7 + (\lambda_5 m + \lambda_6)x_1 + (\lambda_3 S_F + \lambda_4)F$$
$$= A\mu^+x_1 + B\mu^-x_2 + (\lambda_5 m + \lambda_6)x_1 + \lambda_7 + \phi F \qquad (13.192)$$

where

$$A(\lambda) \triangleq \lambda_1(1-\alpha) + \lambda_2\alpha - \lambda_3/Y_{x/s}^+ + \lambda_5 n(1-\alpha)$$

$$B(\lambda) \triangleq \lambda_2 - \lambda_3/Y_{x/s}^-, \phi = \lambda_3 S_F + \lambda_4 \qquad (13.193)$$

For growth-associated product formation, $m = 0$, the specific product formation rate is proportional to the specific growth rate of PBC, $\dot{x}_5 = n\dot{x}_1$, and the amount of product at the final time is maximized by maximizing the cell mass at the final time. When the product formation is both growth- and non-growth-associated, $m > 0$, maximization of the product at the final time is achieved by maximizing the cell mass at all times. Maximization of cell mass at the final time only ($m = 0$) does not guarantee the maximization of product at the final time. However, optimization and analysis of the simple case ($m = 0$) may provide insights to the general case ($m > 0$). Therefore, we consider the simplest case of zero m and constant cell yields and then consider the general problem of nonzero m and variable yields.

13.6.1.2 Constant Yields with Growth-Associated Product Formation, $m = 0$
The adjoint variables must satisfy the following dynamics equations and boundary conditions:

$$\begin{bmatrix} \dot{\lambda}_1 \\ \dot{\lambda}_2 \\ \dot{\lambda}_3 \\ \dot{\lambda}_4 \\ \dot{\lambda}_5 \\ \dot{\lambda}_6 \\ \dot{\lambda}_7 \end{bmatrix} = - \begin{bmatrix} \partial H/\partial x_1 \\ \partial H/\partial x_2 \\ \partial H/\partial x_3 \\ \partial H/\partial x_4 \\ \partial H/\partial x_5 \\ \partial H/\partial x_6 \\ \partial H/\partial x_7 \end{bmatrix} = - \begin{bmatrix} A\mu^+ + \lambda_5 m + \lambda_6 \\ B\mu^- \\ (A\mu_s^+ x_1 + B\mu_s^- x_2)(1/x_4) \\ (A\mu_s^+ x_1 + B\mu_s^- x_2)(-x_3/x_4^2) \\ 0 \\ 0 \\ 0 \end{bmatrix} \qquad (13.194)$$

where $\mu_s^+ = \partial\mu^+/\partial S$ and $\mu_s^- = \partial\mu^-/\partial S$. The final conditions on the adjoint variables are obtained from the transversality conditions:

$$\delta P[\mathbf{x}(t_f)] = \underset{\sim}{n}\delta x_1(t_f) + 0\delta x_2(t_f) + 0\delta x_3(t_f) + 0\delta x_4(t_f) + 0\delta x_5(t_f) + \underset{=}{m}\delta x_6(t_f) + 0\delta x_7(t_f)$$

$$= \boldsymbol{\lambda}^T(t_f)\delta\mathbf{x}(t_f)$$

$$= \underset{\sim1}{\lambda}(t_f)\delta x_1(t_f) + \lambda_2(t_f)\delta x_2(t_f) + \lambda_3(t_f)\delta x_3(t_f) + \lambda_4(t_f)\partial x_4^{=0}(t_f)$$

$$+ \lambda_5(t_f)\delta x_5(t_f) + \underset{\sim6}{\lambda}(t_f)\delta x_6(t_f) + \lambda_7(t_f)\delta x_7(t_f) \qquad (13.195)$$

By matching the coefficients as we did in Chapter 9, we obtain the final condition:

$$[\lambda_1(t_f)\ \lambda_2(t_f)\ \lambda_3(t_f)\ \lambda_4(t_f)\ \lambda_5(t_f)\ \lambda_6(t_f)\ \lambda_7(t_f)] = [n\ 0\ 0\ \beta\ 0\ m\ 0] \qquad (13.196)$$

where β is an unknown constant owing to the terminal condition, $x_4(t_f) = V_{max}$. From Eqs. (13.194) and (13.196), we conclude that

$$\lambda_5(t) = 0, \ \lambda_6(t) = m = 0, \ \lambda_7(t) = 0 \qquad (13.197)$$

Because the differential equations for x_1 through x_4 are independent of x_5 through x_7 in Eq. (13.190), and λ_5 through λ_7 are identically zero from Eq. (13.197), the

independent differential equations are the first through fourth in Eqs. (13.190) and
(13.194). Equation (13.192) reduces to

$$H = A\mu^+ x_1 + B\mu^- x_2 + \phi F \tag{13.198}$$

Because the Hamiltonian is linear with respect to the feed rate F in Eq. (13.198), it
is maximized by picking F according to the sign of the switching function $\phi(t)$, as
follows:

$$F = \begin{cases} F_{\max} & \text{when } \phi > 0 \\ F_{\min} = 0 & \text{when } \phi < 0 \\ F_{\sin} & \text{when } \phi = 0 \text{ over finite time interval(s)}, t_q < t < t_{q+1} \\ F_b = 0 & \text{when } x_4(t) = V_{\max} \end{cases} \tag{13.199}$$

The optimal feed rate profile is any concatenation of $F_{\max}$, $F_{\min} = 0$, $F_{\sin}$, and $F_b = 0$.
When the initial conditions $(X^+ V_0, X^- V_0, SV_0, V_0)$ lie on a singular hyperspace,
the optimal feed rate sequence may lack the maximum and/or minimum feed rate
periods and may be singular from the start until the reactor volume is full. It is also
possible, if the upper limit $F_{\max}$ is not sufficiently large, that the singular feed rate
can reach the upper limit, and therefore, the singular feed rate must be set at $F_{\max}$
before the reactor volume is full. Therefore, a larger pump may be needed to avoid
this situation. If the chosen final time is sufficiently small, then the feed rate may
be set to the maximum to fill the reactor volume completely within the given final
time. Thus, a reasonable final time must be used to avoid this situation. Thus, the
structure of the concatenation of feed rates is affected by the upper and lower limits
on the feed rate, the initial values, and the final time.

13.6.1.2.1 SINGULAR FEED RATE. On the singular arc, $\phi(t) = 0$ over a finite time interval, and therefore, its higher-order time derivatives must also vanish. Thus, to obtain
the singular feed rate, we take sequential time derivatives of the switching function
until the feed rate $F(t)$, or a variable related to it, appears explicitly:

$$\phi = \lambda_3 S_F + \lambda_4 = 0 \tag{13.200}$$

$$\dot{\phi} = -(A\mu_s^+ x_1 + B\mu_s^- x_2)(S_F - x_3/x_4)/x_4 = 0 \quad \Rightarrow \quad A\mu_s^+ x_1 + B\mu_s^- x_2 = 0 \tag{13.201}$$

$$\ddot{\phi} = -(S_F - x_3/x_4)/x_4 \left\{ (A\mu_{ss}^+ x_1 + B\mu_{ss}^- x_2)(dS/dt) - \alpha B(\mu^+/\mu^-)_s (\mu^-)^2 x_1 \right\} = 0$$
$$\Rightarrow (A\mu_{ss}^+ x_1 + B\mu_{ss}^- x_2)(dS/dt) - \alpha B(\mu^+/\mu^-)_s (\mu^-)^2 x_1 = 0 \tag{13.202}$$

The substrate balance equation, Eq. (13.187), may be rearranged to obtain

$$d(SV)/dt = (dS/dt)V + SF = \left[-(\mu^+/Y_{x/s}^+)x_1 - (\mu^-/Y_{x/s}^-)x_2 \right] + S_F F \tag{13.203}$$

which is solved to obtain the singular feed rate,

$$F_{\sin} = \frac{V \, dS/dt + (\mu^+/Y_{x/s}^+)x_1 + (\mu^-/Y_{x/s}^-)x_2}{S_F - S} \tag{13.204}$$

The time rate of change in substrate concentration is obtained by solving simultaneously Eqs. (13.201) and (13.202). These two equations are linear in A and B.

Therefore, nontriviality of both A and B requires that the determinant of the two linear equations vanishes, yielding

$$\frac{dS}{dt} = -\alpha \mu_s^+ \left(\frac{\mu^-}{\mu_s^-}\right)^2 \frac{(\mu^+/\mu^-)_s}{(\mu_s^+/\mu_s^-)_s}\left(\frac{x_1}{x_2}\right) \qquad (13.205)$$

where the subscript s is used to denote the differentiation with respect to the substrate concentration. Substitution of Eq. (13.205) into Eq. (13.204) yields the singular feed rate in feedback mode:

$$F_{\sin} = \frac{-\alpha V \mu_s^+ \left(\mu^-/\mu_s^-\right)^2 \frac{(\mu^+/\mu^-)_s}{(\mu_s^+/\mu_s^-)_s}\left(\frac{x_1}{x_2}\right) + (\mu^+/Y_{x/s}^+)x_1 + (\mu^-/Y_{x/s}^-)x_2}{S_F - S} \qquad (13.206)$$

The generalized Legendre–Clebsch (GLC) condition[44] obtained from Eq. (13.202) is

$$(-1)\frac{\partial}{\partial F}\frac{d^2\phi}{dt^2} = (S_F - S)/x_4 \frac{\partial}{\partial F}[(A\mu_{ss}^+ x_1 + B\mu_{ss}^- x_2)(dS/dt)]$$

$$= (S_F - S)/x_4(A\mu_{ss}^+ x_1 + B\mu_{ss}^- x_2)\frac{\partial}{\partial F}\left[\frac{(S_F - S)F}{V} - \sigma X\right] \geq 0$$

$$\Rightarrow (A\mu_{ss}^+ x_1 + B\mu_{ss}^- x_2) \geq 0 \qquad (13.207)$$

This condition will be used later in identifying regions of singular feed rate.

However, at this point, it is convenient to consider separately two cases: (1) a performance index that does not contain free final time and (2) a performance index that contains free final time.

13.6.1.2.2. FREE FINAL TIME NOT IN THE PERFORMANCE INDEX. The final time is free and does not appear in the performance index so that the Hamiltonian is zero on the optimal path. Therefore, over the singular arc ($\phi = 0$), Eq. (13.198) reduces to

$$H = A\mu^+ x_1 + B\mu^- x_2 = 0 \qquad (13.208)$$

Equations (13.201) and (13.208) are two homogeneous equations in A and B. Therefore, the nontriviality of A and B yields the following relationship:

$$\begin{vmatrix} \mu_s^+ x_1 \mu_s^- x_2 \\ \mu^+ x_1 \mu^- x_2 \end{vmatrix} = 0 \quad \Rightarrow \quad 0 = \mu_s^+ \mu^- - \mu_s^- \mu^+ = (\mu^-)^2(\mu^+/\mu^-)_s$$

$$\Rightarrow (\mu^+/\mu^-)_s = \frac{d(\mu^+/\mu^-)}{dS} = 0 \qquad (13.209)$$

Equation (13.209) states that the ratio of the specific growth rate of PBC to that of PFC μ^+/μ^- *must be maximized on the singular arc*. This can happen at the substrate concentration $S = S_*$ yet to be identified. Therefore, it is apparent that the substrate concentration is kept constant on the singular arc. A better way to identify the singular region is obtained by rearranging Eq. (13.209):

$$\mu_s^+ \mu^- - \mu_s^- \mu^+ = 0$$
$$\Rightarrow \mu_s^+/\mu_s^- = \mu^+/\mu^- > 0 \qquad (13.210)$$

where the inequality is due to the fact that the specific growth rates are positive. To satisfy the inequality, the signs of μ_s^+ and μ_s^- must be the same. Therefore, the regions in which the signs of μ_s^+ and μ_s^- are different are not singular regions. Because the specific rates, μ^+ and μ^-, are assumed to be functions of substrate concentration only, Eq. (13.209) yields

$$\mu^+ = \beta\mu^-, \quad \beta > 0 \tag{13.211}$$

where β is positive owing to Eq. (13.210). Equation (13.211) implies that the specific growth rate of PBC μ^+ must be proportional to that of PFC μ^- over the singular arc. Therefore, the regions not satisfying Eq. (13.210) are not singular regions. Substituting Ax_1 from Eq. (13.201) into Eq. (13.207) yields the equality GLC condition:

$$-Bx_2(\mu_s^+/\mu_s^-)_S[(\mu_s^-)^2/\mu_s^+] = 0 \tag{13.212}$$

Because the substrate concentration is kept constant at $S = S_*$, it follows from Eq. (13.204) that

$$F_{\sin} = \frac{[(\mu^+/Y_{x/s}^+)x_1 + (\mu^-/Y_{x/s}^-)x_2]}{S_F - S_*} \tag{13.213}$$

It is now possible to construct the optimal feed rate profiles that depend on the initial conditions.

Optimal Feed Rate Structure. The general results of this section, Eqs. (13.209) and (13.211), indicate that the singular arc exists at $S = S_*$, which maximizes μ^+/μ^-. This implies that the singular feed rate and the corresponding state and adjoint trajectories are those that maximize μ^+/μ^- at every instant of time. The final values of adjoint variables, Eq. (13.196), are substituted into Eq. (13.198) to obtain the final value of the Hamiltonian,

$$H(t_f) = 0 = A(t_f)\mu^+(t_f)x_1(t_f) + B(t_f)\mu^-(t_f)x_2(t_f) + \phi(t_f)F(t_f)$$
$$= n(1-\alpha)\mu^+(t_f)x_1(t_f) \tag{13.214}$$

which is satisfied by

$$\mu^+(t_f) = 0 \tag{13.215}$$

However, for specific growth rate models of $\mu^+(S)$ and $\mu^-(S)$, $\mu^+(t_f) = 0$ only when the substrate concentration is zero at the final time. Therefore, the limiting nutrient concentration in the culture medium must approach zero at the free final time, and this is only possible if the final time is infinite, unless there is a zero-order term such as a cell death term in the specific growth rate. Because a singular feed rate is preferable to a bang (maximum or minimum) feed rate with the satisfaction of GLC, the optimal feed flow rate profile can now be readily obtained for any given initial state. Let S_* denote the value of the substrate concentration that maximizes $\mu^+/\mu^-(S_*)$ on the singular arc. The initial substrate concentration S_0 can be greater than, less than, or equal to S_*. (1) If $S_0 < S_*$, the first feed rate is $F_{\max}$ until the substrate concentration S reaches S_*, and then the singular feed rate, Eq. (13.213), is applied, which maintains the substrate concentration constant at S_* until the reactor volume is full, unless the singular feed rate exceeds the upper limit, and finally, there is a batch operation, $F_b = 0$, until the substrate concentration is zero; $F_{\max} \to F_{\sin} \to F_b = 0$. (2) If $S_0 > S_*$, a batch period $F_{\min} = 0$ must be used to bring down the substrate concentration to S_*, and the feed rate strategy thereafter is the same as the first case: $S_0 < S_*$, $F_{min} = 0 \rightarrow F_{sin} \rightarrow F_b = 0$. (3) When $S_0 = S_*$, the optimal feed rate sequence is $F_{sin} \rightarrow F_b = 0$. Thus, it is clear that the best initial substrate concentration is $S_0 = S_*$.

Special Case $\mu^- = k\mu^+$, $k > 0$. In this case, Eq. (13.209) is automatically satisfied. The first and second rows of Eq. (13.187) are substituted into the third row and then integrated to yield the following equation:

$$x_1 - \frac{(1-\alpha)}{1/Y_{x/s}^+ - \alpha/Y_{x/s}^-} \left[S_F x_4 - x_3 - \frac{x_2}{Y_{x/s}^-} \right]$$
$$= x_{1,0} - \frac{(1-\alpha)}{1/Y_{x/s}^+ - \alpha/Y_{x/s}^-} \left[S_F x_{4,0} - x_{3,0} - \frac{x_{2,0}}{Y_{x/s}^-} \right] \qquad (13.216)$$

Thus, the state variables are related to each other through Eq. (13.216), and therefore, the dimension is reduced by one. Equations corresponding to (13.187)–(13.213) can also be developed for the reduced system. Substituting $\mu^- = k\mu^+$ into the second row of Eq. (13.187), the first and second rows constitute one differential equation:

$$dx_2/dx_1 = \delta x_2/x_1 + \varepsilon, \quad \delta = k/(1-\alpha), \quad \varepsilon = \alpha/(1-\alpha) \qquad (13.217)$$

The solution is

$$(x_1)^{-\delta}\{x_2 - \varepsilon x_1/(1-\delta)\} = (x_{1,0})^{-\delta}\{x_{2,0} - \varepsilon x_{1,0}/(1-\delta)\} \qquad (13.218)$$

Therefore, the order of dynamic state equations is reduced further by one. Taking the first, fourth, and seventh state variables as the independent state variables, the Hamiltonian is redefined as

$$H = \lambda_1(1-\alpha)\mu^+ x_1 + \lambda_4 F + \lambda_7 \triangleq H_1 + \lambda_4 F + \lambda_7 \qquad (13.219)$$

where $\lambda_7 = 0$ because its time derivative and the final value are zero. The Hamiltonian is zero on the singular arc:

$$H_1 = \lambda_1(1-\alpha)\mu^+ x_1 = 0 \quad \Rightarrow \lambda_1 = 0, \phi = \lambda_4 = 0 \qquad (13.220)$$

Thus, all adjoint variables are zero, which means all and any state can be on a singular arc. Therefore, we can say that *any feed rate sequence (infinite in number) yields the optimal performance index*. This result is not surprising because the performance index does not penalize the final time, and therefore, by taking an infinite time, any feed strategy would give the same result as long as the yield coefficients are constant.

13.6.1.2.3. FREE FINAL TIME IN THE PERFORMANCE INDEX. The final time t_f is an important parameter that affects the optimal feed rate strategy. For example, if the final time t_f is made shorter and shorter, maintaining the substrate concentration at S_* on the singular arc to maximize μ^+/μ^- may not satisfy the final volume constraint for the fixed t_f. Therefore, we can say that the optimal feed rate must vary with time, and so does the substrate concentration S on the singular arc.

Let us consider the maximization of a profit function in the following form:

$$\underset{F(t)}{\text{Max}}\,[P = x_1(t_f) - \varsigma t_f] \qquad (13.221)$$

where ς is a positive constant reflecting the unit reactor operating cost (assumed to be proportional to the final time) relative to the unit price of the PBC mass. The final condition, Eq. (13.196), changes to

$$\underline{\lambda}(t_f)^T = \begin{bmatrix} 1 & 0 & 0 & \beta & 0 & 0 & -\varsigma \end{bmatrix} \tag{13.222}$$

Equations (13.194) and (13.222) show that

$$\lambda_5 = \lambda_6 = 0, \lambda_7(t) = -\varsigma < 0 \tag{13.223}$$

It follows from Eq. (13.192) that over the singular arc,

$$H = 0 = A\mu^+ x_1 + B\mu^- x_2 + \varsigma \quad \Rightarrow A\mu^+ x_1 + B\mu^- x_2 < 0 \tag{13.224}$$

Without any constraints, the singular feed rate is in feedback form (Eq. (13.206)) with a zero value of m if dS/dt can be expressed without the adjoint variables.

Singular Arc. Equation (13.205) shows that the substrate concentration varies with time unless it satisfies $\mu_s^+(S) = 0$ or $(\mu^+/\mu^-)_s(S) = 0$. However, the time derivative of S alone does not provide enough information to predict the optimal feed rate structure or the behavior of substrate concentration. To elucidate the characteristics of the singular arc, we begin with the inequality constraint on the singular arc (13.224). Substitution of A from Eq. (13.201) into (13. 224), together with Eq. (13.202), yields

$$B(\mu^-)^2 \left(\frac{\mu^+}{\mu^-}\right)_s \left(\frac{x_2}{\mu_s^+}\right) = \left(\frac{\dot{S}}{\mu_s^+}\right) \left(\frac{[GLC]x_2}{\alpha x_1}\right) < 0 \tag{13.225}$$

where

$$GLC = (A\mu_{ss}^+ x_1 + B\mu_{ss}^- x_2) \geq 0 \tag{13.226}$$

We obtain another inequality from Inequality (13.225):

$$\text{sign}(dS/dt)\text{sign}(\mu_S^+) < 0 \tag{13.227}$$

Inequality (13.227) shows that the sign of the time derivative of substrate concentration is opposite to that of the substrate derivative of the specific cell growth rate of PBC:

$$\text{sign}[(dS/dt)]\text{sign}(\mu_S^+) < 0 \quad \Rightarrow \begin{bmatrix} dS/dt > 0 & \text{if} & d\mu^+/dS < 0 \\ dS/dt < 0 & \text{if} & d\mu^+/dS > 0 \end{bmatrix} \tag{13.228}$$

Thus, on the singular arc, *the substrate concentration decreases, $dS/dt < 0$, in the region where μ^+ increases with the substrate concentration, $d\mu^+/dS > 0$, while it increases, $dS/dt > 0$, in the region where μ^+ decreases with substrate concentration, $d\mu^+/dS < 0$.* This is an important characteristic in that if the shape of the specific growth rate is known, we can tell if the substrate concentration would increase or decrease during the singular feed rate period. However, it is unknown at this point why the substrate concentration moves in one direction or the other.

The original system of Eq. (13.187) is fourth order owing to the equality condition among state variables. For this reason, taking the first, second, fourth, and fifth rows of Eq. (13.187), the Hamiltonian is redefined as

$$H = [\lambda_1(1-\alpha) + \lambda_2\alpha]\mu^+ x_1 + \lambda_2\mu^- x_2 + \lambda_4 F + \lambda_5 \stackrel{\Delta}{=} H_1 + \lambda_4 F + \lambda_5 \tag{13.229}$$

Time derivatives of λ_1 and λ_2 and the switching function are redefined as

$$-\frac{d\lambda_1}{dt} = [\lambda_1(1-\alpha) + \lambda_2\alpha]\mu^+, \quad -\frac{d\lambda_2}{dt} = \lambda_2\mu^-, \quad \phi = \lambda_4 \quad (13.230)$$

Solving Eq. (13.230) for λ_2, together with the zero final value of λ_2 from Eq. (13.194), yields

$$\lambda_2(t) = \lambda_2(t_f)e^{\mu^-(t_f-t)} = [x_1/(x_2)^2]e^{\mu^-(t_f-t)} > 0 \quad (13.231)$$

Thus, $\lambda_2(t)$ is always positive over all process times. Equation (13.222) is applicable to this redefined Hamiltonian, $\lambda_5(t) = \varsigma > 0$. The final time is free, and therefore, on the singular arc,

$$H_1 = [\lambda_1(1-\alpha) + \lambda_2\alpha]\mu^+x_1 + \lambda_2\mu^-x_2 + \xi = 0 \Rightarrow A\mu^+x_1 + B\mu^-x_2 < 0 \quad (13.232)$$

where $A = [\lambda_1(1-\alpha) + \lambda_2\alpha]$ and $B = \lambda_2 > 0$. Therefore, the following inequality is obtained:

$$[\lambda_1(1-\alpha) + \lambda_2\alpha] = A < 0 \quad \Rightarrow \quad \lambda_1 < 0 \quad (13.233)$$

Equations (13.202) and (13.207) give the following inequality:

$$0 < (A\mu_{ss}^+x_1 + B\mu_{ss}^-x_2) = \frac{\alpha B(\mu^+/\mu^-)_s(\mu^-)^2x_1}{(dS/dt)} \quad (13.234)$$

This and Inequality (13.231) imply that

$$\text{sign}\left(\frac{dS}{dt}\right) = \text{sign}\left[(\mu^+/\mu^-)_s\right] = \text{sign}\left[\frac{d(\mu^+/\mu^-)}{dS}\right] \quad (13.235)$$

Equation (13.235) states that during the singular feed period, *the substrate concentration increases (decreases) with time in the direction of increasing (decreasing) μ^+/μ^-.* Thus, the substrate concentration is regulated by the singular feed rate to increase the ratio, (μ^+/μ^-). Substitution of Eq. (13.235) into Inequality (13.227) yields

$$(\mu^+/\mu^-)_s \cdot \mu_s^+ < 0 \quad (13.236)$$

Taking the sign functions of Inequality (13.236), we obtain

$$\text{sign}\left[(\mu^+/\mu^-)_s\right] = -\text{sign}(\mu_s^+) \quad (13.237)$$

which implies that the sign of μ_s^+ is opposite of the sign of $(\mu^+/\mu^-)_s = d(\mu^+/\mu^-)/dS$. Inequality (13.236) or Eq. (13.237) defines the region in which the ratio (μ^+/μ^-) increases and the specific growth rate of PCB decreases with the substrate concentration.

The first-order time derivative of the switching function is the same as Eq. (13.201). Substituting B from Eq. (13.201) into Eq. (13.232) yields

$$\left(\frac{Ax_1}{\mu_s^-}\right)(\mu^+/\mu^-)_s(\mu^-)^2 = Ax_1(\mu^-)^2\left(\frac{\dot{S}}{\mu_s^-}\right)\left(\frac{(\mu^+/\mu^-)_s}{\dot{S}}\right) < 0 \quad (13.238)$$

Combining Eq. (13.238) with Inequalities (13.233) and (13.234), we obtain

$$\text{sign}\left(\frac{dS}{dt}\right)\text{sign}\left(\frac{d\mu^-}{dS}\right) < 0 \quad (13.239)$$

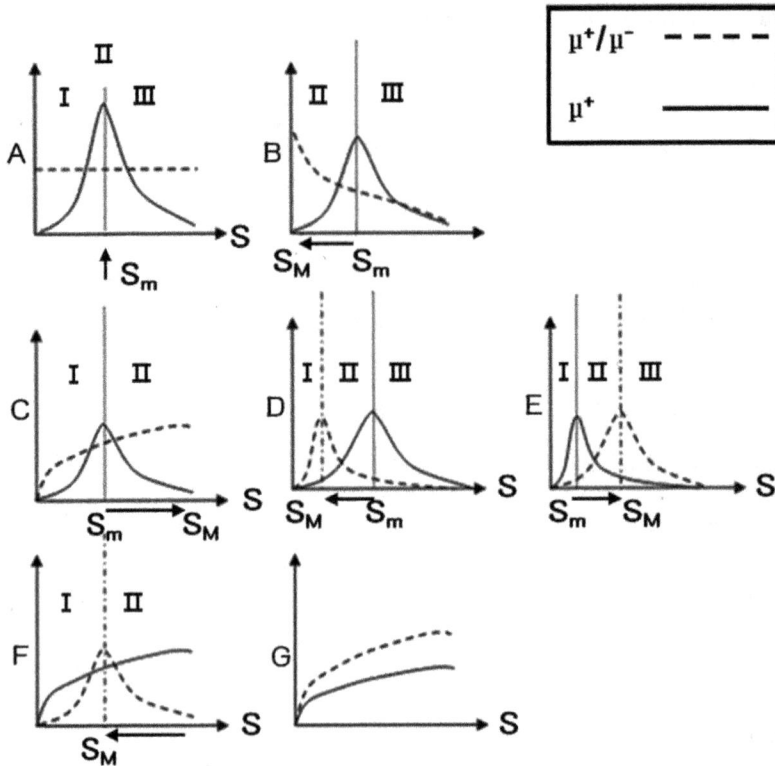

Figure 13.3. Singular regions II and variations in substrate concentrations during the singular interval. The arrows indicate the direction of the change in substrate concentration with time. A = nonmonotonic μ^+ and constant μ^+/μ^-; B = nonmonotonic μ^+ and monotonically decreasing μ^+/μ^-; C = nonmonotonic μ^+ and monotonically increasing μ^+/μ^-; D = nonmonotonic μ^+ and μ^+/μ^- peaking before μ^+_{max}; E = nonmonotonic μ^+ and μ^+/μ^- peaking after μ^+_{max}; F = monotonic μ^+ and nonmonotonic μ^+/μ^-; G = monotonic μ^+ and μ^+/μ^-.

Inequality (13.239) shows that the sign of the time derivative of substrate concentration is opposite to the sign of the substrate derivative of the specific cell growth rate of PFC. In other words, *the substrate concentration should increase (decrease) if the specific growth rate decreases (increases) with the substrate concentration.*

13.6.1.2.4 SINGULAR REGIONS IN TERMS OF SPECIFIC RATES. Substrate concentration S on the singular arc moves with time in the direction of maximizing (μ^+/μ^-). The substrate concentration moves toward maximizing (μ^+/μ^-) away from the direction of maximizing μ^+. As the final time increases, the substrate concentration changes further toward maximizing (μ^+/μ^-). This phenomenon supports that for a free final time case, the singular arc is simply the substrate concentration that maximizes (μ^+/μ^-). The optimal feed rate profiles suggest that the maximum or minimum feed rate is used to bring the process onto the singular arc as soon as possible, and the singular feed rate is used to keep the process on the singular arc until the reactor volume is full. The cell mass of PBC x_1 is maximized during the singular feed rate period.

Figure 13.3 is constructed using Eqs. (13.227), (13.235), and (13.237) and shows all types of singular regions. The regions in which singular feed rate is feasible are denoted by the arrows. The direction of these arrows also indicates whether the substrate concentration remains constant, increases, or decreases with time during the period of singular feed rate. The entire space of the specific rate of PBC μ^+ and the ratio (μ^+/μ^-) versus substrate concentration is divided into regions I, II, and III. The singular region is denoted by region II; except in the case of constant (μ^+/μ^-), it is a vertical line instead of a region.

13.6.1.2.4.1 OPTIMAL FEED RATE SEQUENCES FOR VARIOUS INITIAL CONDITIONS AND VARIATIONS IN SUBSRATE CONCENTRATIONS IN SINGULAR FEED PERIOD. The impor-

tance and very practical utility of Figure 13.3 are illustrated here. Not only can we readily determine by inspection the existence of a singular region but also the approximate ranges of substrate concentration on which the singlar feed rate must be applied and whether the substrate concentration remains constant, increases, or decreases with time during the singular feed rate period. In addition, Figure 13.3 helps us determinc the sequence of feed rates that is needed from various initial conditions.

When (μ^+/μ^-) is constant (Figure 13.3A), *the substrate concentration is held constant at the value that maximizes μ^+, $S = S_m$ during the singular feed period to maximize the specific growth rate μ^+.* In all cases, once the bioreactor volume is full, a batch period follows until the final time is met. If the initial substrate concentration is in region I, $S(0) < S_m$, the optimum feed rate sequence is $F_{max} \to F_{sin} \to F_b = 0$ when the volume is full. If the initial substrate concentration is in region III, $S(0) > S_m$, the optimum sequence is $F_{min} = 0 \to F_{sin} \to F_b = 0$ when the volume is full. Finally, if the initial substrate concentration is at the optimum value that maximizes the specific growth rate μ^+, region II, $S(0) = S_m$, the optimum sequence is $F_{sin} \to F_b = 0$. Hence, the optimal initial substrate concentration is S_m, the substrate concentration that maximizes μ^+.

If (μ^+/μ^-) is a monotonically decreasing function of substrate concentration (Figure 13.3B), *the substrate concentration decreases during the singular feed period to improve the (μ^+/μ^-) at the expense of specific growth rate of PBC μ^+.* If the initial conditions lie in region II, $S(0) < S_m$, the optimum sequence is $F_{max} \to F_{sin} \to F_b = 0$ or $F_{sin} \to F_b = 0$, whereas in region III, $S(0) > S_m$, the optimum sequence is $F_{min} = 0 \to F_{sin} \to F_b = 0$. The optimal initial substrate concentration is near to but less than S_m and must be determined numerically.

When (μ^+/μ^-) is *a monotonically increasing function of substrate concentration* (Figure 13.3C), the optimum sequence is $F_{max} \to F_{sin} \to F_b = 0$ in region I, $S(0) < S_m$, and $F_{min} = 0 \to F_{sin} \to F_b = 0$ in region II, $S(0) > S_m$. The optimal initial substrate concentration is near to but greater than S_m.

We consider next the case of nonmonotonic μ^+ and (μ^+/μ^-). When the substrate concentration corresponding to the peak in the (μ^+/μ^-), S_M, is less than that of the peak in the specific growth rate of PBC μ^+, S_m (Figure 13.3D), $S_M < S_m$, *the substrate concentration starting near S_m decreases toward S_M during the singular feed rate period to improve the rclative rate (μ^+/μ^-) at the expense of the specific growth rate of PBC, μ^+.* Thus, if the initial condition is in region I, the optimum feed rate sequence is $F_{max} \to S_{sin} \to F_b = 0$); in region II, the optimum sequence is $F_{max}($ or $F_{min} = 0) \to F_{sin} \to F_b = 0$, whereas in region III, the optimum sequence is $F_{min} = 0 \to F_{sin} \to F_b = 0$. Conversely, if $S_M > S_m$ (Figure 13.3E), *the substrate concentration starts near S_m and increases toward S_M in the singular feed period to improve the (μ^+/μ^-) at the expense of the specific growth rate of PBC μ^+.* The optimum sequence is $F_{max} \to F_{sin} \to F_b = 0$ if the initial condition lies in region I, it is $F_{min} = 0 \to F_{sin} \to F_b = 0$ in Region III, and in region II, it is F_{max} (or $F_{min} = 0$) $\to F_{sin} \to F_b = 0$ or $F_{sin} \to F_b = 0$. The optimal initial substrate concentration is greater than S_m but less than S_M.

If (μ^+/μ^-) is nonmonotonic but the specific growth rate μ^+ is monotonic (Figure 13.3F), *the substrate concentration decreases with time toward S_M during the period of singular feed rate.* If the initial condition is in region I, the optimum sequence is $F_{max} \to F_{sin} \to F_b = 0$, whereas in region II, it is F_{max} (or $F_{min} = 0$) $\to F_{sin} \to F_b = 0$.

Finally, *if both the specific growth rate of PBC μ^+ and the relative rate (μ^+/μ^-) are monotonic as in* Figure 13.3G, *there is no singular region*, and the optimal feed rate sequence is $F_{max} \to F_b = 0$, or a batch operation.

All these are the result of fixed final time. The singular feed rate increases or decreases the substrate concentration in the reactor to improve the relative specific growth rate (μ^+/μ^-) at the expense of the specific growth rate of PBC (μ^+). Therefore, it is anticipated that if the final time is large, the specific rate of PBC μ^+ is immaterial owing to the ample time available, and instead, the relative rate (μ^+/μ^-) is maximized, while the rate μ^+ would be maximized if the final time were relatively short.

13.6.1.2.5. FIXED FINAL TIME PROBLEMS. Considering a fixed final time problem, the Hamiltonian is a constant value. Therefore, we substitute ς with $-H^*$, the constant value of the Hamiltonian, not known a priori, and Eqs. (13.220)–(13.239) are applicable.

Special Case $\mu^+ = k\mu^-$, $k \neq 0$. Substituting $\mu^+ = k\mu^-$ into Eqs. (13.201) and (13.214) yields

$$\mu_s^+ = \mu_s^- = 0 \tag{13.240}$$

Equation (13.240) defines the singular arc on which the substrate concentration maximizes the specific cell growth rates. However, for Monod-type specific growth rates, μ^+ and μ^- have no maximum points satisfying Eq. (13.240) and therefore *no singular arc exists.*

We now consider specific examples to illustrate the preceding analyses.

EXAMPLE 13.E.4: MONOTONIC SPECIFIC GROWTH RATES AND CONSTANT-YIELD COEF-FICIENTS Let us consider the special case $\mu^+ = k\mu^-$, constant $k \neq 0$, where the specific rates are Monod-type dependent on the limiting nutrient concentration only and the yields are constant. As analyzed earlier, no singular arc exists for a fixed or free final time problem, and the optimal feed rate sequence is F_{max} to fill the reactor followed by $F_b = 0$, a batch period. We shall test this by taking monotonic specific rates:[27]

$$\mu^+ = \frac{0.24S}{0.1 + S}; \mu^- = \frac{0.265S}{0.1 + S}; \quad Y_{x/s}^+ = 0.263, \ Y_{x/s}^- = 0.275, \ m = 0 \tag{13.E.4.1}$$

The performance index to be maximized is the final amount of PBCs, $m = 0$ and $n = 0$ in Eq. (13.188), that is, $P = x_1(t_f)$. The required mass balance equations are four, including PBC, PFC, substrate, and the total mass, x_1 through x_4, in Eq. (13.190).

Owing to the constant yields, we have one algebraic relationship (Eq. (13.216)), thus making this process a third order. We take the first, second, and fourth state variables as the state variables in Eq. (13.190) in addition to the algebraic Eq. (13.216):

$$
\begin{bmatrix} \dot{x}_1 \\ \dot{x}_2 \\ \dot{x}_4 \end{bmatrix} = \begin{bmatrix} (1-\alpha)\mu^+ x_1 \\ \alpha\mu^+ x_1 + \mu^- x_2 \\ 0 \end{bmatrix} + \begin{bmatrix} 0 \\ 0 \\ 1 \end{bmatrix} F, \qquad \begin{array}{c} \underline{x}(t_0) = \underline{x}_0 \\ 0 = F_{\min} \leq F \leq F_{\max} \\ x_4(t_f) = V_{\max} \end{array} \quad (13.E.4.2)
$$

The adjoint equations are the first, second, and fourth equations in Eq. (13.194) with $m = 0$ and $\lambda_6 = 0$:

$$
\begin{bmatrix} \dot{\lambda}_1 \\ \dot{\lambda}_2 \\ \dot{\lambda}_4 \end{bmatrix} = -\begin{bmatrix} \partial H/\partial x_1 \\ \partial H/\partial x_2 \\ \partial H/\partial x_4 \end{bmatrix} = -\begin{bmatrix} A\mu^+ \\ B\mu^- \\ (A\mu_s^+ x_1 + B\mu_s^- x_2)(-x_3/x_4^2) \end{bmatrix},
$$

$$
\begin{bmatrix} \lambda_1(t_f) \\ \lambda_2(t_f) \\ \lambda_3(t_f) \end{bmatrix} = \begin{bmatrix} 1 \\ 0 \\ free \end{bmatrix} \qquad (13.E.4.3)
$$

where x_3 is a function of x_1 and x_4 given by Eq. (13.216), and from Eq. (13.193),

$$
A = \lambda_1(1-\alpha), \ B = \lambda_2 \qquad (13.E.4.4)
$$

The Hamiltonian (Eq. (13.219)) reduces to

$$
H = \lambda_1(1-\alpha)\mu^+ x_1 + \lambda_4 F \qquad (13.E.4.5)
$$

Because the final value is $\lambda_1(t_f) = 1$ and $\lambda_1(t) = \lambda_1(t_f)\exp[\int_t^{t_f}(1-\alpha)\mu^+ d\tau]$, λ_1 is positive for all time $\lambda_1 > 0$; $\lambda_2(t) = \lambda_2(t_f)\int_{t_f}^t \mu^- d\tau = 0$, and as a result, $\dot{\lambda}_4 = [\lambda_1(1-\alpha)\mu_s^+ x_1 + \lambda_2\mu_s^- x_2](x_3/x_4^2) > 0$. Therefore, the switching function $\phi = \lambda_4$ is an increasing function of time; that is, there cannot be a finite time period in which it is identically zero. Therefore, the optimal feed rate is bang-bang with at most one switching time, a batch operation, and no singular feed period.

13.6.1.3 Constant Yields and General Product Formation Rate, $m \neq 0, n \neq 0$

We begin with the case of constant yields for metabolite formation kinetics that are both growth- and non-growth-associated.

13.6.1.3.1 FREE FINAL TIME NOT IN THE PERFORMANCE INDEX. For this situation, the state and adjoint equations and the Hamiltonian are given by Eqs. (13.190), (13.194), and (13.192). On the singular arc, Eqs. (13.201) and (13.202) lead to a feedback singular feed rate with the following time rate of change in substrate concentration:

$$
\frac{dS}{dt} = \mu_s^+ \frac{(\mu^+/\mu^-)_s}{(\mu_s^+/\mu_s^-)_s} \left(\frac{\mu^-}{\mu_s^-}\right)^2 \left[(1-\alpha) - \alpha\left(\frac{x_1}{x_2}\right)\right] \qquad (13.241)
$$

Comparing the case of nonzero m, Eq. (13.241), with the case of zero m, Eq. (13.205), we find that the time rate of change in substrate concentration for $m > 0$ slows down on the singular arc, and the m value does not affect the extent of slowdown.

13.6.1.3.2 FINAL TIME IN THE PERFORMANCE INDEX. As mentioned earlier, the order of the differential Eq. (13.187) is reduced by one owing to the constant yields, which give rise to an algebraic equation, Eq. (13.216). Thus, we retain the first, second, fourth, and fifth state variables and redefine the Hamiltonian:

$$H = [\lambda_1(1-\alpha) + \lambda_2\alpha]\mu^+ x_1 + \lambda_2\mu^- x_2 + \lambda_4 F + \lambda_5 - mx_1 \triangleq H_1 + \lambda_5 - mx_1 + \lambda_4 F$$

(13.242)

Similar to the derivation of existence conditions for the singular arc for the fixed final time, constant yields, and growth-associated metabolite formation, the following two inequalities are derived (derivation not shown):

$$\frac{(\mu^+/\mu^-)_s}{\mu_s^+} < \frac{m}{\lambda_2(\mu^-)^2}(\frac{x_1}{x_2}), \quad \lambda_2 > 0$$

(13.243)

$$\left(\frac{\mu_s^+}{\dot{S}}\right)\left[\frac{(\mu^+/\mu^-)_s}{\mu_s^+} - \frac{(1-\alpha)m}{\alpha\lambda_2(\mu^-)^2}\right] = \frac{GLC}{[x^+\alpha(\mu^-)^2]\lambda_2} > 0$$

(13.244)

where

$$GLC = (\lambda_1(1-\alpha) + \lambda_2\alpha)\mu_{ss}^+ x + \lambda_2\mu_{ss}^- x_2 > 0$$

(13.245)

Let us first consider Eq. (13.243). Because the right-hand side of the inequality is positive, it is sufficient for the left-hand side to be negative, $[(\mu^+/\mu^-)_s]/[\mu_s^+] < 0$. Then the quantity inside the bracket of Inequality (13.244) is negative, leading to the negativity of the remaining term of the left-hand side, $\mu_s^+/\dot{S} < 0$. These are also the conditions for the existence of singular arcs described in Section 13.6.1.2.4 for growth-associated metabolite formation with fixed final time (Eqs. (13.227) and (13.236)). Therefore, A set of Eqs. (13.227) and (13.236) is a limiting case in view of the preceding sufficient condition for the existence of singular arcs.

The unreduced form of the singular feed rate is Eq. (13.206), and Eq. (13.205) is the time rate of substrate concentration variation. However, for the given performance index (Eq. (13.221)), the adjoint variables are given by Eqs. (13.194), (13.222), and (13.223) and the Hamiltonian by Eq. (13.224). Equations (13.224) and (13.201) are solved for B, which is then substituted into Eq. (13.205) to obtain

$$\dot{S} = \mu_s^+ \left(\frac{\mu^-}{\mu_s^-}\right)^2 \frac{(\mu^+/\mu^-)_s}{(\mu_s^+/\mu_s^-)_s}\left(\frac{x_1}{x_2}\right)\left[-\alpha + x_2\frac{m(1-\alpha)}{(mx_1-\varsigma)}\right]$$

(13.246)

The singular feed rate is obtained by substituting Eq. (13.246) into Eq. (13.204):

$$F_{\text{sin}} = \frac{V\mu_s^+ \left(\frac{\mu^-}{\mu_s^-}\right)^2 \frac{(\mu^+/\mu^-)_s}{(\mu_s^+/\mu_s^-)_s}\left(\frac{x_1}{x_2}\right)\left[-\alpha + x_2\frac{m(1-\alpha)}{(mx_1-\varsigma)}\right] + (\mu^+/Y_{x/s}^+)x_1 + (\mu^-/Y_{x/s}^-)x_2}{S_F - S}$$

(13.247)

EXAMPLE 13.E.5: MONOTONIC SPECIFIC GROWTH RATES, GROWTH- AND NON-GROWTH-ASSOCIATED METABOLITE FORMATION Let us consider Monod-type cell growth rates, μ^+ and μ^-, which are not linear to each other:

$$\mu^+ = \frac{aS}{K^+ + S}, \quad \mu^- = \frac{bS}{K^- + S}, \quad m = 0.08, n = 0.1, \alpha = 0.02, a = 0.15,$$

$$b = 0.25, \quad K^+ = 0.2, \quad K^- = 0.3, \quad Y_{X/S}^+ = 0.263, \quad Y_{X/S}^- = 0.275$$

(13.E.5.1)

If $K^+ > K^-$, then μ^+/μ^- increases with the substrate concentration. If the yield coefficients $Y^+_{x/s}$ and $Y^-_{x/s}$ are constant, there is no singular arc, as seen in Figure 13.3g. Because the specific rates are monotonic, there is very little one can do to analyze the switching function behavior a priori. Qualitatively, μ^+ and μ^+/μ^- increase as the substrate concentration increases. Therefore, higher substrate concentration is preferable for the maximum PBC, and so the optimal feed rate sequence is $F_{max} \rightarrow F_{min} = 0$, a batch operation.

Conversely, we have $K^+ < K^-$ so that μ^+ increases but μ^+/μ^- decreases with the substrate concentration. Because the yield coefficients are constant, this case corresponds to region II of Figure 13.3B, in which a singular arc can exist. This means that the competition between μ^+ and μ^+/μ^- entails a singular arc and the substrate concentration decreases with time over the singular arc. Let us see an example, of which simulation results are shown in Figure 13.E.5.1. The initial substrate concentration is 1; μ^+ and μ^+/μ^- are nearly constant until they drop to around 1, and after that, μ^+ decreases, while μ^+/μ^- increases steeply. For this reason, the effective singular arc on which the process operates gives a better performance index than on bang arc. Therefore, we can imagine F_{min} is the acceptable feed rate because it forces the state to be around the singular arc. A proper choice of the second feed rate (bang or singular) is related to convergence of the TPBVP. Figure 13.E.5.1 shows the plots of μ^+, μ^-, and μ^+/μ^- and the optimal time profiles of all state variables. The optimal feed rate is $F_{min} = 0 \rightarrow F_{singular} \rightarrow F_b = 0$, which is consistent with the switching function ϕ, which is positive $\rightarrow$ zero $\rightarrow$ positive. The analysis of the singular arc as described earlier plays a major role in optimizing the singular feed rate problem in this example.

13.6.1.4 Variable-Yield Coefficients, $Y^+_{X/S}(S), Y^-_{X/S}(S)$
When the cell mass yield coefficients are functions of substrate concentration, the analysis is much more complicated, and it is difficult to obtain general results.

13.6.1.4.1. FREE FINAL TIME NOT IN THE PERFORMANCE INDEX. With the performance index of Eq. (13.188), the final values of the adjoint variables are the same as those in Eq. (13.196). The dynamics of the adjoint variables for λ_3 and λ_4 are different and are given subsequently:

$$\dot{\lambda}_3 = -\left[(A\mu^+_s x_1 + B\mu^-_s x_2) + (A_s\mu^+ x_1 + B_s\mu^- x_2)\right]/x_4$$
$$= -\left[(A\mu^+_s x_1 + B\mu^-_s x_2) + \left[(Y^+_{x/s})_s\mu^+ x_1/(Y^+_{x/s})^2 + (Y^-_{x/s})_s\mu^- x_2/(Y^-_{x/s})^2\right]\lambda_3\right]/x_4$$
$$= -C/x_4$$
$$\dot{\lambda}_4 = (-x_3/x_4)\dot{\lambda}_3 \tag{13.248}$$

where

$$C = (A\mu^+_s x^+ + B\mu^-_s x^-) + [\mu^+ x^+(Y^+_{x/s})_s/(Y^+_{x/s})^2 + \mu^- x^-(Y^-_{x/s})_s/(Y^-_{x/s})^2]\lambda_3 \tag{13.249}$$

On a singular arc, Eqs. (13.201) and (13.202) are transformed to

$$\dot{\phi} = -(S_F - S)(C/x_4) = 0 \quad \rightarrow \quad C = 0$$
$$\ddot{\phi} = \left[(A\mu^+_{ss}x_1 + B\mu^-_{ss}x_2) + (A_{ss}\mu^+ x_1 + B_{ss}\mu^- x_2) + 2(A_s\mu^+_s x_1 + B_s\mu^-_s x_2)\right]\dot{S}$$
$$+ \left[(1-\alpha)(A\mu^+)_s\mu^+ x_1 + (B\mu^-)_s(\alpha\mu^+ x_1 + \mu^- x_2)\right] - B\mu^-(\mu^- + \mu^-_s)x_2$$
$$- \left[(1-\alpha)(A\mu^+ - m) + \alpha B\mu^-\right](\mu^+ + \mu^+_s)x_1 = E\dot{S} + G = 0 \tag{13.250}$$

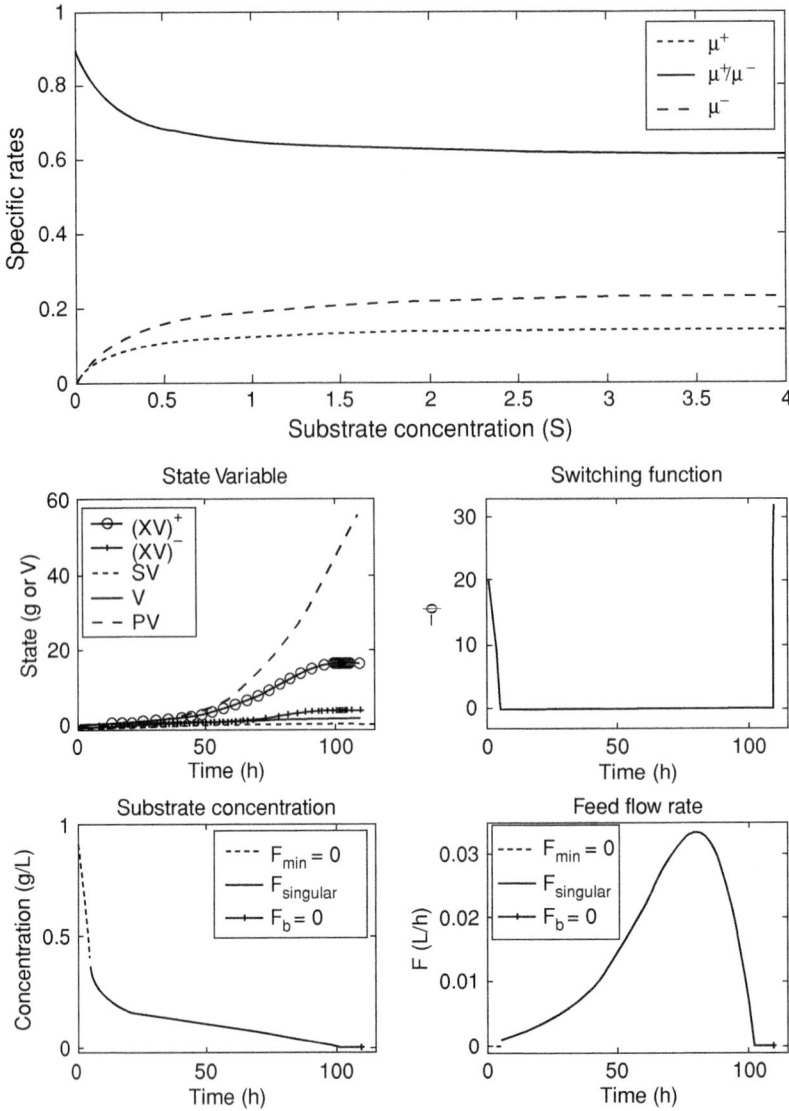

Figure 13.E.5.1. Time profiles of the simulation results. The optimal feed rate sequence is $F_{min} = 0 \to F_{singular} \to F_b = 0$. The initial states and operational parameters are, respectively, $[x_1, x_2, x_3, x_4](t_0) = [0.1 \text{ g}, 0 \text{ g}, 0.5 \text{ g}, 0.5 \text{ L}]$, $[S_F, F_{max}, F_{min}, V_{max}, t_f, \alpha, m, n] = [50 \text{ g/L}, 0.5 \text{ L/hr}, 0 \text{ L/hr}, 2 \text{ L}, 110 \text{ hr}, 0.02, 0.08, 0.1]$.

where

$$E = (A\mu_{ss}^+ x_1 + B\mu_{ss}^- x_2) + (A_{ss}\mu^+ x_1 + B_{ss}\mu^- x_2) + 2(A_s\mu_s^+ x_1 + B_s\mu_s^- x_2) \quad (13.251)$$

and

$$G = (1-\alpha)(A\mu^+)_s \mu^+ x_1 + (B\mu^-)_s (\alpha\mu^+ x_1 + \mu^- x_2) - B\mu^-(\mu^- + \mu_s^-)x_2$$
$$- \left[(1-\alpha)(A\mu^+ - m) + \alpha B\mu^-\right](\mu^+ + \mu_s^+)x_1 \quad (13.252)$$

From Eq. (13.208) and $\dot{\phi} = 0$ of Eq. (13.250), A and B can be expressed as follows:

$$\begin{bmatrix} A \\ B \end{bmatrix} = \begin{bmatrix} -[\mu^- H + \mu_s^- m x_1]/(Dx_1) \\ [\mu^+ H + \mu_s^+ m x_1]/(Dx_2) \end{bmatrix} \tag{13.253}$$

where

$$H = [\mu^+ x^+ (Y_{x/s}^+)_s / (Y_{x/s}^+)^2 + \mu^- x^- (Y_{x/s}^-)_s / (Y_{x/s}^-)^2] \lambda_3 \tag{13.254}$$

and

$$D = (\mu_s^+ \mu^- - \mu^+ \mu_s^-) \tag{13.255}$$

Substituting A and B from Eq. (13.253) into $\dot{\phi} = 0$ of Eq. (13.250) and solving for $\dot{S}$ obtains

$$\dot{S} = -(G/E)(x_1, x_2, S, \lambda_3) \tag{13.256}$$

where λ_3 is an unknown constant. Substitution of Eq. (13.256) into Eq. (13.204) yields the singular feed rate, but it contains an unknown constant λ_3,

$$F_{\sin} = \frac{[-(G/E)x_4 + (\mu^+/Y_{x/s}^+)x_1 + (\mu^-/Y_{x/s}^-)x_2]}{S_F - S} \tag{13.257}$$

which is a function of S, x_1, x_2 and a constant value of λ_3, which is not known a priori. Therefore, a feedback form of a singular feed flow rate with one unknown constant parameter is obtained. Hence, an iterative solution is required.

13.6.1.4.2. FREE FINAL TIME THAT APPEARS IN THE PERFORMANCE INDEX. Equations (13.222)–(13.224) are applicable to the performance index of Eq. (13.221). Solving Eq. (13.224) and $\dot{\phi} = 0$ of Eq. (13.250) for A and B,

$$\begin{bmatrix} A \\ B \end{bmatrix} = \begin{bmatrix} -[\mu^- H + \mu_s^- (mx^+ - \varsigma)]/(Dx) \\ [\mu^+ H + \mu_s^+ (mx^+ - \varsigma)]/(Dx_2) \end{bmatrix} \tag{13.258}$$

where A and B are functions of x_3/x_4, x_1, x_2 and two constant parameters, ς and λ_3, which are not known a priori. Substituting A and B from Eq. (13.258) into $\dot{\phi} = 0$ of Eq. (13.250), the singular feed rate with two unknown constant parameters, ς and λ_3, is obtained. The time rate of change in substrate concentration and the singular feed rate are the same forms as Eqs. (13.256) and (13.257), respectively.

Although the performance index does not contain the fixed final time t_f, replacing ς with the negative constant Hamiltonian value $-H^*$, we see that the singular feed rate form is the same as the preceding case, having two unknown constant parameters $-H^*$ and λ_3, which can be obtained by numerical iteration.

EXAMPLE 13.E.6: VARIABLE-YIELD COEFFICIENTS If the yields $Y_{x/s}^+$ and $Y_{x/s}^-$ are not constant but vary with the substrate concentration, it is difficult to show the existence of a singular arc in a form that is independent of adjoint variables (analysis not shown). Nevertheless, the limiting case, constant yields, can give an inference to the existence of singular arcs for this nonconstant-yield case, although it may not be completely consistent with the constant-yield case.

The plasmid-bearing methylotroph cells free of substrate inhibition in a continuous culture have been analyzed.[38] The plasmid-free cells are subject to substrate inhibition. The performance index to be considered is to maximize PBC in the fed-batch reactor at a final time. Specific growth and substrate consumption rates are functions of substrate concentration only:

$$\mu^+ = \frac{0.9S}{0.8 + S}; \mu^- = \frac{1.2S}{0.4 + S + 0.4S^2}; \sigma = S/(0.1 + S) \qquad (13.E.6.1)$$

where the substrate consumption rates of PBC and PFC are the same. Optimal singular control structure is a concatenation of maximum, minimum, and singular control. However, only with the necessary conditions deduced from the optimality condition, it is not simple to predict the sequence. This example deals with a general case, in which specific rates are independent of each other so that it is more difficult to predict the characteristics of the singular arc and the optimal feed rate strategy. However, the limiting case characterized by Figure 13.E.6.1 can show the feature of singular arcs, although the results are based on constant-yield coefficients. Qualitatively, we can say that the substrate concentration moves in the direction of increasing the ratio μ^+/μ^- at the expense of μ^+ on the singular arc. However, for this system, both μ^+/μ^- and μ^+ increase with respect to substrate concentration. This means that there is no competition between the value of μ^+ and that of μ^+/μ^-, which is the key characteristic of a singular arc, as mentioned previously in the case of constant yield. In view of the limiting case, there is no singular arc over all the substrate concentration region. For these reasons, we can say that the optimal operation is bang-bang, $F_{max} \rightarrow F_b = 0$, a batch operation.

EXAMPLE 13.E.7: NONMONOTONIC SPECIFIC GROWTH RATES AND CONSTANT-YIELD COEFFICIENTS Let us consider nonmonotonic specific rates μ^+ and μ^- with constant yields, as follows:

$$\mu^+ = \frac{S}{1 + 2S + S^2}, \mu^- = \frac{2S}{1 + S + S^2}, Y_{x/s}^+ = 0.25, Y_{x/s}^- = 0.28, m = 0.5 \quad (13.E.7.1)$$

As shown in specific rate behavior in Figure 13.E.7.1, Inequalities (13.225) and (13.234) are satisfied over all substrate concentrations. Therefore, according to the singular arc conditions, if m is zero, all substrate concentration areas can be candidates for a singular arc. However, for the substrate concentration below 1, μ^+/μ^- increases and μ^- decreases much more rapidly as compared to the substrate concentrations over 1. Therefore, we predict that the optimal singular region is the substrate concentration below 1, even though m is not zero. The initial substrate concentration is 2. Hence, the first feed is $F_{min} = 0$ to reduce the substrate concentration so as to reach the singular arc as quickly as possible. The second feed, F_{sin}, continues up to the fixed full reactor volume, followed by a batch operation, $F_b = 0$, to the fixed final time ($F_{min} = 0 \rightarrow F_{sin} \rightarrow F_b = 0$). Figure 13.E.7.1 shows the consistent switching function behaviorϕ, negative $\rightarrow$ zero $\rightarrow$ negative. As predicted, the substrate concentration on a singular arc decreases away from 1, at which μ^+ is maximum. Therefore, we can say that the singular arc criteria with zero m infer to some degree the singular arc behavior with nonzero m.

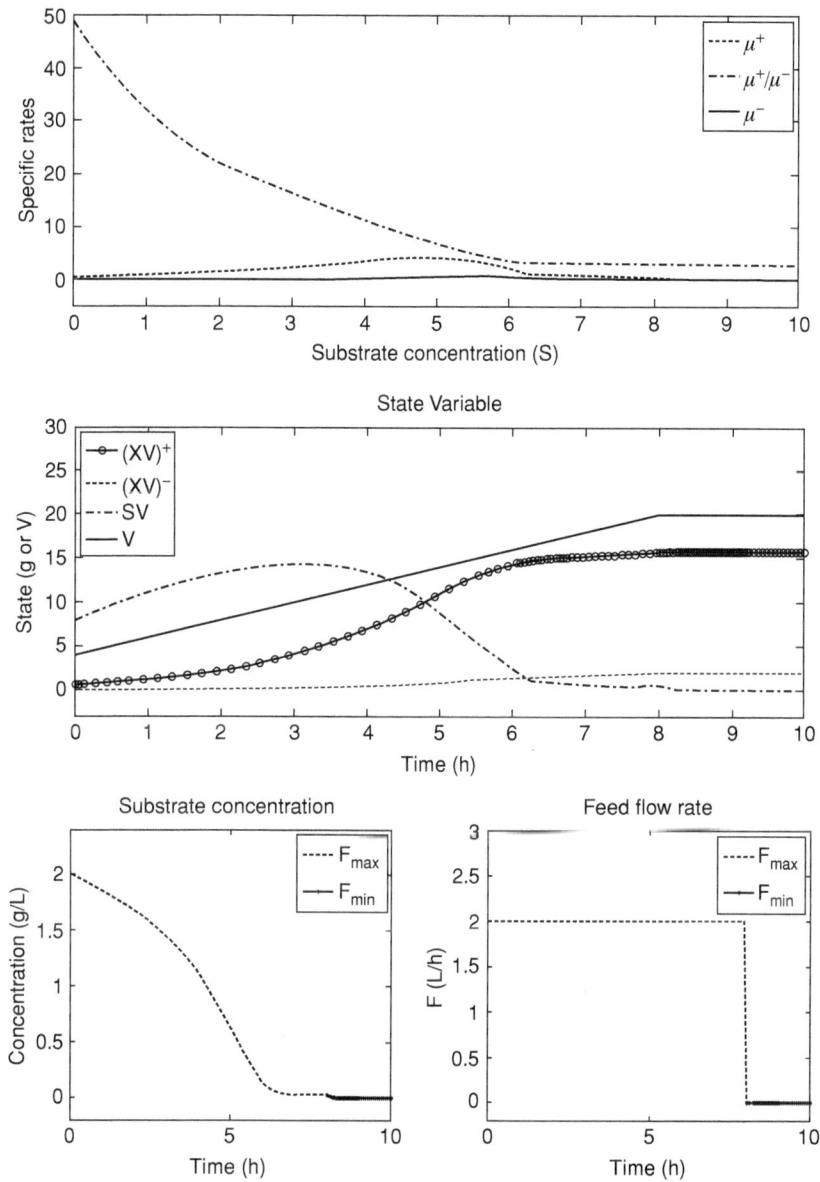

Figure 13.E.6.1. Time-profiles of the simulation results of Example 13.E6. The optimal feed rate sequence is $F_{max} \to F_b = 0$. The initial states and operation parameters are, respectively, $[x_1, x_2, x_3, x_4](t = 0) = [0.65\,\text{g}, 0\,\text{g}, 8\,\text{g}, 4\,\text{L}]$ $[S_F, F_{max}, F_{min}, V_{max}, t_f, \alpha] = [2\,\text{g/L}, 2\,\text{L/hr}, 0\,\text{L/hr}, 20\,\text{L}, 10\,\text{hr}, 0.02]$.

13.6.2 Recombinant Cells with Plasmid Instability and Subject to Cell Death

The death rates for PFC and PBC can be different. Therefore, we take up the problem of optimizing the fed-batch culture of recombinant cells with plasmid instability and with different death rates. Two types of singular arcs are elucidated, and the optimal policies over these singular arcs are explored. These findings are important practically, owing to revelation of qualitative information of the singular arc. Even though most fed-batch fermentation is known to have first-order singularity (so that

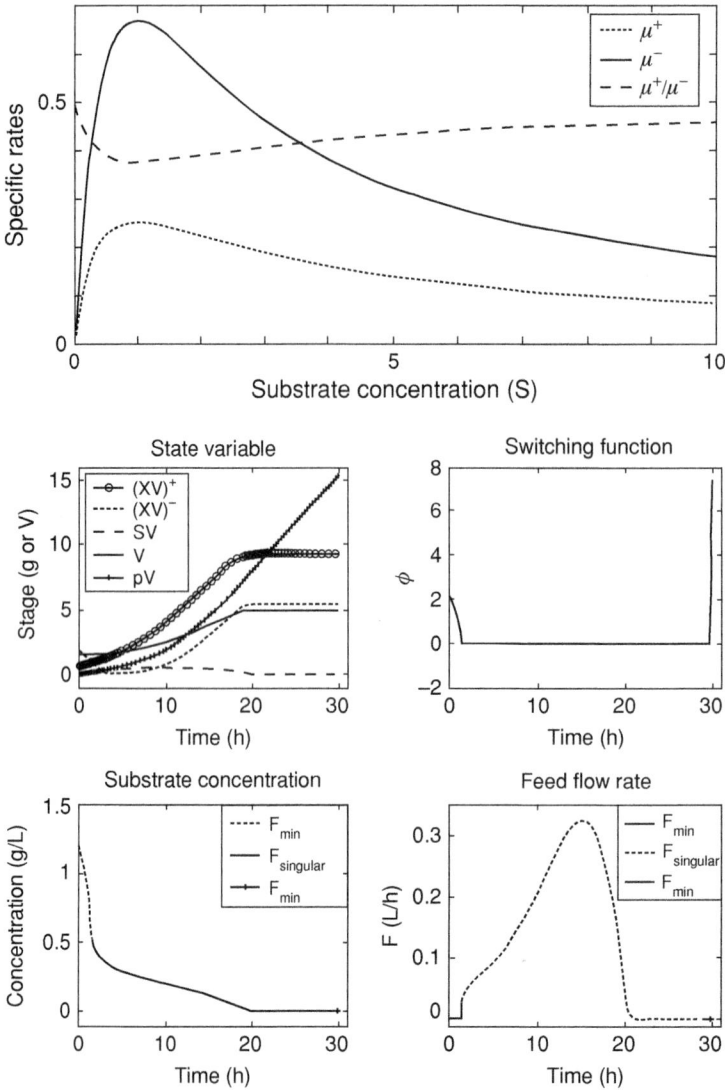

Figure 13.E.7.1. Time profiles of the simulation results. The optimal feed rate sequence is $F_{min} = 0 \rightarrow F_{singular} \rightarrow F_b = 0$. The initial states and operation parameters are, respectively, $[x_1, x_2, x_3, x_4](t_0) = [0.65$ g, 0 g, 2 g, 1.5 L$]$, $[S_F, F_{max}, F_{min}, V_{max}, t_f, \alpha, m, n] = [15$ g/L, 2 L/hr, 0 L/hr, 5 L, 30 hr, 0.02, 0.08, 0.1$]$.

the singular arc is determined from the switching function and its first- and second-order derivatives), this study shows that a singular arc with second-order singularity is possible for a recombinant cell process if PFC and PBC are subjected to death and their specific growth rates are Monod type and proportional to each other.

The concentration of PBC, the plasmid copy number, plasmid stability, and gene expression efficiency affect the rate and yield of recombinant products, and these factors depend on the nutrient concentration.[39] Thus, the optimal fed-batch operation involving manipulation of feed rates to control the nutrient concentration and the reactor environment is a powerful fermentation method and has been used widely to produce many recombinant proteins in high concentration or productivity.[47–50]

Loss of plasmid during fermentation is a major drawback, for which two major reasons are known: plasmid instability[50,51] and depression of the growth rate of PBC.[53] Several operational methods have been suggested to improve the plasmid stability in nonselective media,[45,54–56,58] and a fed-batch operation is a preferred method to produce recombinant products.[59] Experimental results show that low substrate concentration (mainly carbon source) leads to high plasmid stability.[48,55,60–62] Although PFC grow faster than PBC in a nonselective medium of high substrate concentration,[63] when the concentration of limiting carbon source is low, the PFC are more susceptible to starvation of the limiting carbon source than the PBC.[55,59,64] This phenomenon plays a selective pressure on the PFC population and has been used to improve the level of the PBC formation in fed-batch processes,[55] where the carbon source (glucose) is added periodically into the reactor, thus causing starvation between glucose additions. This operation uses the host auxotrophic property and generates a substandard transport mechanism for the limiting nutrient. This approach is very attractive owing to possible applications to many microorganisms.

However, this approach may be adequate only for low cell density fermentation, in which the low sustained concentration of carbon source (glucose) yields a low yield of PBC and is inferior because a high density is a prerequisite to obtaining high product density. Thus, there is a need for an optimal feeding strategy to increase both cell productivity and plasmid stability.

The problem of determining the optimal feed rate profile for fed-batch culture is a singular control problem. However, no study of singular optimal control of the recombinant cell culture has been reported. A model proposed by Cheng et al.[54] utilizes a death rate for the PFC, which occurs only at a low range of substrate concentration, making it possible to have two different singular arcs, one for the low substrate concentration and another for the rest of the substrate concentration. In the region of high substrate concentration, the cell productivity is favorable, but the plasmid instability increases with time, while the region low in substrate concentration favors the plasmid stability but disfavors the productivity. In this section, the characteristics of these two singular arcs are analyzed, which can provide important information essential in obtaining numerical solutions for the optimal feed rate strategies of fed-batch fermentation. Numerical solutions are provided.

13.6.2.1 Problem Formulation

For simplicity and as a first step in analyzing recombinant cells with plasmid instability, we follow the traditional approach of lumping recombinant cells into two groups: one with plasmids (PBC) and the other without plasmids (PFC).[64] A more sophisticated model involving cells with different copy numbers of plasmid would be much more desirable and realistic. However, the resulting high-order model makes it computationally difficult and a generalization almost impossible. The simplified model could reveal a first approximation to fed-batch optimization of recombinant cell processes, and the results should shed light on the case of more sophisticated models. Therefore, we consider here the lumped model of PBC and PFC.

It is usually assumed that the rate of reversion of PBC to PFC is proportional to the growth rate of PBC, $r_{X^-} = \mu^- x_2 = \alpha r_{X^+} = \alpha \mu^+ x_1$, where α is a fraction of PBC that reverts[64] to PFC. Cell death occurs normally when the substrate concentration is low, and PFC is known to be more susceptible than PBC to the starvation of the limiting nutrient, leading to a higher death rate for PFC.

As with nonrecombinant cells, recombinant cells may be characterized by the specific rates of cell growth for PBC, μ^+, and PFC, μ^-, the substrate consumption, σ, product formation, π, and any intermediate formation. Only one feed rate F is assumed. The dynamic behavior is expressed by unsteady state mass balance equations for PBC, PFC, substrate, product, and total mass:

$$
\begin{bmatrix} \dot{x}_1 \\ \dot{x}_2 \\ \dot{x}_3 \\ \dot{x}_4 \\ \dot{x}_5 \end{bmatrix} = \begin{bmatrix} (1-\alpha)\mu^+ x_1 \\ \alpha\mu^+ x_1 + (\mu^- - k_d)x_2 \\ -(\mu^+/Y_{x/s}^+)x_1 - (\mu^-/Y_{x/s}^-)x_2 \\ 0 \\ 1 \end{bmatrix} + \begin{bmatrix} 0 \\ 0 \\ S_F \\ 1 \\ 0 \end{bmatrix} F, \quad \begin{bmatrix} x_1(0) \\ x_2(0) \\ x_3(0) \\ x_4(0) \\ x_5(0) \end{bmatrix} = \begin{bmatrix} x_{10} \\ x_{20} \\ S_0 V_0 \\ V_0 \\ 0 \end{bmatrix}
$$

$$
x_4(t_f) = V_{max} k_d = \begin{bmatrix} \text{a constant when } S < S_c \\ 0 \text{ when } S > S_c \end{bmatrix}
$$

$$
0 = F_{min} \le F \le F_{max} \tag{13.259}
$$

where $x_1^+ = X^+ V$, $x_2^- = X^- V$, $x_3 = SV$ and $x_4 = V$ and the superscripts (+) and (−) refer to PBC and PFC, respectively. The balance on PV is depend only on S and x_1, $d(PV)/dt = \pi X^+ V$, and therefore not included in the state equations. The constant death rate of PFC is denoted by k_d, which is active below a critical substrate concentration S_c. An additional state variable, $\dot{x}_5 = 1$ and $x_5(0) = 0$, is introduced to handle an explicit dependence of the performance index on the final time, t_f. The performance index is defined as

$$
J = \underset{F(t)}{\text{Max}} \; P[\mathbf{x}(t_f)] \tag{13.260}
$$

According to PMP,[43] the Hamiltonian is defined as

$$
H = \lambda^T \mathbf{f} = [\lambda_1(1-\alpha) + \lambda_2\alpha - \lambda_3/Y_{x/s}^+]\mu^+ x_1 + (\lambda_2 - \lambda_3/Y_{x/s}^-)\mu^- x_2 - \lambda_2 k_d x_2
$$
$$
+ \lambda_5 + (\lambda_3 S_F + \lambda_4)F \stackrel{\Delta}{=} A\mu^+ x_1 + B\mu^- x_2 - \lambda_2 k_d x_2 + \lambda_5 + \phi F \tag{13.261}
$$

where

$$
A = (1-\alpha)\lambda_1 + \alpha\lambda_2 - \lambda_3/Y_{x/s}^+
$$
$$
B = \lambda_2 - \lambda_3/Y_{x/s}^- \tag{13.262}
$$

and the switching function is

$$
\phi = S_F \lambda_3 + \lambda_4 \tag{13.263}
$$

The adjoint equations are

$$
\begin{bmatrix} \dot{\lambda}_1 \\ \dot{\lambda}_2 \\ \dot{\lambda}_3 \\ \dot{\lambda}_4 \\ \dot{\lambda}_5 \end{bmatrix} = -\frac{\partial H}{\partial \mathbf{x}} = -\begin{bmatrix} A\mu^+ \\ B\mu^- - k_d\lambda_2 \\ (A\mu_s^+ x_1 + B\mu_s^- x_2)(1/x_4) \\ (A\mu_s^+ x_1 + B\mu_s^- x_2)(-x_3/x_4^2) \\ 0 \end{bmatrix}, \quad \begin{bmatrix} \lambda_1(t_f) \\ \lambda_2(t_f) \\ \lambda_3(t_f) \\ \lambda_4(t_f) \\ \lambda_5(t_f) \end{bmatrix} = \begin{bmatrix} \partial P/\partial x_1(t_f) \\ \partial P/\partial x_2(t_f) \\ \partial P/\partial x_3(t_f) \\ \eta \\ \partial P/\partial x_5(t_f) \end{bmatrix}
$$
$$
\tag{13.264}
$$

where $\mu_s^+ \triangleq \partial\mu^+/\partial S$, $\mu_s^- \triangleq \partial\mu^-/\partial S$, and where η is an unknown constant owing to the corresponding state variable whose final value is fixed, $x_4(t_f) = V_{max}$. It is clear from Eq. (13.264) that $\lambda_5(t) = \lambda_5(t_f) = \partial P/\partial x_5(t_f)$ is a constant.

13.6.2.2 Constant Cell Mass Yield Coefficients, $Y_{x/s}^+$ and $Y_{x/s}^-$

We begin with a limiting case that is simpler to analyze. We consider first the case of constant cell mass yield coefficients for both PBC and PFC.

13.6.2.2.1 PERFORMANCE INDEX DEPENDENT ON FREE FINAL TIME. Because the final time does not appear in the performance index, the auxiliary state variable, x_6, is not needed. Also λ_5 is a constant. Hence, we work with four state and four adjoint variables. Assuming that the specific cell growth rates depend only on one limiting nutrient concentration and that cell mass yield coefficients are constants, the adjoint variables must satisfy the following ordinary differential equations and the final values according to PMP:

$$\begin{bmatrix} \dot{\lambda}_1 \\ \dot{\lambda}_2 \\ \dot{\lambda}_3 \\ \dot{\lambda}_4 \end{bmatrix} = -\frac{\partial H}{\partial \mathbf{x}} = -\begin{bmatrix} A\mu^+ \\ B\mu^- - k_d\lambda_2 \\ (A\mu_s^+ x_1 + B\mu_s^- x_2)(1/x_4) \\ (A\mu_s^+ x_1 + B\mu_s^- x_2)(-x_3/x_4^2) \end{bmatrix}, \begin{bmatrix} \lambda_1(t_f) \\ \lambda_2(t_f) \\ \lambda_3(t_f) \\ \lambda_4(t_f) \end{bmatrix} = \begin{bmatrix} \partial P/\partial x_1(t_f) \\ \partial P/\partial x_2(t_f) \\ \partial P/\partial x_3(t_f) \\ \eta \end{bmatrix}$$

$$(13.265)$$

Because the Hamiltonian is linear with respect to the feed rate F in Eq. (13.261), the Hamiltonian is minimized by picking F according to the sign of its coefficient, the switching function $\phi(t)$, as follows:

$$F = \begin{cases} F_{max} & \text{when } \phi = S_F\lambda_3 + \lambda_4 > 0 \\ F_{min} = 0 & \text{when } \phi = S_F\lambda_3 + \lambda_4 < 0 \\ F_{sin} & \text{when } \phi == S_F\lambda_3 + \lambda_4 = 0 \text{ over finite time interval(s), } t_q < t < t_{q+1} \\ F_b = 0 & \text{when } x_4(t) = V_{max} \end{cases}$$

$$(13.266)$$

The optimal feed rate profile is any concatenation of F_{max}, $F_{min} = 0$, F_{sin}, and $F_b = 0$. When the initial conditions $(X_0 V_0, S_0 V_0, V_0)$ lie on the singular hyperspace, it is possible for the optimal feed rate to be singular from the start, until the reactor volume is full. It is also possible, if the upper limit F_{max} is small, that the singular feed rate can reach the upper limit, and therefore, the singular feed rate must be set at F_{max} before the reactor volume is full. If the chosen final time is sufficiently small, then it is possible for the feed rate to be the maximum to fill the reactor volume completely within the given final time. Thus, the shape of the concatenation of feed rates is affected by upper and lower limits on the feed rate, the initial values, and the final time.

On the singular arc, $\phi(t) = 0$ over the finite time interval, and therefore, its higher-order time derivatives must also vanish (up to the second-order derivative if the order of singularity is one). Thus, to obtain the singular feed rate, we take sequential time derivatives of the switching function until the feed rate $F(t)$ appears explicitly:

$$\phi = S_F\lambda_3 + \lambda_4 = 0 \qquad (13.267)$$

$$\dot{\phi} = S_F\dot{\lambda}_3 + \dot{\lambda}_4 = 0 \quad \Rightarrow \quad A\mu_s^+ x_1 + B\mu_s^- x_2 = 0 \qquad (13.268)$$

$$\ddot{\phi} = -(S_F - S)/x_4 \left\{ \begin{array}{l} (A\mu_{ss}^+ x_1 + B\mu_{ss}^- x_2)(dS/dt) - \alpha B(\mu^+/\mu^-)_s(\mu^-)^2 x_1 \\ +k_d(\lambda_2\alpha\mu_s^+ x_1 + (\lambda_3/Y_{x/s}^-)\mu_s^- x_2) \end{array} \right\} = 0$$

$$(13.269)$$

Because the term dS/dt in Eq. (13.269) is related to the feed rate through the substrate balance equation (13.259), no higher derivative needs to be taken:

$$\dot{x}_3 = d(SV)/dt = V\,dS/dt + SF = \left[-(\mu^+/Y_{x/s}^+)x_1 - (\mu^-/Y_{x/s}^-)x_2 \right] + S_F F$$

$$\Rightarrow F_{\sin} = \frac{V[dS/dt + (\mu^+/Y_{x/s}^+)x_1 + (\mu^-/Y_{x/s}^-)x_2]}{S_F - S} \qquad (13.270)$$

The GLC condition[44] obtained from Eq. (13.269) is

$$(-1)\frac{\partial}{\partial F}\frac{d^2\phi}{dt^2} = (S_F - S)\frac{x_1}{x_4}\frac{\partial}{\partial F}[(A\mu_{ss}^+ x_1 + B\mu_{ss}^- x_2)(dS/dt)]$$

$$= (S_F - S)\frac{x_1}{x_4}[(A\mu_{ss}^+ x_1 + B\mu_{ss}^- x_2)]$$

$$\times \frac{\partial}{\partial F}\left[\frac{\left(S_F - \frac{x_3}{x_4}\right)F}{x_4} - \left(\frac{\mu^+}{Y_{x/s}^+}\right)x_1 - \left(\frac{\mu^-}{Y_{x/s}^-}\right)x_2 \right] \geq 0$$

$$\Rightarrow (A\mu_{ss}^+ x_1 + B\mu_{ss}^- x_2) \geq 0 \qquad (13.271)$$

13.6.2.2.2 PERFORMANCE INDEX INDEPENDENT OF THE AMOUNT OF SUBSTRATE AT THE FINAL TIME. When the performance index is independent of the final amount of substrate, $x_3(t_f)$, the final value of the third adjoint variable is zero, $\lambda_3(t_f) = 0$. The optimal feed rate structure of this problem may have degenerate cases for certain initial conditions, the magnitudes of t_f and $F_{\max}$, and functional forms of μ, σ, and π. However, we consider a case in which the magnitude of the maximum feed rate is large enough so that the singular feed rate $F_{\sin}$ does not reach $F_{\max}$ and therefore is implemented without running into a magnitude constraint. This can be realized by a proper selection of a pump capacity to increase $F_{\max}$. It is known that any $F_{\sin}$ satisfying GLC is preferable to bang-bang types of feed rates. Accordingly, once $F_{\sin}$ takes place, it is unnecessary to change the feed rate to a bang ($F_{\max}$ or $F_{\min}$) to meet the fixed volume constraint (for first-order singularity). Therefore, we assume that the sequence $(F_{\max} \rightarrow$ or $F_{\min} = 0) \rightarrow F_{\sin} \rightarrow F_b = 0$ is, in general, the optimal feed rate strategy.

To gain further information, we look at Eqs. (13.265) and (13.268) to obtain

$$-(A\mu_s^+ x_1 + B\mu_s^- x_2)x_1/x_4 = \dot{\lambda}_3 = \frac{\dot{\phi}}{S_F - S} \qquad (13.272)$$

The switching function is zero on the singular arc ($\phi = 0$, $\dot{\phi} = 0$) and is negative, $\phi < 0$, in the region of $F_b = 0$, which follows the singular feed rate $F_{\sin}$. Therefore, $\dot{\lambda}_3 > 0$ over the same region. Because λ_3 is zero at the final time, $\lambda_3(t_f) = 0$, and Eqs. (13.265) and (13.268) show that λ_3 is constant over the singular arc, and λ_3 is a negative constant over the singular arc and must increase from the junction point between F_{sin} and the final $F_b = 0$ to meet the zero final value. Therefore, we can make the following assertion.

For the structure of $(F_{max}$ or $F_{min} = 0) \rightarrow F_{singular} \rightarrow F_b = 0$, if the performance index is independent of $x_3(t_f)$ and there are no points satisfying $\dot{\phi}(t^*) = 0$ in the region of $F_b = 0$ following F_{sin}, λ_3 is negative over the region of $F_{sin} \rightarrow F_b = 0$.

Potential performance indices include maximization of PBC at the final time, its productivity, the ratio of PBC and PFC, and so on. These performance indices are favorable to PBC and unfavorable to PFC, leading to $\lambda_1(t_f) < 0$ and $\lambda_2(t_f) > 0$. The second ordinary differential equation (ODE) of Eq. (13.265) has the solution in the following integral form on the assumption that the performance index is independent of $x_3(t_f)$ so that $\lambda_3(t_f) = 0$:

$$\lambda_2(t) = e^{-\int_{t_f}^{t} p d\tau} \int_{t}^{t_f} e^{\int_{t_f}^{t} p d\tau} (-q) d\varsigma + \lambda_2(t_f) \tag{13.273}$$

where $p = \mu^- - k_d$ and $q = (\mu^-/Y_{x/s}^-)\lambda_3$. Because $\lambda_2(t_f)$ is positive and q is negative over the singular arc, $\lambda_2(t)$ *is positive over the singular arc*. Rewriting the first ODE of Eq. (13.265),

$$\dot{\lambda}_1 = -\lambda_1(1-\alpha)\mu^+ - (\lambda_2\alpha - \lambda_3/Y_{x/s}^+)\mu^+ = p\lambda_1 + q \tag{13.274}$$

where $p = -\mu^+(1-\alpha)$ and $q = -\mu^+(\lambda_2\alpha - \lambda_3/Y_{x/s}^-)$. On the singular arc, p is always negative. If $q = 0$, $\lambda_1(t)$ approaches 0 asymptotically. However, q is always negative over the same region. This means that $\lambda_1(t)$ is less than the negative value of $\lambda_1(t)$ when $q = 0$ for all times over the singular arc. Therefore, we can say that $\lambda_1(t)$ is always negative over a singular arc. Correspondingly, B is always positive over the singular arc.

In summary, $\lambda_5(t) = 0$ and $\lambda_3(t) < 0$ over all times, based on the performance index with fixed final time and assertion 1, respectively. In addition, for the performance index favorable to PBC and unfavorable to PFC, $\lambda_1(t) < 0$ and $\lambda_2(t) > 0$ over a singular arc. The assumptions stated here are realistic for the system.

13.6.2.2.3 ANALYSIS OF SINGULAR ARC. On the basis of the assumptions and results stated earlier, we now look over the feature of the singular arc. The Hamiltonian is negative from PMP. Over the singular arc where $\phi = 0$ and because $\lambda_5(t) = 0$, we conclude that

$$A\mu^+ x_1 + B\mu^- x_2 < \lambda_2 k_d x_2 \tag{13.275}$$

Because λ_2 is positive over the singular arc, the sufficient condition for Eq. (13.275) is

$$A\mu^+ x_1 + B\mu^- x_2 < 0 \tag{13.276}$$

Substituting Eq. (13.268) into Eq. (13.276) and recalling that B is positive over the singular arc,

$$(Bx_2/\mu_s^+)(\mu^+/\mu^-)_s < 0 \quad \Rightarrow \quad \text{sign}(\mu_s^+) = -\text{sign}(\mu^+/\mu^-)_s \tag{13.277}$$

Before going through a further analysis of the singular arc, let us restrict the form of specific growth rates μ^+ and μ^-. As stated in the introduction, Section 13.6.2, the process uses the difference in death rates of PBC and PFC. At very low substrate concentrations, the PFC go through death, while the PBC do not. At the low range of substrate concentrations, no inhibition takes place, and both specific growth rates μ^+ and μ^- increase with the substrate concentration so that μ_s^+ and μ_s^- are both positive over the affected ranges (very low substrate concentration). Then, $\text{sign}(\mu^+/\mu^-)_s$ is negative from Eq. (13.277) because $\text{sign}(\mu_s^+)$ is positive. This means that if substrate concentration increases with time ($\dot{S} > 0$), μ^+ increases and μ^+/μ^- decreases correspondingly, but if $\dot{S} < 0$, μ^+ decreases and μ^+/μ^- increases. Although it is not easy to elucidate a definite sign of $\dot{S}$, we can intuitively say that $\dot{S} < 0$ on a singular arc so that the substrate concentration decreases with time in the direction of the increasing ratio of specific rates (μ^+/μ^-) gradually at the expense of μ^+. It is this increasing μ^+/μ^- that is favorable to the amount of PBC, even though low μ^+ and decreasing substrate concentration force us to take more time until the process reaches the final reactor volume. This can be represented by the following:

$$\frac{dS}{dt} < 0, \quad \frac{dS}{dt}\frac{d\mu^+}{dS} = \frac{d\mu^+}{dt} < 0, \quad \frac{dS}{dt}\frac{d(\mu^+/\mu^-)}{dS} = \frac{d(\mu^+/\mu^-)}{dt} > 0 \quad (13.278)$$

On a singular arc, the substrate concentration decreases with time in the direction of increasing the ratio of specific rates (μ^+/μ^-) gradually at the expense of specific rates μ^+ and μ^-.

13.6.2.2.4. FREE FINAL TIME NOT IN THE PERFORMANCE INDEX.
Because the performance index is not dependent on the final time, $\lambda_5(t) = \lambda_5(t_f) = 0$. In addition, the Hamiltonian is zero for this case, and therefore, on a singular arc,

$$A\mu^+x_1 + B\mu^-x_2 - \lambda_2 k_d x_2 = 0 \quad (13.279)$$

Equations (13.268), (13.269), and (13.279) are linear in λ_1, λ_2, and λ_3. The matrix form of the three equations is $\mathbf{R}[\lambda_1, \lambda_2, \lambda_3]^T = \mathbf{0}$. The determinant $\mathbf{R}$ is zero due to a nontriviality of adjoint variables, leading to the time derivative of the substrate concentration that depends on state variables only. When incorporated into Eq. (13.270), the result constitutes a feedback form of a singular feed rate:

$$F_{\text{sin}} = \frac{V\frac{\mu^-(\mu^+/\mu^-)_s(\mu^-)^2 - k_d\mu_s^+}{(\mu_s^+/\mu_s^-)_s(\mu_s^-)^2} + (\mu^+/Y_{x/s}^+)x_1 + (\mu^-/Y_{x/s}^-)x_2}{S_F - S} \quad (13.280)$$

13.6.2.2.5 A SPECIAL CASE, $\mu^-(S) = \eta\mu^+(S)$.
A special case in which the specific growth rates of PBC and PFC are proportional to each other, $\mu^-(S) = \eta\mu^+(S)$, and cell mass yields are constant, $Y_{x/s}^+$ and $Y_{x/s}^-$, requires special treatment.

Two different performance indices will be treated: (I) a performance index, which depends on the free final time, and (II) a performance index independent of the free final time.

I. Free Final Time in the Performance Index. Substituting $\mu^-(S) = \eta\mu^+(S)$, where η is a constant and a Monod form of μ^+, into ϕ and $\dot{\phi}$ (Eqs. (13.268) and (13.269)), yields the following:

$$Ax_1 + \eta Bx_2 = 0 \quad (13.281)$$

$$(\alpha\lambda_2)x_1 + (\eta\lambda_3/Y^-)x_2 = 0 \quad (13.282)$$

Because the singular feed rate does not appear in the second-order time derivative of the switching function $\ddot{\phi}$, we take a third-order time derivative of Eq. (13.282) and set it to zero, $\dddot{\phi} = 0$:

$$(\alpha\lambda_2)x_1\left[(1-\alpha)\mu^+ - (\eta\mu^+ - k_d)\right] + (\eta\lambda_3/Y_{X/S}^-)\left[2\alpha x_1\mu^+ + (\eta\mu^+ - k_d)x_2\right] = 0 \tag{13.283}$$

Equations (13.281)–(13.283) are linear in adjoint variables, λ_1, λ_2, and λ_3. To retain the nontriviality of the adjoint variables, the determinant of the matrix originating from Eqs. (13.281)–(13.283) must vanish:

$$2\alpha\mu^+ x_1 + \left[(2\eta + \alpha - 1)\mu^+ - 2k_d\right]x_2 = 0 \tag{13.284}$$

which is the singular arc. Because the singular feed rate does not appear in the third-order time derivative $\dddot{\phi} = 0$, we take a fourth-order time derivative again and substitute the substrate balance equation (13.259) for $\dot{S}$:

$$\begin{aligned}
\phi^{(4)} &= \alpha x_1 \lambda_1 / x_2 \{-\alpha(1-\alpha)(\mu^+)^2 x_1 - [2\alpha x_1 + (2\beta + \alpha - 1)x_2]\mu_S^+ \dot{S}\} \\
&= \alpha x_1 \lambda_1 / x_2 \left[\begin{array}{l} -[2\alpha x_1 + (2\beta + \alpha - 1)x_2]\mu_S^+ \{-(\mu^+/Y_{X/S}^+)x_1 - (\mu^-/Y_{X/S}^-)x_2\}/V \\ -\alpha(1-\alpha)(\mu^+)^2 x_1 - [2\alpha x_1 + (2\beta + \alpha - 1)x_2]\mu_S^+ (S_F - S)F/V \end{array}\right]
\end{aligned} \tag{13.285}$$

In Eq. (13.285), the feed rate appears linearly, and therefore, it belongs to the singular control problems. Additionally, the order of singularity, defined as q, in which $2q$ is the number of time derivatives of the switching function ϕ, in which the feed rate appears linearly, is 1 ($2q = 2$) in most cases, but this problem has the order of singularity of 2 ($2q = 4$). The singular feed rate is obtained from Eq. (13.285) by solving for the feed rate:

$$F_{\text{sin}} = \frac{[-2\alpha x_1 + (1 - 2\beta - \alpha)x_2]\mu_S^+\{(\mu^+/Y_{X/S}^+)x_1 + (\mu^-/Y_{X/S}^-)x_2\} - \alpha(1-\alpha)(\mu^+)^2 x_1 V}{[2\alpha x_1 + (2\beta + \alpha - 1)x_2]\mu_S^+ (S_F - S)} \tag{13.286}$$

Equation (13.284) represents a singular arc, a function of state variables only, which implies, once state variables are determined at an instant of time, that the optimal feed rate at that time is determined to be either bang or singular. We assume that a singular feed rate satisfying the GLC condition in combination with bang-bang yields a better result than bang-bang with singular. If the initial state is not on the singular arc, the first feed rate could be bang-bang with a number of switchings until the state is on the singular arc. Then, a singular feed rate continues until the reactor volume is full, when a batch period, $F_b = 0$, takes over until the final time. For the given final time, the structure bang-singular feed rate with any switching time may not have the final reactor volume constraint reached. In this case, bang control can be added. If the initial state is on the singular arc, the singular feed begins first until the reactor volume constraint is matched, and then $F_b = 0$ to the final time.

    II. Free Final Time Not in the Performance Index. Because the performance index is independent of the final time, $\lambda_5(t) = 0$ from Eq. (13.264). In addition, the condition $\mu^-(S) = \eta\mu^+(S)$ and Eq. (13.281) lead to $\lambda_2(t) = 0$ on the singular arc from Eq. (13.278). However, the fact that $\lambda_2(t) = 0$ on the singular arc implies that $\lambda_1(t) = \lambda_3(t) = \lambda_4(t) = 0$ from Eqs. (13.267), (13.281), and (13.282); that is, all adjoint variables on the singular arc are zero. This trivial case means that all types of feed rate structures satisfy the necessary conditions for optimality; any feed rate structure is optimal.

13.6.2.3. Feed Rate Policy for Maximizing PBC and Plasmid Stability
The process is divided into two types, depending on the substrate concentration S. Plasmid stability is improved when the process is operated in the low S region. However, the operations in the region for a given final time entail inputs of small amounts of substrate, resulting in small amounts of PBC, as dictated by the mass balance. Therefore, operations in the region of the relatively high S are indispensable for maximizing the final cell mass. The need to operate the process on a higher S region becomes greater as the final reactor volume constraint $V_{\max}$ is high, the final time t_f is short, and the feed substrate concentration S_F is high. Because a singular control with the satisfaction of the GLC condition is known to improve the performance index beyond that which can be achieved by the bang-bang control alone, two types of singular arcs are anticipated, depending on S. Conclusively, the optimal feed rate strategy is a concatenation of bang and two types of singular feed rates.

EXAMPLE 13.E.8: β-GALATOSIDASE FROM GLUCOSE   Cheng et al.[54] proposed a kinetic model for producing β-galactosidase from *S. cerevisiae*. The specific rates and kinetic parameters are as follows:

$$\mu^+ = 0.24S/(0.1 + S), \quad \mu^- = 0.265S/(0.1 + S), \quad \pi = (Y_{p/s}^+/Y_{x/s}^+)\mu^+, \quad Y_{x/s}^+ = 0.263,$$
$$Y_{x/s}^- = 0.275, \quad Y_{p/s}^+ = 0.144, \quad \alpha = 0.08, \quad k_d = 0.025 \text{ when } S < 0.1, \quad k_d = 0.0$$
$$\text{when } S \geq 0.1 \qquad\qquad (13.\text{E}.8.1)$$

They proposed and operated the fed-batch with periodic pulse feedings and allowed depletion of glucose between pulses. This case study applies the singular feed strategy instead and confirms the improvement of the performance index and the maximization of metabolite at the final time.

There are two distinct specific rates, depending on substrate concentration ranges, owing to the discontinuous death term $k_d = 0.025[U(S) - U(S - 0.1)]$. The death occurs at a constant rate 0.025 when the substrate concentration is less than 0.1, and no death occurs when the substrate concentration is greater than 0.1. The metabolite production rate is assumed to be directly proportional to the PBC formation rate (Eq. (13.E.8.1)). The original mass balances are fourth order, but constant-yield coefficients $Y_{p/s}^+$, $Y_{x/s}^+$, and $Y_{x/s}^-$ allow us to reduce the order by one. The first through third rows of Eq. (13.259) yield the following equation with the initial values:

$$x_1^+ - \frac{(1-\alpha)}{1/Y_{x/s}^+ - \alpha/Y_{x/s}^-}\left[S_F x_4 - x_3 - \frac{x_2^-}{Y_{x/s}^-}\right]$$
$$= x_{1,0} - \frac{(1-\alpha)}{1/Y_{x/s}^+ - \alpha/Y_{x/s}^-}\left[S_F x_{4,0} - x_{3,0} - \frac{x_{2,0}}{Y_{x/s}^-}\right] \qquad (13.\text{E}.8.2)$$

The four state variables are related to each other by the preceding algebraic equation, and therefore, the dimension is reduced by one. Substituting $\mu^- = k\mu^+$ into the second row of Eq. (13.259), the first and second rows constitute one differential equation,

$$dx_2^-/dx_1^+ = \varpi x_2^-/x_1^+ + \tau, \qquad \varpi = k/(1-\alpha), \quad \tau = \alpha/(1-\alpha) \quad (13.E.8.3)$$

which yields a solution in terms of the initial states:

$$(x_1^+)^{-(\varpi-1)}\{x_2^- - [\tau/(1-\varpi)]x_1^+\} = (x_{1,0}^+)^{-(\varpi-1)}\{x_{2,0}^- - [\tau/(1-\varpi)]x_{1,0}^+\} \quad (13.E.8.4)$$

Thus, the dimension is further reduced by one, and the original mass balances reduce to a second-order process:

$$[S_F, t_f, V_{max}] = [200\,g/L, 90\,hr, 4.729\,L]$$
$$[x_{10}, x_{20}, x_{30}, x_{40}, x_{50}] = [1.322, 0, 12.2, 3.05, 0]$$

Taking the first and fourth rows of Eq. (13.259), the Hamiltonian is redefined as

$$H = \lambda_1(1-\alpha)\mu^+ x_1 + \lambda_4 F \qquad (13.E.8.5)$$

The Hamiltonian is zero on the singular arc:

$$\phi = \lambda_4 = 0 \qquad (13.E.8.6)$$

$$\dot{\phi} = \dot{\lambda}_4 = \lambda_1(1-\alpha)\mu_s^+ \left(-x_1 x_2/x_4^2\right) = 0 \qquad (13.E.8.7)$$

There are two solutions to Eq. (13.E.8.7): $\lambda_1 = 0$ or $\mu_s^+ = 0$. However, $\mu_s^+ \neq 0$, as seen in Eq. (13.E.8.1), and therefore, $\lambda_1 = 0$. This means that the two independent adjoint variables λ_1 and λ_4 are zero on the singular arc. This is a trivial case, and therefore, any feed rate is optimal on a singular arc. The optimal feed rate is a concatenation of singular and bang feed rates. Therefore, for this example, when $S \geq 0.1$, the optimal feed rate is a bang-bang feed rate. When $S < 0.1$, k_d is not zero, and the singular arc and the singular feed rate are Eqs. (13.284) and (13.270), respectively.

Parameters S_F and the final time t_f are 200 g/L and 90 hrs, respectively, and the initial inoculum states $[(XV)^+, (XV)^-, SV, V, PV]$ are [1.3222, 0, 12.2, 3.05, 0]. Similar to Cheng et al.,[54] the first simulation is to feed the substrate after four hours following depletion of glucose, and this strategy is repeated up to the final time of 90 hrs. The results are shown in Figure 13.E.8.1. The final amounts of PBC and metabolite are 61.47 and 35.79, respectively, while the ratio of PBC/PFC is maintained around 80 percent.

To examine the superiority of the singular feed rate strategy, all parameters and initial inoculum states must be the same as the first simulation of the addition–depletion feeding strategy. Considering the feed strategy, the first feed rate is F_{min} so that the state can be on a singular arc as soon as possible. $F_{singular}$ continues until the reactor is full, V_{max}, and then a batch period takes place until the final time, t_f. The simulation results are shown in Figures 13.E.8.2 and 13.E.8.3. The final amounts of PBC and metabolite are 74.32 (vs. 61.47, an improvement of 20.9%) and 43.44 (vs. 35.79, an improvement of 21.4%), while the ratio of PBC/PFC is maintained over 85 percent (vs. 80%, an improvement of 6.3%). Consequently, we can say that the application of the singular feed rate strategy is much better than the repeated addition–depletion feed strategy in improving the plasmid stability and the final amount of metabolite.

Glucose starvation between additions

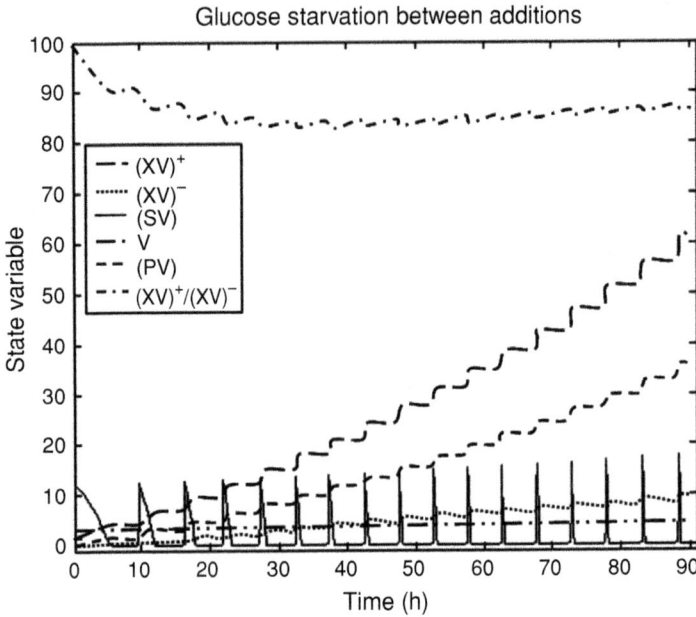

Figure 13.E.8.1. Simulated fed-batch fermentation in nonselective medium, in which glucose is added up to 4 g/L with 4 hrs of glucose starvation between additions: $[S_F, t_f, V_{max}] = [200 \text{ g/L}, 90 \text{ hr}, 4.729 \text{ L}]$, $[x_{10}, x_{20}, x_{30}, x_{40}, x_{50}] = [1.322, 0, 12.2, 3.05, 0]$.

Singular feed of glucose

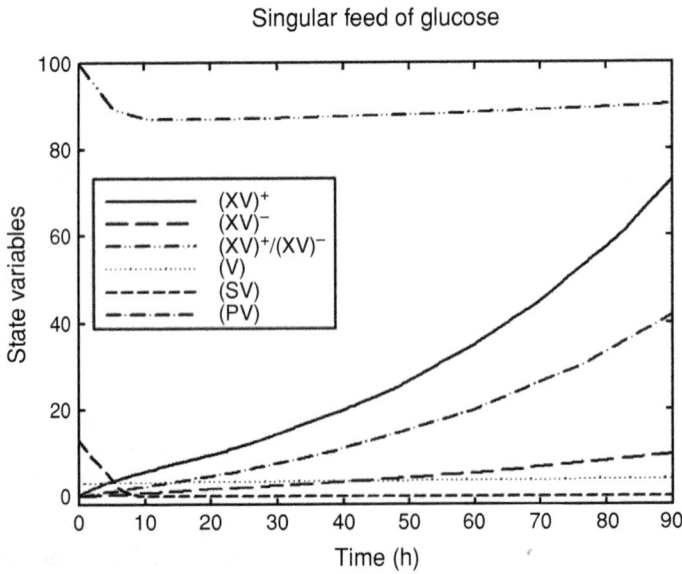

Figure 13.E.8.2. Simulated fed-batch fermentation in nonselective medium, in which glucose is added following the optimal singular feed rate strategy: $[S_F, t_f, V_{max}] = [200 \text{ g/L}, 90 \text{ hr}, 4.729 \text{ L}]$, $[x_{10}, x_{20}, x_{30}, x_{40}, x_{50}] = [1.322, 0, 12.2, 3.05, 0]$.

Substrate concentration on F_{sin} and Feed rate

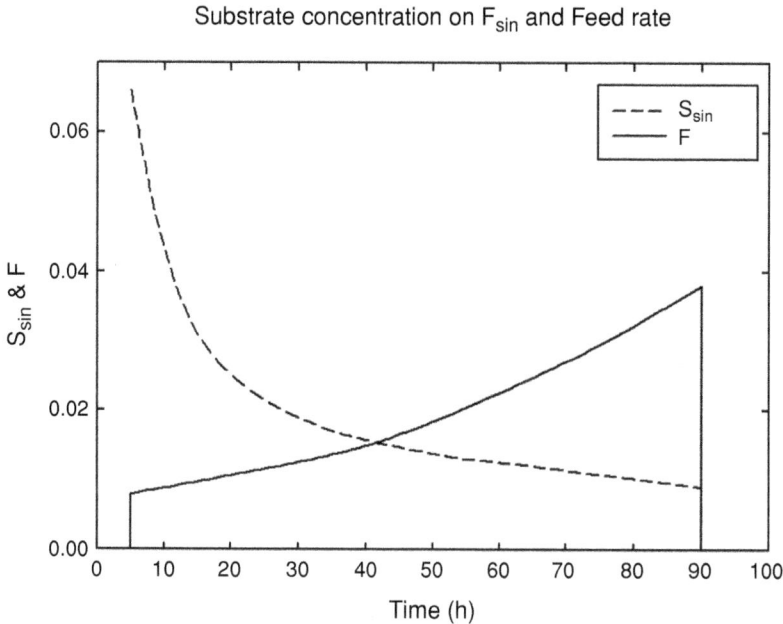

Figure 13.E.8.3. Optimal singular feed rate and substrate concentration profiles.

13.7 Higher-Order Models

When the process to be optimized is complex and requires a large number (more than five) of mass balance equations, it is difficult to determine the structure of the optimal feed rate profile, and there is very little hope of obtaining any analytical results. Therefore, it may be necessary to resort to a purely numerical procedure. The PMP and singular control theory provide that the optimal feed rate is a concatenation of periods of maximum, minimum, and singular feed rates. The exact sequence and the number of occurrences of individual periods are not known. In addition, the singular feed rate expression contains $(n - 4)$ adjoint variables for free final time problems and $(n - 3)$ for fixed final time problems, and therefore, any process that is higher than fourth order would include at least one adjoint variable, which in turn makes any analytical approach to the solution very difficult. There is a purely numerical approach using the substrate concentration profiles as the manipulated variables, instead of the feed rate. By so doing, the problem becomes nonsingular, and therefore, any gradient search method can be used to obtain a numerical solution (open-loop solution) rather than feedback (closed-loop solution).

In Chapter 9, we treated the nonsingular approach in detail, including the selection of substrate concentrations as the manipulated variables, a transformation of a singular problem into a nonsingular problem, solution to the optimal substrate concentration profiles, and finally, if desired, the corresponding optimal feed rates.

13.7.1 Animal Cell Cultures

Animal cell cultures are more complex than the microbial cell cultures and, therefore, are modeled by at least five mass balance equations. Hence, optimization of animal cell cultures is more complex and even numerically difficult to solve. Therefore, there has not been much literature on rigorous treatments of optimization of the feed rate strategy. However, more and more applications of fed-batch operations are used to produce biologics from animal cells.

Genetic engineering techniques have been applied to develop highly productive cell lines and to maximize cell culture longevity and highly specific secretion rates. For optimal operation of fed-batch cell cultures, the nutrient concentration and culture environmental conditions are manipulated.[66,67] The cell viability and operational period dictate the performance of many mammalian cell fed-batch cultures. Glucose and glutamine are the main carbon and nitrogen sources, and their concentrations are critical to the production of monoclonal antibodies.[68,69]

Owing to the complexity of animal cell metabolisms and poor understanding of the intracellular factors that affect the product formation and secretion rates, few studies have dealt with the monitoring and control of mammalian cell fed-batch operations. Structured models have been proposed to develop feeding strategies[70–72] and simulation, and intermittent[73] and continuous[74] feedings to maximize cell growth have been reported.

Dynamic programming was used to calculate the optimal feed rate profile to maximize the monoclonal antibody[75] production using a sixth-order model. Owing to the high-order process and the nature of dynamic programming requiring extensive computation, the numerical difficulty led to a suboptimal policy. A comparison of three different methods of computation was reported:[75] the unidirectional singular control algorithm,[76] dynamic programming, and a modified nonsingular method.[76] It is instructive to go over the study.

A sixth-order model[75] of monoclonal antibody production dynamics by hybridoma cells involved in the study is

$$\frac{dX_v}{dt} = (\eta - k_d)X_v - DX_v$$

$$\frac{dGlc}{dt} = D(Glc_{in} - Glc) - Q_{glc}X_v$$

$$\frac{dGlt_e}{dt} = D(Glt_{in} - Glt_e) - K_{tr}(Glt_e - Glt_i)X_v$$

$$\frac{dGlt_i}{dt} = K_{tr}((Glt_e - Glt_i)X_v - Q_{glt_i}X_v \qquad (13.287)$$

$$\frac{dMAb}{dt} = Q_{AMb}X_v - DMAb$$

$$\frac{dV}{dt} = F,$$

where X_v, Glc, Glt, and MAb stand for the concentrations of viable cells, glucose, glutamine, and monoclonal antibody, respectively, and the subscripts e and i stand for the external and internal, respectively. The rate expressions and parameters are as follows:

$$\eta = \mu_{\max}\frac{Glc}{K_{glc} + Glc}\frac{Gln_i}{K_{Gln} + Gln_i},$$

$$k_d = k_{d\max}\frac{K_{dgln}}{(K_{dGln} + Gln_i)}, \mu = \eta - k_d$$

Table 13.2 *Parameter values for antibody production model*

Parameters	Numerical values
η_{max}	1.13
K_{glc}	1.4
K_{gln}	1.5
$K_{d\,max}$	0.78
K_{dgln}	0.02
$Y_{X_v/glc}$	0.87
$Y_{X_v/gln}$	1.95
K_{tr}	4.81
α_0	2.6
K_μ	0.001

$$\begin{aligned}
Q_{Glc} &= \eta/Y_{X_v/Glc} = \sigma_1, \\
Q_{Gln_i} &= \eta/Y_{X_v/Gln} = \sigma_2, \\
Q_{MAb} &= \frac{\alpha_0}{K_\mu + \eta}\eta = \pi
\end{aligned} \tag{13.288}$$

The numerical values of the kinetic parameters are given in Table 13.2.

Because the rate of transfer of glutamine across the cell membrane is thought to be very high as compared to its consumption rate in the cell, one can make a simplifying assumption that the internal glutamine concentration is approximately equal to the external concentration.[75] Under this assumption, the equations for the internal and external glutamine may be combined into one so that we can drop the subscripts on glutamine. A standard form of Eq. (13.287) is as follows:

$$\begin{bmatrix} d(X_vV)/dt \\ d(GlcV)/dt \\ d(GltV)/dt \\ d(MAbV)/dt \\ d(V)/dt \end{bmatrix} = \begin{bmatrix} dx_1/dt \\ dx_2/dt \\ dx_3/dt \\ dx_4/dt \\ dx_5/dt \end{bmatrix} = \begin{bmatrix} \dot{x}_1 \\ \dot{x}_2 \\ \dot{x}_3 \\ \dot{x}_4 \\ \dot{x}_5 \end{bmatrix} = \begin{bmatrix} \mu x_1 \\ S_FF - \sigma_1 x_1 \\ N_FF - \sigma_2 x_1 \\ \pi x_1 \\ F \end{bmatrix} \tag{13.289}$$

where $[x_1, x_2, x_3, x_4, x_5, S, N] = [X_vV, SV, NV, MAbV, V, Glc, Glt]$ so that $S = x_2/x_5$, $N = x_3/x_5$, and $[\mu, \sigma_1, \sigma_2, \pi] = [\eta - k_d, Q_{Glc}, Q_{Gln}, Q_{MAb}]$.

The objective is to maximize the concentration of MAb at the end of final time, which is free:

$$\underset{F}{\text{Max}}\,[x_4(t_f)] \tag{13.290}$$

Although there are five state variables for this problem, and therefore, the problem appears to be fifth order, a constant ratio of $\sigma_2/\sigma_1 (Q_{gln}/Q_{glc})$ leads to a reduction to a fourth-order system. Equation (13.291) indicates that the ratio of the specific substrate consumption rate of glucose to that of glutamine is constant:

$$\frac{\sigma_2}{\sigma_1} = \frac{Y_{X_v/Glc}}{Y_{X_v/Gln}} = \frac{0.87}{1.95} = 0.446 \tag{13.291}$$

Therefore, the concentrations of glucose and glutamine are not independent of each other but related by the following stoichiometric relationship, which is obtained by combining Eq. (13.291) with Eq. (13.289),

$$\frac{d(GltV)}{dt} = N_F F - \sigma_2 x_1 = N_F F - 0.446\sigma_1 x_1 = N_F \frac{dV}{dt} - 0.446 \left[S_F \frac{dV}{dt} - \frac{d(GlcV)}{dt} \right]$$

$$(13.292)$$

which is integrated from $t = 0$ to t to obtain

$$(Glt)(V) - (Glt)_0 (V)_0$$
$$= N_F(V - V_0) - 0.446\{S_F(V - V_0) - [(Glc)(V) - (Glc)_0(V)_0]\} \quad (13.293)$$

According to Eq. (13.293), the amount of glutamine is a function of the amount of glucose and the volume. Therefore, one can pick a new set of state variables without the glutamine; cells, glucose, antibody, and the total mass; or $X_v V$, SV, PV, and V :

$$
\begin{bmatrix}
d(X_v V)/dt \\
d(GlcV)/dt \\
d(MAbV)/dt \\
dV/dt
\end{bmatrix}
=
\begin{bmatrix}
d(X_v V)/dt \\
d(SV)/dt \\
d(PV)/dt \\
d(V)/dt
\end{bmatrix}
=
\begin{bmatrix}
dx_1/dt \\
dx_2/dt \\
dx_4/dt \\
dx_5/dt
\end{bmatrix}
=
\begin{bmatrix}
\mu x_1 \\
S_F F - \sigma_1 x_1 \\
\pi x_1 \\
F
\end{bmatrix}
\quad (13.294)
$$

Therefore, $S = SV/V = x_2/x_5$ and $\partial()/\partial x_i = [\partial()/\partial S](\partial S/\partial x_i)$.

The Hamiltonian for this process is

$$H = (\lambda_1 \mu - \lambda_2 \sigma_1 + \lambda_4 \pi)x_1 + (\lambda_2 S_F + \lambda_5)F \stackrel{\Delta}{=} H_1 x_1 + \phi F,$$

$$H_1 \stackrel{\Delta}{=} \lambda_1 \mu - \lambda_2 \sigma_1 + \lambda_4 \pi, \phi \stackrel{\Delta}{=} (\lambda_2 S_F + \lambda_5) \quad (13.295)$$

The adjoint variables must satisfy the following:

$$
-
\begin{bmatrix}
\dot{\lambda}_1 \\
\dot{\lambda}_2 \\
\dot{\lambda}_4 \\
\dot{\lambda}_5
\end{bmatrix}
=
\begin{bmatrix}
\partial H/\partial x_1 \\
\partial H/\partial x_2 \\
\partial H/\partial x_4 \\
\partial H/\partial x_5
\end{bmatrix}
=
\begin{bmatrix}
\lambda_1 \mu - \lambda_2 \sigma_1 + \lambda_4 \pi \\
(\lambda_1 \mu' - \lambda_2 \sigma_1' + \lambda_4 \pi')x_1/x_5 \\
0 \\
-(\lambda_1 \mu' - \lambda_2 \sigma_1' + \lambda_4 \pi')(x_1/x_5)(x_2/x_5)
\end{bmatrix}
\quad (13.296)
$$

where the prime is used to denote the partial derivative with respect to the total amount of glucose, $()' = \partial()/\partial S$. The final conditions are as follows:

$$[\lambda_1(t_f), \lambda_2(t_f), \lambda_4(t_f), \lambda_5(t_f)] = [0, 0, 1, a] \quad (13.297)$$

where a is an unknown constant. It is clear from Eqs. (13.296) and (13.297) that $\lambda_4 = 1$.

Because the Hamiltonian is linear in F, the feed rate must take the following form:

$$
F =
\begin{cases}
F_{max} & \text{if } \phi > 0 \ \phi = \lambda_2 S_F + \lambda_5 \\
F_{min} = 0 & \text{if } \phi < 0 \\
F_{sin} & \text{if } \phi = 0 \text{ over finite time interval(s)} \\
F_b = 0 & \text{if } V = V_{max}
\end{cases}
\quad (13.298)
$$

The singular feed rate is determined from $\phi, \dot{\phi},$ and $\ddot{\phi}$:

$$\phi = S_F \lambda_2 + \lambda_5 = 0$$
$$\dot{\phi} = S_F \dot{\lambda}_2 + \dot{\lambda}_5 = [(S - S_F)(\lambda_1 \mu' - \lambda_2 \sigma_1' + \pi')]X_v = 0 \Rightarrow \lambda_1 \mu' - \lambda_2 \sigma_1' + \pi' = 0$$
$$\ddot{\phi} = S_F \ddot{\lambda}_2 + \ddot{\lambda}_5 = [(\dot{\lambda}_1 \mu' - \dot{\lambda}_2 \sigma_1')X_v](S - S_F) + (\lambda_1 \mu'' - \lambda_2 \sigma_1'' + \pi'')\dot{S}X_v(S - S_F) = 0$$
$$(13.299)$$

Solving $\ddot{\phi}$ for $\dot{S}$ and equating to Eq. (13.296), we obtain

$$\dot{S} = \frac{(\dot{\lambda}_2 \sigma_1' - \dot{\lambda}_1 \mu')}{(\lambda_1 \mu'' - \lambda_2 \sigma_1'' + \pi'')} = \frac{-(\lambda_1 \mu - \lambda_2 \sigma_1 + \pi)\mu'}{(\lambda_1 \mu'' - \lambda_2 \sigma_1'' + \pi'')} = \frac{(S_F - S)\,F - \sigma_1 x_1}{x_5} \quad (13.300)$$

Finally, solving Eq. (13.300) for the singular feed rate, we obtain

$$F_{\sin} = \left[\sigma_1 x_1 - \frac{(\lambda_1 \mu - \lambda_2 \sigma_1 + \pi)\mu' x_5}{(\lambda_1 \mu'' - \lambda_2 \sigma_1'' + \pi'')}\right] / (S_F - S) \quad (13.301)$$

The singular feed rate expression given by Eq. (13.301) is a function of state and adjoint variables. Therefore, it requires knowledge of both state and adjoint variables for evaluation. As in Chapter 12, we carry out further analysis by analyzing the adjoint equations, Eq. (13.296), with the switching functions of Eq. (13.299):

$$-\begin{bmatrix} \dot{\lambda}_1 \\ \dot{\lambda}_2 \\ \dot{\lambda}_4 \\ \dot{\lambda}_5 \end{bmatrix} = \begin{bmatrix} \lambda_1 \mu - \lambda_2 \sigma_1 + \pi \\ (\lambda_1 \mu' - \lambda_2 \sigma_2' + \pi')x_1/x_5 \\ 0 \\ -(\lambda_1 \mu' - \lambda_2 \sigma_2' + \pi')(x_1/x_5)(x_2/x_5) \end{bmatrix} = \begin{bmatrix} \lambda_1 \mu + \pi \\ 0 \\ 0 \\ 0 \end{bmatrix} \quad (13.302)$$

According to Eq. (13.302), λ_2 and λ_5 are constants in the singular interval, and therefore, the switching function, $\phi = S_F \lambda_2 + \lambda_5$, is a constant and does not change sign, and there is no singular interval, suggesting that the optimal feed rate sequence is a bang-bang type, *a batch process*. This finding is not too surprising because every one of the specific rate expressions proposed in the preceding model is monotonic, favoring the highest possible concentrations of glucose and glutamine to maximize the rate.

13.7.2 Transformation of Singular Problems to Nonsingular Problems

A general transformation to convert a feed rate optimization problem into the problem of an optimal substrate concentration profile is based on the amounts of substrate consumed (a form of mass balance) in places of the normal substrate balances. This transformation[4,77,78] can be used for processes with multiple feed rates, and there is no limit on the order of process models. Let us consider the penicillin fermentation model presented in Chapter 6:

Mass Balance Equations[39]
Viable cell $d(AV)/dt = (\mu - k_d)AV$
Nonviable fraction $d(A_u V)/dt = k_d AV$
Substrate $d(SV)/dt = FS_F - \sigma AV$ (13.303)
Penicillin $d(PV)/dt = \pi AV - k_h PV$
Overall $dV/dt = F$
Total cell mass $d(XV)/dt = \mu AV = d(AV)/dt + d(A_u V)/dt$

Specific Rates

Growth $\mu = \mu_m/(K + S)$

Penicillin formation $\pi = k_P S/[K_P + S(1 + S/K_I)]$

Cell degradation and differentiation $k_d = k_2/(L + S)$

Maintenance $m = m_1 S/(K_m + S)$

Branching $k_b = vS/(K + S)$

Substrate consumption $\sigma = \mu/Y_{X/S} + \pi/Y_{P/S} + m$

Because the total cell mass $XV = AX + A_u X$ and nonviable cell mass $A_u V = x_2$ are not involved in the other mass balance equations, we can work with the balance equations of viable cells $(AV = x_1)$, glucose $(SV = x_3)$, penicillin $(PV = x_4)$, and the total mass $(V = x_5)$ (constant density assumption). In terms of state variables, the state equations are as follows:

$$\frac{dx_1}{dt} = (\mu[x_3/x_5] - k_d)x_1 \quad dx_3/dt = FS_F - \sigma[x_3/x_5]x_1$$
$$dx_4/dt = \pi[x_3/x_5]x_1 - k_h x_4 \quad dx_5/dt = F \tag{13.304}$$

To eliminate F in the substrate balance equation (13.304), we introduce a new state variable that represents the total amount of substrate consumed, which is the amount present initially plus the amount added minus the amount remaining in the reactor:

$$x_6 \overset{\Delta}{=} S_0 V_0 + S_F(V - V_0) - SV = x_{30} + S_F(x_5 - V_0) - x_3 \tag{13.305}$$

Then, the time derivative of Eq. (13.303), dx_6/dt, should represent the rate of consumption of substrate:

$$\sigma XV = \frac{dx_6}{dt} = \sigma x_1 = S_F \frac{dx_5}{dt} - \frac{dx_3}{dt} \tag{13.306}$$

Equation (13.306) is in fact another form of substrate balance equation (13.306). Thus, we can rewrite Eq. (13.306) as

$$\frac{dx_6}{dt} = \sigma x_1 \tag{13.307}$$

The initial condition on x_6, the amount consumed initially, is obviously zero:

$$x_6(0) = S_0 V_0 + S_F(V_{t=0} - V_0) - SV_{t=0} = S_0 V_0 + S_F(V_0 - V_0) - S_0 V_0 = 0 \tag{13.308}$$

It is also clear that the overall balance need not be included because it does not appear explicitly in four component balance equations:

$$\frac{dx_1}{dt} = [\mu(S) - k_d]x_1 \quad x_1(0) = A_0 V_0$$
$$\frac{dx_3}{dt} = FS_F - \sigma(S)x_1 \quad x_3(0) = S_0 V_0$$
$$\frac{dx_4}{dt} = \pi(S)x_1 - k_h x_4 \quad x_4(0) = 0 \tag{13.309}$$
$$\frac{dx_5}{dt} = F \quad x_5(0) = V_0$$
$$\frac{dx_6}{dt} = \sigma(S)x_1 \quad x_6(0) = 0$$

Having chosen the substrate concentration profile instead of the feed flow rate as the manipulated variable, we can inspect Eq. (13.309) to remove the third and fifth state x_3 and x_5:

$$\frac{dx_1}{dt} = [\mu(S) - k_d]x_1 \quad x_1(0) = A_0V_0$$
$$\frac{dx_4}{dt} = \pi(S)x_1 - k_h x_4 \quad x_4(0) = 0 \qquad (13.310)$$
$$\frac{dx_6}{dt} = \sigma(S)x_1 \qquad x_6(0) = 0$$

The performance index is the total amount of penicillin at the end of the final time, which can be either free or fixed:

$$\underset{S(t)}{\text{Max}}[P = x_4(t_f)] \qquad (13.311)$$

The volume constraint is obtained from Eq. (13.305):

$$V - V_{\max} = \frac{x_6 + V_0(S_F - S_0)}{(S_F - S)} - V_{\max} = g(x_6, S) - V_{\max} \le 0 \qquad (13.312)$$

Equation (13.312) is a constraint on the state and manipulated variables. Once we obtain the optimum substrate concentration profile and therefore dS^*/dt, the corresponding optimum feed rate $F^*(t)$ is obtained from Eq. (13.304):

$$F^*(t) = [\sigma^*(XV)^* + V^* dS^*/dt]/(S_F - S^*) \qquad (13.313)$$

The reactor volume is obtained from Eq. (13.305):

$$V^* = [x_6^* + V_0(S_F - S_0)]/(S_F - S^*) \qquad (13.314)$$

Therefore, the optimal feed rate is obtained by substituting Eq. (13.314) into Eq. (13.313):

$$F^*(t) = \left\{ \sigma^* x_1 + [x_6^* + V_0 \frac{(S_F - S_0)}{(S_F - S^*)}](dS^*/dt) \right\} / (S_F - S^*) \qquad (13.315)$$

According to Eq. (13.315), the optimal feed rate profile is obtained from the optimal time profiles of substrate concentration, the time derivative of substrate concentration, the profile of the total amount of cell mass, and the amount of substrate consumed. Because the volume is a monotonic function, $dV/dt = F \ge 0$, the constraint on volume, $V(t) \le V_{\max}$, is equivalent to the terminal constraint, $V(t_f) = V_{\max}$, or

$$[x_6(t_f) + V_0(S_F - S_0)]/[S_F - S(t_f)] = V_{\max} \qquad (13.316)$$

The constraint on the manipulated variable is

$$S(0) = S_0, S_{\min} \le S(t) \le S_{\max} \qquad (13.317)$$

where $S_{\min}$ and $S_{\max}$ are to be determined from the magnitude constraints on the feed rate:

$$0 = F_{\min} \le F \le F_{\max} \qquad (13.318)$$

The solution to the problem posed by Eqs. (13.310), (13.311), (13.312), and (13.316) is obtained via PMP, according to which the Hamiltonian is maximized:

$$\underset{S(t)}{\text{Max}} H = [\lambda_1(\mu - k_d) + \lambda_4 \pi + \lambda_6 \sigma)]x_1 - \lambda_4 k_h x_4 + \eta[g(x_6, S) - V_{\max}]$$
$$= [\lambda_1(\mu - k_d) + \lambda_4 \pi + \lambda_6 \sigma)]x_1 - \lambda_4 k_h x_4 + \eta[h(x_6, S)] \qquad (13.319)$$

where $\eta(t)$ is a Lagrangian multiplier. We note that $S(t)$ appears nonlinearly. Thus, the singular problem with the feed rate $F(t)$ as the manipulated variable is avoided, and a nonsingular problem is formulated with the substrate concentration $S(t)$ as the manipulated variable. The Hamiltonian is maximized:

$$\frac{\partial H}{\partial S} = \frac{\partial[\lambda_1(\mu - k_d) + \lambda_4\pi + \lambda_6\sigma)]x_1 - \lambda_4 k_h x_4 + \eta h]}{\partial S} = 0 \qquad (13.320)$$

where

$$\eta \begin{cases} = 0 \text{ when } h < 0, \text{ off equality control boundary} \\ \leq 0 \text{ when } h = 0, \text{ on equality control boundary} \end{cases} \qquad (13.321)$$

Therefore, off the equality control boundary, the adjoint variables and control function must satisfy

$$\frac{d}{dt}\begin{pmatrix} \lambda_1 \\ \lambda_4 \\ \lambda_6 \end{pmatrix} = -\begin{pmatrix} \partial H/\partial x_1 \\ \partial H/\partial x_4 \\ \partial H/\partial x_6 \end{pmatrix} = -\begin{pmatrix} \lambda_1(\mu - k_d) + \lambda_4\pi + \lambda_6\sigma \\ -\lambda_4 k_h \\ 0 \end{pmatrix} \qquad (13.322)$$

$$\partial H/\partial S = 0 = \lambda_1 \partial\mu/\partial S + \lambda_4 \partial\pi/\partial S + \lambda_6 \partial\sigma/\partial S \qquad (13.323)$$

On the control constraint boundary, the corresponding adjoint equations and control function are given by

$$\frac{d}{dt}\begin{pmatrix} \lambda_1 \\ \lambda_4 \\ \lambda_6 \end{pmatrix} = -\begin{pmatrix} \partial H/\partial x_1 \\ \partial H/\partial x_4 \\ \partial H/\partial x_6 \end{pmatrix} = -\begin{pmatrix} \lambda_1(\mu - k_d) + \lambda_4\pi + \lambda_6\sigma \\ -\lambda_4 k_h \\ 0 \end{pmatrix} \begin{pmatrix} \lambda_1(t_f) \\ \lambda_4(t_f) \\ \lambda_6(t_f) \end{pmatrix} = \begin{pmatrix} 0 \\ 1 \\ 0 \end{pmatrix}$$
$$(13.324)$$

It is obvious from Eq. (13.324) that $\lambda_6(t) = 0$. Therefore, the Hamiltonian reduces to

$$\underset{S(t)}{\text{Max }} H = \lambda_1(\mu - k_d)x_1 + \lambda_4(\pi x_1 - k_h x_4) \qquad (13.325)$$

According to Eq. (13.325), the total growth rate $dx_1/dt = (\mu - k_d)x_1$ is time weighted by $\lambda_1(t)$, and the product formation rate $dx_4/dt = (\pi x_1 - k_h x_4)$ is time weighted by $\lambda_4(t)$; the sum of the two is to be maximized by varying the substrate concentration. The weighing factors are the adjoint variables:

$$\frac{d}{dt}\begin{bmatrix} \lambda_1 \\ \lambda_4 \end{bmatrix} = -\begin{bmatrix} \lambda_1(\mu - k_d) + \lambda_4\pi \\ -\lambda_4 k_h \end{bmatrix} \qquad (13.326)$$

Because it is simple to integrate λ_4,

$$\lambda_4(t) = \exp^{k_h(t-t_f)} \qquad (13.327)$$

Inspection of Eq. (13.325) shows that the penicillin formation rate is weighted lightly in the early stage of fermentation and weighted more and more toward the final time. Maximizing the Hamiltonian, we obtain

$$\partial H/\partial S = \lambda_1 \partial\mu/\partial S + \lambda_4 \partial\pi/\partial S = 0 \qquad (13.328)$$

13.7.3 Transformation of Singular Problems with Multiple Feed Rates

Consider the problem of manipulating two feed streams, for example, the fed-batch culture of the PHB:[77]

$$\frac{dx_1}{dt} = \frac{d(XV)}{dt} = \mu(S_1, S_2)XV = \mu(S_1, S_2)x_1 \tag{13.329}$$

$$\frac{dx_2}{dt} = \frac{d(S_1 V)}{dt} = S_{1F}F_1 - \sigma_1(S_1, S_2)XV = S_{1F}F_1 - \sigma_1(S_1, S_2)x_1 \tag{13.330}$$

$$\frac{dx_3}{dt} = \frac{d(S_2 V)}{dt} = S_{2F}F_2 - \sigma_2(S_1, S_2)XV = S_{2F}F_2 - \sigma_2(S_1, S_2)x_1 \tag{13.331}$$

$$\frac{dx_4}{dt} = \frac{d(PV)}{dt} = \pi(S_1, S_2, P)XV = \pi(S_1, S_2, P)x_1 \tag{13.332}$$

$$\frac{d(x_5 \rho)}{dt} = \frac{d(V\rho)}{dt} = F_1 \rho_1 + F_2 \rho_2 \tag{13.333}$$

In this PHB model, S_1 and S_2 represent the concentrations of glucose and ammonium chloride, respectively, P the concentration of PHB, X the concentration of active cell mass (cell concentration − PHB concentration), V the fermentor volume, S_{1F} and S_{2F} the feed concentrations of glucose and ammonia, respectively, F_1 and F_2 the glucose and ammonium chloride feed rates, and ρ, ρ_1, and ρ_2 the densities of the culture, the glucose feed, and the ammonium chloride feed, respectively. It is assumed in general that the densities are the same, $\rho = \rho_1 = \rho_2$, so that Eq. (13.333) reduces to

$$\frac{dx_5}{dt} = \frac{dV}{dt} = F_1 + F_2 \tag{13.334}$$

Because there are two feed streams, glucose and ammonium chloride, we introduce the amounts of glucose and ammonium chloride consumed, x_6 and x_7 – the amount present initially plus the amount added minus the amount remaining:

$$x_6(t) \triangleq S_{10}V_0 + S_{1F}\int_0^t F_1 d\tau - S_1 V(t) = x_2(0) - x_2(t) + S_{1F}\int_0^t F_1 d\tau \tag{13.335}$$

$$x_7(t) \triangleq S_{20}V_0 + S_{2F}\int_0^t F_2 d\tau - S_2 V(t) = x_3(0) - x_3(t) + S_{2F}\int_0^t F_2 d\tau \tag{13.336}$$

Then, the time derivatives of x_6 and x_7 are equal to the consumption of substrates S_1 and S_2:

$$\frac{dx_6}{dt} = \sigma_1(S_1, S_2, P)x_1 \tag{13.337}$$

$$\frac{dx_7}{dt} = \sigma_2(S_1, S_2, P)x_1 \tag{13.338}$$

Conversely, differentiation of Eqs. (13.335) and (13.336) yields

$$\frac{dx_6}{dt} = S_{1F}F_1 - \frac{dx_2}{dt} = S_{1F}F_1 - (S_{1F}F_1 - \sigma_1 x_1) = \sigma_1 x_1$$

$$\frac{dx_7}{dt} = S_{2F}F_2 - \frac{dx_3}{dt} = S_{2F}F_2 - (S_{2F}F_2 - \sigma_2 x_1) = \sigma_2 x_1 \qquad (13.339)$$

Equation (13.339) confirms that by introducing the amounts of substrate consumed (Eqs. (13.335) and (13.336)), we can replace the two substrate balance equations (Eqs. (13.330) and (13.331)), yielding a new set of balance equations:

$$\frac{dx_1}{dt} = \mu(S_1, S_2)x_1 \quad x_1(0) = x_{10} \qquad (13.340)$$

$$\frac{dx_6}{dt} = \sigma_1(S_1, S_2)x_1 \quad x_6(0) = 0 \qquad (13.341)$$

$$\frac{dx_7}{dt} = \sigma_2(S_1, S_2)x_1 \quad x_7(0) = 0 \qquad (13.342)$$

$$\frac{dx_4}{dt} = \pi(S_1, S_2, P)x_1 \quad x_4(0) = x_{40} \qquad (13.343)$$

The performance index is maximized using the concentration profiles of $S_1(t)$ and $S_2(t)$:

$$\underset{S_1(t), S_2(t)}{\text{Max}} \, P[x_4(t_f)] \qquad (13.344)$$

The Hamiltonian to be maximized is

$$\underset{S(t)}{\text{Max}} \, H = \lambda_1 \mu x_1 + \lambda_6 \sigma_1 x_1 + \lambda_7 \sigma_2 x_1 + \lambda_4 \pi x_1 + \eta[g(x_4, S) - V_{\max}]$$

$$= [\lambda_1 \mu + \lambda_6 \sigma_1 + \lambda_7 \sigma_2 + \lambda_4 \pi]x_1 + \eta(t)h(x_4, S) \qquad (13.345)$$

For this problem, both $S_1(t)$ and $S_2(t)$ appear nonlinearly in the specific rates, and therefore, this problem is nonsingular with respect to $S_1(t)$ and $S_2(t)$ so that

$$\frac{\partial H}{\partial S_1} = 0 \quad \text{and} \quad \frac{\partial H}{\partial S_2} = 0 \qquad (13.346)$$

The optimal numerical solutions, $S_1^*(t)$ and $S_2^*(t)$, can be obtained using the steepest ascent method, as described in Chapter 10. With the optimal substrate concentrations, one can construct a control strategy[78] in which the measured substrate concentration profiles are forced to follow the optimal paths, $S_1^*(t)$ and $S_2^*(t)$, by manipulating the feed rates, $F_1(t)$ and $F_2(t)$.

If it is desired to obtain the optimal feed rates, F_1^* and F_2^*, from the preceding optimal $S_1^*(t)$ and $S_2^*(t)$, we first collect the data: $x_1^*(t) = (XV)^*, x_6^*(t), x_7^*(t)$, and $x_4^*(t)$. Then, expressions for F_1^* and F_2^* are obtained from the substrate balances (Eqs. (13.330) and (13.333)):

$$\frac{d(S_1 V)}{dt} = \frac{dS_1}{dt}V + S_1 \frac{dV}{dt} = \frac{dS_1}{dt}V + S_1(F_1 + F_2) = S_{1F}F_1 - \sigma_1 x_1 \qquad (13.347)$$

or

$$- (S_{1F} - S_1)F_1 + S_1 F_2 = -\frac{dS_1}{dt}V - \sigma_1 x_1 \qquad (13.348)$$

Likewise, from Eq. (13.334),

$$S_2F_1 - (S_{2F} - S_2)F_2 = -\frac{dS_2}{dt}V - \sigma_2 x_1 \tag{13.349}$$

Solving Eqs. (13.348) and (13.349) for F_1 and F_2,

$$F_1 = \frac{[(S_{2F} - S_2)\dot{S}_1 + S_1\dot{S}_2]V + [(S_{2F} - S_2)\sigma_1 + S_1\sigma_2]x_1}{(S_{1F} - S_1)(S_{2F} - S_2) - S_1S_2} \triangleq \frac{N_{11}V + N_{12}}{D} \tag{13.350}$$

and

$$F_2 = \frac{[(S_{1F} - S_1)\dot{S}_2 + S_2\dot{S}_1]V + [(S_{1F} - S_1)\sigma_2 + S_2\sigma_1]x_1}{(S_{1F} - S_1)(S_{2F} - S_2) - S_1S_2} \triangleq \frac{N_{21}V + N_{22}}{D} \tag{13.351}$$

where

$$N_{11} \triangleq [(S_{2F} - S_2)\dot{S}_1 + S_1\dot{S}_2], N_{12} \triangleq [(S_{2F} - S_2)\sigma_1 + S_1\sigma_2]x_1$$
$$N_{21} \triangleq [(S_{1F} - S_1)\dot{S}_2 + S_2\dot{S}_1], N_{22} \triangleq [(S_{1F} - S_1)\sigma_2 + S_2\sigma_1]x_1 \tag{13.352}$$

These two feed rates are the functions of the culture volume, $V(t)$, which is also the function $F_1(t)$ and $F_2(t)$, $V(t) = \int_0^t (F_1 + F_2)d\tau$. Hence, we have two coupled integral equations. We can solve Eqs. (13.350) and (13.351) for V, and by equating them, we obtain the relationship between F_1 and F_2:

$$F_1 = \frac{N_{11}}{N_{21}}F_2 + \frac{N_{12}N_{21} - N_{11}N_{22}}{DN_{21}} \triangleq \alpha F_2 + \beta \tag{13.353}$$

Therefore,

$$F_1 + F_2 = \alpha F_2 + \beta + F_2 = (1 + \alpha)F_2 + \beta \tag{13.354}$$

Substitution of Eq. (13.354) into (13.349) yields

$$F_2 = \frac{N_{11}}{D\alpha}\int_0^t [(1 + \alpha)F_2 + \beta]d\tau + \frac{N_{12}}{D\alpha} - \frac{\beta}{\alpha} \tag{13.355}$$

which is a special form of Volterra equation of the second kind:

$$F_2(t) = f(t) + \int_0^t K(\tau)F_2(\tau)d\tau \tag{13.356}$$

Once we solve this equation numerically, we obtain the numerical values of $F_2(t)$. Obviously, $F_1(t)$ is obtained using Eq. (13.353). Numerical schemes to solve the Volterra equation of the second kind are available.[81]

REFERENCES

1. Guthke, R., and Knorre, W. A. 1981. Optimal substrate profile for antibiotic fermentations. *Biotechnology and Bioengineering* 23: 2771–2777.
2. Alvarez, J., and Alvarez, J. 1998. Analysis and control of fermentation processes by optimal and geometric methods, in *Proceedings of 1998 Automatic Control Conference*, p. 1112. American Automatic Control Council.
3. Fishman, V. M., and Biryukov, V. V. 1974. Kinetics model of secondary metabolite production and its use in computation of optimal conditions. *Biotechnology and Bioengineering Symposium* 4: 647–662.
4. Modak, J. M., and Lim, H. C. 1989. Simple nonsingular control approach to fed-batch fermentation optimization. *Biotechnology and Bioengineering* **33**: 11–15.

5. Yamane, T., Kume, T., Sada, E., and Takamatsu, T. 1977. A simple optimization technique for fed-batch culture: Kinetic studies on fed-batch cultures (VI). *Journal of Fermentation Technology* 55: 587–598.
6. San, K.-Y., and Stephanopoulos, G. 1989. Optimization of fed-batch penicillin fermentation: A case of singular optimal control with state constraints *Biotechnology and Bioengineering* 34: 72–78.
7. Menawat, A., Mutharasan, R., and Coughanowr, D. R. 1987. Singular optimal control strategy for a fed-batch bioreactor: Numerical approach. *American Institute of Chemical Engineers Journal* 33: 776–783.
8. Modak, J. M. 1993. Choice of control variable for optimization of fed-batch fermentation. *Chemical Engineering Journal* 52: B59–B69.
9. Jayant, A., and Pushpavanam, S. 1998. Optimization of a biochemical fed-batch reactor: Transition from a nonsingular to a singular problem. *Industrial and Engineering Chemistry Research* 37: 4314–4321.
10. Sin, H. S., and Lim, H. C. 2007. Maximization of metabolite in fed-batch cultures: Sufficient conditions for singular arc and optimal feed rate profiles. *Biochemical Engineering Journal* 37: 62–74.
11. Ohno, H., Nakanishi, E., and Takamatsu, T. 1976. Optimal control of a semi-batch fermentation. *Biotechnology and Bioengineering* 18: 847–864.
12. Shin, H. S., and Lim, H. C. 2007. Optimization of metabolite production in fed-batch cultures: Use of sufficiency and characteristics of singular arc and properties of adjoint vectors in numerical computations. *Industrial and Engineering Chemistry Research* 46: 2526–2534.
13. Aiba, S., Shoda, M., and Nagatani, M. 1968. Kinetics of product inhibition in alcohol fermentation. *Biotechnology and Bioengineering* 10: 845–864.
14. Hong, J. 1986. Optimal substrate feeding policy for a fed-batch fermentation with substrate and product inhibition kinetics. *Biotechnology and Bioengineering* 28: 1421–1431.
15. Ingram, L. O., Conway, T., Clark, D. P., Sewell, G. W., and Preston, J. F. 1987. Genetic engineering of ethanol production in *Escherichia coli*. *Applied Environmental Microbiology* 53: 2420–2425.
16. Imanaka, T. 1986. Application of recombinant DNA technology to the production of useful biomaterials. *Advances in Biochemical Engineering/Biotechnology* 33: 1–27.
17. Graumann, K., and Premstaller, A. 2006. Manufacturing of recombinant therapeutic proteins in microbial systems. *Biotechnology Journal* 1: 164–186.
18. Macauley-Patrick, S., Fazenda, M. L., McNeil, B., and Harvey, L. M. 2005. Heterogonous protein production using the *Pichia pastoris* expression system. *Yeast* 22: 249–270.
19. Georgiou, G. 1988. Optimizing the production of recombinant proteins in microorganisms. *American Institute of Chemical Engineers Journal* 34: 1233–1248.
20. Anderson, T. F., and Lustbader, E. 1975. Inheritability of plasmids and population dynamics of cultured cells. *Proceedings of the National Academy of Sciences of the United States of America* 72: 4085–4089.
21. Meacock, P. A., and Cohen, S. N. 1980. Partitioning of bacterial plasmids during cell division: A cis-acting locus that accomplishes stable plasmid inheritance. *Cell* 20: 529–542.
22. Zund, P., and Lebek G. 1980. Generation time-prolonging R plasmids: Correlation between increases in the generation time of *Escherichia coli* caused by R plasmids and their molecular size. *Plasmid* 3: 65–69.
23. Son, K. H., Jang, J. H., and Kim, J. H. 1987. Effect of temperature on plasmid stability and expression of cloned cellulase gene in a recombinant *Bacillus megaterium*. *Biotechnology Letters* 9: 821–824.

24. Aiba, S., and Koizumi, J. 1984. Effects of temperature on plasmid stability and penicillinase productivity in a transformant of *Bacillus stearothermophilus. Biotechnology and Bioengineering* 26: 1026–1031.
25. Koizumi, J., Monden, Y., and Aiba, S. 1985. Effects of temperature and dilution rate on the copy number of recombinant plasmid in continuous culture of *Bacillus stearothermophilus* (pLPl1). *Biotechnology and Bioengineering* 27: 721–728.
26. Jones, S. A., and Melling, J. 1984. Persistence of pBR322-related plasmids in *Escherichia coli* grown ichemostat cultures. *FEMS Microbiology Letters* 22: 239–243.
27. Adams, C. W., and Hatfield, G. W. 1984. Effects of promoter strengths and growth conditions on copy number of transcription-fusion vectors. *Journal of Biological Chemistry* 259: 399–403
28. Weber, A. E., and San, K. Y. 1987. Persistence and expression of the plasmid pBR322 in *Escherichia coli* K-12 cultured in complex media. *Biotechnology Letters* 9: 757–760.
29. Imanaka, T., and Aiba, S. 1981. A perspective on the application of genetic engineering: Stability of recombinant plasmid. *Annals of the New York Academy of Science* 369: 1–14.
30. Dwivedi, C. P., Imanaka, T., and Aiba, S. 1982. Instability of plasmid-harboring strain of *E coli* in continuous culture. *Biotechnology and Bioengineering* 24: 1465–1468.
31. Lee S. B., and Bailey, J. E. 1984. Analysis of growth rate effects on productivity of recombinant *Escherichia coli* populations using molecular mechanism models. *Biotechnology and Bioengineering* 26: 66–73.
32. Bailey, J. E., Hjortso, M., Lee, S. B., and Srienc, F. 1983. Kinetics of product formation and plasmid segregation in recombinant microbial populations. *Annals of the New York Academy of Science* 413: 71–87.
33. Warner, P. J., Higgins, I. J., and Drozd, J. W. 1980. Conjugative transfer of antibiotic resistance to methylotrophic bacteria. *FEMS Microbiology Letters* 7: 181–185.
34. Jimenez, A., and Davies, J. 1980. Expression of a transposable antibiotic resistance element in *Saccharomyces. Nature* 287: 869–871.
35. Rawlings, D. E., Pretorius, I., and Woods, D. R. 1984. Construction of arsenic-resistant *Thiobacillus ferrooxidans* recombinant plasmids and the expression of autotrophic plasmid genes in a heterotrophic cell-free system. *Journal of Biotechnology* 1: 129–133.
36. Webster, T. D., and Dickson, R. C. 1983. Direct selection of *Saccharomyces cerevisiae* resistant to the antibiotic G418 following transformation with a DNA vector carrying the kanamycin-resistance gene of Tn903. *Gene* 26: 243–252.
37. Ryder, D. F., and DiBiasio, D. 1984. An operational strategy for unstable recombinant DNA cultures. *Biotechnology and Bioengineering* 26: 942–947.
38. Parulekar, S. J., Chang, Y. K., Modak, J. M., and Lim, H. C. 1986. Analysis of continuous cultures of recombinant methylotrophs. *Biotechnology and Bioengineering* 29: 911–923.
39. Chittur, V. 1989. Modeling and optimization of fedbatch penicillin fermentation, Ph.D. thesis, Purdue University.
40. Gupta, J. C., and Mukherjee, K. J. 2002. Stability studies of recombinant *Saccharomyces cerevisiae* in the presence of varying selection pressure. *Biotechnology and Bioengineering* 78: 475–488.
41. Park, S., and Ramirez, W. F. 1988. Optimal production of secreted protein in fed-batch reactors. *American Institute of Chemical Engineering Journal* 34: 1550–1558.

42. Lee, J., and Ramirez, W. F. 1992. Mathematical modeling of induced foreign protein production by recombinant bacteria. *Biotechnology and Bioengineering* 39: 635–646.
43. Pontryagin, L. S., Boltyanskii, V. G., Gamkrelidze, R. V., and Mishchenko, E. F. 1962. *The Mathematical Theory of Optimal Processes.* Interscience.
44. Kelley, H. J. 1965. A transformation approach to singular subarcs in optimal trajectory and control problems. *SIAM Journal of Control and Optimization* 2: 234–241.
45. Kumar, P. K. R., Maschke, H.-E., Friehs, K., and Schugerl, K. 1991. Strategies for improving plasmid stability in genetically modified bacteria in bioreactors. *TIBTECH* 9: 279–284.
46. Wang, F. S., and Cheng, W. M. 1999. Simultaneous optimization of feeding rate and operation parameters for fed-batch fermentation processes. *Biotechnology Progress* 15: 949–952.
47. Shi, X. M., Jiang, Y., and Chen, F. 2002. High-yield production of lutein by the green microalga *Chlorella protothecoides* in heterotrophic fed-batch culture. *Biotechnology Progress* 18: 723–727.
48. Yee, L., and Blanch, H. W. 1992. Recombinant protein expression in high cell density fed-batch cultures of *Escherichia coli*. *Bio/Technology* 10: 1550–1556.
49. Skolpap, W., Scharer, J. M., Douglas, P. L., and Moo-Young, M. 2004. Fed-batch optimization of α-amylase and protease-producing *Bacillus subtilis* using Markov chain methods. *Biotechnology and Bioengineering* 86: 706–717.
50. Anderson, T. F., and Lustbader, E. 1975. Inheritability of plasmids and population dynamics of cultured cells. *Proceedings of the National Academy of Sciences of the United States of America* 72: 4085–4089.
51. Meacock, P. A., and Cohen, S. N. 1980. Partitioning of bacterial plasmids during cell division: A cis-acting locus that accomplishes stable plasmid inheritance. *Cell* 20: 529–542.
52. Zund, P., and Lebek, G. 1980. Generation time-prolonging R plasmids: Correlation between increases in the generation time of *Escherichia coli* caused by R plasmids and their molecular size. *Plasmid* 3: 65–69.
53. Altıntas, M. M., Ülgen, K., Kırdar, B., Önsan, Z. I., and Oliver, S. 2003. Optimal substrate feeding policy for fed-batch cultures of *S. cerevisiae, YPG* expressing bifunctional fusion protein displaying amylolytic activities. *Enzyme and Microbial Technology* 33: 262–269.
54. Cheng, C., Huang, Y. L., and Yang, S. T. 1997. A novel feeding strategy for enhanced plasmid stability and protein production in recombinant yeast fed-batch fermentation. *Biotechnology and Bioengineering* 56: 23–31.
55. Gupta, J. C., Pandey, G., and Mukherjee, K. J. 2001. Two-stage cultivation of recombinant *Saccharomyces cerevisiae* to enhance plasmid stability under non-selective conditions: Experimental study and modeling. *Enzyme and Microbial Technology* 28: 89–99.
56. Hempel, C., Erb, R. W., Deckwer, W. D., and Hecht, D. V. 1998. Plasmid stability of recombinant *Pseudomonas sp.* B13 FR1 pFRC20p in continuous culture. *Biotechnology and Bioengineering* 57: 62–70.
57. Raj, A. E., Sathish, H. S., Kumar, S., Kumar, U., Misra, M. C., Ghildyal, N. P., and Karanth, N. G. 2002. High-cell-density fermentation of recombinant *Saccharomyces cerevisiae* using glycerol. *Biotechnology Progress* 18: 1130–1132.
58. Zhang, Y., Taiming, L., and Liu, J. 2003. Low temperature and glucose enhanced T7 RNA polymerase-based plasmid stability for increasing expression of glucagon-like peptide-2 in *Escherichia coli*. *Protein Expression and Purification* 29: 132–139.

59. Oh, G., Moo-Young, M., and Chistit, Y. 1998. Automated fed-batch culture of recombinant *Saccharomyces cerevisiae* based on on-line monitored maximum substrate uptake rate. *Biochemical Engineering Journal* 1: 211–217.

60. Bravo, S., Mahn, A., and Shene, C. 2000. Effect of feeding strategy on *Zymomonas mobilis* CP4 fed-batch fermentations and mathematical modeling of the system. *Applied Microbiology and Biotechnology* 54: 487–493.

61. Xu, P., Thomas, A., and Gilson, C. D. 1996. Combined use of three methods for high concentration ethanol production by *Saccharomyces cerevisiae*. *Biotechnology Letters* 18: 1439–1440.

62. Lin, C.-S., and Lim, H. C. 1992. Utilization of dynamic responses of recombinant cells to improve CSTBR operations. *Chemical Engineering Communications* 118: 265–278.

63. Zhang, X., Xia, Z., Zhao, B., and Cen, P. 2002. Enhancement of plasmid stability and protein productivity using multi-pulse, fed-batch culture of recombinant *Saccharomyces cerevisiae*. *Biotechnology Letters* 24: 995–998.

64. Imanaka, T., and Aiba, S. 1981. A perspective on the application of genetic engineering: Stability of recombinant plasmid. *Annals of the New York Academy of Science* 369: 1–14.

65. Bibila, T. A., and Robinson, D. K. 1995. In pursuit of the optimal batch process for monoclonal antibody production. *Biotechnology Progress* 11: 1–13.

66. Jo, E.-C., Park, H.-J., Kim, D.-I., and Moon, H. M. 1993. Repeated fed-batch culture of hybridoma cells in nutrient-fortified high-density medium. *Biotechnology and Bioengineering* 42: 129–137.

67. Omasa, T., Ishimoto, M., Higashiyama, K. I., Shioya, S., and Suga, K. I. 1992. The enhancement of specific antibody production rate in glucose and glutamine controlled fed batch culture. *Cytotechnology* 8: 75–84.

68. Truskey, G. A., Nicolakis, D. P., Haberman, A., and Swartz, R. W. 1990. Kinetic studies and unstructured models of lymphocyte metabolism in fed batch culture. *Biotechnology and Bioengineering* 36: 797–807.

69. Barford, J. P., Phillips, P. J., and Harbour, C. 1992. Simulation of animal cell metabolism. *Cytotechnology* 10: 63–74.

70. Barford, J. P., Phillips, P. J., and Harbour, C. 1992. Enhancement of productivity by yield improvements using simulation techniques, in *Animal Cell Technology: Products of Today, Prospects of Tomorrow*, ed. Spier, R. E., Griffiths, J. B., and Berthold, W., p. 397. Butter-Heinemann.

71. Batt, B. C., and Kompala, D. S. 1989. A structured kinetic modeling framework for the dynamics of hybridoma growth and monoclonal antibody production in continuous suspension cultures. *Biotechnology and Bioengineering* 34: 515–531.

72. Hansen, H. A., Madsen, N. M., and Emborg, C. 1993. An evaluation of fed-batch cultivation methods for mammalian cells based on model simulation. *Bioprocess Engineering* 9: 205–213.

73. Noe, W., Schorn, P., and Berthold, W. 1994. Fed batch strategies for mammalian cell cultures, in *Animal Cell Technology: Products of Today, Prospects of Tomorrow*, ed. Spier, R. E., Griffiths, J. B., and Berthold, W., p. 413. Butter-Heinemann.

74. de Trembley, M., Perrier, M., Chavarie, C., and Archambault, J. 1993. Optimization of fed-batch culture of hybridoma cells using dynamic programming: Single and multi feed cases. *Bioprocess Engineering* 7: 229–234.

75. Lee, J.-H., Lim, H. C., Yoo, Y. J., and Park, Y. H. 1999. Optimization of feed rate profile for the monoclonal antibody production. *Bioprocess Engineering* 20: 137–146.

76. Lim, H. C., Tayeb, Y. J., Modak, J. M., and Bonte, P. 1984. Computational algorithms for optimal feed rates for a class of fed-batch fermentation: Numerical

results for penicillin and cell mass production. *Biotechnology and Bioengineering* 28: 1408–1420.

77. Shin, H. S., and Lim, H. C. 2008. Optimal fed-batch operation for recombinant cells with segregational plasmid instability. *Chemical Engineering Communications* 195: 1122–1143.

78. Merriam, C. W. III. 1964. *Optimization theory and the design of feedback control systems*, McGraw-Hill.

79. Lee, J. H., Lim, H. C. and Hong, J. 1997. Application of nonsingular transformation to on-line optimal control of poly-hyroxybutyrate fermentation. *Journal of Biotechnology* 55: 135–150.

80. Lee, J. H., Lim, H. C. and Kim, S. I. 2001. A nonsingular optimization approach to the feed rate profile optimization of fedbatch cultures. *Bioprocess and Biosystems Engineering* 24: 115–121.

81. Arfken, G. 1985. *Mathematical methods for physicists*, Academic Press.

14 Simple Adaptive Optimization

In previous chapters, optimizations were carried out using known models with fixed parameter values. The assumption was that the model and its parameter values remain practically invariant during the entire course of bioreactor operation. This assumption may not hold in reality. First, because the model used is only an approximation of real cellular processes, it may not represent well the process over the entire operational time period. The rate-limiting step may change during the course of operation, for example, the rate-limiting step in the cell growth phase may not be the same as that in the product formation phase. To overcome the shortcomings of a fixed model with fixed values of parameters, it may be necessary to allow the parameter values to vary or even to alter the model during the course of the operational time period to obtain a model that better fits the experimental data.

The objective of adaptive optimization is to optimize the process under uncertainties in model and/or parameter values. In this chapter, we consider only simple adaptive optimization schemes with a fixed model with adjustable parameters; the parameter values are updated after one run using the experimental data generated, and the optimization is repeated and implemented in the subsequent run. This process is repeated run after run (cycle-to-cycle, off-line optimization). Alternatively, within one run, the operational time may be divided into a number of time intervals, and the parameter values may be updated from one time interval to the next time interval during the course of one run using the experimental data generated in previous time intervals, and the optimization may be repeated from one time interval to the next with the updated parameter values and implemented in the subsequent time interval (on-line adaptive optimization). This process may be repeated from one interval to the next, until the entire time interval is covered.

These optimization methods yield the optimal feed rate profile as a function of time. This is an open-loop policy, but a feedback law (closed loop) reduces the sensitivity in the presence of uncertainties or disturbances. Thus, it is desirable to obtain a feedback control instead of the open-loop policy. However, as we have seen in Chapters 12 and 13, it is difficult to obtain the optimum feed rate in feedback mode, unless the model is low in order (fewer than five independent dynamic equations) and the final time is free. When the final time is fixed (given), the number of independent dynamic equations must be fewer than four. When it is not possible to obtain the optimum feed rate in feedback form, it is desirable to apply adaptive optimization

393

so that the model is updated and the process is optimized repeatedly to overcome this shortcoming.

Dynamic programming and a linear predictive regression analysis[1] were applied[2-3] to a fed-batch culture of *Brevibacterium divaricatum* for glutamic acid production with ethanol as the feed substrate. A high-density fermentation of *Candida utilis*[4] was carried out using the optimal substrate feed rate, which maximizes the specific growth rate based on an on-line estimation technique with a moving data window that tracks the time-varying parameters and estimates the cell mass.

An on-line adaptive optimal control algorithm was developed[5] and applied to fed-batch fermentation of *Streptomyces* C5, in which the objective was to maximize the cell mass. Using a dynamic model for the substrate, cell mass and fermentor volume were used to determine the optimal control policy that was to maintain, during the singular period, the substrate concentration constant at a value at which the specific growth rate was maximum. This is a simple cell mass maximization problem for a third-order process, which was dealt with in sufficient detail in Chapter 12. An extended Kalman filter (EKF) was used to estimate the state variables along with the parameters in the model.

An adaptive optimization algorithm for fed-batch culture of *Bacillus subtilis* for maximization of α-amylase production was developed[6] using a recursive least squares method with exponential weighting to estimate the parameters in the model. Instead of the recursive least squares method, an EKF was used[7] to estimate the required state variables and model parameters.[6] We consider in this chapter a simple intuitive approach to adaptive optimizations and refer to the literature for more sophisticated adaptive optimization methods.

14.1 Off-Line Cycle-to-Cycle (Sequential) Optimization

Off-line cycle-to-cycle optimization, sometimes referred to as *run-to-run iterative optimization*, is an intuitive off-line optimization scheme in which the parameters in the model are reestimated (or a new model with adjustable parameters) using the experimental data obtained in the previous run and optimization is performed using the updated model parameters to obtain the optimal feed rate profile, which is then applied to the next run. This process is repeated successively from run to run. This off-line as well as the on-line adaptive optimization schemes are outlined in Figure 14.1. The only difference between the two is that the off-line optimization scheme is applied from one run to the next, whereas the on-line adaptive optimization is applied within one run by applying the scheme on a number of consecutive time intervals in the entire time period.

The off-line cycle-to-cycle optimization is an obvious and practical approach that can be implemented readily. A pertinent question to be asked concerns convergence. Intuitively, it would appear that if the model is functionally appropriate to mirror most important parts of cell physiology, this procedure should improve the result successively and asymptotically approach the optimum in a few runs. However, if the model is functionally inadequate, the procedure may not successively converge to the optimal solution. This aspect will be shown in an example to follow. This off-line, cycle-to-cycle iterative scheme can be summarized by the following steps:

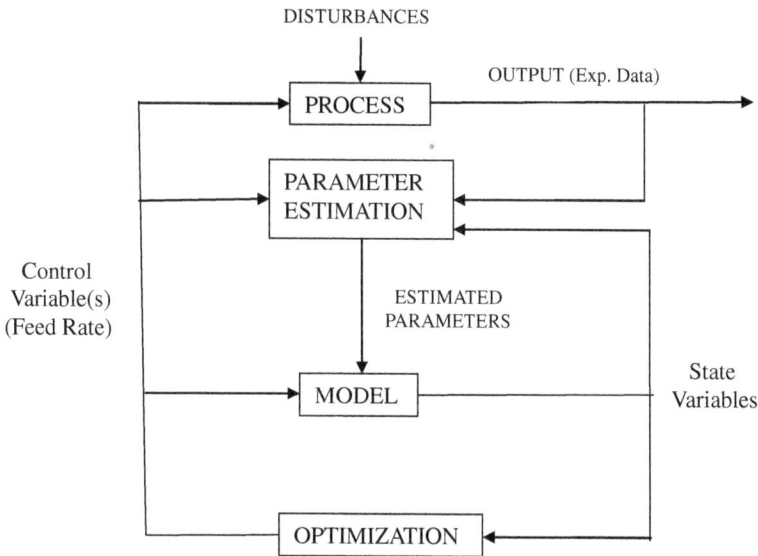

Figure 14.1. Cycle-to-cycle and on-line adaptive optimization.

1. Develop a dynamic model based on up-to-date knowledge of the process.
2. Make a run to generate experimental data and fit the model to the data by estimating the parameters in the model (off-line parameter estimation).
3. Using the model obtained in step 2, optimize the feed rate profile (off-line optimization).
4. Apply the optimal feed rate profile obtained in step 3 for the next run to generate experimental data (implementation).
5. Repeat steps 3 and 4 until no appreciable improvement is obtained (iteration).

This intuitive and obvious approach to improving the process successively from one run to the next is illustrated with examples of penicillin production[9] by *Penicillium chrysogenum* and invertase production[10] by a recombinant strain of *Saccharomyces cerevisiae*. The cycle-to-cycle optimization scheme has been applied to (1) simulation studies in which the "data" generated using a fairly complex model to represent the process are used for parameter estimation in a simplified model, which is then optimized to obtain the feed rate profile, and (2) experimental studies in which the data generated experimentally in one run are used to update the parameter values in the model, which is then optimized to determine the optimal feed rate profile for the next run. This procedure is repeated to reach the optimum and to maintain it there for subsequent runs.

14.1.1 Off-Line Cycle-to-Cycle Optimization of Penicillin Production

Both simulation and experimental studies were made to show the effectiveness of off-line adaptive optimization of penicillin fermentation. The simulation studies were made by rigorously optimizing (impulse response optimization) the updated model by Pontryagin's maximum principle (PMP) and also by a parameter optimization method using the "experimental data" generated by a complex model. Experimental

Table 14.1. *Cagney's differential state model for penicillin fermentation*

	Differential state model of Cagney[8]
Mass balance equations	
Growing tips	$d(A_0V)/dt = k_bA_1V - k_{A_0/A_1}A_0V$
Penicillin-producing fraction	$d(A_1V)/dt = \mu A_0V - k_bA_1V + k_{A_0/A_1}A_0V - k_dA_1V$
Nonviable fraction	$d(A_2V)/dt = k_dA_1V$
Substrate	$d(GV)/dt = FG_F - (\mu/Y_{X/S})(A_0V) - (\pi/Y_{P/S} + m)(A_1V)$
Penicillin	$d(PV)/dt = \pi A_1V - k_hPV$
Overall	$dV/dt = F$
Total cell mass	$d(XV)/dt = d(A_0V)/dt + d(A_1V)/dt + d(A_2V)/dt = \mu A_0V$
Specific rate expressions	
Growth	$\mu = \mu_m/(K + G)$
Penicillin formation	$\pi = k_pG/[K_P + G(1 + G/K_I)]$
Cell degradation	$k_d = k_2/(L + G)$
Cell branching	$k_{A_0/A_1} = k_1/(L + G)$
Maintenance	$m = m_1G/(K_m + G)$
Branching	$k_b = \nu G/(K + G)$
Substrate consumption for cell growth	$\sigma = \mu/Y_{X/S}$
Substrate consumption for penicillin formation	$\eta = \pi/Y_{P/S}$

optimization studies are performed using actual experimental data to estimate the model parameters and optimize the updated model to obtain the optimal feed rate profile, which is implemented in a subsequent run.

14.1.1.1 Simulation Studies

Simulation studies are made with a complex differentiation state model proposed by Cagney[8] for the strain of *P. chrysogenum* (C455.0/3) growing on a semidefined medium. This model is used to generate "simulated data" for the fractions of non-viable cells, penicillin-producing cells, and growing-tip cells and the concentrations of substrate, cell, and penicillin G. The simulated data are then used to fit a simpler model proposed by Chittur.[9]

Cagney's model is sixth order and composed of six mass balance equations for concentrations of penicillin and substrate, the culture volume, and three hyphal fractions: a nonviable cell fraction, a penicillin-producing cell fraction, and a growing-tip cell fraction. The dynamic model equations are given in Table 14.1. This model is capable of fitting experimental data well. Cagney's model was used to generate "experimental data."

A simpler, de facto fourth-order model of Chittur[9] consists of five mass balances by lumping together the penicillin-producing cell fraction and the viable cell fraction and allowing the nonviable fraction to be dependent only on the viable cell fraction. Thus, there are four independent mass balance equations. The model equations are given in Table 14.2. Chittur's model was used for model parameter estimation and also for optimization.

Table 14.2. *Chittur's simplified model for penicillin fermentation*[9]

	Mass balance equations
Mass balance species	
Viable cell	$d(AV)/dt = (\mu - k_d)AV$
Nonviable fraction	$d(A_u V)/dt = k_d AV$
Substrate	$d(GV)/dt = FG_F - (\mu/Y_{X/S} + \pi/Y_{P/S} + m)AV$
Penicillin	$d(PV)/dt = \pi AV - k_h PV$
Overall	$dV/dt = F$
Total cell mass	$d(XV)/dt = \mu AV = d(AV)/dt + d(A_u V)/dt$
Specific rate expressions	
Growth	$\mu = \mu_m/(K + G)$
Penicillin formation	$\pi = k_p G/[K_P + G(1 + G/K_I)]$
Cell degradation and differentiation	$k_d = k_2/(L + G)$
Maintenance	$m = m_1 G/(K_m + G)$
Branching	$k_b = vG/(K + G)$
Substrate consumption for cell growth	$\sigma = \mu/Y_{X/S}$
Substrate consumption for penicillin formation	$\eta = \pi/Y_{P/S}$

There are 14 parameters in Cagney's complex model, whereas 12 parameters are present in Chittur's simplified model. The former is used to generate the data, whereas the latter is used to model the data generated by the former. A sensitivity analysis[10] of the 12 parameters revealed that the parameters K_m, k_h, L, and $Y_{P/S}$ are least sensitive, and therefore, these are kept constant at nominal values. In addition, $Y_{X/S}$ was also kept constant because it has a generally accepted value. A 5 percent random error was added to the data generated by Cagney's model. The remaining seven parameters were then estimated from the data by means of an iterative least squares parameter estimation scheme.[10]

The optimization of the simpler de facto fourth-order model of Chittur with the estimated parameter values is carried out using two different approaches: (1) the impulse response optimization (the theoretical optimum feed rate profile) and (2) the parameter optimization. The former approach is based on the theoretical optimal feed rate profile that consisted of a period of the maximum flow rate, followed by a period of the minimum flow rate (a batch period), a period of the singular flow rate, and a batch period. The latter approach[10] is based on an empirical feed rate that can approximate the optimal feed rate expression by the following form with adjustable parameters:

$$F_s(\tau) = a_0 + a_1\tau + a_2\tau^2 + \cdots + a_n\tau^n + \alpha \exp(\beta\tau), \quad \tau = t - t_s \quad (14.1)$$

where a_0 through a_n, α, and β are constant parameters that must be optimally determined from the experimental data, τ is the time elapsed from the start of the singular interval, t_s, and n is the order of the polynomial. The idea is to optimize the performance index by searching for the best parameter values in Eq. (14.1). Because of the exponential function and the polynomial form, a third- or fourth-order polynomial turns out to be adequate in approximating the optimal feed rate, which requires extensive computational effort, and its sequence cannot be determined readily if the process is modeled by more than four dynamic equations.

Table 14.3. *Simulated cycle-to-cycle adaptive optimization of fed-batch penicillin fermentation*[10]

Run	Cycle	Optimization method	Parameter estimation method	Performance index (g)
1	1	Parameter	Iterative least squares	45.02
2	1	Impulse response	Iterative least squares	44.13
3	2	Impulse response	Iterative least squares	56.29

The experimental data obtained from the first run (run 1) are used to estimate the best parameter values in the model, and this updated model is then optimized by searching for the best parameter values (run 1, parameter optimization) in Eq. (14.1). This updated model is optimized rigorously (impulse optimization) using PMP (run 2). This newly determined feed rate profile is implemented experimentally in a subsequent run (run 3) to obtain new set of experimental results, which are then used to estimate the best parameter values again. This process is repeated until no appreciable changes are observed in the performance index. The results are summarized in Table 14.3.[10]

Runs 1 and 2 show that there is no appreciable difference in the result between the impulse response optimization and the parameter optimization using the parameterized function of Eq. (14.1). In other words, the parameter optimization using the approximated feed flow rate expression of Eq. (14.1) yields practically the same result as the rigorous optimization, which requires considerably more effort. Therefore, if the system order is high or the application is on-line and requires a rapid computation, the approximation based on the functional form of Eq. (14.1) is practical and effective. Runs 2 and 3 show that the iterative optimization applied to the second run (run 3) yields the anticipated improvement over the first run (run 1).

14.1.1.2 Experimental Studies

Experimental studies[9] of fed-batch penicillin fermentation by *P. chrysogenum* using a computer-interfaced fermentor with an on-line filtration device[11] to estimate the viable and nonviable cells were carried out. The de facto fourth-order model[2] (a simplified model) shown in Table 14.2[9] is used for optimization. The experimental details are available elsewhere.[9] The experimental data are generated starting from arbitrary initial conditions and a substrate feed profile for the first run and are used in estimating the parameters in the simple model. The model with updated parameter values is optimized to determine the optimal feed rate profile consisting of a period of maximum flow rate, a period of minimum flow rate (a batch period), a singular feed rate period (intermediate flow rate profile), and a final batch period, when the culture volume is full. This optimum feed profile is then applied in the second run, and the data thus generated are used to reestimate the parameters in the model. This updated model is optimized again to obtain a new optimal feed rate profile, which is applied in the subsequent run. This process is presumed to continue for subsequent runs. To make certain that the procedure converges, another set of runs starting from near the optimum was run. The results are summarized in Table 14.4.[9]

Run 1 in (I) was made using an arbitrary feed profile to generate experimental data at various times during the run; concentrations of glucose, penicillin and cell fractions, and the reactor volume profile. As anticipated, the amount of penicillin G

Table 14.4. *Experimental cycle-to-cycle adaptive optimization of fed-batch penicillin fermentation*[9]

	Run 1	Run 2, 1st optimization	Run 3, 2nd optimization	Run 4, 3rd optimization
		I. Runs from arbitrary point		
Amount predicted, g		41.4	61.8	
Amount obtained, g	30	53.9	63.8	
		II. Runs from near optimum		
Amount predicted, g		55.0	67.1	68.4
Amount obtained, g		64.6	67.8	68.1

produced was 30 g, far from the optimum value that was obtained by two successive sequential optimizations, 63.7 g. These data are then used to estimate the parameters in the simplified model. The model with the estimated parameter values is then used for optimization to obtain the optimal feed rate profile. Then, the newly calculated optimum profile is used for the next run (first iterative optimization run), obtaining a completely new set of experimental data and yielding 53.9 g of penicillin G. The data obtained from the first experimental optimization run were used to reestimate the adjustable parameters in the model, and optimization was carried out to obtain the optimal feed rate profile. The newly determined optimal feed rate profile was used to carry out the next experimental run (second iterative optimization run), obtaining 63.7 g of penicillin and generating new sets of experimental data to be used for the subsequent run. The results from (I) indicate that the sequential optimization using the simplified model converges to the optimum in two iterations when the optimization is started from an arbitrary point, far from the optimum.

The results from (II) show that the iterative optimization starting near the optimum does not diverge but remains close to the optimum. Thus, the proposed run-to-run iterative approach worked well in rapidly approaching the optimum when started from an arbitrary point, and once the optimum is approached, it does not diverge. The details can be found elsewhere.[9]

14.1.2 Experimental Off-Line Cycle-to-Cycle Optimization of Invertase Production

Studies similar to the preceding are made with a recombinant strain of *S. cerevisiae* SEY2102/pRB58[12] cloned for invertase. The yeast 2 μm based plasmid pRB58 is introduced into the host SEY2102 to study the expression of the *SUC2* gene in yeast. It has been reported that the *SUC2* promoter is very tightly controlled by the glucose concentration in the medium and is fully repressed above 2 g/L.

Experimental optimization studies are made using two models: (1) the Modak–Patkar model,[13,14] which ignores the ethanol inhibition of both invertase formation and cell growth (Table 14.5), and (2) a modified form of the Modak–Patkar model[13,14] that accounts for the ethanol inhibition (Table 14.6). The details of these two models are given in Tables 14.5 and 14.6, respectively.

The inhibitions by ethanol both on the growth rate of cells on glucose and on the invertase formation rate are modeled by the same factor, $(1 - E/k_G) \geq 0$, a linearly

Table 14.5. *Modak–Patkar model for aerobic yeast,* Saccharomyces cerevisiae[13,14]

	Mass balance equations
Mass balance species	
Cell	$d(XV)/dt = (\mu_G + \mu_E)XV$
Glucose	$d(GV)/dt = FS_F - \sigma XV$
Ethanol	$d(EV)/dt = (\pi_E - \eta)XV$
Invertase	$d(IXV)/dt = (\pi - k_d I)XV$
Overall	$dV/dt = F$
Specific rate expressions	
Growth	
On glucose	$\mu_G = (k_1 G + k_2 G^2)/(k_3 + k_4 G + G^2)$
On ethanol	$\mu_E = k_5 E/[(k_6 + k_7\sigma + E)(1 + k_8 E)]$
Fraction of glucose fermented	$R = (1 + k_9 G^n)/(k_{10} + k_9 G^n)$
Ethanol production rate	$\pi_E = Y_{E/G}^F \sigma R$
Ethanol consumption rate	$\eta = \mu_E/Y_{X/E}^R$
Cellular yield	$Y_X = (1 - R)Y_{X/G}^R + RY_{X/G}^F$
Invertase formation	$\pi = k_P G/[K_P + G(1 + G/K_I)]$
Substrate consumption for cell growth	$\sigma = \mu_G/Y_X$

Note: X = cell concentration; V = reactor volume; G = glucose concentration; E = ethanol concentration; P = amount of invertase per cell amount; F = glucose feed rate.

decreasing function of ethanol concentration. This technique has been widely used to model the ethanol inhibition effect.

The results of two cycles of experimental studies with the Modak–Patkar model that ignores ethanol inhibition are given in Table 14.7 as C1 and C2, while the results

Table 14.6. *A modified Modak–Patkar model for* Saccharomyces cerevisiae *(inhibition by ethanol of both invertase formation and cell growth on glucose)*[13,14]

	Mass balance equations
Mass balance species	
Cell	$d(XV)/dt = (\mu_G + \mu_E)XV$
Glucose	$d(GV)/dt = FS_F - \sigma XV$
Ethanol	$d(EV)/dt = (\pi_E - \eta)XV$
Invertase	$d(IXV)/dt = (\pi - k_d I)XV$
Overall	$dV/dt = F$
Specific rate expressions	
Growth	
On glucose	$\mu_G = \dfrac{(k_1 G + k_2 G^2)}{(k_3 + k_4 G + G^2)}(1 - E/k_E)$
On ethanol	$\mu_E = k_5 E/[(k_6 + k_7\sigma + E)(1 + k_8 E)]$
Fraction of glucose fermented	$R = (1 + k_9 G^n)/(k_{10} + k_9 G^n)$
Ethanol production rate	$\pi_E = Y_{E/G}^F \sigma R$
Ethanol consumption rate	$\eta = \mu_E/Y_{X/E}^R$
Cellular yield	$Y_X = (1 - R)Y_{X/G}^R + RY_{X/G}^E$
Invertase formation	$\pi = k_P G(1 - E/k_E)/[K_P + G(1 + G/K_I)]$
Substrate consumption for cell growth	$\sigma = \mu_G/Y_X$
	$(1 - E/k_E) \geq 0$

Note: X = cell concentration; V = reactor volume; G = glucose concentration; E = ethanol concentration; P = amount of invertase per cell amount; F = glucose feed rate.

Table 14.7. *Experimental on-line and cycle-to-cycle optimization results with Modak–Patkar model (inhibition by ethanol ignored)*[10]

Run	$(PXV)_f$	t_f (hrs)	$(XV)_f$ (g)	E_f (g/L)	$(PXV^{cell})_f$	$(PXV^{sup})_f$	$(SIA)_f$ (gDCW)	$(SIA^{cell})_f$ (gDCW)	$(SIA^{sup})_f$ (gDCW)
C1	3.88E5	18.8	42.1	12.2	3.08E5	8.00E4	7067	5600	1467
C2	5.19E5	19.0	52.6	12.0	3.96E5	1.23E5	9867	7533	2333

Note: $(PXV)_f$ = total amount of invertase at the final time; $(XV)_f$ = total amount of cell mass at the final time; t_f = final time; E_f = ethanol concentration at the final time; $(PXV^{sup})_f$ = total amount of intracellular (periplasmic space) invertase at the final time; $(PXV^{sup})_f$ = total amount of extracellular invertase at the final time; $(SIA)_f$ = specific invertase activity at the final time; $(SIA^{cell})_f$ = specific intracellular invertase activity at the final time; $(SIA^{sup})_f$ = specific extracellular invertase activity at the final time; C1 = cycle-to-cycle adaptive optimization run 1; C2 = cycle-to cycle adaptive optimization run 2.

with the modified Modak–Patkar model, which accounts for ethanol inhibition, are given in Table 14.8 as C1 and C2.[10] The details of experimental conditions and protocols are available elsewhere.[10] The results of the optimization with a model that accounts for ethanol inhibition yield almost twice as much invertase (Table 14.8) as those obtained with the model that ignores ethanol inhibition, clearly demonstrating the need to work with a functionally correct model. The high ethanol concentration (10–12 g/L at the final time) inhibited the invertase formation rate and is probably the reason for low production of invertase for the optimization with the Modak–Patkar model that does not take into account the inhibitory effect of ethanol.

The total amounts of invertase obtained in two cycles of off-line cycle-to-cycle optimization with the Modak–Patkar model are 3.9E5 units for the first cycle and 5.2E5 units for the second cycle. The improvement of the second cycle over the first is about 34 percent. Conversely, when the ethanol inhibition is accounted for in the modified model of Modak–Patkar, the amounts of invertase obtained increased almost twofold over those obtained with the model that ignored the inhibition effect (Table 14.8), to 7.6E5 units for the first cycle and 12.6E5 units for the second cycle, an

Table 14.8. *Experimental cycle-to-cycle and on-line adaptive optimization results with a modified Modak–Patkar model (inhibition by ethanol of both invertase formation and cell growth on glucose)*[2,10,17]

Run	$(PXV)_f$	t_f (hrs)	$(XV)_f$ (g)	E_f (g/L)	$(PXV^{cell})_f$	$(PXV^{sup})_f$	$(SIA)_f$ (gDCW)	$(SIA^{cell})_f$ (gDCW)	$(SIA^{sup})_f$ (gDCW)
C1	7.60E5	13.0	48.1	3.75	5.86E5	1.75E5	15800	12200	3630
C2	12.6E5	25.0	54.3	1.71	8.82E5	3.78E5	23200	16200	6970
A1	13.4E5	22.7	54.0	1.33	1.00E6	3.39E5	24800	18500	6270
A2	12.9E5	25.7	51.3	2.05	9.56E5	3.34E5	25100	18600	6500

Note: $(PXV)_f$ = total amount of invertase at the final time; $(XV)_f$ = total amount of cell mass at the final time; t_f = final time; E_f = ethanol concentration at the final time; $(PXV^{sup})_f$ = total amount of intracellular (periplasmic space) invertase at the final time; $(PXV^{sup})_f$ = total amount of extracellular invertase at the final time; $(SIA)_f$ = specific invertase activity at the final time; $(SIA^{cell})_f$ = specific intracellular invertase activity at the final time; $(SIA^{sup})_f$ = specific extracellular invertase activity at the final time; C1 = cycle-to-cycle adaptive optimization run 1; C2 = cycle-to cycle adaptive optimization run 2; A1 = on-line adaptive optimization run 1; A2 = on-line adaptive optimization run 2.

improvement of 66 percent. It is clear that the optimization should be performed with a functionally correct model. It is interesting to note that the amounts of invertase stored in the periplasmic space are almost three times those excreted out of the cells.

For comparison purposes, the results of off-line cycle-to-cycle optimization are given along with those of on-line adaptive optimization. One can project using the results of Table 14.8 that it would take about three cycles to match one run of on-line adaptive optimization. In other words, the on-line adaptive optimization in one run can match the results obtained by three runs of off-line cycle-to-cycle optimization.

The results of optimization with the modified Modak–Patkar model are shown in Table 14.8. The second run-to-run optimization yielded 1.26E5 units of invertase, almost three times more in the periplasmic space than in the extracellular space when the ethanol concentration was kept low. When comparing the results of Table 14.7 with those of Table 14.8, it is clear that the use of the modified Modak–Patkar model results in twice as much invertase as obtained with the use of the Modak–Patkar model and that this is achieved mainly by suppressing the formation of ethanol. In fact, the ethanol concentration at the final time obtained with the modified model is only about 10 to 20 percent of that obtained with the unmodified model. It is well known that ethanol concentrations in excess of 2 g/L inhibit invertase formation as well as cell growth, and therefore, ethanol concentrations should be held below this level. Because the Modak–Patkar model does not account for the inhibitory effect of ethanol, the optimization procedure ignores the detrimental effects of high ethanol concentration.

The details of these optimization studies are available elsewhere.[10] The successive optimal feed rate profiles and concentration profiles of various species are also available in the literature.[10]

14.2 On-Line Adaptive Optimization

While off-line adaptive optimization may yield satisfactory results after a few runs, a more desirable approach would be to apply an on-line adaptive optimization to improve the results within the same run. Obviously, a number of requirements must be met. First, one must be able to measure, within a reasonable time, the state variables, that is, various species concentrations, or if not, one must have the capability of predicting difficult-to-measure state variables. Second, the computations involved in optimization must be completed within a short time scale. The reason for these requirements become apparent if we look into what is involved in an on-line adaptive optimization scheme.

On-line adaptive optimization begins first by breaking up the total fermentation time period into N intervals and implementing the iterative optimization procedure in each interval. Using a fed-batch model with a number of adjustable parameters, the experimental data obtained in the first interval are used to estimate the parameters in the model, and the optimization is carried out using the updated parameter values to determine the optimal feed rate profile. The portion of the calculated optimal feed rate profile corresponding to the second interval is applied in the second interval. The experimental data thus generated from the second time interval (together with or without the experimental data obtained in the first interval) are used to once again update the adjustable parameters in the fed-batch culture model, and a new optimal feed rate profile is determined for the updated model. The portion of the calculated optimal feed rate profile corresponding to the third interval is then applied in the third interval; the experimental data thus obtained (with or without previous time interval data) are used to update once again the adjustable parameters in the model, and a new optimal feed rate profile is calculated and implemented in the next interval. This procedure is repeated until the entire N intervals are covered.

In the preceding approach, it may be more advantageous to consider the experimental data generated up to jth intervals instead of just the data obtained in the jth interval alone. It is also conceivable to use a moving windows approach and use only the data generated in the last few intervals. The data generated in the last few intervals may be more pertinent than the entire data because the fermentation may go through different stages, such as growth and production phases, and the data generated during the growth phase (idiophase) may not be as pertinent as the data generated in the production phase (protophase).

An adaptive feature incorporating an on-line parameter estimation and reoptimization should greatly improve the performance of bioreactors while the requirement for a highly sophisticated model would be relaxed as the model parameter values are frequently updated using experimental data. The time necessary to reestimate the key parameter values and to determine the corresponding optimal feed rate profile may be short or long relative to the time of subintervals used. If it is short, then the calculated feed rate profile may be applied in the next interval. However, if the time is long, then the feed rate profile computed in the previous interval should be applied while the computation is being carried out. Thus, there may be an overlapping period in this approach. As stated previously, the fermentation time period is divided into N time intervals, and the parameter estimation is carried out using the data generated during one or more intervals. Thus, the parameter estimation and optimization are performed repeatedly for each time interval, until the final time is reached, while the fermentation is being carried out.

A more detailed description of the on-line adaptive optimization procedure as well as some variations and results are given elsewhere.[9,10] Simulation and experimental results were obtained using the approach outlined here.

14.2.1 Simulation Studies of On-Line Adaptive Optimization of Penicillin Production

We demonstrate the efficacy of on-line adaptive optimization through physical examples of penicillin production by *P. chrysogenum*. Owing to the time delay for off-line measurements, samples can only be taken hourly. Furthermore, a two-hour delay is required for off-line analysis, and an additional hour is required for estimation and optimization calculations. The total delay from the last measurement to the feed rate update is therefore three hours. The total fermentation time for the fed-batch experiments is between 14 and 26 hours for this system. Therefore, the number and duration of the time intervals is limited. For the experiments conducted, the first time interval of adaptive optimization did not include parameter estimation. This is

Table 14.9. *Simulated on-line adaptive optimization of penicillin fermentation*[10]

Problem type	Optimization method	Parameter estimation method	Parameter estimation initial condition	Performance index
Adaptive	Impulse response	ILS	True conditions	56.92
Adaptive	Parameter	ILS	True conditions	56.82
Adaptive	Impulse response	ILS	Measured conditions with exp. errors	47.75
Adaptive	Impulse response	EKF	Estimated future IC	56.57
Adaptive	Parameter	EKF	Estimated future IC	56.44

Note: ILS = iterative least squares; EKF = extended Kalman filter.

because an appropriate number of data points is necessary to begin the parameter estimation, which, coupled with the three-hour time delay, would result in a large lag before the adaptive optimization can begin. By updating the feed rate during the first time interval, corrections are made for deviations in the initial conditions from the expected values that were used to determine the initial feed rate profile.

For simulation studies, a 5 percent random error was added to the data generated from Cagney's model. To overcome measurement errors, an EKF was implemented for the simultaneous estimation of parameters and states. The EKF estimated future initial conditions, which were used to initiate optimization for each interval, and estimated initial conditions were used for parameter estimation.

The optimization at each interval was initialized by estimating the initial conditions of the interval by integrating from the last data point using the updated parameters. The EKF algorithm successfully estimated the parameters and the initial conditions for each interval for the optimization algorithm. The simulation results are given in Table 14.9.

The simulation studies[10] show that the parameterized feed rate (Eq. (14.1)) is practically indistinguishable from the optimum feed rate (impulse response) in terms of the final penicillin concentration. Thus, the parameterized feed rate can be used to reduce the computational time considerably. It is also noted that the iterative least squares method used to estimate the parameters in the model worked well, except when the measured conditions with added experimental error were used; the estimations of parameters and states by the EKF also worked well.

14.2.2 Experimental On-Line Adaptive Optimization of Invertase Production

The experimental optimization studies are made using two models: one that ignores and the other that accounts for the ethanol effect, that is, (1) the Modak–Patkar model[13,14] (Table 14.5) and (2) a modified form of the Modak–Patkar model[13,14] (Table 14.6). The experimental results[10] of on-line optimization of invertase fermentation are shown in Table 14.10. As noted for the cycle-to-cycle optimization, it is clear that the use of the modified Modak–Patkar model results in almost a twofold increase in invertase as compared to the case of the Modak–Patkar model, and this is achieved mainly by suppressing the formation of ethanol. In fact, the concentration of ethanol at the final time obtained with the modified model is only

Table 14.10. *Experimental on-line adaptive optimization of invertase production*[10]

Model used	$(PXV)_f$	t_f (hrs)	$(XV)_f$ (g)	E_f (g/L)	$(PXV^{cell})_f$	$(PXV^{sup})_f$
Modified Modak–Patkar	13.4E6	22.7	54.0	1.33	10.0E5	3.39E5
Modified Modak–Patkar	12.9E6	25.7	51.3	2.05	9.56E5	3.34E5
Modak–Patkar	6.16E5	19.9	47.8	10.5	5.03E5	1.13E5

Note: $(PXV)_f$ = total amount of invertase at the final time; $(XV)_f$ = total amount of cell mass at the final time; t_f = final time; E_f = ethanol concentration at the final time; $(PXV^{sup})_f$ = total amount of intracellular (periplasmic space) invertase at the final time; $(PXV^{sup})_f$ = total amount of extracellular invertase at the final time.

about 10 to 20 percent of that obtained with the unmodified model. Once again, this clearly demonstrates the need to use a functionally correct model in modeling and optimization of fed-batch processes. The details of these optimization studies are available elsewhere.[10] The successive optimal feed rate profiles and concentration profiles of various species are also available in the literature.[10]

In this chapter, a simple intuitive approach to an adaptive optimization scheme is used to demonstrate that such a scheme can be successfully used to carry out off-line cycle-to-cycle and on-line adaptive optimizations of invertase and penicillin production. Although many sophisticated adaptive optimization methods are available in the literature, the proposed intuitive and simple method is practical and effective.

REFERENCES

1. Kishimoto, M., Yoshida, T., and Taguchi, H. 1981. Simulation of fed-batch culture for glutamic acid production with ethanol feeding by use of regression analysis. *Journal of Fermentation Technology* 59: 43–48.
2. Kishimoto, M., Sawano, T., Yoshida, T., and Taguchi, H. 1982. Optimization of a fed-batch culture by statistical analysis, in *IFAC Workshop on Modelling and Control of Biotechnical Processes, Proceedings*, p. 161, Pergamon Press.
3. Kishimoto, M., Yoshida, T., and Taguchi, H. 1981. On-line optimal control of fed-batch culture of glutamic acid production. *Journal of Fermentation Technology* 59: 125–129.
4. Huang, H.-P., Chang, L.-L., and Chao, Y.-C. 1990. Experimental study of on-line optimal substrate feed for a fed-batch culture. *Chemical Engineering Communication* 94: 105–118.
5. Schlasner, S. M., Strohl, W. R., and Woo, W.-K. 1987. On-line adaptive optimal control of a fed-batch fermentation of *Streptomyces* C5, in *Proceedings of the American Control Conference*, p. 687. American Automatic Control Council.
6. Staniskis, J., and Levisauskas, D. 1983. An adaptive control algorithm for fed-batch culture. *Biotechnology and Bioengineering* 26: 419–425.
7. Yoo, Y. J., Hong, J., and Hatch, R. T. 1985. Sequential estimation of states and kinetic parameters and optimization of fermentation processes, in *Proceedings of the American Control Conference*, p. 866. American Automatic Control Council.
8. Cagney, J. W. 1984. Experimental investigation of a differential state model for fed-batch penicillin fermentation. PhD diss., Purdue University.
9. Chittur, V. K. 1989. Modeling and optimization of the fed-batch penicillin fermentation. PhD dissertation, Purdue University.
10. Hansen, J. 1996. On-line adaptive optimization of fed-batch fermentations. PhD dissertation, University of California, Irvine.

11. Thomas, D. C., Chittur, V. K., Cagney, J. W., and Lim, H. C. 1985. On-line estimation of mycelial cell mass concentrations with a computer-interfaced filtration probe. *Biotechnology and Bioengineering* 27: 729–742.

12. Emr, S., Schekman, R., Fessel, M., and Thorner, J. 1983. An Mfα1-SUC2 (α-factor-invertase) gene fusion for study of protein localization and gene expression in yeast. *Proceedings of the National Academy of Sciences of the United States of America* 80: 7080–7084.

13. Modak, J. M. 1985. A theoretical and experimental optimization of fed-batch fermentation processes. PhD dissertation, Purdue University.

14. Patkar, A., Seo, J.-H., and Lim, H. C. 1993. Modeling and optimization of cloned invertase expression in *Saccharomyces cerevisiae*. *Biotechnology and Bioengineering* 41: 1066–1074.

15 Measurements, Estimation, and Control

To optimize bioreactor operations and to operate them effectively, it is essential to be able to monitor and control the bioreactor operating conditions. It is most desirable to have reliable in situ, on-line sensors and measurement devices that can supply measurements of various variables and parameters without a significant time delay so that the information provided by these devices can be used to implement optimization strategies to improve the productivity and efficiency of processes. These devices can be broadly classified as in situ, on-line, and off-line.

This on-line data acquisition would allow on-line optimization and control of bioreactors. For the purposes of process optimization and control, it is desirable to have reliable in situ sensors with a short response time, and because they come into direct contact with the culture medium, they must withstand the sterilization process. When it is not possible to sterilize the sensors, on-line measurement via sampling of the culture medium or culture gas stream can be utilized. Thus, shorter sampling lines that can maintain sterility are required for the liquid and gas sample lines.

The purpose of process control is to manipulate the control variables: (1) to maintain the desired output either at a constant desired value or to force it to follow a desired time profile by suppressing the influence of external disturbances, (2) to stabilize the processes, and (3) to optimize the process performances such as yield, productivity, or profit. The questions associated with the design of a control system include the objective(s), variables to be measured, variables to be manipulated, pairing of the measured and manipulated variables, control system structures, and controller tunings.

Various control techniques are used to implement optimal and semioptimal strategies for fed-batch fermentations. In this chapter, we consider various forms of indirect and direct controls. Indirect feedback controls are based on certain parameters such as carbon dioxide evolution rate (CER), dissolved oxygen (DO), pH, respiratory quotient (RQ), specific growth rate of cells, and substrate concentration. These parameters are indicative of some measure of fed-batch fermentation and are normally held at a constant value that is considered to be the best from a priori information.

The optimal strategy developed in previous chapters is either an open-loop time profile (exponential profiles or numerical values) or a closed-loop (feedback control) feed rate requiring a number of measurements such as the culture volume and concentrations of cells, substrates, and product, X, S, P, and V. The open-loop time profiles of feed rate provide the time-variant set point values for the entire fermentation period but are subject to error because there is no self-corrective action based on the measurements of outputs. Although the closed-loop (feedback) control is desirable, it is not always possible to obtain the feedback control, and implementation requires measurements of process variables.

15.1 Measurements of Process Variables and Parameters

Various fed-batch process variables and parameters are classified into physical and chemical variables.

15.1.1 Physical Properties

Physical properties that can be monitored continuously include agitation speed, gas (aeration) and liquid (medium) flow rates, pressure, temperature, broth viscosity and density, and bioreactor culture volume and mass.

Agitator (*shaft*) *speed* is measured outside of the vessel, thus avoiding problems associated with sterilization using a variety of tachometers such as continuous-current and alternating-current electromagnetic tachometers, synchronous and asynchronous tachometers, impulsion tachometers, variable-reluctance sensors, Foucault current sensors, and optical tachometers.

Aeration (*gas flow*) *rate* is usually measured by a thermal mass flow meter in which gas flows through a capillary tube around which a low-power heating coil is spooled, and both ends of the tube are connected to a Wheatstone bridge. Gas flow causes a temperature variation that is proportional to the mass flow rate.

Medium supply (*liquid flow rate*) *rate* is measured by classical instrumentation such as turbine flow meters, depression flow meters, vortex flow meters, electromagnetic flow meters, and Coriolis effect flow meters.

The *pressure* inside the bioreactor is normally monitored by diaphragm gauges with and without a transducer. Pressure monitoring is important during sterilization and also for providing positive pressure for fermentation processes involving volatile substrates. Some fermentation is also carried out under pressure to enhance oxygen transfer. The principle of pressure measurement is based on the deformation of differently shaped solids under the influence of a pressure difference between the inlet and outlet sides. Bourdon tubes of various shapes, such as C shaped, twisted, and helical, are used.

For *temperature* measurements, thermistors, semiconductor devices that exhibit slightly nonlinear changes in resistance to temperature changes, are most commonly used. Other temperature sensors are platinum resistance sensors and thermocouples.

Broth viscosity measurements are usually made with coaxial cylinder viscosimeters such as Couette viscosimeters with a rotating outer cylinder and Searle viscosimeters with a rotating inner cylinder. On-line measurements of broth viscosity are useful for controlling fermentation processes. Viscosities of broths of microorganisms have been reported: *Aspergillus niger* by a tube viscosimeter,[1] *Hansenula polymorpha* by a capillary tube method,[2] *Penicillium chrysogenum* by a slot-type viscosimeter,[3] and *Aureobasidium pullulans* by an impeller-type[4] viscosimeter.

Biomass concentration is measured by optical sensors, filtration methods, calorimetric techniques, viscosity methods, electrochemical methods, and acoustic methods. Optical sensors can be broadly classified as nephelometric methods or fluorescence. Most of the nephelometric methods are ineffective for high cell mass concentrations. Normally concentrated biomass samples are diluted to less than 1 g/L, and the optical density of the diluted samples is measured and compared to a standard calibration curve to estimate the cell concentration of the original sample. For continuous operation, recycling of diluted broth (for effective measurement at low concentration) can be avoided by a flow-through device that reduces the effective light path.[5] Turbidity probes are used for high cell concentration measurements and are commercially available.[6,7] For filamentous organisms such as *P. chrysogenum*, filtration devices[9-13] were developed in which filtration of fermentation broth takes place and the accumulation of filtration cake can be monitored over a short time period to estimate the concentration of cells with semiempirical correlations.

Dissolved oxygen concentrations are measured by electrochemical methods.[14] There are two types of electrodes: polarographic and galvanic.[15] The galvanic electrodes are most commonly used for small fermentors and are made up of a lead anode and a silver cathode and employ potassium hydroxide, chloride, bicarbonate, or acetate as electrolytes. The polarographic electrodes are in greater demand and are commonly used in biotechnical and fermentation industries. The polarographic electrodes are made of an Ag-AgCl anode and a platinum cathode with potassium chloride and measure the partial pressure of the dissolved oxygen and not the dissolved oxygen concentration. At equilibrium, the partial pressure of dissolved oxygen P_{O_2} as sensed by the probe is equal to mole fraction of oxygen in the gas phase Y_{O_2} times the total pressure P_T:

$$P_{O_2} = Y_{O_2} \times P_T \qquad (15.1)$$

The actual readings are expressed as percentage saturation with air at atmospheric pressure so that 100 percent dissolved oxygen implies a partial pressure of 160 mmHg. Thus, water at 25°C and 760 mmHg pressure saturated with air contains 8.4 mg O_2 in 1 L.

15.1.2 Chemical Properties

15.1.2.1 Carbon Dioxide Evolution Rate

Assimilation of carbon source by microorganisms is always accompanied by the evolution of carbon dioxide, and therefore, the carbon source feed rate based on CER allows regulation of the concentration of the carbon source in media. The idea behind the use of CER is to take advantage of the fact that CER responds rapidly to various inputs that cause dynamic changes in microbial processes and is also a readily measurable parameter so that normally sluggish control based on a slowly responding measurement, say, cell concentration, can be made fast by relying on the fast-responding CER. In this way, the feedback control is rapid without much dynamic delay and results in superior performance. In that sense, this is a form of feedforward control.

CER is computed by applying a mass balance on the inlet and outlet gas streams. Under regulated pH, the inlet and outlet concentrations of carbon dioxide are measured by an infrared CO_2 analyzer or mass spectrometer, and the inlet and outlet gas flow rates are also measured. The carbon dioxide evolution rate is simply the difference between the outlet flow rate of carbon dioxide and the inlet carbon dioxide supply rate. Because the inlet carbon dioxide concentration is fairly constant, if the aeration rate is held constant, the partial pressure of carbon dioxide in the exit gas may be used to feedback-control the substrate feed rates.

Oxygen uptake rate (OUR) is computed by applying a mass balance on the inlet and outlet gas streams. Under regulated pH, the inlet and outlet concentrations of oxygen are measured by a paramagnetic O_2 analyzer or mass spectrometer, and the inlet and outlet gas flow rates are also measured. The OUR is simply the difference between the outlet flow rate of oxygen and the oxygen supply rate.

RQ as a control parameter was suggested in 1959 for yeast production because then, feedback control of a substrate feed rate based on RQ was used extensively for baker's yeast production. Four types of yeast metabolism (glucose oxidation, aerobic fermentation yielding ethanol, oxidation of both glucose and ethanol, and oxidation of ethanol) are known to be related to certain ranges of RQ values. Therefore, by carefully regulating the range of RQ values, it is possible to minimize ethanol (by-product) formation and maximize the yeast cell yield from glucose by a proper feeding strategy of glucose.

The RQ is the ratio of the CER to the OUR:

$$RQ = CER/OUR = r_{CO_2}/r_{O_2} \qquad (15.2)$$

where r_{CO_2} and r_{O_2} are the rate of formation of carbon dioxide and the rate of consumption of oxygen, respectively. Both CER and OUR are calculated from the measurements of carbon dioxide and oxygen mass flow rates in the inlet and outlet air flows. Initially, the glucose feed rate was manually changed stepwise[16] so as to maintain the RQ value within the range 1.0–1.2 throughout the course of yeast fermentation. A linear relationship between RQ and the ethanol generation rate, r_{EtOH}, was established:[17]

$$r_{EtOH} = r_{CO_2} - (RQ)_0 r_{O_2} \qquad (15.3)$$

where r_{EtOH} is the rate of formation of ethanol and $(RQ)_0$ is the RQ value in the absence of ethanol production, which is in the range of 0.95–1.05. However, it was found[18] that the control of RQ alone was not satisfactory in experimental production of yeast from molasses, and it was proposed to modify the volumetric feed rate of molasses according to the following formula:

$$F(t) = \frac{\mu X V}{Y_{X/S}}[1 - K_p\{r_{CO_2} - (RQ)_0 r_{O_2}\}] \qquad (15.4)$$

where K_p is the proportional controller constant. Use of this formula was effective in overcoming problems of oxygen starvation, molasses quality, and variations in inoculums. The aeration and agitation rates are increased or decreased to make the right-hand side constant and, therefore, the DO constant. When the total oxygen uptake is very high owing to high cell density, aeration and agitation may not be enough to control the DO with air. In this situation, one may have to use oxygen-enriched air or pure oxygen.

15.1.2.2 Off-Gas Analyses

To estimate the gas exchange rates, CER, OUR, and RQ, it is necessary to be able to analyze the inlet and outlet gases for carbon dioxide and oxygen. In the past, off-gas analyses were made using a paramagnetic oxygen analyzer and an infrared carbon dioxide analyzer. With availability of relatively inexpensive mass spectrometers, it is now common practice to use a mass spectrometer multiplexed to a number of bioreactors to sequentially analyze not only oxygen and carbon dioxide but also other volatile components in a number of off-gases from various bioreactors.

15.1.2.3 pH Probes[19]

The voltage difference between a measuring electrode and a reference electrode is measured, which varies with the activity of H^+ ions according to the Nernst equation:

$$E = E_0 + (RT/F) \ln a_{H^+} \qquad (15.5)$$

where R is the gas constant, F is the Faraday constant, and a_{H^+} is the activity of H^+ ion concentration, which, in dilute solutions, can be taken as equal to the H^+ ion concentration. For bioprocesses, pH probes combining the reference electrode and the measuring glass electrode are used. The glass electrode (Ag/AgCl in chloride-containing buffer) and the reference electrode (Ag/AgCl in saturated KCl solution) are combined in the combined pH electrode.

15.1.2.4 Redox Potential

Although redox systems play an important part in living systems,[20,21] redox potential does not define any specific property of the culture, and interpretation is subject to some questions. However, in anaerobic cultures, it replaces the DO as an indicator of electron acceptors, and it also does so in aerobic cultures in which the DO is extremely low so that the conventional DO probe cannot be used. The redox electrode of a conducting wire (usually platinum) immersed in the culture exchanges electrons with the oxidation–reduction until equilibrium is reached, at which time the wire acquires a potential E:

$$E_h = E_0 + (RT/nF) \ln a_{Ox}/a_{Red} \qquad (15.6)$$

Because of the difficulty of interpreting the redox potential measurements, examples have been limited to the conversion of sorbitol to sorbose,[22] amino acid optimization,[22,23] and batch culture of *Escherichia coli*.[24]

15.1.2.5 Conductivity and Ionic Probes

Conductivity is an overall measurement of the ionic content of the liquid phase, and specific ion contents are measured with ion-sensitive electrodes. The general principle is based on the potential difference between a reference electrode and an ion-sensitive electrode, which is usually a metal electrode in a reference solution and isolated from the solution by a membrane that is permeable only to the ion of interest. The potential that develops is proportional to the logarithm of the activity (therefore, approximately concentration) of the ion. A number of ion probes are available.[25]

15.1.3 Culture Conditions

In this section, we deal with biochemical sensors that are designed to measure biological variables such as concentrations of substrates and metabolites and biomass concentration. These data are crucially important for process control of bioreactor operations. However, these devices are not readily available for large-scale bioreactors, although they are used in research bioreactor operations, especially off-line. On-line applications of these devices for large-scale bioreactor operations have been very limited, and we look to the future for wider realization.

15.1.3.1 Enzyme and Microbial Electrodes

The enzymes, whole cells, fragments of plant and animal cells, and antibodies/antigens are chosen on the basis of selectivity as catalysts that convert a substrate A to a product B with a simultaneous modification of some property that is detected by a transducer, which in turn converts it into an electrical signal, which is then measured to determine the specific concentration. Examples include glucose oxidase in glucose analyzers;[26–28] immobilized cells[29] of *Brevibacterium lactofermentum*; L-amino acid oxidases for amino acids, including lysine,[30] glutamate,[31] glutamine,[32] aspartate,[33] and aspartame;[34] and organic acid analyzers including lactic acid,[35–37] acetic acid,[38] formic acid,[39] and antibiotics such as penicillin,[40] cephalosporin,[41] and nystatin.[42]

15.1.3.2 Biomass Measurements

Biomass measurements are critically important, and therefore numerous techniques for making these measurements have been proposed, including broth turbidity, direct and electronic counting, fluorescence, light scattering, and luminescence. Extensive coverage is available elsewhere,[43] and therefore, only a brief description is given here.

Nephelometry and turbidimetry are two methods used to evaluate the turbidity of a microbial suspension. Nephelometry[44] is based on the measurement of scattered light, which is proportional to the cell mass. This method is particularly good for low concentrations of cell mass. Turbidity[45] devices measure either the transmitted light or the optical density and are good for high cell concentrations.

Electronic counting is done, for example, in a Coulter counter by monitoring the effect of microorganisms on an electric field as the microorganisms cross the field. The microbial cells suspended in the growth medium are forced to flow in single file through a small aperture, across which an electric field is applied from a constant current source. Microbial cells are relatively nonconducting so that as they cross through the field, the electrical resistance within the aperture increases and the voltage drop increases. The frequency of the pulse is the number of cells and the magnitude of the pulse is proportional to the size of the cells. Flow cytometry is an optical device built on a similar principle.

15.1.3.3 Gas–Liquid Oxygen Transfer

In aerobic processes, oxygen is one of the key substrates, and owing to its low solubility in aqueous solutions, oxygen transfer rate from the gas phase to the liquid phase is a critical factor that determines the oxidative metabolism of the cells. The oxygen transfer rate depends on the oxygen mass transfer coefficient and the driving force, the difference between the saturated oxygen concentration and the oxygen concentration in the liquid. The oxygen transfer rate coefficient depends on the liquid suspension properties, aeration rate, agitation rate, pressure, and temperature, while the driving force can be altered by using oxygen-enriched air or pure oxygen instead of ambient air.

There are a number of methods of determining the oxygen transfer coefficients.[46] These methods include static and dynamic measurement methods as well as dynamic oxygen balance methods around the fermentor.

The oxygen balance around a fed-batch bioreactor is

$$\frac{d[OV]}{dt} = O\frac{dV}{dt} + \frac{dO}{dt}V = k_L aV(O^* - O) - Q_o XV \qquad (15.7)$$

where $O, O^*, V, k_L a, Q_o$, and X are the oxygen concentration in the liquid, the saturated oxygen concentration in equilibrium with the gas phase, the bioreactor volume, the volumetric mass transfer coefficient, the specific oxygen uptake rate, and cell concentration, respectively. DO expressed in percentage saturation[47] is

$$\frac{d[(DO)V]}{V dt} = \frac{d(DO)}{dt} + (DO)\frac{d\ln V}{dt} = k_L a(100 - DO) - 376.48 Q_o X \qquad (15.8)$$

where an assumption that the oxygen concentration on each side of the gas–liquid interface of a dilute aqueous solution is related by Henry's law. The mass transfer coefficient determination using Eq. (15.8) requires the time derivatives of DO and $\ln V$ or $(DO)V$, both of which are subject to numerical error. Thus, we need to have another way of determining the mass transfer coefficient. In addition, there have been several different empirical correlations[48] for the volumetric mass transfer coefficient, $k_L a$. One of the practical empirical correlations involves the agitation and aeration rates as the pressure and temperature of the fermentors are usually maintained constant during the course of operation. Thus, for the purposes of process control for industrial applications in which aeration is the manipulated variable, the practical correlation for the oxygen transfer coefficient should depend on the aeration rate, while for research-type fermentors for which both the aeration and agitation rates can serve as the manipulated variables, the practical correlation would involve both the aeration and agitation rates.

15.2 Estimation Techniques

For the purposes of control and optimization of bioreactors, it is essential to be able to measure on-line many key physiological parameters. Unfortunately, such measurements are difficult, if not impossible, for many of the process parameters. Without such measurements, it is not possible to implement the control and optimization of bioreactors, which are based on key parameters and therefore require information on-line. Therefore, there is an urgent need to supply the information by estimating these parameters from other parameters that are simpler and easier to measure. In this section, we present estimation schemes for key parameters and states that can be used to estimate the physiological states of bioreactors.

15.2.1 Macroscopic Balances

Some indirect measurements may be combined with other measurements that can provide physically significant information on bioreactor performance. The material balance method depends heavily on the gas exchange conditions in a bioreactor, and with availability of continuous gas analyzers, such as paramagnetic oxygen analyzers, infrared carbon dioxide analyzers, and mass spectrometers, it is possible to complete elemental balances (C, H, N, and O) and relate various gas exchange rates such as OUR and CER to physiologically important bioreactor parameters.

There are two variations to the material balance methods, the first of which[49,50] is based on the concept of conservation of mass and the overall chemical reaction stoichiometry. To illustrate the procedure, we take a simple example of growing cells without formation of a metabolite. We consider conversion of a substrate with oxygen and a nitrogen source to the cell mass and formation of carbon dioxide and water by the following overall chemical reaction:

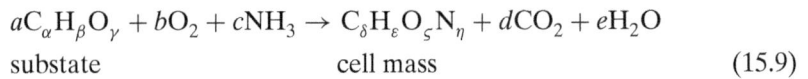

$$aC_\alpha H_\beta O_\gamma + bO_2 + cNH_3 \rightarrow C_\delta H_\varepsilon O_\varsigma N_\eta + dCO_2 + eH_2O$$

$$\text{substate} \qquad\qquad\qquad \text{cell mass} \qquad\qquad\qquad (15.9)$$

In this stoichiometric representation, the stoichiometric coefficient for cell mass has been normalized to unity and all chemical formulas (α through η) are assumed known and constant, although the chemical composition of cells is known to be affected by growth rates and the nature and composition of the medium used. The five stoichiometric coefficients (a through e) are unknown at this point, and the objective is to determine these five coefficients experimentally, which requires five equations. Four elemental (C, H, N, and O) balances provide four equations:

$$
\begin{array}{lll}
\text{Carbon} & \alpha a - d = \delta \\
\text{Hydogen} & \beta a + 3c - 2e = \varepsilon \\
\text{Nitrogen} & c = \eta & (15.10) \\
\text{Oxygen} & \gamma a + 2b - 2d - e = \varsigma
\end{array}
$$

therefore, an additional equation is needed. We note from the preceding stoichiometric equation (15.9) that

$$\text{CER/OUR} = d/b \qquad\qquad (15.11)$$

Therefore, we need to calculate the CER and the OUR from on-line monitoring of the gas exchange rates. Thus, Eqs. (15.10) and (15.11) provide five equations in five unknowns, a through e. Solving for the unknowns, we obtain

$$a = \frac{\text{CER/OUR}}{\alpha} \frac{(\varsigma - \varepsilon + 3\eta)}{[(\text{CER/OUR}/\alpha)(\gamma - \beta) + 2(1 - \text{CER/OUR})]}$$

$$b = \frac{(\varsigma - \varepsilon + 3\eta)}{[(\text{CER/OUR}/\alpha)(\gamma - \beta) + 2(1 - \text{CER/OUR})]}$$

$$c = \eta \qquad\qquad (15.12)$$

$$d = \frac{\text{CER/OUR}(\varsigma - \varepsilon + 3\eta)}{[(\text{CER/OUR}/\alpha)(\gamma - \beta) + 2(1 - \text{CER/OUR})]}$$

$$e = 3\eta - \varepsilon + \frac{(\varsigma - \varepsilon + 3\eta)(\beta\text{CER/OUR}/\alpha)}{[(\text{CER/OUR}/\alpha)(\gamma - \beta) + 2(1 - \text{CER/OUR})]}$$

Once we determine the stoichiometric coefficients, we can calculate various rates from the stoichiometric equation (15.9):

$$\begin{array}{lll} \text{Substrate consumption rate} & \text{SCR} = \sigma XV = \text{OUR}/b \\ \text{Ammonia uptake rate} & \text{AUR} = (c/d)\text{CER} & (15.13) \\ \text{Cell growth rate} & \text{CGR} = \mu XV = \text{CER}/d \end{array}$$

Thus, this approach allows us to estimate the rates that are difficult to measure on-line from the off-gas exchange rates. For fed-batch cultures, it is also possible to estimate on-line the substrate and cell concentrations that are sometimes difficult to measure on-line using the mass balance on substrate and cell mass:

Cell balance

$$\frac{d(XV)}{dt} = \mu XV = \text{CER}/d \Rightarrow X(t)V(t) = X(0)V(0) + \int_0^t (\text{CER}/d)d\theta \quad (15.14)$$

Substrate balance

$$\frac{d(SV)}{dt} = FS_F - \sigma XV = FS_F - \text{CER}(a/d)$$

$$\Rightarrow S(t)V(t) = S(0)V(0) + S_F \int_0^t F d\theta - \int_0^t \text{CER}(a/d)d\theta \quad (15.15)$$

Overall balance

$$\frac{d(V\rho_F)}{dt} = F\rho \Rightarrow V(t)\rho_F(t) - V(0)\rho_F(0) = \int_0^t F\rho d\theta \quad (15.16)$$

Under the assumption of equal and constant densities, $d(\rho_F = \rho)/dt = 0$,

$$V(t) = V(0) + \int_0^t F d\theta \quad (15.17)$$

Therefore, the substrate and cell concentrations are

$$X(t) = \frac{[X(0)V(0) + \int_0^t (\text{CER}/d)d\theta]}{V(0) + \int_0^t F d\theta} \quad (15.18)$$

and

$$S(t) = \frac{S(0)V(0) + S_F \int_0^t F d\theta - \int_0^t \text{CER}(a/d)d\theta}{V(0) + \int_0^t F d\theta} \quad (15.19)$$

Thus, by monitoring the gas exchange rates and the volume changes, it is possible to estimate the concentrations of cells and substrates and various rates such as CGR, AUR, OUR, and SCR.

This approach can be extended to bioreactor operations in which one or more substrates and one or more products are involved. Consider a case of one product given by the following stoichiometric equation:

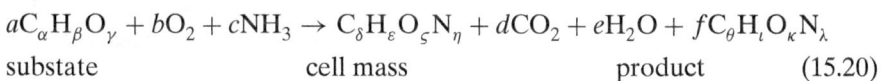

$$a\text{C}_\alpha\text{H}_\beta\text{O}_\gamma + b\text{O}_2 + c\text{NH}_3 \rightarrow \text{C}_\delta\text{H}_\varepsilon\text{O}_\varsigma\text{N}_\eta + d\text{CO}_2 + e\text{H}_2\text{O} + f\text{C}_\theta\text{H}_\iota\text{O}_\kappa\text{N}_\lambda$$

$$\text{substate} \qquad\qquad\qquad \text{cell mass} \qquad\qquad\qquad \text{product} \qquad (15.20)$$

There are six unknown stoichiometric coefficients, *a* through *f*. The elemental balances provide four equations:

$$
\begin{aligned}
\text{Carbon} \quad & \alpha a = \delta + \theta f + e \\
\text{Hydrogen} \quad & \beta a + 3c = \varepsilon + 2e + \iota f \\
\text{Nitrogen} \quad & c = \eta + \lambda f \\
\text{Oxygen} \quad & \gamma a + 2b = \varsigma + 2d + e + \kappa f
\end{aligned}
\tag{15.21}
$$

Two more equations are needed to solve for six unknown stoichiometric coefficients *a* through *f*. As earlier, the gas exchange information provides one equation:

$$
\text{CER/OUR} = d/b \tag{15.11}
$$

One more equation is needed. Among the possible measurements are concentrations of some species such as the substrate, nitrogen, and product. Although some of these species concentrations may be measurable, it is actually the time derivatives that must be monitored. Nevertheless, this method was applied to batch glutamic acid fermentation[51] by *Brevibacterium flavum* with glucose concentration as the additional measurement and the growth of *Hansenula polymorpha*[52] on methanol.

The second method was used to estimate biomass concentration and growth rate through on-line material balance of one chemical component and a mathematical kinetic model that relates the growth rate to the chemical species. Therefore, the accuracy of the method depends heavily on the validity and accuracy of the mathematical model. Using OUR and its relationship to biomass and growth rate of biomass, this method was applied successfully to batch cultures of *Thermoactinomyces* sp. and *Streptomyces* sp. but less successfully to *Saccharomyces cerevisiae*.

15.2.2 Mathematical Estimation Techniques

The balancing methods described suffer from the inaccuracy of available instrumentation. The errors in primary measurements are often large, and these errors can lead to profound effects on the accuracy of estimation. As seen, the gas exchange rates, OUR and CER, are integrated over a time period so that errors in gas exchange rate can compound the errors of on-line estimation of biomass from the off-line assay values as fermentation progresses, and these errors can become unacceptably large. Therefore, it was necessary from time to time to reinitialize the biomass concentration in the midst of the fermentation period,[53] and it became necessary to employ a noise filtration algorithm to improve the reliability of the estimated values.

The deterministic and stochastic estimation techniques were treated in Chapter 6 for estimating parameters. In this chapter, the main emphasis is on estimating those state variables that are difficult, if not impossible, to measure on-line.

In certain situations, some of the state variables are not directly measurable but are observable (can be calculated from the measurements of other state variables). Therefore, it may be necessary to estimate the state variables that are not directly measurable before carrying out the estimation of parameters in the model. In fact, it is possible to simultaneously predict the unmeasurable state variables and estimation of the parameters.

When a process is linear and a model is available, Kalman[54–55] filters are very powerful tools that can estimate not only the unknown parameters in the model but also the state variables that are not possible to measure on-line. However, this technique requires knowledge of certain stochastic properties of measurement and disturbance noises. For nonlinear systems, this technique is applied to linearized models and is known as the extended Kalman filter (EKF).[56] This is a recursive least squares estimator and has been applied to estimate nonlinear bioreactor state variables and kinetic model parameters.

15.2.2.1 Extended Kalman Filter

We begin with a set of mass balance equations for fed-batch cultures in the form of a general nonlinear dynamic model:

Mass balance model

$$\dot{\mathbf{x}} = \mathbf{f}(\mathbf{x}(t), t) + \mathbf{G}(\mathbf{x}(t), t)\mathbf{w}(t) \tag{15.22}$$

where $\mathbf{f}$ is a nonlinear vector function of state vector, $\mathbf{x}(t)$ and t, $\mathbf{G}$ is a matrix function of $\mathbf{x}(t)$ and t, and $\mathbf{w}(t)$ is a random disturbance vector to account for those effects that are not included explicitly in the mass balance equations. We note that the fed-batch mass balance equations in vector form given in Chapter 8 were

$$\dot{\mathbf{x}} = \mathbf{f}[\mathbf{x}(t), \mathbf{u}(t)] \tag{15.23}$$

Therefore, the inputs (feed rates) are imbedded in Eq. (15.23) as time functions and the disturbance function represents all other factors that affect the state of fed-batch operation but are unaccounted for in the fed-batch mass balance equations:

Measurement model

$$\mathbf{z}(t_i) = \mathbf{h}(\mathbf{x}(t_i), t_i) + \boldsymbol{\nu}(t_i) \tag{15.24}$$

where $\mathbf{h}$ is a nonlinear vector function of $\mathbf{x}(t_i)$ and $\boldsymbol{\nu}(t_i)$ is a random error vector in the measurement of $\mathbf{z}(t_i)$. There are assumptions on the random vectors: $\mathbf{w}(t)$ and $\boldsymbol{\nu}(t)$ are zero-mean white noises (frequency independent) with covariance matrices $\mathbf{Q}(t)$ and $\mathbf{R}(t)$, respectively. The noises $\mathbf{w}(t)$ and $\boldsymbol{\nu}(t)$ are uncorrelated with each other and also uncorrelated with the initial conditions, $\mathbf{x}(0)$. The mean and covariance of $\mathbf{x}(0)$ are designated as $\mathbf{m}(0)$ and $\mathbf{P_X}(0)$, respectively.

The EKF problem may be stated as follows: given a measurement sequence $Z(k) = \{\mathbf{z}(t_0), \mathbf{z}(t_1), \mathbf{z}(t_2), \dots, \mathbf{z}(t_k)\}$ for the model given by Eqs. (15.22) and (15.24), find an estimator to provide unbiased, minimum-variance estimates of $\mathbf{x}(t)$, denoted by $\mathbf{x}(t|t_k)$ patterned after a linear Kalman filter and yielding small errors between the true state and the estimated state, $\tilde{\mathbf{x}}(t|t_k) = \mathbf{x}(t) - \hat{\mathbf{x}}(t|t_k)$. The convention on the notation is that when $t > t_k$, the estimation is referred to as prediction, when $t = t_k$, as filtering, and when $t < t_k$, as smoothing. The EKF estimation is described by the following set of filtering equations.

15.2.2.1.1. PREDICTOR. A prediction of estimate is produced by integration of the following differential equations:

$$\dot{\hat{x}}(t|t_k) = \mathbf{f}(\mathbf{x}(t|t_k), t) \tag{15.25}$$

15.2.2.1.2. CORRECTOR. A current state estimate is obtained from the past value by a correction, the difference between the corrupted measurement vector $\mathbf{z}(t_k)$ and the estimated noise-free measurement $\mathbf{h}(\hat{\mathbf{x}}(t_k|t_{k-1}, t_k)]$ multiplied by Kalman gain matrix $\mathbf{K}(t_k)$:

$$\hat{\mathbf{x}}(t_k|t_k) = \hat{\mathbf{x}}(t_k|t_{k-1}) + \mathbf{K}(t_k)[\mathbf{z}(t_k) - \mathbf{h}(\hat{\mathbf{x}}(t_k|t_{k-1}, t_k)], \hat{\mathbf{x}}(t_0|t_0) = m_0 \qquad (15.26)$$

where the Kalman gain matrix is given by

$$\mathbf{K}(t_k) = \mathbf{P}(t_k|t_{k-1})\mathbf{H}^T(t_k)[\mathbf{H}(t_k)\mathbf{P}(t_k|t_{k-1})\mathbf{H}^T(t_k) + \mathbf{R}(t_k)]^{-1},$$
$$\mathbf{H}(t_k) = [\partial\mathbf{h}/\partial\mathbf{x}]_{\mathbf{x}=\hat{\mathbf{x}}(t_k|t_{k-1})} \qquad (15.27)$$

where the error covariance matrix is

$$\mathbf{P}(t_k|t_k) = [\mathbf{I} - \mathbf{K}(t_k)\mathbf{H}(t_k)]\mathbf{P}(t_k|t_{k-1}), \mathbf{P}(t_0|t_0) = \mathbf{P}_{\mathbf{x}_0} \qquad (15.28)$$

Error covariance prediction is

$$\mathbf{P}(t_k|t_{k-1}) = \mathbf{\Phi}(t_k, t_{k-1})\mathbf{P}(t_{k-1}, t_{k-1})\mathbf{\Phi}^T(t_k, t_{k-1}) + \mathbf{Q}(t_{k-1}) \qquad (15.29)$$

where the transition matrix (the fundamental matrix) $\mathbf{\Phi}$ is given by

$$\frac{\partial\mathbf{\Phi}}{\partial t} = \mathbf{F(t)}\mathbf{\Phi}(t, t_k), \mathbf{F}(t) = [\partial\mathbf{f}/\partial\mathbf{x}]_{\mathbf{x}=\hat{\mathbf{x}}(t_k|t_{k-1})} \qquad (15.30)$$

The estimation begins by obtaining the error covariance matrix from Eqs. (15.28) and (15.29), and the Kalman gain matrix is calculated from Eq. (15.27). The current noise-free predicted value $\hat{\mathbf{x}}(t_k|t_k)$ is obtained from the previous value $\hat{\mathbf{x}}(t_k|t_{k-1})$ by adding the product of the estimated noise $\mathbf{z}(t_k) - \mathbf{h}(\hat{\mathbf{x}}(t_k|t_{k-1}, t_k)$ to the Kalman gain according to Eq. (15.26).

A computational algorithm to determine $\hat{\mathbf{x}}(t_1|t_1)$, $\mathbf{P}(t_1|t_1)$, and $\mathbf{K}(t_1)$ for the first interval $[t_0, t_1]$ is as follows:

1. Calculate $\mathbf{\Phi}(t_1, t_0)$ using Eq. (15.30) with the initial conditions $\hat{\mathbf{x}}(t_0|t_0) = m_0$.
2. Calculate $\mathbf{P}(t_1|t_0)$ using Eq. (15.29) with initial conditions $\mathbf{P}(t_0|t_0) = \mathbf{P}_0$ and $\mathbf{Q}(t_0)$.
3. Calculate $\hat{\mathbf{x}}(t_1|t_0)$ by integrating Eq. (15.25) with the initial conditions $\hat{\mathbf{x}}(t_0|t_0) = m_0$.
4. Calculate $\mathbf{K}(t_1)$ and $\mathbf{H}(t_1)$ using Eq. (15.27), where $\mathbf{P}(t_1|t_0)$ was determined in step 2 and $\mathbf{R}(t_1)$ must be computed.
5. Update the state estimate and covariance matrix, $\hat{\mathbf{x}}(t_1|t_1)$ and $\mathbf{P}(t_1|t_1)$, at time t_1 using Eqs. (15.26) and (15.28), where $\mathbf{K}(t_1)$ and $\mathbf{H}(t_1)$ were calculated in step 4.

A computational algorithm to determine $\hat{\mathbf{x}}(t_k|t_k)$, $\mathbf{P}(t_k|t_k)$, and $\mathbf{K}(t_k)$ for the first interval $[t_{k-1}, t_k]$ is as follows:

1. Calculate $\mathbf{\Phi}(t_k, t_{k-1})$ using Eq. (15.30) with initial conditions $\hat{\mathbf{x}}(t_{k-1}|t_{k-1})$.
2. Calculate $\mathbf{P}(t_k|t_{k-1})$ using Eq. (15.29) with initial conditions $\mathbf{P}(t_{k-1}|t_{k-1}) = \mathbf{P}_{k-1}$ and $\mathbf{Q}(t_{k-1})$.
3. Calculate $\hat{\mathbf{x}}(t_k|t_{k-1})$ by integrating Eq. (15.25) with the initial conditions $\hat{\mathbf{x}}(t_{k-1}|t_{k-1})$. Calculate $\mathbf{K}(t_k)$ and $\mathbf{H}(t_k)$ using Eq. (15.27), where $\mathbf{P}(t_k|t_{k-1})$ was determined in step 2 and $\mathbf{R}(t_k)$ must be computed.
4. Update the state estimate and covariance matrix, $\hat{\mathbf{x}}(t_k|t_k)$ and $\mathbf{P}(t_k|t_k)$, at time t_1 using Eqs. (15.26) and (15.28), where $\mathbf{K}(t_k)$ and $\mathbf{H}(t_k)$ were calculated in step 4.

EXAMPLE 15.E.1: PARAMETER ESTIMATION USING EXTENDED KALMAN FILTER Following is an example of parameter estimation using EKF[57] of fed-batch fermentation for poly$-\beta-$hydroxybutyricacid (PHB) by *Alcaligenes eutrophus*. The mass balance model[58] for the fermentation consists of a set of ordinary differential equations:

$$\dot{X}_1 = \mu \cdot X_1 \qquad X_1(0) = X_{10} \tag{15.E1.1}$$

$$\dot{X}_2 = S_{1f} \cdot F_1 - \sigma_1 \cdot X_1 \qquad X_2(0) = X_{20} \tag{15.E1.2}$$

$$\dot{X}_3 = S_{2f} \cdot F_1 - \sigma_2 \cdot X_1 \qquad X_3(0) = X_{30} \tag{15.E1.3}$$

$$\dot{X}_4 = \pi \cdot X_1 \qquad X_4(0) = 0 \tag{15.E1.4}$$

$$\dot{X}_5 = F_1 + F_2 \qquad X_5(0) = X_{50}$$
$$X_5(t_f) = X_{5f} = V_f \tag{15.E1.5}$$

$$\mu = \mu_m \left(\frac{S_1}{K_G + S_1 + S_1^2/K_{GI}} \right) \left(\frac{S_2}{K_N + S_2 + S_1^2/K_{NI}} \right) \tag{15.E1.6}$$

$$\pi = \pi_m \left(1 - \frac{P/X}{(P/X)_m} \right) \left(\frac{S_1}{K_{PG} + S_1 + S_1^2/K_{PGI}} \right) \left(\frac{S_2 + V_P}{K_{PN} + S_2 + S_2^2/K_{PNI}} \right) \tag{15.E1.7}$$

$$\sigma_1 = \left(\frac{\mu}{Y_{R/C}} + \frac{\mu}{Y_{P/C}} + m_e \right) \tag{15.E1.8}$$

$$\sigma_2 = \frac{\mu}{Y_{R/N}} \tag{15.E1.9}$$

where $X_1 = XV$ is the active cell mass (total cell mass, PHB), $X_2 = S_1V$ is the total amount of glucose, $X_3 = S_2V$ is the total amount of ammonia, $X_4 = PV$ is the total amount of PHB, $X_5 = V$ is the bioreactor volume, P/X is the PHB fraction, F_1 is the glucose feed rate, F_2 is the ammonia feed rate, $(P/X)_m$ is the maximum PHB fraction, m_e is the maintenance energy, $Y_{P/C}$, $Y_{R/C}$, and $Y_{R/N}$ are the yield coefficients, V_p is the PHB production rate constant, μ, σ_1, σ_2, and π are specific rates of cell growth, glucose consumption, ammonia consumption, and product formation, respectively, and K_G, K_{GI}, K_N, K_{NI}, K_{PG}, K_{PGI}, K_{PN}, and K_{PNI} are kinetic constants.

Experiments were performed with arbitrary feed rates (see Figure 15.E.1.1a), while keeping the glucose concentration between 10 and 20 g/L and the ammonium chloride concentration between 0 and 0.5 g/L. Figure 15.E.1.1a shows the feed rates and fermentor volume, whereas Figures 15.E.1.1b and 15.E.1.1c show the experimentally measured data for parameter estimation.

The parameters $Y_{P/C}$, $Y_{R/C}$, $Y_{R/N}$, K_G, K_{GI}, K_N, K_{NI}, K_{PG}, K_{PGI}, K_{PN}, K_{PNI}, $(P/X)_m$, m_e, and V_p were estimated using the EKF. State variables and parameters are simultaneously estimated by the EKF by augmenting the state variables with the parameters to be estimated. To apply the algorithm of the EKF, the appropriate problem formulation was first carried out. The state and measurement vectors were defined as follows:

$$\mathbf{x}^{\mathbf{T}} = [X\ S_1\ S_2\ P Y_{P/C}\ Y_{R/C}\ Y_{R/N}\ K_G\ K_{GI}\ K_N\ K_{NI}\ K_{PG}\ K_{PGI}\ K_{PN}\ K_{PNI}\ (P/X)_m\ m_e\ V_p]$$
$$\mathbf{z}^{\mathbf{T}} = [X\ S_1\ S_2\ P]$$
$$\mathbf{h}^{\mathbf{T}} = [x_1\ x_2\ x_3\ x_4]$$

$$\mathbf{H} = \begin{bmatrix} 1 & 0 & 0 & 0 & 0 & 0 & 0 & 0 & 0 & 0 & 0 & 0 & 0 & 0 & 0 & 0 & 0 & 0 \\ 0 & 1 & 0 & 0 & 0 & 0 & 0 & 0 & 0 & 0 & 0 & 0 & 0 & 0 & 0 & 0 & 0 & 0 \\ 0 & 0 & 1 & 0 & 0 & 0 & 0 & 0 & 0 & 0 & 0 & 0 & 0 & 0 & 0 & 0 & 0 & 0 \\ 0 & 0 & 0 & 1 & 0 & 0 & 0 & 0 & 0 & 0 & 0 & 0 & 0 & 0 & 0 & 0 & 0 & 0 \end{bmatrix}$$

The system model is defined as

$$\mathbf{f}(\mathbf{x}(t),\ D(t),\ t) = \mathbf{a}(\mathbf{x}) + \mathbf{b}D_1(t) + \mathbf{c}D_1(t),$$
$$\mathbf{x}(t_0) = \mathbf{x}_0,\ D_1(t) = F_1(t)/V(t),\ D_2(t) = F_2(t)/V(t)$$
$$\mathbf{a}^{\mathbf{T}} = [\mu x_1\ -\sigma_1 x_1\ -\sigma_2 x_1\ \pi x_1\ 0\ 0\ 0\ 0\ 0\ 0\ 0\ 0\ 0\ 0\ 0\ 0\ 0\ 0]$$
$$\mathbf{b}^{\mathbf{T}} = [-x_1\ (S_{1f} - x_2)\ (S_{2f} - x_3)\ -x_4\ 0\ 0\ 0\ 0\ 0\ 0\ 0\ 0\ 0\ 0\ 0\ 0\ 0\ 0]$$
$$\mathbf{c}^{\mathbf{T}} = [-x_1\ -x_2\ -x_3\ -x_4\ 0\ 0\ 0\ 0\ 0\ 0\ 0\ 0\ 0\ 0\ 0\ 0\ 0\ 0]$$

The EKF algorithm was applied to the problem formulated along with measured data shown in Figure 15.E.1.1, and the estimated parameters are shown in Table 15.E.1.1.

15.3 Feedback Control Systems

A standard feedback control system diagram is shown in Figure 15.1. The purpose of control is to manipulate the control variables (inputs to the process that can be manipulated) (1) to maintain the desired outputs at desired constant values (the *regulation problem*) or force the output to follow desired time profiles (the *servo problem*) in the presence of external disturbances, (2) to stabilize unstable or potentially unstable processes (the *stabilization problem*), or (3) to optimize performance as defined by certain measures such as yield, productivity, or profit (the *optimization problem*). These objectives must be met under certain constraints such as (1) operational constraints, (2) safety constraints, (3) environmental regulations, and (4) limited resources. Thus, questions to be answered prior to design of control systems include the following: (1) What is the objective of control? (2) What variables are to be controlled? (3) What variables are to be manipulated? (4) What variables are to be measured? (5) Which control variable is to be paired with which measured output? and (6) Which disturbances are to be measured and coupled with which control variables? Once these questions are answered, then there are problems of

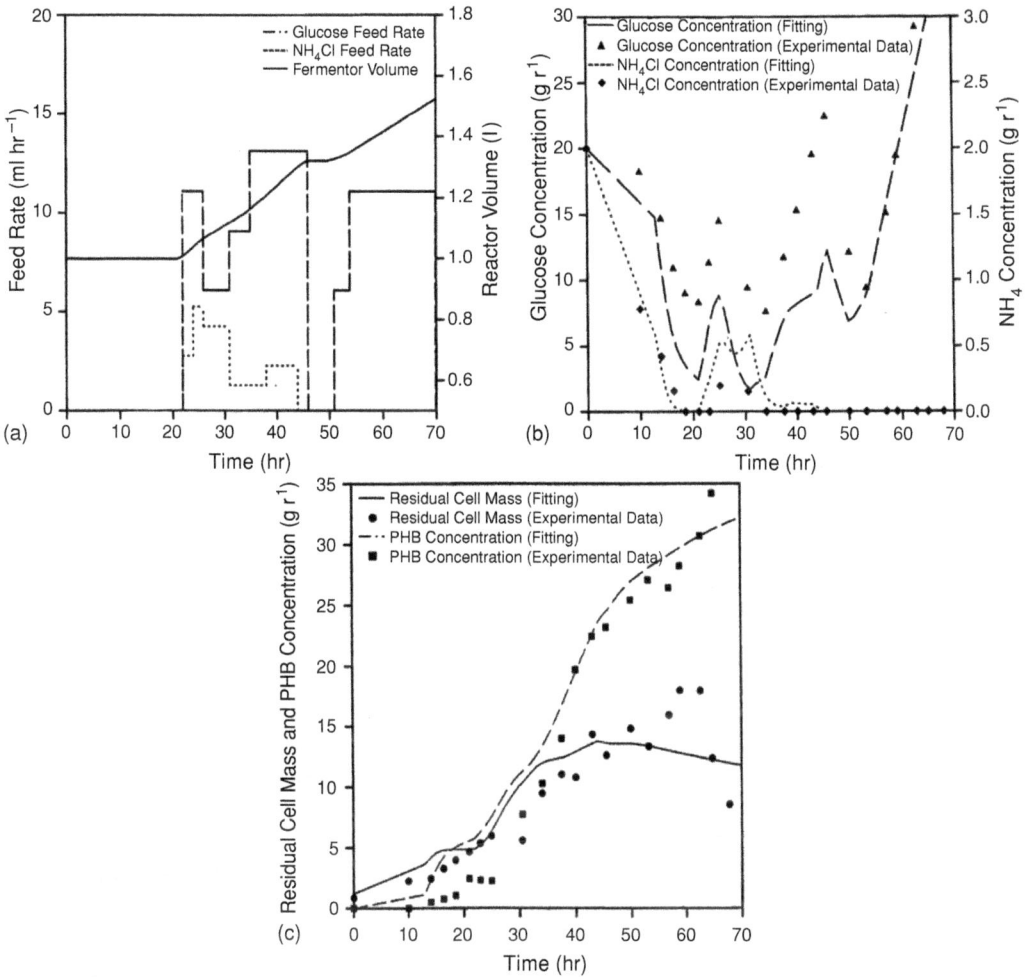

Figure 15.E.1.1. (a) Profiles of glucose and NH_4Cl feed rates and fermentor volumes. (b) Profiles of glucose and NH_4Cl concentrations. (c) Profiles of active cell mass and PHB concentrations.

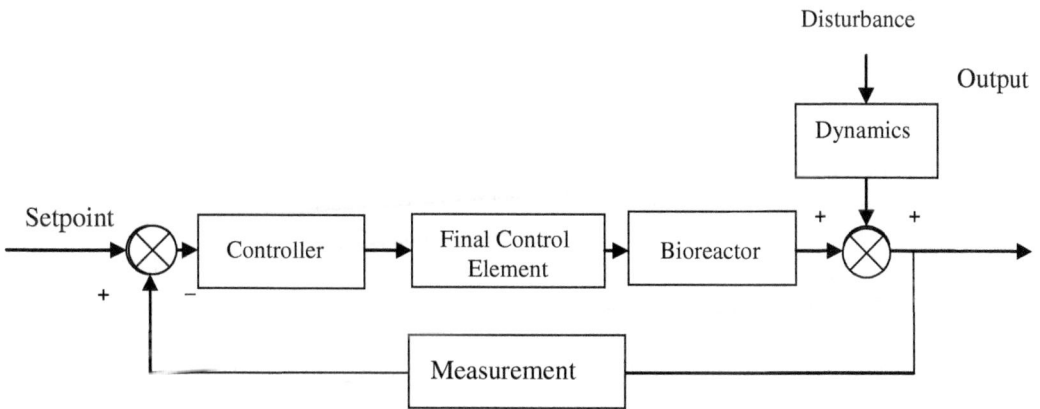

Figure 15.1. A standard feedback control system.

Table 15.E.1.1. *Parameter estimation by extended Kalman filter*[54]

Parameter	EKF estimate
$(P/X)_m$	0.85
K_G	8.11
K_{GI}	17.43
K_N	0.59
K_{NI}	1.5
K_{PG}	8.0
K_{PGI}	80
K_{PN}	0.024
K_{PNI}	2.5
m_e	0.01 ($= 0$ when $S_1 = 0$)
μ_m	0.80
π_m	0.88
V_P	0.0095
$Y_{P/C}$	0.47
$Y_{R/C}$	0.45
$Y_{R/N}$	2.11

(1) what types of controllers should be selected (proportional (P), proportional plus integral (PI), or proportional plus integral and derivative (PID)), (2) how these controllers should be tuned, and (3) how the system should be optimized (constant values or variable profiles). It is not appropriate to discuss these questions here, but the details are readily available elsewhere.[59]

We shall here take a simple example to answer the preceding questions. Consider the problem of maintaining in the bioreactor the temperature (T) and the dissolved oxygen concentration (DO) at some desired values (T_d, DO_d) by manipulating the bioreactor coolant flow rate (F_c) and the aeration rate (R_a) in the presence of disturbances such as the inlet coolant temperature (T_{ci}) and changes in microbial growth rate, which can alter the oxygen demand rate. The statement of this problem answers questions 1 and 3. Obviously, T and DO should be measured, answering question 2. The question of which measured variables should be coupled with which manipulated variables is intuitively obvious in this example; that is, T should be paired with F_c and DO with R_a. In some situations, the question of proper pairing is not intuitively answered but rather must be determined by a systematic method based on the concept of a relative gain array.[59]

A classical feedback control system diagram is shown in Figure 15.2. The outputs to be maintained constant are T and DO of the bioreactor, which are compared with the set points to generate the errors ε_1 and ε_2, which are then put through the controllers G_{c1} and G_{c2}, which may be P, PI, or PID, which generates the input to the aeration motor and the coolant flow rate valve to manipulate the control variables R_a and F_c.

The PID controller is described by the following equation:

$$m = m_s + K_c \left(\varepsilon + \frac{1}{\tau_I} \int_0^t \varepsilon d\theta + \tau_D \frac{d\varepsilon}{dt} \right) \tag{15.31}$$

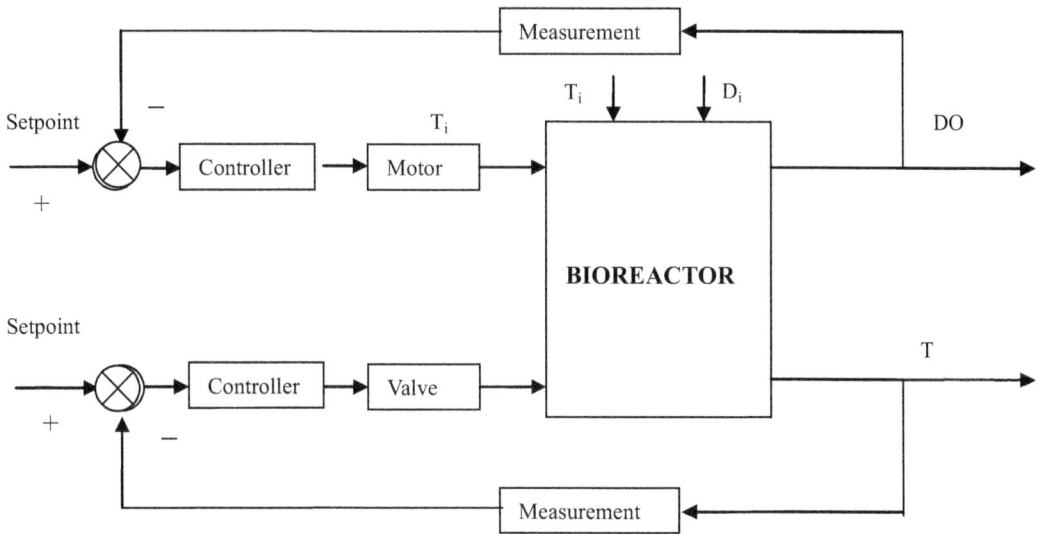

Figure 15.2. Two input–two output feedback.

where m_s is the steady state value of m, K_c is the proportional gain, τ_I is the integral time constant, and τ_D is the derivative time constant. These are adjustable parameters that require tuning for satisfactory performance of the resulting control system.

15.3.1 Single-Loop Control

Examples of common single-loop controls are flow rate control, feed rate control, pressure control, temperature control, dissolved oxygen control, and pH control.

15.3.1.1 Flow Rate Control Loop

These are characterized by fast responses in seconds and essentially no time delay. Most disturbances are high-frequency noises owing to stream turbulence, valve changes, and pump vibrations. Therefore, PI controllers without the derivative mode are usually used.

15.3.1.2 Gas Pressure Control

For gas pressure control, PI controllers are normally used with a small amount of integral action. Owing to a short time constant associated with pressure control as compared to other process time constants, no derivative action is needed.

15.3.1.3 Temperature Control

Owing to a variety of heat transfer equipment and processes, no general guideline can be assigned. Because of time delays and high-order processes with large time constants, PID controllers with a small amount of integral action are used to provide the speed of action.

15.3.1.4 pH Control Systems

Control systems for pH are highly nonlinear, and therefore, at times, pH is difficult to control, requiring addition of both acid and base.

15.3.1.5 Dissolved Oxygen Control

This can be accomplished either by the aeration rate, the agitator speed, or a combination of the two. For industrial-scale fermentors, the manipulation of the agitator speed is costly, and therefore, the agitator speed is normally fixed and the DO concentration is controlled by varying the aeration rate. A PI controller is used for this purpose. Research-scale fermentors are regulated by a cascade control system in which the inner loop involves the measurement of agitation speed and the outer loop has the desired DO value as its set point. DO measurements have been used successfully to control the rate of addition of substrate for more than 50 years. An increase (decrease) in DO level is an indication of a decrease (increase) in the substrate consumption rate. DO probes that are sterilizable, reliable, and accurate are now readily available for feedback control of the substrate feed rate. There still remain some unanswered questions. Should the DO level be maintained constant (and at what value) during the entire course of fermentation, or should it be varied? Should the entire operation be optimized rigorously, as we did in Chapter 13?

We begin with the differential mass balance on DO:

$$\frac{d\{V[DO]\}}{dt} = k_l a V \{[DO]^* - [DO]\} - q_{O_2}(VX) \tag{15.32}$$

where [DO] is the dissolved oxygen concentration; [DO]* is the liquid-phase DO concentration, which is in equilibrium with the balk gas phase; $k_l a$ is the mass transfer coefficient times the gas–liquid interfacial area per culture volume; and q_{O_2} represents the specific oxygen uptake rate. Expanding the left-hand side of Eq. (15.32), we obtain

$$\frac{d\{V[DO]\}}{dt} = V\frac{d[DO]}{dt} + [DO]F = k_l a V\{[DO]^* - [DO]\} - q_{O_2}(VX)$$

$$\Rightarrow \frac{1}{(F/V + k_l a)}\frac{d[DO]}{dt} + [DO] = \frac{1}{(F/V + k_l a)}(k_l a[DO]^* - q_{O_2}X) \tag{15.33}$$

Equation (15.33) suggests that the time constant for this first-order process is $1/(F/V + k_l a)$ and that the process gains are $-X/(F/V + k_l a)$ and $k_l a/(F/V + k_l a)$ for q_{O_2} and [DO]*, respectively. Therefore, the DO response is faster with larger F/V and $k_l a$. The DO level increases with the increase in the air-saturated DO, DO*, and the oxygen transfer rate, $k_l a$, and decreases with the cell concentration and the oxygen uptake rate, q_{O_2}. Instead of air, if pure oxygen is used, the oxygen-saturated DO, $[DO]^*_{O_2}$, would be almost five times higher than the air-saturated DO, $[DO]^*$, and therefore, the driving force for DO is almost fivefold. Thus, pure oxygen or oxygen-enriched air provides an advantage over ambient air. When the dilution rate $(F/V)(t)$ is much smaller than $k_L a$, the time constant is inversely proportional to the mass transfer coefficient $k_L a$.

To maintain DO constant at a desired value, the steady state value from Eq. (15.33) is

$$[DO] = \frac{1}{(F/V + k_l a)}(k_l a[DO]^* - q_{O_2}X) \tag{15.34}$$

Because the desired value of DO is fixed, the right-hand side of Eq. (15.34) must be adjusted in the presence of increasing cell concentration X and time-variant oxygen uptake rate q_{O_2}. Thus, some parameters on the right-hand side must be varied to hold

DO constant. The mass transfer coefficient per unit culture volume, $k_l a$, depends on aeration and agitation rates. Thus, the aeration and agitation rates are increased or decreased to make the right-hand side constant and, therefore, DO constant. When the total oxygen uptake is very high owing to high cell density, aeration and agitation may not be enough to control the DO with air. In this situation, one may have to use oxygen-enriched air or pure oxygen. A number of membrane devices can enrich the oxygen content of ambient air.

15.3.2 Controller Selection and Tuning Methods

Subsequent to the selection of outputs, measured variables, and manipulated variables and their pairing with appropriate outputs, it is necessary to specify the controller types and controller parameter values.

15.3.2.1 Controller Type Selections

There are no fixed criteria to use in specifying the controller type: P, PI, proportional and derivative (PD), PID, or programmed controllers. General characteristics of responses of controlled processes to load disturbances may be taken into consideration in choosing the controller type:

1. When no offset (the difference between the desired and the actual final response) is desired, an integral action is required. PD control results in the shortest time to reach steady state with the least oscillation at the smallest maximum deviation, but at the expense of offset, and is very sensitive to measurement noises.
2. PI action results in no offset but at the expense of a higher maximum deviation, a longer period of oscillation, and a longer time for the oscillation to cease.
3. PID action eliminates the offset and lowers the maximum deviation. It also eliminates some of the oscillation that occurs in PI control.

15.3.2.1.1. CONTROLLER TUNING METHODS. Each controller has settings (adjustable parameters) that need to be tuned to obtain a satisfactory response. These adjustable parameters are the proportional gain (K_c), the integral time constant (τ_I), and the derivative time constant (τ_D), as shown in the PID controller in Eq. (15.31). Although the manufacturers of bioreactor instrumentation supply the controller settings, on-site tuning may be needed to meet individual requirements. Widely used tuning methods include the *ultimate gain method* and the *process reaction curve method*.

The ultimate gain method, also referred to as loop tuning or the continuous cycling method,[60] is based on a continuous cycling intentionally caused by increasing the proportional gain, K_C, while the integral (largest possible τ_I) and derivative mode ($\tau_D \cong 0$) are inoperative. The value of K_C that caused the continuous oscillation is referred to as the *ultimate gain K_U*, and the period of the sustained oscillation is referred to as the *ultimate period P_U*. Ziegler–Nichols tuning is based on these values of K_U and P_U to give approximately a quarter decay ratio. It turns out that these settings were conservative, and modified settings[61] were proposed. The original and modified Ziegler–Nichols settings are given in Table 15.1.

Although the tuning method recommended here is simple and rapid, there are a number of disadvantages, which include (1) the time-consuming process to obtain K_U, especially if the process dynamics are slow; (2) that the tuning method may cause

Table 15.1. *Original*[53] *and modified*[56]
Ziegler–Nichols controller settings

Controller	K_C	τ_I	τ_D
Original Z–N			
P	$0.5\,K_U$		
PI	$0.45\,K_U$	$P_U/1.2$	
PID	$0.6K_U$	$P_U/2$	$P_U/8$
Modified Z–N			
PID	$0.33\,K_U$	$P_U/2$	$P_U/3$
PID	$0.2\,K_U$	$P_U/3$	$P_U/2$

instability that can result in lost productivity or poor product quality; and (3) that
the method relies on a complete process operation. Thus, other authors proposed
an open-loop tuning procedure based on the process reaction curve. A small step
change of magnitude M is introduced in the manipulated variable of the open control
loop (by introducing a temporary step change in the set point), and the response of
the output variable is recorded against time, the process reaction curve, from which
two parameters are determined: S, the slope of the tangent line drawn through the
inflection point, and T_d, the time at which the tangent line intersects the time axis, as
shown in Figure 15.3. The Ziegler–Nichols tuning constants are given in Table 15.2,
in which $S^* = S/M$ is the normalized slope by the size of the input.

Observing that the step response of most processing units, including the process,
the final control element (valve or motor), and the measuring element, can be
approximated by the response of a first-order transfer function with a time delay,

$$\bar{Y}_m(s)/\bar{C}(s) = G_p G_f G_m = \frac{K_p e^{-T_d s}}{\tau_p s + 1} \tag{15.35}$$

new controller settings were proposed,[62] as shown in Table 15.3. The model parameters in the preceding equation can be approximated by

$$K_p = B/M$$
$$\tau_p = B/S \tag{15.36}$$

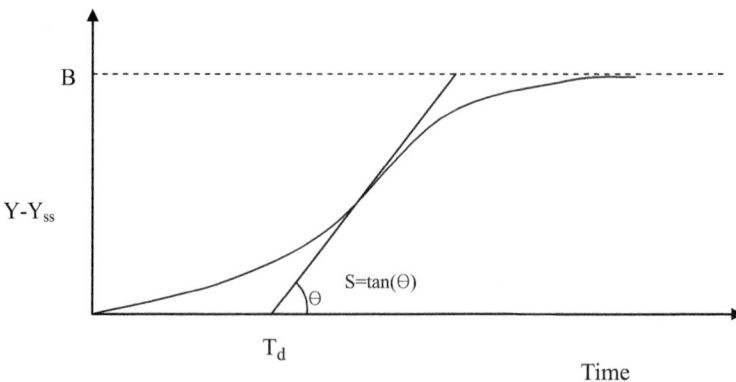

Figure 15.3. A process reaction curve.

Table 15.2. *Controller settings based on process reaction curve*[53]

Controller	K_c	τ_I	τ_D
P	$1/T_d S^*$		
PI	$0.9/T_d S^*$	$3.33\, T_d$	
PID	$1.2/T_d S^*$	$2\, T_d$	$T_d/2$

The settings given in Table 15.3 are to give quarter decay ratios, a minimum offset, and a minimum area under the load response curve.

15.3.3 Multiple-Loop Control

Many control systems contain multiple loops to achieve better control than a single loop can provide. Perhaps the most common multiple-loop control systems include *cascade control systems* and *feedforward–feedback control systems.*

15.3.3.1 Feedforward–Feedback Control

The basic idea behind the feedforward control is that it is not necessary to wait until the disturbances actually affect the output; rather, one can measure the disturbances (T_i and O_i) and apply corrective actions in anticipation of the expected effects. Thus, it is a better and faster acting control than the feedback control. However, to implement the feedforward control, the disturbances must be recognized and measurable, and the effect of the disturbances on the output must be known. Feedforward control is rarely used alone but rather is used in combination with the usual feedback control because the potential errors caused by the imperfect knowledge of their effects on the output cannot be corrected. A combination of the feedback and feedforward control scheme is shown in Figure 15.4.

15.3.3.2 Cascade Control

The cascade control scheme involves only one manipulated variable but has two loops with two measurements. The idea is to measure an intermediate variable to initiate control actions before the output is affected, not to wait until the effect of disturbances affects the output. In this sense, a control action is applied in anticipation

Table 15.3. *Cohen and Coon controller settings*[57]

Controller	K_c	τ_I	τ_D
P	$\dfrac{1}{K}\dfrac{\tau_p}{T_d}\left(1+\dfrac{T_d}{3\tau_p}\right)$		
PI	$\dfrac{1}{K}\dfrac{\tau_p}{T_d}\left(0.9+\dfrac{T_d}{12\tau_p}\right)$	$T_d\dfrac{30+3T_d/\tau_p}{9+20T_d/\tau_p}$	
PID	$\dfrac{1}{K}\dfrac{\tau_p}{T_d}\left(\dfrac{4}{3}+\dfrac{T_d}{12\tau_p}\right)$	$T_d\dfrac{32+6T_d/\tau_p}{13+8T_d/\tau_p}$	$T_d\dfrac{4}{12+2T_d/\tau_p}$

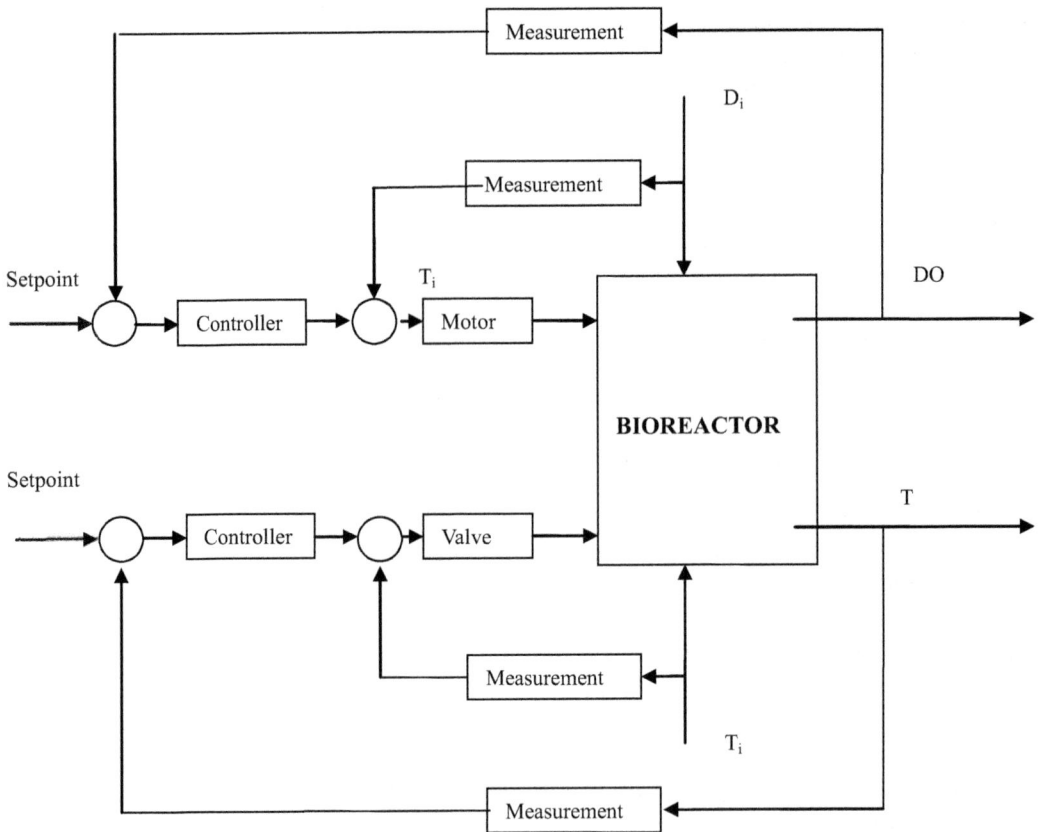

Figure 15.4. A feedforward–feedback control.

of the eventual effect on the output. Thus, the objective is similar to the feedforward control scheme, but the implementation is done differently in a feedback manner. This is illustrated in Figure 15.5 for the control of dissolved oxygen concentration. There are two loops: the inner loop (called a *slave*), involving the measurement of aeration rate, and the outer loop (called a *master*), involving the measurement of DO concentration. The output of the DO controller is used as the set point for the aeration loop. By doing this, the control action can take place earlier without having to wait for the effect of the disturbance to appear in the output. For small-scale bioreactors that are built with adjustable agitator speeds, there is also a cascade control system consisting of an inner loop involving the measurement of agitation speed and an outer loop involving the measurement of DO concentration. These two cascade loops are shown in parallel. Initially, the agitator speed system is used, and only when it reaches the limit is the aeration system used. However, for large bioreactors, the regulation of agitator speed is costly, and only the aeration rate regulation is used.

15.3.3.3 Adaptive Control
More often, the process characteristics are not known precisely or change with time as living cells go through different life cycles, and therefore, the operating conditions, including the controller parameters and set points, may have to be changed during

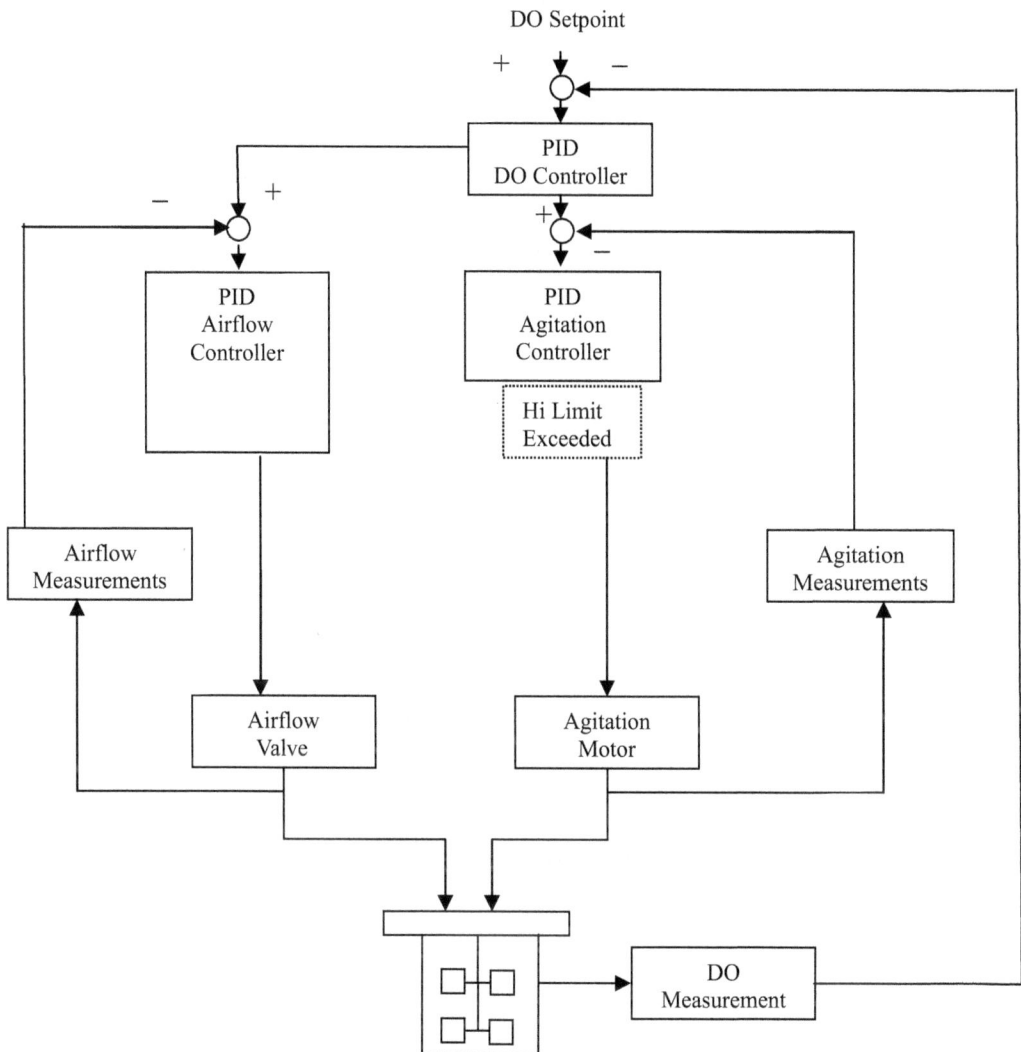

Figure 15.5. A cascade control.

the fermentation period. Adaptive control schemes adjust the controller parameters automatically to compensate for variations in the process characteristics. As shown in Figure 15.6, a typical adaptive control system consists of two loops, one to identify the changes in the process (identification) from the input and output data and another to carry out the loop optimization (optimizer) to reset the controller parameters.

There are two classes of adaptive control: *a programmed adaptive control*, for processes whose changes can be either measured or anticipated so that the controller settings can be adjusted systematically based on the measured or anticipated process changes, and a *self-tuning controller*,[63] for processes whose changes cannot be measured or predicted so that the adaptive control is implemented in a feedback manner on-line through computer control.

As shown in Figure 15.6, the process identification scheme estimates the process parameter values in the process model using the input–output data acquired on-line and supplies these parameter values to the optimization scheme, which determines

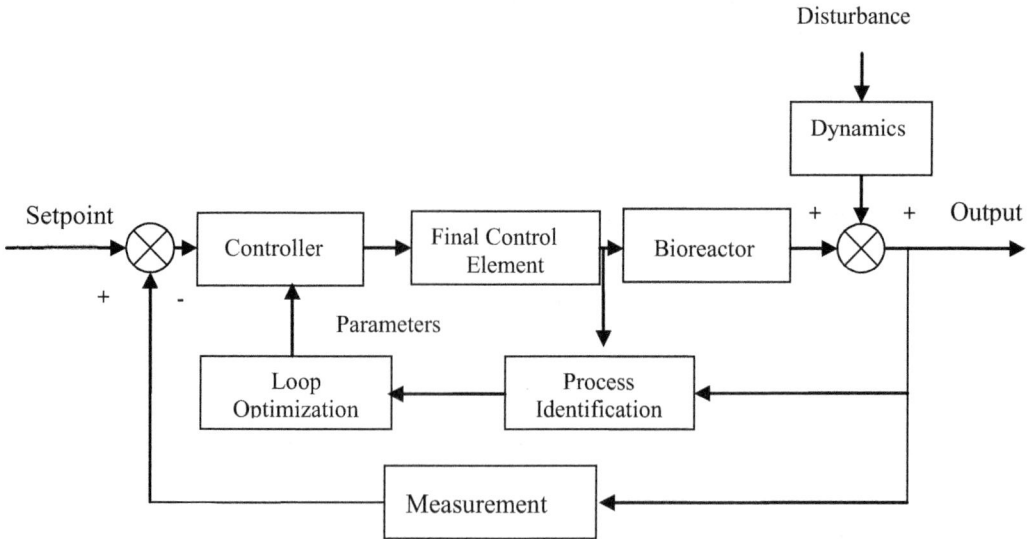

Figure 15.6. An adaptive control.

the optimum controller parameter settings based on the new process parameter values, and the controller settings in the feedback loop are automatically adjusted by the loop optimizer. In this self-tuning or self-adaptive scheme, most often, an external forcing function that excites the process most effectively is introduced to obtain the best process model parameters.

15.4 Indirect Feedback Control

Indirect feedback control refers to the situation in which a parameter that is related to the overall performance of fed-batch fermentation is measured and feedback is controlled to achieve the purpose. Slow dynamics associated with microbial processes should be taken into account so that rapidly responding and readily measurable parameters are selected so that control action can be applied in anticipation of the slow dynamic response.

15.4.1 Carbon Dioxide Evolution Rate

Assimilation of a carbon source by microorganisms is always accompanied by the evolution of carbon dioxide, and therefore, the carbon source feed rate based on CER allows regulation of the concentration of the carbon source in media. The idea behind the use of CER is to take advantage of the fact that CER responds rapidly to various inputs that cause dynamic changes in microbial processes and is also a readily measurable parameter so that normally sluggish control based on a slowly responding measurement, say, the cell concentration, which is usually slow in response, can be made fast by relying on the fast-responding CER. In this way, the feedback control is rapid without much dynamic delay and results in a superior and better performance.[61] In that sense, this is a form of feedforward control.

CER is computed by applying a mass balance on the inlet and outlet gas streams. Under regulated pH, the inlet and outlet concentrations of carbon dioxide are measured by an infrared CO_2 analyzer or a mass spectrometer, and the inlet and outlet gas flow rates are also measured. The carbon dioxide evolution rate is simply the difference between the outlet rate of carbon dioxide and the inlet carbon dioxide supply rate.

15.4.2 Specific Growth Rates

In the literature, many authors have developed control algorithms based on maintaining the specific growth rate constant throughout the fermentation, although we have shown in Chapter 12 that even if the cell mass is the product, keeping the specific growth rate at the maximum value is not optimal unless the cell mass yield coefficient is constant. When the cell mass yield coefficient is a function of the limiting substrate concentration, the optimal policy initially favors the specific growth rate but gradually favors the yield, and thus, the substrate concentration starts near where specific growth rate is maximum but slowly changes to favor the yield at the expense of the specific rate.

The specific growth rate μ has to be either calculated on-line using the definition or estimated using a readily measurable parameter or estimated using the Kalman filter or EKF. The specific growth rate may be obtained from the cell mass balance,

$$\frac{d(XV)}{dt} = \mu XV \Rightarrow \frac{1}{XV}\frac{d(XV)}{dt} = \frac{d(\ln XV)}{dt} = \mu \qquad (15.37)$$

or

$$\ln[(XV)_t/(XV)_{t-\Delta t}] = \int_{t-\Delta t}^{t} \mu(S)d\tau = \overline{(\mu)}\Delta t = \mu\Delta t \qquad (15.38)$$

where $\overline{\mu}$ is the mean value of μ so that when the substrate concentration is approximately constant in the exponential growth phase, $\overline{\mu} = \mu$, the specific growth rate is equal to the temporal slope of the natural logarithm of exponentially growing cell mass concentration:

$$\mu(t) = [\ln(XV)_t - \ln(XV)_{t-\Delta t}]/\Delta t \qquad (15.39)$$

The current value of the specific growth rate may be obtained from the finite difference of the slope of the plot of $\ln(XV)$ versus time.

15.5 Optimal Control

As we have seen in Chapters 12 and 13, the equation-based model optimization results in an optimal feed rate profile that is a concatenation of maximum, minimum, and singular feed rates and is therefore open loop, that is, a function of time. Only for low-order processes and under certain conditions can we obtain an analytical form of the singular feed rate in terms of the state variables (X, S, P, and V), and therefore, the optimal feed rate is in closed-loop, that is, feedback, mode. For processes modeled by more than four mass balance equations, it is not possible to obtain analytical solutions (closed loop); instead, numerical solutions for the feed rates that are functions of time are obtained.

15.5.1 Optimal Open-Loop Control

The numerical solutions of feed rates are called *open loop* because no information about the current status of the process (state variables) is fed back for implementing the optimal feed rate. Only when the process can be modeled by four or fewer mass balance equations under certain conditions[1] can the optimal feed rate be obtained in analytical form involving the state variables and specific rates, and these feed rates are called *closed loop* (feedback) because the current status of the process is fed back for implementation.

The optimal open-loop solution is computed numerically for a particular set of initial conditions and process models with known specific rates. Consequently, the solution depends heavily on the initial conditions and the process parameter values. Therefore, errors in the initial conditions and system parameters can lead to inferior results. The disadvantages of an open-loop optimization and control are well known as there is no means of accounting for the uncertainties and disturbances. The open-loop feed rate profile simply provides numerical values of the feed rate as a function of time. Therefore, the open-loop feed rate profile provides only a time-variant set point to be followed, regardless of the outcome of the applied feed rate. Conversely, a closed-loop (feedback control) feed rate attenuates errors in system parameter values and is independent of the initial conditions by feeding back the current status of the process.

Thus, one has no choice but to implement the optimal open-loop feed rate. To account for the uncertainties in the model parameter values, error in initial conditions, and unknown disturbances, the open-loop policy may be implemented in conjunction with an adaptive scheme. In other words, the optimal open-loop feed rate may be applied in cycle-to-cycle or on-line adaptive optimization schemes. This approach is covered in detail in Chapter 14.

15.5.2 Optimal Closed-Loop (Feedback) Control

The optimal feed rate is a concatenation of minimum, maximum, and singular feed rate periods. The exact sequence and the times at which the feed rate switches from one mode to another must be computed numerically. When the process to be optimized is fourth order at most, the optimization covered in Chapters 12 and 13 leads to an optimal singular feed rate that is in closed-loop form, requiring the measurements of state variables $S, X, P,$ and V and knowledge of specific rates $\sigma, \mu,$ and π. Thus, one must be able to measure the state variables $S, X, P,$ and V and calculate the specific rates $\sigma, \mu,$ and π. If it is not possible to measure one or more of the state variables or to calculate $\sigma, \mu,$ and π, then one has to be able to estimate the difficult-to-measure state variables and specific rates. Then, it is possible to implement the singular closed-loop feed rate in a feedback scheme. The optimal feed rate profile is a concatenation of maximum, F_{max}, minimum, $F_{min} = 0$ (batch), and singular, F_{sin} and $F_b = 0$, feed rates. Therefore, in addition to the singular feed rate, one has to be able to implement in feedback mode (in terms of state variables) the switching times between two different modes of feed rates: between two extremes, $F_{min} \to F_{max}$ and $F_{max} \to F_{min}$, between extreme and singular, $F_{max} \to F_{sin}$ and $F_{min} \to F_{sin}$, and between singular and batch, $F_{sin} \to F_b = 0$. Then, a complete feedback control scheme can be established. The measurements and estimation schemes for the state variables are covered earlier in this chapter. However, we must consider switching times in feedback manner, that is, in terms of state variables. Therefore, we consider the periods of singular feed rate (interior singular arc) and boundary control (boundary arc).

The switching between the singular mode and batch mode, $F_{sin} \to F_b = 0$, is initiated when the bioreactor volume is full, $V = V_{max}$. Therefore, by monitoring the bioreactor volume (a state variable) V to reach the maximum value, the switching takes place, $F_b = 0$. In general, for processes described (Eq. (12.8) or (13.8)) and the performance given by Eq. (12.9) or Eq. (13.10), the boundary control is given by $dh/dt = dx_3/dt = F_b = 0$. This is a feedback control as it applies when one of the state variables, $x_3 = V$, reaches the boundary $V = V_{max}$.

Switching between two extremes, $F_{min} \to F_{max}$ and $F_{max} \to F_{min}$, and between extreme and singular, $F_{max} \to F_{sin}$ and $F_{min} \to F_{sin}$, is usually obtained numerically as a function of time, and feedback realization requires obtaining hypersurfaces (switching curves for two-dimensional problems, switching surfaces for three-dimensional problems, switching volumes for four-dimensional problems, etc.). The task of obtaining these hypersurfaces is extremely difficult as it requires repeated numerical solutions covering the entirety of initial conditions and imbedding the numerically obtained times into the state space. Therefore, interested readers are referred to the literature[65] dealing with a low-order model. At any rate, for low-order models (second or third order), it is possible to obtain, under certain conditions, analytical solutions of switching curves and switching surfaces. In general, at this stage of development, a combination of open-loop and closed-loop optimum solutions is practical.

REFERENCES

1. Blakebrough, N., McManamey, W. J., and Tart, K. R. 1978. Rheological measurements on *Aspergillus niger* fermentation systems. *Journal of Applied Chemistry and Biotechnology* 28: 453–461.
2. Perley, C. R., Swartz, J. R., and Cooney, C. L. 1979. Measurements of cell concentration with a continuous viscosimeter. *Biotechnology and Bioengineering* 21: 519–513.
3. Langer, G., and Werner, U. 1981. Measurements of viscosity of suspensions in different viscosimeter flows and stirring systems. *German Chemical Engineering* 4: 226–241.
4. Kemblowski, Z., Kristiansen, B., and Alayi, O. 1985. On-line rheometer for fermentation liquids. *Biotechnology Letters* 7: 803–808.
5. Lee, C., and Lim, H. 1980. New device for continuously monitoring the optical density of concentrated microbial cultures. *Biotechnology and Bioengineering* 22: 639–642.
6. Monitek, Mösenbroicher Weg 200, D-2000 Düsseldorf 30, Germany.
7. Aquasant Measuring Technique Ltd, Hauptstrasse 22/4416 Bubendorf Switzerland.
8. Lee, Y. T., and Tsao, G. T. 1979. Dissolved oxygen electrodes. *Advances in Biochemical Engineering* 13: 35–86.

9. Nestaas, E., and Wang, D. I. C. 1981. A new sensor, the "filtration probe," for quantitative characterization of the penicillin fermentation. I. Mycelial morphology and culture activity. *Biotechnology and Bioengineering* 23: 2803–2813.

10. Nestaas, E., and Wang, D. I. C. 1981. A new sensor, the "filtration probe," for quantitative characterization of the penicillin fermentation. II. The monitor of mycelial growth. *Biotechnology and Bioengineering* 23: 2815–2824.

11. Nestaas, E., and Wang, D. I. C. 1983. A new sensor, the "filtration probe," for quantitative characterization of the penicillin fermentation. III. An automatically operating probe. *Biotechnology and Bioengineering* 25: 1981–1987.

12. Thomas, D. C., Chittur, Y. K., Cagney, J. W., and Lim, H. C. 1985. On-line estimation of mycelial cell mass with a computer interfaced filtration probe. *Biotechnology and Bioengineering* 27: 729–742.

13. Reuss, M., Boelcke, C., Lenz, R., and Peckmann, U. 1987. A new automatic sampling device for determination of filtration characteristics of biosuspensions and coupling of analyzers with industrial fermentation processes. *BTF-Biotech-Forum* 4: 2–12.

14. Hitchman, M. L. 1978. *Measurement of Dissolved Oxygen*. John Wiley.

15. Van Hemert, P., Kilburn, D. D., Righelato, R. C., and Van Wezei, A. L. 1969. A steam-sterilizable electrode of galvanic type for the measurement of dissolved oxygen. *Biotechnology and Bioengineering* 11: 549–560.

16. Aiba, S., Nagai, S., and Nishizawa, Y. 1976. Fed batch culture of *Saccharomyces cerevisiae*: A perspective of computer control to enhance the productivity in baker's yeast cultivation. *Biotechnology and Bioengineering* 18: 1001–1016.

17. Wang, H. Y., Cooney, C. L., and Wang, D. I. C. 1977. Computer-aided baker's yeast fermentations. *Biotechnology and Bioengineering* 19: 69–86.

18. Wang, H. Y., Cooney, C. L., and Wang, D. I. C. 1979. Computer control of bakers' yeast production. *Biotechnology and Bioengineering* 21: 975–995.

19. Wetcott, C. C. 1978. *pH Measurements*. Academic Press.

20. Dahod, S. K. 1982. Redox potential as a better substitute for dissolved oxygen in fermentation process control. *Biotechnology and Bioengineering* 24: 2123–2125.

21. Kjaergaard, L. 1977. The redox potential: Its use and control in biotechnology. *Advances in Biochemical Engineering* 7: 131–149.

22. Akashi, K., Ikeda, S., Shibai, H., Kobayashi, K., and Hirose, Y. 1978. Determination of redox potential levels critical for cell respiration and suitable for L-leucine production. *Biotechnology and Bioengineering* 20: 27–41.

23. Radjai, M. K., Hatch, R. T., and Cadman, T. W. 1984. Optimization of amino acid production by automatic self-tuning digital control by redox potential. *Biotechnology and Bioengineering* 14: 657–679.

24. Thompson, B. G., and Gerson, D. F. 1985. Electrochemical control of redox potential in batch cultures of *Escherichia coli*. *Biotechnology and Bioengineering* 27: 1512–1515.

25. Fiechter, A., Meiners, H., and Sukatsch, D. S. 1982. *Biologische regulation und prozeßführung in handbuch der biotechnologie*, ed. Präve, P., et al., p. 173. Akademische Verlagsgesellschaft.

26. Yellow Springs Instruments International, Yellow Springs, Ohio.

27. Bradley, J., Anderson, P. A., Dear, A. M., Ashby, R. E., and Turner, A. P. F. 1988. Glucose biosensors for the study and control of bakers compressed yeast production. Fourth International Conference on Computer Applications in Fermentation, Cambridge, MA. Sept. 25–29.

28. *Electrode with Interchangeable Enzyme GLUC1*. 1988. Taccusel.

29. Hakuma, M., Obana, H., Yasuda, T., Karube, I., and Suzuki, S. 1980. Amperometric determination of total assimilable sugars in fermentation broths with use of immobilized whole cells. *Enzyme Microbial Technology* 2: 234–238.

30. Tran, M. D., Romette, J. L., and Thomas, D. 1983. An enzyme electrode for specific determination of L-lysine: A real-time control sensor. *Biotechnology and Bioengineering* 25: 329–340.

31. Yamauchi, H., Kusakabe, H., Midorikawa, Y., Fujishima, T., and Kuninaka, A. 1984. Enzyme electrode for specific determination of L-glutamate. Third European Congress on Biotechnology, Munich, Germany. Sept. 10–14.

32. Wollenburger, U., Scheller, F. W., Bömer, A., Passarge, M., and Müller, H. G. 1989. A specific enzyme electrode for L-glutamate development and applications. *Biosensors* 4: 381–391.

33. Fatibello-Filho, O., Suleiman, A., and Guilbaut, G. 1989. Enzyme electrode for the determination of aspartate. *Biosensors* 4: 313–321.

34. Renneberg, R., Riedel, K., and Scheller, F. 1985. Microbial sensor for aspartame. *Applied Microbiology Biotechnology* 21: 180–181.

35. Mascini, M., Mosocne, D., Palleschi, G., and Pilloton, R. 1988. In-line determination of metabolites and milk components with electrochemical biosensors. *Analytica Chimica Acta* 213: 101–110.

36. Se'chaud, F., Penguin, S., Coulet, P., and Bardeletti, G. 1989. Fast and reliable organic acid determination in fermented milk using an enzyme-electrode based analyzer. *Process Biochemistry* 33: 33–38.

37. Microzym L, Sétric Génie Industriel, Toulouse, France.

38. Karube, I., Tamiya, E., Sode, K., Yokoyama, K., Kitakawa, Y., Suzuki, H., and Asano, Y. 1988. Application of microbiological sensors in fermentation process. *Analytica Chimica Acta* 213: 69–77.

39. Matsunaga, T., Karube, I., and Suzuki, S. 1980. A specific microbial sensor for formic acid. *European Journal of Applied Microbiology and Biotechnology* 10: 235–242.

40. Enfors, S. O., and Nilsson, H. 1979. Design and response characteristics of an enzyme electrode for measurement of penicillin in fermentation broth. *Enzyme Microbial Technology* 1: 260–264.

41. Matsumoto, K., Seijo, H., Watanabe, T., Kaube, I., Satoh, I., and Suzuki, S. 1979. Immobilized whole cell-based flow-type sensor for cephalosporins. *Analytica Chimica Acta* 105: 429–432.

42. Karube, I. 1986. Microbial sensors for process and environmental control. *ACS Symposium Series, Fundamentals and Applications of Chemical Sensors* 309: 330–348.

43. Pons, M. N. 1991. *Measurements of Biological Variables in Bioprocess Monitoring and Control*, ed. Pons, M. N. Hanser.

44. Meschner, K. 1984. An automated nephelometric system for evaluation of the growth of bacterial cultures. *Analytica Chimica Acta* 163: 85–90.

45. Cox, R. P., Miller, M., Nielson, M., and Thomsen, J. K. 1989. Continuous turbidimetric measurements of microbial cell density in bioreactors using a light emitting diode and a photodiode. *Journal of Microbiological Methods* 10: 25–51.

46. Ensari, S., and Lim, H. C. 2003. Apparent effects of operational variables on continuous culture of *Corynebacteriun lactofermentum*. *Process Biochemistry* 38: 1531–1538.

47. Heinzle, E., and Dunn, I. J. 1991. Methods and instruments in fermentation gas analysis. *Biotechnology* 4: 30–74.

48. Moo-Young, M., and Blanch, H. W. 1981. Design of biochemical reactors: Mass transfer criteria for simple and complex systems. *Advances in Biochemical Engineering* 19: 1–69.

49. Cooney, C. L., Wang, H. Y., and Wang, D. I. C. 1977. Computer-aided material balancing for prediction of fermentation parameters. *Biotechnology and Bioengineering* 19: 55–67.

50. Wang, H. Y., Cooney, C. L., and Wang, D. I. C. 1977. Computer-aided baker's yeast fermentations. *Biotechnology and Bioengineering* 19: 69–86.

51. Constantinides, A., and Shao, P. 1981. Material balancing applied to the prediction of glutamic acid production and cell mass formation. *Annals of the New York Academy of Science* 369: 167–180.

52. Swartz, J. R., and Cooney, C. L. 1979. Indirect fermentation measurements as a basis for control. *Biotechnology and Bioengineering Symposium* 9: 95–101.

53. Wang, H. Y., Cooney, C. L., and Wang, D. I. C. 1977. Computer-aided baker's yeast fermentations. *Biotechnology and Bioengineering* 19: 69–86.

54. Kalman, R. E. 1960. A new approach to linear filtering and prediction problems. *Journal of Basic Engineering* 82: 35–45.

55. Seinfeld, J. H., Gavalas, G. R., and Hwang, M. 1969. Control of nonlinear stochastic systems. *Industrial Engineering Chemistry Fundamentals* 8: 257–262.

56. Seinfeld, J. H. 1970. Optimal stochastic control of nonlinear systems. *American Institute of Chemical Engineering Journal* 16: 1016–1022.

57. Lee, J. H., Lim, H. C., and Hong, J. 1997. Application of nonsingular transformation to on-line optimal control of poly-β-hydroxybutyrate fermentation. *Journal of Biotechnology* 55: 135–150.

58. Lee, J. H., Lee, Y. W., and Yoo, Y. J. 1992. A simulation study of two-stage fed-batch culture for optimization and control of PHB production. *Korean Journal of Applied Microbiology and Biotechnology* 20: 668–676.

59. Stephanopoulos, G. 1984. *Chemical Process Control: Introduction to Theory and Practice*. PTR Prentice Hall.

60. Ziegler, J. G., and Nichols, N. B. 1942. Optimum settings for automatic controllers. *Transactions of the ASME* 64: 759–768.

61. Perry, R. H., and Green, D., eds. 1984. *Perry's Chemical Engineer's Handbook*. 6th ed. McGraw-Hill.

62. Cohen, G. H., and Coon, G. A. 1953. Theoretical consideration of related control. *Transactions of the ASME* 75: 827–834.

63. Astrom, K. J., and Wittenmark, B. 1988. *Adaptive Control Systems*. Addison-Wesley.

64. Chang, Y. K., and Lim, H. C. 1990. Fast inferential adaptive optimization of continuous yeast culture based on carbon dioxide evolution rate. *Bioengineering and Biotechnology* 3: 8–14.

65. Modak, J. M., and Lim, H. C. 1987. Feedback optimization of fed-batch fermentation. *Biotechnology and Bioengineering* 30: 518–540.

16 Feasibility Assessment and Implementable Feed Rates

In previous chapters on optimization, we saw that once a process model in the form of mass balance equations is developed, it can be optimized using, among other methods, Pontryagin's maximum principle (PMP), or when a statistical model such as a neural network model is available, it can be optimized using a method of optimization that is most appropriate for the model.

As discussed in Chapter 4, various physical and chemical phenomena favor fed-batch operations. For physical reasons, such as the need to use auxotrophic mutants, attainment of high cell and metabolite concentrations, and alleviation of high viscosity, the use of fed-batch is obvious and intuitive, whereas for chemical reasons, such as substrate inhibition, glucose effects, and catabolite repressions, the use of fed-batch operation requires recognition of *nonmonotonic specific rates* such as cell growth, substrate consumption, and product formation, μ, σ, and π. In other words, at least one of the specific rates must exhibit a maximum with respect to the substrate concentration.

In this chapter, we go over a strategy of quickly assessing if a fed-batch operation is superior. In previous chapters dealing with the optimizations of cell mass, metabolite, and recombinant cell products (Chapters 12 and 13), the mass balance equation models required specific rates, in particular, the dependence on substrate and/or product concentration of specific rates of cell growth, substrate consumption, and product formation, μ, σ, and π. Indeed, the sufficient conditions for potential advantage of fed-batch operations over other forms of reactor operation are that one or more specific rates must be a nonmonotonic function of substrate and/or product concentration, exhibiting a maximum. Intuitively, it is clear that when one or more of the specific rates is nonmonotonic, one can manipulate the substrate concentration to maximize the specific rates. Looking from another angle, the nonmonotonic nature of the specific rates implies that the yield coefficients either are constants or vary with the substrate concentration. Therefore, it is essential to investigate the dependence of specific rates on substrate and/or product concentration. It is also intuitively clear that if all specific rates are monotonic, that is to say, they continuously increase and approach asymptotically saturation levels with the substrate concentration, then there is no obvious kinetic advantage to fed-batch operation, other than the physical advantages stated earlier, such as the high cell density that can be obtained with a fed-batch operation or the need to alleviate a high-viscosity effect during the course of bioreactor operation.

16.1 Estimation of Specific Rates

The kinetic information, in particular, the specific rates of cell formation, product formation, and substrate consumption, μ, π, and σ, is essential to characterize various reactor operations and is usually obtained from batch, continuous, or fed-batch operations. However, because the ultimate purpose is to assess the advantage of a fed-batch operation, it is best to generate the kinetic information using the fed-batch operation. However, batch operations including shake-flask cultures are simple to perform and can be used to expedite the time required to assess the feasibility of fed-batch operation. Fed-batch operation is a dynamic operation, and kinetic information under a dynamic situation would be more appropriate than steady state operations of a continuous culture. Chemical reactions are usually well defined so that the data from any type of reactor (batch, semi-batch, or continuous) can be interchangeably used. However, biological processes with living cells not only involve an extremely large number of reactions but are also so complex that mathematical models built around a limited number of key reactions are incapable of predicting biological processes over a wide range of process operation. Therefore, the data obtained in one reactor operating under conditions of limited range are not capable of predicting the behavior over the range beyond what is used to obtain the data, let alone the operation in other reactor types. In other words, the reactor data obtained from one type of reactor do not necessarily represent well the reactor operation of other types. Hence, the kinetic information obtained from one type of reactor operation, say, a steady state continuous reactor operation, does not necessarily represent well that of another reactor operation, say, a fed-batch operation. Therefore, it is best to obtain the kinetic information from a fed-batch operation. However, preliminary information can be obtained quickly and conveniently from a large number of batch operations such as shake-flask cultures. Therefore, it is convenient to begin with flask cultures.

16.1.1 Shake-Flask and Batch Experiments

The kinetic information can be obtained quickly from a large number of shake-flask experiments, and the data can be analyzed and used to obtain semiquantitative estimates of the dependence on substrate concentration of the specific rates of cell growth, substrate consumption, and product formation, $\mu(S)$, $\sigma(S)$, and $\pi(S)$. In the course of normal process development, a large number of shake-flask cultures are usually run, and this information is available anyway. In Chapter 8, we covered in some detail the methods of obtaining the specific rates as functions of substrate concentration or both the substrate and product concentrations. Therefore, this information is not repeated here.

If the flask cultures indicate that none of the specific rates exhibit a maximum at a finite substrate concentration, that is, they are monotonic, then there is no apparent chemical advantage to fed-batch operation other than the physical grounds mentioned earlier. However, in many situations, steady state continuous operations have been found to be ineffective for metabolite production and are also subject to contamination and therefore are very seldom used industrially, with the exception of single-cell protein productions, certain beer-making processes, and wastewater treatments.

We can summarize the steps to assess the feasibility of using fed-batch operations. In general, we can classify fermentation processes to (1) processes whose fed-batch operation can be modeled by mass balances of key components such as the cell mass, substrate, product, and intermediates, and (2) processes for which we are not able to complete mass balances. For the former, we need to develop a model based on mass balances of key components with known dependence of specific rates on the substrate concentration or substrate and product concentrations so that it can be optimized using a technique such as PMP. For the latter, we cannot complete the mass balances for the key components, and therefore, only qualitative information is needed in the early stage of process development, and a statistical technique, such as a neural network, can be used for modeling and optimization.

We begin with shake-flask studies:

1. Carry out a number of shake-flask experiments in which the initial substrate concentration is varied over a wide range and the specific rates are determined during the exponential growth phase to assess the dependence of specific rates on substrate concentration or substrate and product concentrations.
2. Estimate the specific rates as detailed in Chapter 8. Obtain their dependence on substrate concentration or substrate and product concentrations. If none of the specific rates show a maximum with respect to substrate concentration, that is, they are monotonic (case G in Figure 12.11), then a fed-batch operation is not advantageous, unless there are physical reasons that favor fed-batch operations, such as the need to obtain a high cell density operation or to avoid high-viscosity culture. If one or more of specific rates is found to be dependent on the initial substrate concentration, then additional shake-flask experiments should be planned around the substrate concentration at which the specific rates peak.

When a fed-batch operation is found to be feasible, then more carefully planned experiments may be carried out to obtain quantitative information on specific rates by running controlled (dissolved oxygen, pH, temperature, etc.) fed-batch operations in which the substrate feed rate is varied over a range around the maximum rate and the corresponding time profiles of cell, product, and substrate concentrations are recorded.

16.1.2 Fed-Batch Operations

Once the fed-batch operation is found to be feasible, it should be carried out in the range of substrate concentration around the maximum specific rate using the method discussed in detail in Chapter 8.

The method of parameter estimation was covered in detail in Chapter 6. Therefore, it is not repeated here. When there are some state variables that are impossible or difficult to measure, one can use the estimation scheme presented there (e.g., the extended Kalman filter). In this way, not only the difficult-to-measure state variables but also the parameters in the model can be estimated.

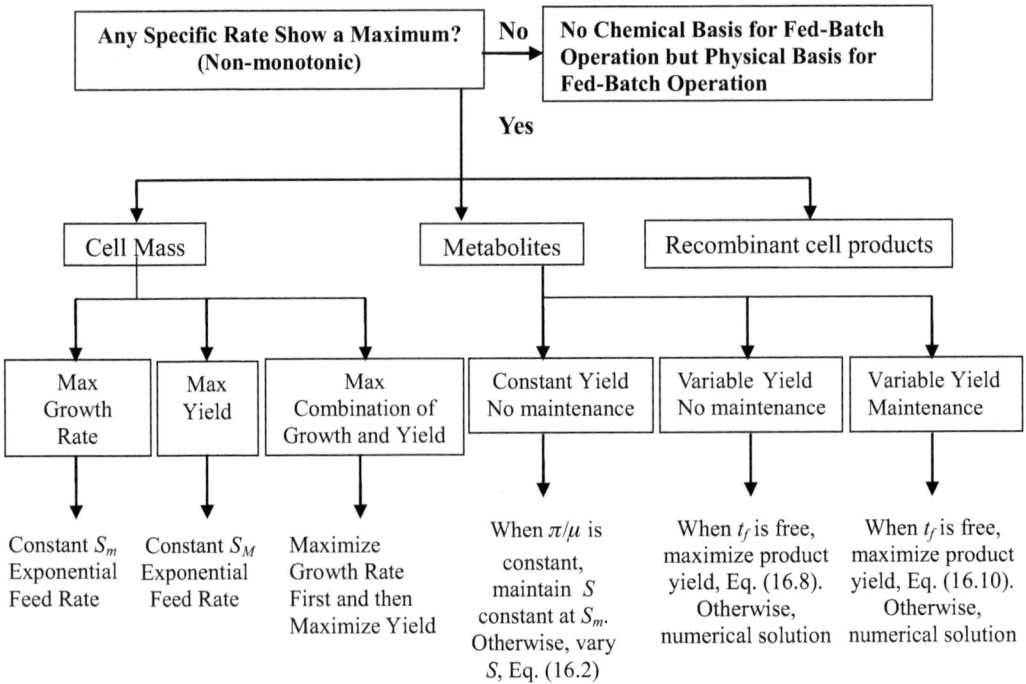

Figure 16.1. Flow chart of optimal–suboptimal feed rate policies.

Once the specific rates are determined as a function of substrate concentration alone or substrate plus product concentrations, the entire time profiles of concentrations of substrate, cell mass, and product are used with the model presented in Chapter 6 to see if the model with the specific rates fits the experimental time profiles. If not, the parameters in the specific rates are adjusted to fit the time profiles. It should be noted here that once a model that fits the experimental data is obtained, its ability to predict outcomes under different experimental conditions should be tested.

16.2 Sequential Approach to Feasibility Assessment

We now consider the important task of determining if a fed-batch culture operation is feasible and a step-by-step procedure to optimize the fed-batch operation to produce, first, cell masses and, second, metabolites. A schematic diagram is given in Figure 16.1. First, we must confirm nonmonotonic specific rates of cell growth, substrate consumption, and product formation with respect to substrate and/or product concentration, that is, at least one of the specific rates must show a maximum.

If none of the specific rates show a maximum, there is no chemical basis for fed-batch operation, and one must consider physical phenomena that favor fed-batch operation, as presented in Chapter 4. A number of physical factors favor fed-batch operation such as the necessity to utilize auxotrophic mutants, to have high-density cell and metabolite concentrations, to extend the operational time, to alleviate high viscosity, to make up for lost water by evaporation, and to obtain better plasmid stability of recombinant cells.

If at least one specific rate shows a maximum with respect to the substrate concentration or substrate and product concentrations, there is a chemical basis that favors fed-batch operation. Therefore, we proceed to a step-by-step procedure to obtain the optimal or suboptimal feed rate strategy. Given in the following are brief notes to identify the existence of singular regions and a qualitative description of how the substrate concentration ought to be varied during the singular period:

1. For cell mass production, using the specific growth rate μ and the yield coefficient $Y_{X/S} = \mu/\sigma$, we identify the appropriate case (cases A–G) in Figure 12.11, which is duplicated here as Figure 16.2, and identify the singular region in the substrate concentration to determine the range over which the substrate concentration should be varied or held constant.

    For example, consider case D (both μ and $Y_{X/S}$ are nonmonotonic); the substrate concentration should start at S_m (at which μ is maximum) and should decrease toward S_Y (at which $Y_{X/S}$ is maximum). The time rate of the change in substrate concentration is not precisely known and must be determined, if desired, by the numerical procedure presented in Chapter 12. This policy would favor cell growth initially and, eventually, the cell mass yield.

2. For metabolite production, singular regions are not well defined, except that provided by the limited sufficient conditions listed in Table 13.1, which are replicated here as Table 16.1.

    Using the notation of a single hat to denote the first partial derivative, $(\ )^\wedge = \partial (\ )/\partial \ln(S_F - S) + \partial (\ )/\partial \ln P$, it is clear that when $\hat{\pi} < 0\ \hat{\sigma} > 0$, $\hat{\mu} < 0$ and $\hat{\sigma} < 0$, $\hat{\mu} > 0$ and $\hat{\pi} > 0$, there is no singular region. This implies that there is no kinetic advantage to fed-batch operation. Only when $\hat{\sigma}$, $\hat{\mu}$ and $\hat{\pi}$ are all positive or all negative and the constraints listed in Table 16.1 (numbers 1, 2, 7, and 8) are met are there are singular regions. By calculating $\sigma, \mu, \pi, \hat{\sigma}, \hat{\mu}, \hat{\pi}$, $\hat{\hat{\sigma}}, \hat{\hat{\mu}}$, and $\hat{\hat{\pi}}$ as functions of substrate concentration, one can check against the constraints listed in Table 16.1, where the second partial derivatives are defined as $(\ )^{\wedge\wedge} = \partial^2(\ )/[\partial \ln(S_F - S)]^2 + \partial^2(\ )/(\partial \ln P)^2$. However, in most cases, it may be necessary to resort to the numerical procedures given in Chapter 13.

3. For recombinant cell processes, assuming that the simple model used in Chapter 13 holds, one can assess the region of singularity by referring to Figure 13.3 and by considering the specific growth rate of the plasmid-containing cells, μ^+, and the ratio of specific growth rates of plasmid-containing cells to plasmid free cells, μ^+/μ^-. The procedure is the same as that for cell mass production. If a precise solution is desired, the numerical procedure given in Chapter 13 should be followed.

For those recombinant product processes that cannot be described by the model presented in Chapter 13, it is necessary to resort to a numerical procedure to determine the optimal feed rate profiles as there are no rule-of-thumb methods to identify the singular regions.

When it is not possible to develop a model based on mass balance equations for the key elements, the process optimization based on an equation-based mathematical model is not applicable, and one must resort to a statistical approach such as statistical

Table 16.1. *Sufficient conditions for existence of singular arc for metabolite production*

$\hat{\sigma}$	$\hat{\mu}$	$\hat{\pi}$	Constraints	Singular arc	No.
+	+	+	$\hat{\sigma}/\sigma = \hat{\mu}/\mu > \hat{\pi}/\pi > 0$ or $\hat{\sigma}/\sigma > \hat{\mu}/\mu \geq \hat{\pi}/\pi > 0$	Yes	1
+	+	+	$\hat{\sigma}/\sigma \geq \hat{\mu}/\mu \geq \hat{\pi}/\pi > 0$, $\hat{\hat{\sigma}}/\hat{\sigma} \geq \hat{\hat{\mu}}/\hat{\mu} > \hat{\hat{\pi}}/\hat{\pi}$, except $\hat{\sigma}/\sigma = \hat{\mu}/\mu = \hat{\pi}/\pi$ and $\hat{\hat{\sigma}}/\hat{\sigma} = \hat{\hat{\mu}}/\hat{\mu} = \hat{\hat{\pi}}/\hat{\pi}$	Yes	2
+	+	+	$0 > \hat{\sigma}/\sigma \leq \hat{\mu}/\mu \leq \hat{\pi}/\pi$ or $\hat{\hat{\sigma}}/\hat{\sigma} \leq \hat{\hat{\mu}}/\hat{\mu} \leq \hat{\hat{\pi}}/\hat{\pi}$, except, $\hat{\sigma}/\sigma = \hat{\mu}/\mu = \hat{\pi}/\pi$ and $\hat{\hat{\sigma}}/\hat{\sigma} = \hat{\hat{\mu}}/\hat{\mu} = \hat{\hat{\pi}}/\hat{\pi}$	No	3
+	+	+	$0 < \hat{\sigma}/\sigma \leq \hat{\pi}/\pi \leq \hat{\mu}/\mu$ or $\hat{\hat{\sigma}}/\hat{\sigma} \leq \hat{\hat{\pi}}/\hat{\pi} \leq \hat{\hat{\mu}}/\hat{\mu}$, except $\hat{\sigma}/\sigma = \hat{\mu}/\mu = \hat{\pi}/\pi$ and $\hat{\hat{\sigma}}/\hat{\sigma} = \hat{\hat{\mu}}/\hat{\mu} = \hat{\hat{\pi}}/\hat{\pi}$	No	4
−	−	−	$\hat{\pi}/\pi \leq \hat{\mu}/\mu \leq \hat{\sigma}/\sigma < 0$ or $\hat{\hat{\pi}}/\hat{\pi} \leq \hat{\hat{\mu}}/\hat{\mu} \leq \hat{\hat{\sigma}}/\hat{\sigma}$, except $\hat{\sigma}/\sigma = \hat{\mu}/\mu = \hat{\pi}/\pi$ and $\hat{\hat{\sigma}}/\hat{\sigma} = \hat{\hat{\mu}}/\hat{\mu} = \hat{\hat{\pi}}/\hat{\pi}$	No	5
−	−	−	$\hat{\mu}/\mu \leq \hat{\pi}/\pi \leq \hat{\sigma}/\sigma < 0$ or $\hat{\hat{\mu}}/\hat{\mu} \leq \hat{\hat{\pi}}/\hat{\pi} \leq \hat{\hat{\sigma}}/\hat{\sigma}$, except $\hat{\sigma}/\sigma = \hat{\mu}/\mu = \hat{\pi}/\pi$ and $\hat{\hat{\sigma}}/\hat{\sigma} = \hat{\hat{\mu}}/\hat{\mu} = \hat{\hat{\pi}}/\hat{\pi}$	No	6
−	−	−	$0 > \hat{\pi}/\pi \geq \hat{\mu}/\mu > \hat{\sigma}/\sigma$ or $0 > \hat{\pi}/\pi > \hat{\mu}/\mu = \hat{\sigma}/\sigma$	Yes	7
−	−	−	$0 > \hat{\pi}/\pi \geq \hat{\sigma}/\sigma \geq \hat{\mu}/\mu > 0$, $\hat{\hat{\sigma}}/\hat{\sigma} \geq \hat{\hat{\mu}}/\hat{\mu} > \hat{\hat{\pi}}/\hat{\pi}$, except $\hat{\sigma}/\sigma = \hat{\mu}/\mu = \hat{\pi}/\pi$ and $\hat{\hat{\sigma}}/\hat{\sigma} = \hat{\hat{\mu}}/\hat{\mu} = \hat{\hat{\pi}}/\hat{\pi}$	Yes	8
+	−	−		No	9
−	+	+		No	10

design or a neural network approach. Also, there are situations in which one can write mass balance equations for some components, while it is not possible to write mass balances for other key components. In these situations, one can apply a *hybrid neural network* approach. The neural network approach to modeling fed-batch culture was covered in Chapter 7. Once a neural network model is developed, the feed rate profile may be optimized using various optimization techniques.

The preceding and yet-to-be-presented general characteristics of feed rate profiles can be utilized to generate easily implementable suboptimal feed rate profiles. They can also be used to reduce greatly the computational burden in numerical computations involved in determining the optimal feed rate profiles as presented in Chapters 12 and 13.

After implementing a few cycles of the optimal feed rate profiles to optimize off-line fed-batch operations, the next logical sequence would be on-line adaptive optimization, as detailed in Chapter 14 and as summarized here:

1. Start a cycle-to-cycle optimization, as detailed in Chapter 14. Carry out a pilot plant study using the optimum open-loop or feedback feed rate obtained in Chapters 12 and 13. Obtain measurements of concentration profiles of cells, substrate, and produc t– the feed rate profiles.
2. Estimate the parameters in the model using the data obtained in step 1, and then determine the optimal feed rate using the updated model.
3. Implement the optimal feed rate determined in step 2.

4. If on-line optimization can be implemented, start on-line adaptive optimization, as presented in Chapter 14.

16.3 Implementable Optimal–Suboptimal Feed Rates

As we have seen in Chapters 12 and 13, in most cases, optimization criteria for various performance indices turn out to be maximization of specific growth rates or yield coefficients, or a combination of the two. For example, for cell mass maximization with a constant-yield coefficient, the optimal feed rate maximizes the specific growth rate by maintaining the substrate concentration constant at the value at which the specific growth rate is at its maximum. Hence, we determine the substrate concentration that maximizes the specific growth rate and apply an exponential feed rate to maintain the substrate concentration or apply a feedback control scheme to maintain the substrate concentration constant.

Throughout Chapters 12 and 13, we learned that the optimal feed rate profile is a concatenation of maximum, minimum, and singular feed rates. Although not theoretically proven, it is intuitively clear that the maximum and minimum feed rates are used to transfer as quickly as possible bioreactors to the singular arcs so that the singular feed rates can be used to carry out the optimization of a given performance index. Thus, heuristically, we can argue or propose a conjecture that the minimum and maximum feed rates are used to bring the state of bioreactors on the singular arc as soon as possible. Then, we can construct readily implementable optimal–suboptimal feed rates. In other words, if we can determine the initial conditions that place bioreactors on the singular arc, it would not be necessary to use the maximum, minimum, or a combination of the two, and instead the feed rate would be singular from the start, until the bioreactor is full, perhaps followed by a batch operation until the final time or conditions are met. It is the conjecture on which we can construct the optimal–suboptimal feed rates that can be implemented readily.

16.3.1 Cell Mass as Product

If the cell mass is the product, such as yeasts, or the intracellular components occupy fixed fractions of cell mass, then the optimal feed rate maximizes the specific growth rate μ for the case of a constant-yield coefficient or a weighted sum of the specific growth rate and the yield coefficient for the case of variable yield. Thus, we can determine the substrate concentrations that maximize the specific growth rate or the yield coefficient. To implement the feed rate to maintain the substrate concentration at a specific value that maximizes the specific growth rate or the yield coefficient, it is important to be able to measure the substrate concentration so that a feedback control scheme may be implemented. If on-line measurement is not possible, either an estimation scheme is needed or an open loop exponential feed rate can be implemented.

16.3.1.1 Specific Growth Rate Optimization
As treated extensively in Chapter 12, when the yield coefficient is constant, the optimal feed rate profiles for maximization of cell mass at the final time, maximum

cellular productivity, and minimum time problems are all exponential and maximize the specific growth rate, $\partial\mu/\partial S = 0$, by maintaining the substrate concentration constant, $\mu(S_m) = \mu_{max}$:

$$F_{sin} = \frac{\sigma(S_m)XV}{S_F - S_m} = \frac{\sigma(S_m)(XV)_0}{S_F - S_m}\exp[\mu(S_m)t] = \alpha_m\exp(\mu_{max}t) \quad (16.1)$$

Therefore, if it is possible to have an on-line substrate concentration measurement device, one can deploy a feedback control system based on the measurement of substrate concentration. An estimation scheme for specific growth rate in conjunction with a control scheme to maximize the specific growth rate may be used. It is also important to choose the best initial substrate concentration $S(0) = S_m$ corresponding to the maximum specific growth rate $\mu(S_m) = \mu_{max}$ so that the optimal feed rate (Eq. (16.1)) is used to maintain the substrate concentration at S_m for the entire operational time period.

16.3.1.2 Yield Coefficient Optimization

When the yield coefficient is a function of substrate concentration, the optimal feed rate profile during the singular period is semiexponential and varies the substrate concentration to maximize a weighted sum of the specific growth rate and cell mass yield coefficient. This conclusion holds for cell mass maximization, maximum cellular productivity, and minimum time problems. When the final time is relatively small, the optimal feed rate favors the specific growth rate, while for a long final time, it favors the yield coefficient. Thus, one may choose a performance index that takes into account the cost of operation (which is assumed to be proportional to the final time) that can be used to determine the weighting factor.

When the maintenance cost is insignificant relative to the product price, then we should pick the feed rate that maximizes the yield coefficient, $\partial Y_{X/S}/\partial S = 0$, $Y_{X/S}(S_Y) = Y_{X/S,max}$, as a suboptimal but readily implementable policy by maintaining the substrate concentration constant at the value that maximizes the yield coefficient, S_Y. Thus, one would choose the initial substrate concentration of $S(0) = S_Y$ and regulate the feed rate, which is exponential, to maintain the substrate concentration constant at S_Y:

$$F_{sin} = \frac{\sigma(S_Y)XV}{S_F - S_Y} = \frac{\sigma(S_Y)(XV)_0}{S_F - S_Y}\exp[\mu(S_Y)t] = \alpha_Y\exp[\mu(S_Y)t] \quad (16.2)$$

Once again, if one is able to measure on-line or is capable of reliably predicting the total cell mass, XV, a feedback control of the feed rate is possible. Thus, knowledge of the functional dependence of the specific growth rate and yield coefficient on substrate concentration is critical in implementing the suboptimal policies. It should be noted that the exponential rate may be less than or equal to the maximum specific growth rate, depending on whether $S_Y \neq S_m$ or $S_Y = S_m$.

16.3.1.3 Optimization of Specific Growth Rate and Yield Coefficient

When the cost of maintaining the reactor operation relative to the price of the product is significant, operational time should not be long, and therefore, it is important to maximize the rate. The feed rate that maximizes the specific growth rate, $\partial\mu/\partial S = 0$,

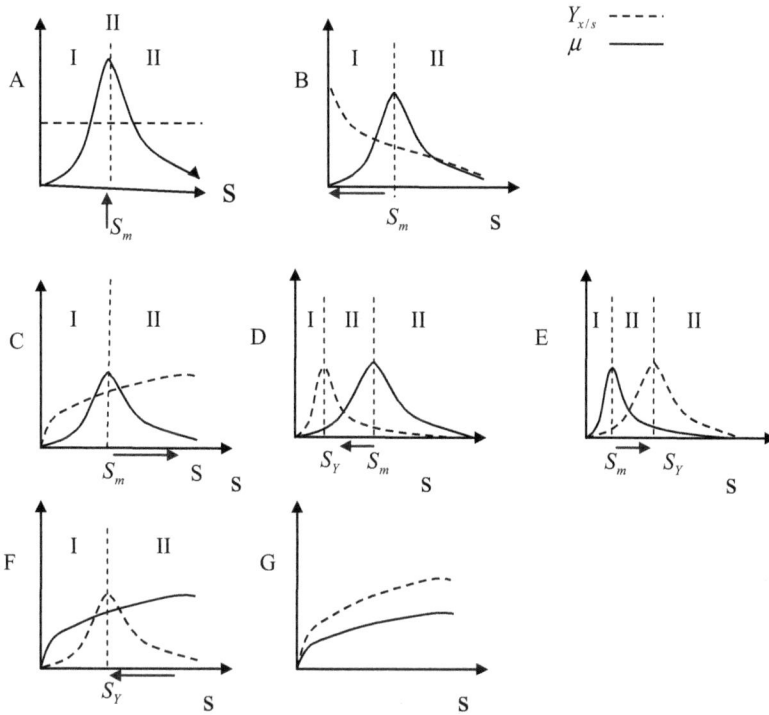

Figure 16.2. Singular regions (II) and variations in substrate concentration during singular interval for cell mass with fixed time and constant-yield coefficients. Solid arrows indicate the direction of the change in S with respect to time (same as Figure 12.11).

by maintaining the substrate concentration constant at the value that maximizes the specific growth rate, $[\partial\mu/\partial S]_{S=S_m} = 0$, can be chosen as the readily implementable policy of specific growth rate maximization. Of course, the initial substrate concentration must be S_m so that the feed rate is once again an exponential function:

$$F_{sin} = \frac{\sigma(S_m)XV}{S_F - S_m} = \frac{\sigma(S_m)(XV)_0 \exp[\mu(S_m)t]}{S_F - S_m} = \alpha_m \exp(\mu_{max}t) \quad (16.1)$$

Conversely, if the maintenance cost relative to the product price is insignificant, the operation time length is insignificant. Therefore, the yield should be maximized by maintaining the substrate concentration at S_Y, $[\partial Y_{X/S}/\partial S]_{S=S_Y} = 0$. The initial substrate concentration should be $S(0) = S_Y$, and the feed rate should maximize the yield. The feed rate to maintain the substrate concentration constant at $S = S_Y$ is an exponential function that differs in the preexponential factor and exponent from Eq. (16.1),

$$F_{sin} = \frac{\sigma(S_Y)XV}{S_F - S_Y} = \frac{\sigma(S_Y)(XV)_0 \exp[\mu(S_Y)t]}{S_F - S_Y} = \alpha_Y \exp[\mu(S_Y)t] \quad (16.2)$$

which can be an open-loop exponential feed or a feedback if XV can be measured or estimated on-line.

Suboptimal but easily implantable feed rate profiles can be generated based on Figure 12.11, which is duplicated here as Figure 16.2, in which the singular regions are characterized by specific growth rates and yield coefficients and the variations in substrate concentrations (increase or decrease) with time are indicated.

1. When the specific growth rate is nonmonotonic and the yield coefficient is constant (Figure 16.2A), the substrate concentration is held constant during the singular feed period at the value S_m, at which the specific growth rate is at its maximum. Therefore, one would choose the initial substrate concentration to be S_m and regulate the feed rate to maintain the substrate concentration constant at S_m. The feed rate is exponential, and one can apply a closed-loop feedback control based on Eq. (12.58) if it is possible to measure on-line the total cell mass, XV:

$$F_{\text{sin}} = \frac{\mu_m XV}{Y_{X/S}(S_F - S_m)} = \frac{\sigma_m XV}{(S_F - S_m)}$$

If it is not possible to measure on-line, then an open-loop exponential feed rate policy can be used:

$$F_{\text{sin}} = \frac{\sigma(S_m)(XV)_0 \exp(\mu_{\max}t)}{(S_F - S_m)} \exp(\mu_{\max}t) = \alpha_m \exp(\mu_{\max}t)$$

2. When the specific growth rate is nonmonotonic and the yield coefficient is a monotonically decreasing function of substrate concentration (Figure 16.2B), the initial substrate concentration should be chosen close to S_m, and the feed rate should be regulated to decrease the substrate so that the cell mass yield is increased at the expense of decreasing specific growth rate. The rate at which the substrate concentration should be decreased is not precisely known.
3. When the yield coefficient is a monotonically increasing function of substrate concentration (Figure 16.2C), the initial substrate concentration should be chosen to be at S_m, and the feed rate should be adjusted to increase the substrate concentration to improve the cell mass yield at the expense of specific growth rate. The rate at which the substrate concentration should be increased is not precisely known.
4. When both μ and Y are nonmonotonic and the peak in the yield S_Y occurs at a lower substrate concentration than that S_m in the specific growth rate (Figure 16.2D), $S_Y < S_m$, the initial substrate concentration should be S_m, and the feed rate should be regulated to decrease the substrate concentration from S_m to S_Y during the course of fermentation.
5. When both μ and Y are nonmonotonic and the peak in the yield occurs at a higher substrate concentration than that in the specific growth rate (Figure 16.2E), $S_Y > S_m$, the initial substrate concentration should be S_m, and the feed rate should be regulated to increase the substrate concentration from S_m to S_Y during the course of fermentation.
6. When the specific growth rate is a monotonically increasing function of substrate concentration, while the yield coefficient is nonmonotonic (Figure 16.2F), the initial substrate concentration should be as high as possible, and the feed rate should be regulated to decrease the substrate concentration to S_Y to improve the yield.

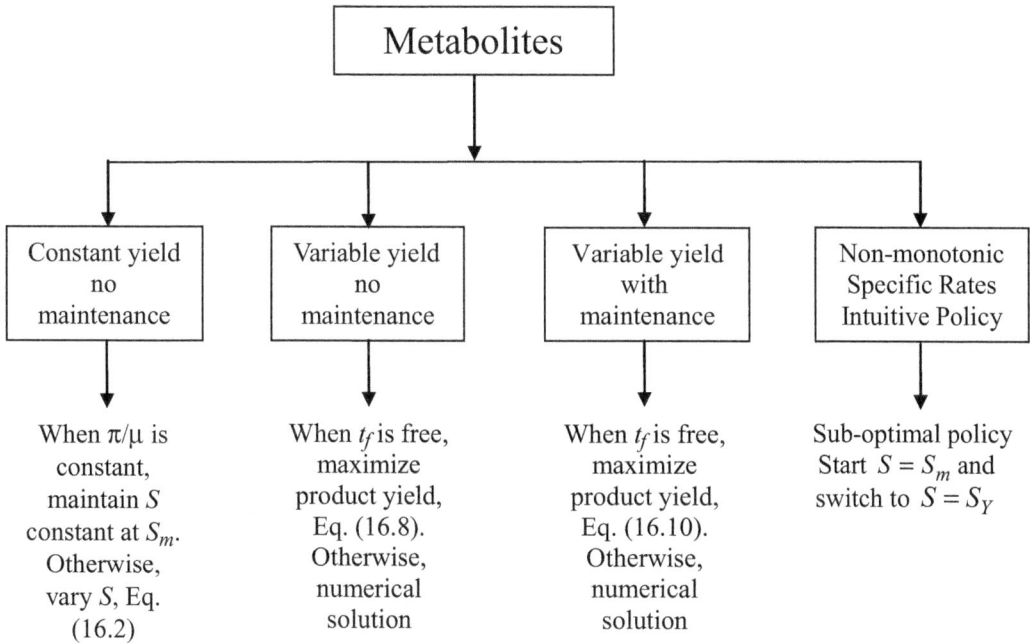

Figure 16.3. Flow chart for optimal/suboptimal feed rates for metabolite production.

7. When both the specific growth rate and the cell mass yield are monotonically increasing functions of substrate concentration (Figure 16.2G), no singular region exists, and a batch operation would be best.

In all of the preceding situations in which the substrate concentration should be regulated from S_m to S_Y or from S_Y to S_m, the rates at which the change, dS/dt, must take place are not known precisely, and an engineering judgment is necessary. Thus, the preceding strategies are suboptimal.

16.3.2 Metabolites as Product

When the product is a metabolite, the process is more complicated so that it is modeled by four or more mass balance equations, so that only for simple cases can we develop simple strategies, and no simple optimal–suboptimal strategy can be made in a general situation. Figure 16.3 represents a roadmap to optimal/suboptimal feed rates for simple processes that can be modeled by four mass balance equations.

16.3.2.1 Constant-Yield Coefficients, $Y_{X/S}$ and $Y_{P/S}$, and without Maintenance, $m = 0$

16.3.2.1.1 PERFORMANCE INDEX INDEPENDENT OF FREE FINAL TIME, $P[x(t_f)]$. When the final time t_f is fixed to meet a production schedule and the performance index is independent of final time, $\partial P/\partial t_f = 0$, we have two situations. First, the limiting case is when *the specific growth rate is proportional to the specific product formation rate*, $(\pi/\mu) = a$; in this case, one should set the initial substrate concentration to the value, $S(0) = S_m$, at which the specific rates are maximum, $[\partial \mu/\partial S]_{S=S_m} = 0$, and the feed rate should be regulated to maintain the substrate concentration constant at $S = S_m$, until the reactor is full, followed by a batch period, $F_b = 0$, to the final time. The singular feed rate is obtained from the substrate balance equation:

$$F_{sin} = \frac{\sigma(S_m)(XV)}{S_F - S_m} = \frac{\sigma(S_m)(XV)_0 \exp[\mu(S_m)t]}{S_F - S_m} = \alpha_m \exp(\mu_{max}t) \quad (16.1)$$

If it is possible to measure on-line the total cell mass XV, one can apply a closed-loop feedback control, $F_{sin} = [\sigma(S_m)/(S_F - S_m)]XV$. If it is not possible to measure it on-line, then an open-loop exponential policy, $F_{sin} = \alpha_m \exp(\mu_{max}t)$, can be used because only the initial amount of cell mass and the maximum specific growth rate are required. A general case requires that the substrate concentration be varied according to

$$\frac{dS}{dt} = \frac{(\pi/\mu)'}{(\pi'/\mu')'}\mu'\left(\frac{\mu}{\mu'}\right)^2 = f(S) \quad (13.71)$$

Integration of Eq. (13.68) yields

$$S(t) = S(t_s) + \int_{t_s}^{t} f(\lambda)d\lambda \quad (16.3)$$

where t_s is the time at which the singular period begins and $S(t_s)$ is not known a priori. Therefore, we have a choice here: to numerically solve the optimization problem, as outlined in Chapter 13, or to apply a suboptimal policy by picking the initial substrate concentration to maximize the specific rate, $S(0) = S_m = S(t_s)$, and apply the singular substrate concentration given by Eq. (16.3).

16.3.2.1.2 PERFORMANCE INDEX DEPENDENT ON FREE FINAL TIME, $P[x(t_f), t_f]$, MINIMUM TIME PROBLEM.

No simple feed rate policy or law can be obtained in a general case, and only extremely simple cases can yield simple feed rates. For special cases in which both μ and π peak at the same substrate concentration $S = S_m$, that is, $(\partial \pi/\partial S)_{S=S_m} = 0$ and $(\partial \mu/\partial S)_{S=S_m} = 0$, or the specific product formation rate is proportional to the specific cell growth rate $\pi = a\mu$ and, therefore, they both peak at the same substrate concentration, $S = S_m$, it is much easier to obtain a simple solution because we can choose the initial values of substrate concentration ($S_0 = S_m$) and regulate the feed rate given by Eq. (16.4), an exponential function, until $V = V_{max}$:

$$F_{sin} = \frac{\sigma(S_m)XV}{S_F - S_m} = \frac{(\mu(S_m)/Y_{X/S} + \pi(S_m)/Y_{P/S} + m)XV}{S_F - S_m} = a_{mm}\exp(\mu_{max}t) \quad (16.4)$$

Then a batch period takes over until $t = t_f$, at which point the final condition is met.

16.3.2.1.3 MAXIMUM PRODUCTIVITY PROBLEM.

As in the case of the preceding minimum time problem, a special case arises when the specific product formation rate is proportional to the specific growth rate, $\pi = a\mu$. In this case, the substrate concentration is maintained constant, $S = S_m$, at which the specific rates are at maximum, $\mu'(S_m) = \pi'(S_m)$. Equation (16.1) is the singular feed rate in feedback mode. The initial substrate concentration is $S(0) = S_m$.

16.3.2.2 Variable-Yield Coefficients $Y_{X/S}(S)$ and $Y_{P/S}(S)$ and without Maintenance

When the yield coefficients $Y_{X/S}$ and $Y_{P/S}$ are functions of substrate concentration S, we can come up with simple feed rate strategies for limiting cases.

16.3.2.2.1 FREE FINAL TIME. When the final time is free, and if the performance index weighs the product only, the singular feed rate maximizes the product yield coefficient, Eq. (13.124), which reduces to

$$\text{Max}_{S^*}\left[\left(\frac{\pi}{\sigma}\right)\right] = \text{Max}_{S^*}[(Y_{P/S}] \tag{16.5}$$

Thus, the singular feed rate maximizes the specific product formation rate relative to the specific substrate consumption rate or the product yield coefficient $Y_{P/S} = \pi/\sigma$. To maximize the variable product yield coefficient, we set $\partial Y_{P/S}/\partial S = 0$ to obtain the substrate concentration that maximizes $Y_{P/S} = \pi/\sigma$ and denote that concentration as $S^* = S_{P/S}$. The optimal initial substrate concentration is $S(0) = S_{P/S}$, and we apply the feed rate, an exponential rate, to maintain the substrate concentration constant at $S^* = S_{P/S}$, until the bioreactor is full. Therefore, the singular feed rate is exponential:

$$F_{\text{sin}} = \frac{\sigma(S_{P/S})XV}{S_F - S_{P/S}} = \frac{\sigma(S_{P/S})\alpha \exp[\mu(S_{P/S})t]}{S_F - S_{P/S}} = \alpha_{P/S} \exp[\mu(S_{P/S})t] \tag{16.6}$$

16.3.2.2.2 FIXED FINAL TIME. When the final time is fixed, we have one less equation to work with, and therefore, it is difficult to assess the characteristics of the singular feed rate and come up with a simple rule. Therefore, one has to make a simulation study in which a known sequence of feed rates (e.g., a period of singular feed rate followed by a batch period) is used to determine the best initial substrate concentration that maximizes the performance index, $F_{\text{sin}}(0, t_{\text{full}}) \to F_b = 0 \ (t_{\text{full}}, t_f)$, where t_{full} becomes known by integration of the differential equations, while t_f is given.

16.3.2.3 Variable-Yield Coefficients and with Maintenance Requirement

When the yield coefficient is a function of substrate concentration and there is a maintenance requirement, $m \neq 0$, we need to look at various performance indices.

16.3.2.3.1 FREE FINAL TIME, t_f. In case 1, performance index depends on only $x_2(t_f)$: this case corresponds to maximization of the product at the final time so that $P = x_2(t_f)$. Then, $\partial P/\partial x_1(t_f) = 0$ and $\partial P/\partial x_2(t_f) = 1$, and Eq. (13.131) reduces to

$$\text{Max}_{S^*}\left[\left(\frac{\pi}{\sigma}\right)\right] = \text{Max}_{S^*}[(Y_{P/S}] \tag{16.7}$$

Thus, the singular feed rate maximizes the specific product formation rate relative to the specific substrate consumption rate or the yield coefficient $Y_{P/S} = \pi/\sigma$. Therefore, we choose the initial substrate concentration to be $S(0) = S^*$, $[\partial Y_{P/S}/\partial S]_{S=S^*} = 0$ and apply the singular feed rate that maintains the substrate concentration at $S(t) = S^*$, until the bioreactor is full, followed by a batch period, $F_b = 0$:

$$F_{\text{sin}} = \frac{\sigma(S^*)XV}{S_F - S^*} = \frac{(\mu(S^*)/Y_{X/S} + \pi(S^*)/Y_{P/S} + m)XV}{S_F - S^*} \tag{16.8}$$

In case 2, the performance index depends on $x_1(t_f)$ and $x_2(t_f)$: this is the case in which both the cell mass and metabolite are valuable so that the performance index weighs the total amounts of both the cell mass and metabolite, such as $P = [\alpha x_1(t_f) + x_2(t_f)]$, where $\alpha \ll 1$ is the value of cell mass relative to that of metabolite. In this case, the weighing factors are $\partial P/\partial x_1(t_f) = \alpha$ and $\partial P/\partial x_2(t_f) = 1$. Thus, the singular feed rate maximizes the weighted sum of the cell mass yield $(\mu/\sigma) = Y_{X/S}$ and the product yield $(\pi/\sigma) = Y_{P/S}$. Equation (13.131) becomes

$$\underset{S^{*1}}{\text{Max}}\left[\alpha\left(\frac{\mu}{\sigma}\right) + \left(\frac{\pi}{\sigma}\right)\right] = \underset{S^{*1}}{\text{Max}}(\alpha Y_{X/S} + Y_{P/S}), \quad \left[\frac{\partial(\alpha Y_{X/S} + Y_{P/S})}{\partial S}\right]_{S^{*1}} = 0 \quad (16.9)$$

Equation (16.9) states that the singular feed rate maximizes the weighted sum of cell and product yields. We do know that the substrate concentration must be held at this value, $S = S^{*1}$, throughout the singular feed rate period. Therefore, we choose the initial substrate concentration as $S(0) = S^{*1}$ and apply the singular feed rate (exponential) that is obtained from the substrate balance equation, Eq. (13.4), until the bioreactor is full:

$$F_{\text{sin}} = \frac{\sigma(S^{1*})XV}{S_F - S^{1*}} = \frac{(\mu(S^{1*})/Y_{X/S} + \pi(S^{1*})/Y_{P/S} + m)XV}{S_F - S^{1*}} \quad (16.10)$$

16.3.2.3.2 MAXIMUM PRODUCTIVITY. A special case arises when the specific product formation rate is proportional to the specific growth rate, $(\pi/\sigma) = \alpha(\mu/\sigma)$. Then, the right-hand side of Eq. (13.178) vanishes and the substrate concentration is maintained constant, $S = S_m$, at which the specific growth rate and product formation rates are maxima. The feed rate required is obtained from Eq. (13.4):

$$F_{\text{sin}} = \frac{\sigma(S_m)XV}{S_F - S_m} = \frac{(\mu(S_m)/Y_{X/S} + \pi(S_m)/Y_{P/S} + m)XV}{S_F - S_m} \quad (16.11)$$

16.3.2.4 Intuitive Suboptimal Policy for Nonmonotonic Specific Rates

When both the specific growth rate and specific product formation rates are nonmonotonic, that is, showing maxima, a suboptimal feed rate profile would be first to maintain the substrate concentration that maximizes the specific growth rate, $S = S_m$, and then shifts the substrate concentration that maximizes the specific product formation rate, $S = S_M$. When one would shift the substrate concentration is not known and requires a simulation study. The initial substrate concentration should be $S(0) = S_m$, and the feed rate would be an exponential function or in feedback mode if the total cell mass can be measured on-line or predicted,

$$F_{\text{sin}} = \frac{\sigma(S_m)XV}{S_F - S_m} = \frac{(\mu(S_m)/Y_{X/S} + \pi(S_m)/Y_{P/S} + m)XV}{S_F - S_m} \quad (16.12)$$

followed by the singular feed rate that maximizes the specific product formation rate, $S = S_M$, which is also an exponential function or in feedback mode if the total cell mass can be measured on-line or predicted:

$$F_{\text{sin}} = \frac{\sigma(S_M)XV}{S_F - S_M} = \frac{(\mu(S_M)/Y_{X/S} + \pi(S_M)/Y_{P/S} + m)XV}{S_F - S_M} \quad (16.13)$$

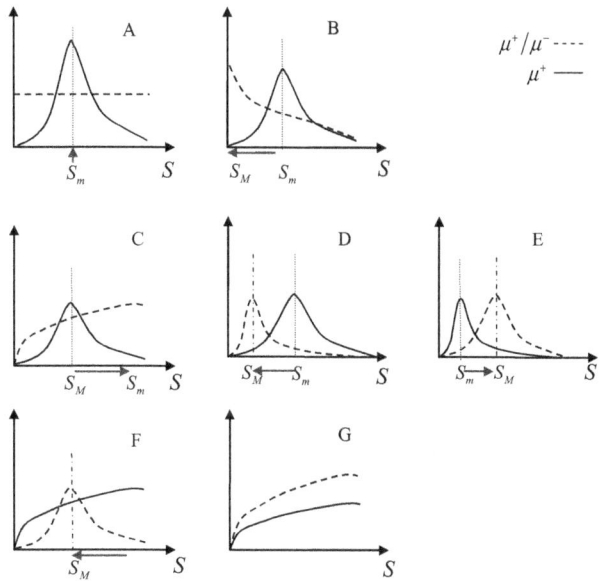

Figure 16.4. Singular regions (II) and variations in substrate concentration during singular interval for recombinant cell products. Solid arrows indicate the direction of the change in S with respect to time.

16.3.3 Recombinant Cell Products

Very few reported works deal with recombinant cell product optimization. We deal here with the case of constant yields and growth-associated products.

16.3.3.1 Constant Yields and Growth-Associated Product Formation

16.3.3.1.1 FREE FINAL TIME. The singular arc is simply the substrate concentration that maximizes (μ^+/μ^-). Therefore, the initial substrate concentration should be that which maximizes μ^+/μ^-, $[\partial(\mu^+/\mu^-)/\partial S]_{S=S_R} = 0$, that is, $S(0) = S_R$. The singular feed rate is used to keep the substrate concentration constant at $S(t) = S_R$, until the reactor volume is full. The singular feed rate is an exponential function, or it can be in feedback mode if a continuous on-line measurement or prediction is available:

$$F_{\sin} = \frac{\sigma(S_R)(XV)}{S_F - S_R} = \frac{\sigma(S_R)(XV)_0 \exp[\mu(S_R)t]}{S_F - S_R} \qquad (16.14)$$

16.3.3.1.2 FIXED FINAL TIME. When the final time is fixed, the yield coefficients are constant, and the product formation is growth associated, the singular regions can be classified in terms of specific rate μ^+ and the relative specific rates (μ^+/μ^-), as observed in Figure 13.3, which is duplicated here as Figure 16.4. The regions in which a singular feed rate is feasible are denoted by the arrows. The direction of these arrows also indicates whether the substrate concentration remains constant, increases, or decreases with time during the period of singular feed rate. The entire space is divided into three regions: I, II, and III. The singular region is denoted by region II, except in the case of constant (μ^+/μ^-), when it is a vertical line instead of a region.

1. When (μ^+/μ^-) is constant (Figure 16.4A), the substrate concentration is held constant at S_m during the singular feed period to maximize the specific growth

rate μ^+. The optimal initial substrate concentration is $S(0) = S_m$. In region II, the optimum sequence is singular, until the volume is full, when a batch period takes over, $F_{sin} \to F_b = 0$. The singular feed rate is exponential if a continuous measurement of total plasmid bearing cells (PBC) is possible, or the exponential function is,

$$F_{sin} = \sigma(S_m)X^+V/(S_F - S_m) = \sigma(S_m)(X^+V)_0 \exp[\mu^+(S_m)t]/(S_F - S_m)$$

(16.15)

2. If (μ^+/μ^-) is a monotonically decreasing function of substrate concentration (Figure 16.4B), the substrate concentration decreases during the singular feed rate period to improve the (μ^+/μ^-) at the expense of the specific growth rate of PBC μ^+. The optimal initial substrate concentration is near to but less than S_m, and the substrate concentration must decrease with time, which must be determined numerically.

3. When (μ^+/μ^-) is a monotonically increasing function of substrate concentration (Figure 16.4C), the optimal initial substrate concentration is near to but greater than S_m and must increase with time during the singular period.

4. We consider next the case of nonmonotonic μ^+ and (μ^+/μ^-). When the concentration corresponding to the peak in the (μ^+/μ^-), S_M, is less than that of the peak S_m in the specific growth rate of PBC, μ^+ (Figure 16.4D), $S_M < S_m$, the substrate concentration starting near S_m decreases toward S_M during the singular feed rate period to improve the relative rate (μ^+/μ^-) at the expense of the specific growth rate of PBC, μ^+.

5. When the reverse situation arises (Figure 16.4E), $S_M > S_m$, the initial substrate concentration should start near S_m, and the feed rate should be regulated to increase the substrate concentration toward S_M, improving the (μ^+/μ^-) at the expense of the specific growth rate μ^+.

6. If (μ^+/μ^-) is nonmonotonic but the specific growth rate μ^+ is monotonic (Figure 16.4F), the initial substrate concentration starts very high but decreases with time toward S_M during the period of singular feed rate.

7. Finally, if both the specific growth rate μ^+ and (μ^+/μ^-) are monotonic, as in Figure 16.4G, there is no singular region, and the optimal feed rate sequence is $F_{max} \to F_b = 0$, or a batch operation.

Index